Embedded Memories for Nano-Scale VLSIs

Series on Integrated Circuits and Systems

Series Editor: Anantha Chandrakasan
Massachusetts Institute of Technology
Cambridge, Massachusetts

Embedded Memories for Nano-Scale VLSIs
Kevin Zhang (Ed.)
ISBN 978-0-387-88496-7

Carbon Nanotube Electronics
Ali Javey and Jing Kong (Eds.)
ISBN 978-0-387-36833-7

Wafer Level 3-D ICs Process Technology
Chuan Seng Tan, Ronald J. Gutmann, and L. Rafael Reif (Eds.)
ISBN 978-0-387-76532-7

Adaptive Techniques for Dynamic Processor Optimization: Theory and Practice
Alice Wang and Samuel Naffziger (Eds.)
ISBN 978-0-387-76471-9

mm-Wave Silicon Technology: 60 GHz and Beyond
Ali M. Niknejad and Hossein Hashemi (Eds.)
ISBN 978-0-387-76558-7

Ultra Wideband: Circuits, Transceivers, and Systems
Ranjit Gharpurey and Peter Kinget (Eds.)
ISBN 978-0-387-37238-9

Creating Assertion-Based IP
Harry D. Foster and Adam C. Krolnik
ISBN 978-0-387-36641-8

Design for Manufacturability and Statistical Design: A Constructive Approach
Michael Orshansky, Sani R. Nassif, and Duane Boning
ISBN 978-0-387-30928-6

Low Power Methodology Manual: For System-on-Chip Design
Michael Keating, David Flynn, Rob Aitken, Alan Gibbons, and Kaijian Shi
ISBN 978-0-387-71818-7

Modern Circuit Placement: Best Practices and Results
Gi-Joon Nam and Jason Cong
ISBN 978-0-387-36837-5

CMOS Biotechnology
Hakho Lee, Donhee Ham and Robert M. Westervelt
ISBN 978-0-387-36836-8

Continued after index

Kevin Zhang
Editor

Embedded Memories for Nano-Scale VLSIs

Editor
Kevin Zhang
Intel Corporation
2501 NW. 229th Ave.
Hillsboro, OR 97124
USA
kevin.zhang@intel.com

ISBN 978-0-387-88496-7 e-ISBN 978-0-387-88497-4
DOI 10.1007/978-0-387-88497-4

Library of Congress Control Number: 2008936472

Printed on acid-free paper

springer.com

Contents

Contributors

John Barth IBM, Essex Junction, Vermont, jbarth@us.ibm.com

Anantha P. Chandrakasan Massachusetts Institute of Technology Cambridge, MA, USA

Jeffrey S. Cross Tokyo Institute of Technology, 2-12-1 Ookayama, Meguro-ku, Tokyo 152-8550, Japan, cross.j.aa@m.titech.ac.jp

Hideto Hidaka MCU Technology Division, Renesas Technology Corporation, 4-1 Mizuhara, Itami, 664-0005, Japan, hidaka.hideto@renesas.com

Shoichiro Kawashima Fujitsu Microelectronics Limited, System Micro Division, 1-1 Kamikodanaka 4-chome, Nakahara-ku, Kawasaki, 211-8588, Japan kawashima@jp.fujitsu.com

Donghyun Kim KAIST

Rob A. Rutenbar Electrical and Computer Engineering, Carnegie Mellon University, Pittsburgh, PA, USA, rutenbar@ece.cmu.edu

Amith Singhee IBM, Thomas J. Watson Research Center, Yorktown Heights, NY, USA

Naveen Verma Massachusetts Institute of Technology, Cambridge, MA, USA, nverma@mit.edu

Hiroyuki Yamauchi Fukuoka Institute of Technology, Fukuoka, Japan

Hoi Jun Yoo KAIST

Kevin Zhang Intel Corporation, Hillsboro, OR, USA

Chapter 1
Introduction

Kevin Zhang

Advancement of semiconductor technology has driven the rapid growth of very large scale integrated (VLSI) systems for increasingly broad applications, including high-end and mobile computing, consumer electronics such as 3D gaming, multi-function or smart phone, and various set-top players and ubiquitous sensor and medical devices. To meet the increasing demand for higher performance and lower power consumption in many different system applications, it is often required to have a large amount of on-die or embedded memory to support the need of data bandwidth in a system. The varieties of embedded memory in a given system have also become increasingly more complex, ranging from static to dynamic and volatile to nonvolatile.

Among embedded memories, six-transistor (6T)-based static random access memory (SRAM) continues to play a pivotal role in nearly all VLSI systems due to its superior speed and full compatibility with logic process technology. But as the technology scaling continues, SRAM design is facing severe challenge in maintaining sufficient cell stability margin under relentless area scaling. Meanwhile, rapid expansion in mobile application, including new emerging application in sensor and medical devices, requires far more aggressive voltage scaling to meet very stringent power constraint. Many innovative circuit topologies and techniques have been extensively explored in recent years to address these challenges.

Dynamic random access memory (DRAM) has long been an important semiconductor memory for its well-balanced performance and density. With increasing demand for on-die dense memory, one-transistor and one-capacitor (1T1C)-based DRAM has found varieties of embedded applications in providing the memory bandwidth for system-on-chip (SOC) applications. With increasing amount of on-die cache memory for high-end computing and graphics application, embedded DRAM (eDRAM) is becoming a viable alternative to SRAM for large on-die memory. To meet product requirements for eDRAM while addressing continuous technology scaling, many new memory circuit design technologies, which are often

---

K. Zhang (✉)
Intel Corporation, 2501 NE 229th Ave, Hillsboro, OR 97124, USA
e-mail: kevin.zhang@intel.com

K. Zhang (ed.), *Embedded Memories for Nano-Scale VLSIs*, Series on Integrated Circuits and Systems, DOI 10.1007/978-0-387-88497-4_1,

drastically different from commodity DRAM design, have to be developed to substantially improve the eDRAM performance while keeping the overall power consumption at minimum.

Solid-state nonvolatile memory (NVM) has played an increasingly important role in both computing and consumer electronics. Many new applications in most recent consumer electronics and automobiles have further broadened the embedded application for NVM. Among various NVM technologies, floating-gate-based NOR flash has been the early technology choice for embedded logic applications. With technology scaling challenges in the floating-gate technologies, including the increasing need for integrating NVM along with more advanced logic transistors, varieties of NVM technologies have been extensively explored, including alternative technology based on charge-trapping mechanism (Fig. 1.1). More efficient circuit design techniques for embedded flash also have to be explored to achieve optimal product goals.

With increasing demand of NVM for further scaling of the semiconductor technology, several emerging memory technologies have drawn increasingly more attention, including magnetic RAM (MRAM), phase-change RAM (PRAM), and ferroelectric RAM (FeRAM). These new technologies not only address some of the fundamental scaling limits in the traditional solid-state memories, but also have brought new electrical characteristics in the nonvolatile memories on top of the random accessing capability. For example, MRAM can offer significant speed improvement over traditional floating-gate memory, which could open up whole new applications. FeRAM can operate at lower voltage and consume ultra low power, which has already made it into "smart-card" marketplace today. These new memory technologies also require a new set of circuit topologies and sensing techniques to maximize the technology benefits, in comparison to the traditional NVM design.

With rapid downward scaling of the feature size of memory device by technology and drastic upward scaling of number of storage elements per unit area, process-induced variation in memory has become increasingly important for both memory technology and circuit design. Statistical design methodology has now

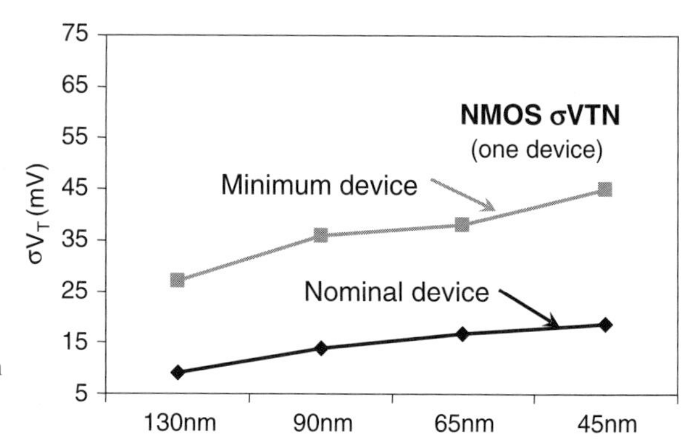

Fig. 1.1 Transistor variation trend with technology scaling [1]

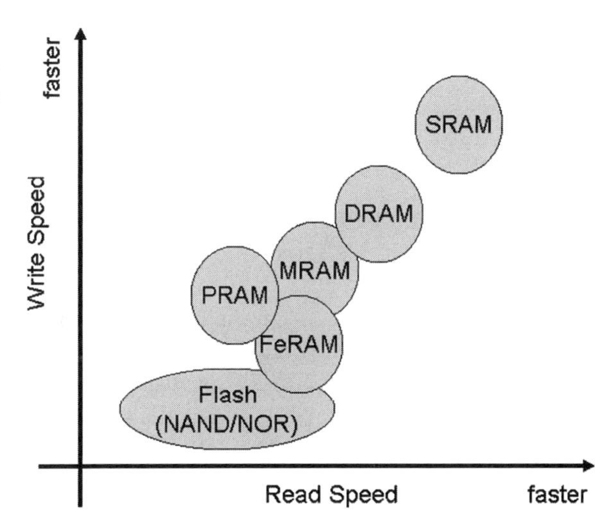

Fig. 1.2 Relative performance among different types of embedded memories

become essential in developing reliable memory for high-volume manufacturing. The required statistical modeling and optimization capability has grown far beyond the memory cell to comprehend many sensitive peripheral circuits in the entire memory block, such as critical signal development paths. Advanced statistical design techniques are clearly required in today's memory design.

In traditional memory field, there is often a clear technical boundary between different kinds of memory technology, e.g., SRAM and DRAM, volatile and non-volatile. With growing demand for on-die memory to meet the need of future VLSI system design, it is very important to take a broader view of overall memory options in order to make the best design tradeoff in achieving optimal system-level power and performance. Figure 1.2 illustrates the potential tradeoff among these different memories. With this in mind, this book intends to provide a state-of-the-art view on most recent advancements of memory technologies across different technical disciplines. By combining these different memories together in one place, it should help readers to gain a much broadened view on embedded memory technology for future applications. Each chapter of the book is written by a set of leading experts from both industry and academia to cover a wide spectrum of key memory technologies along with most significant technical topics in each area, ranging from key technical challenges to technology and design solutions. The book is organized as follows:

1.1 Chapter 2: Embedded Memory Architecture for Low-Power Application Processor, by Hoi Jun Yoo

In this chapter, an overview on embedded memory architecture for varieties of mobile applications is provided. Several real product examples from advanced application processors are analyzed with focus on how to optimize the memory architecture to achieve low-power and high-performance goal. The chapter intends to provide readers an architectural view on the role of embedded memory in mobile applications.

1.2 Chapter 3: Embedded SRAM Design in Nanometer-Scale Technologies, by Hiroyuki Yamauchi

This chapter discusses key design challenges facing today's SRAM design in nano-scale CMOS technologies. It provides a broad coverage on latest technology and design solutions to address SRAM scaling challenges in meeting power, density, and performance goal for product applications. A tradeoff for each technology and design solution is thoroughly discussed.

1.3 Chapter 4: Ultra Low Voltage SRAM Design, by Naveen Verma and Anantha P. Chandrakasan

In this chapter, an emerging family of SRAM design is introduced for ultra-low-voltage operation in highly energy-constrained applications such as sensor and medical devices. Many state-of-the-art circuit technologies are discussed for achieving very aggressive voltage-scaling target. Several advanced design implementations for reliable sub-threshold operation are provided.

1.4 Chapter 5: Embedded DRAM in Nano-Scale Technologies, by John Barth

This chapter describes the state-of-the-art eDRAM design technologies for varieties of applications, including both consumer electronics and high-performance computing in microprocessors. Array architecture and circuit techniques are explored to achieve a balanced and robust design based on high-performance logic process technologies.

1.5 Chapter 6: Embedded Flash Memory, by Hideto Hidaka

This chapter provides a very comprehensive view on the state of embedded flash memory technology in today's industry, including process technology, product application, and future trend. Several key technology options and their tradeoffs are discussed. Product design examples for micro-controller unit (MCU) are analyzed down to circuit implementation level.

1.6 Chapter 7: Embedded Magnetic RAM, by Hideto Hidaka

Magnetic RAM has become a key candidate for new applications in nonvolatile applications. This chapter introduces both key technology and circuit design elements associated with this new technology. The future application and market trend for MRAM are also discussed.

1.7 Chapter 8: FeRAM, by Shoichiro Kawashima and Jeffrey S. Cross

This chapter introduces the latest material, device, and circuit advancement in ferroelectric RAM (FeRAM). With excellent write-time, random accessing capability, and compatibility with logic process, FeRAM has penetrated into several application areas. Real product examples are provided along with future trend of the technology.

1.8 Chapter 9: Statistical Blockade: Estimating Rare Event Statistics for Memories, by Amith Singhee and Rob A. Rutenbar

This chapter introduces a comprehensive statistical design methodology that is essential in today's memory design. The core of this methodology is called statistical blockade and it combines Monte Carlo simulation, machine learning, and extreme value theory in effectively predicting rare failure ($>$ 5 sigma) event. Real design examples are used to illustrate the benefit of the methodology in memory design and optimization.

Reference

1. K. Kuhn, "Reducing Variation in Advanced Logic Technologies: Approaches to Process and Design for Manufacturability of Nanoscale CMOS," IEEE IEDM Tech Digest, pp. 471–474, Dec. 2007.

Chapter 2
Embedded Memory Architecture for Low-Power Application Processor

Hoi Jun Yoo and Donghyun Kim

2.1 Memory Hierarchy

2.1.1 Introduction

Currently, the state-of-the-art high-end processors operate at 3–4 GHz frequency whereas even the fastest off-chip memory operates at just around 600 MHz [1–6]. In decades, along with advances in processor technology, the speed gap between processors and memories has become intolerably large [7], and this speed gap has driven the processor designers to introduce a memory hierarchy into the processor architecture. For processors, it is ideal to have indefinitely large memory with no access latencies [8]. However, implementing large-capacity memory with fast operation speed is infeasible due to the physical limitations of the electrical circuits. Thus, the capacity is usually traded off with the operation speed in memory designs. For example, on-chip L1 caches are able to operate as fast as the state-of-the-art processor cores but have at most few kilobytes capacity. On the other hand, off-chip DRAMs are capable of storing few gigabytes though their operation frequencies are just around hundreds of megahertz.

The memory hierarchy is an arrangement of different types of memories with different capacities and operation speeds to approximate the ideal memory behavior in a cost-efficient way. The idea of memory hierarchy comes from observing two common characteristics of the memory accesses in the wide range of programs, namely temporal locality and spatial locality. When a program accesses a certain data address repeatedly for a while, it is temporal locality. Spatial locality means that the memory accesses occur within a small region of memory for a short duration. Due to these localities, embedding a small but fast memory is sufficient to provide a processor with frequently required data for a short period of time. However,

H.J. Yoo (✉)
KAIST

K. Zhang (ed.), *Embedded Memories for Nano-Scale VLSIs*, Series on Integrated Circuits and Systems, DOI 10.1007/978-0-387-88497-4_2,

large-capacity memory to store the entire working set of a program and other necessary data such as the operating system is also necessary. In this case, the former is usually an L1 cache and the latter is generally realized by external DRAMs or hard disk drives in conventional computer systems. Since the speed difference between these two memories is at least more than four orders of magnitude, more levels of the memory hierarchy are required to hide and reduce long access latencies resulting from the small number of levels in the memory hierarchy. In typical computer systems, more than four levels of the memory hierarchy are widely adopted, and a memory at the higher level is realized as a smaller and faster memory than those of the lower levels. Figure 2.1 describes typical arrangement of the memory hierarchy.

2.1.2 Advantages of the Memory Hierarchy

The advantage of adopting the memory hierarchy is threefold. The first advantage is to reduce cost of implementing a memory system. In many cases, faster memories are more expensive than slower memories. For example, SRAMs require higher cost per unit storage capacity than DRAMs because a 6-transistor cell in the SRAMs consumes more silicon area than a single transistor cell of the DRAMs. Similarly, DRAMs are more costly than hard disk drives or flash memories for the same capacity. Flash memory cells consume less silicon area and platters of the hard disk drives are much cheaper than silicon die in a mass production. A combination of different types of memories in the memory system enables a trade-off between performance and cost. By storing infrequently accessed data in the slow but low-cost memories, the overall system cost can be reduced.

The second advantage is an improved performance. Without the memory hierarchy, a processor should directly access the lowest level memory that operates very slowly and contains all required data. In this case, every memory access results in

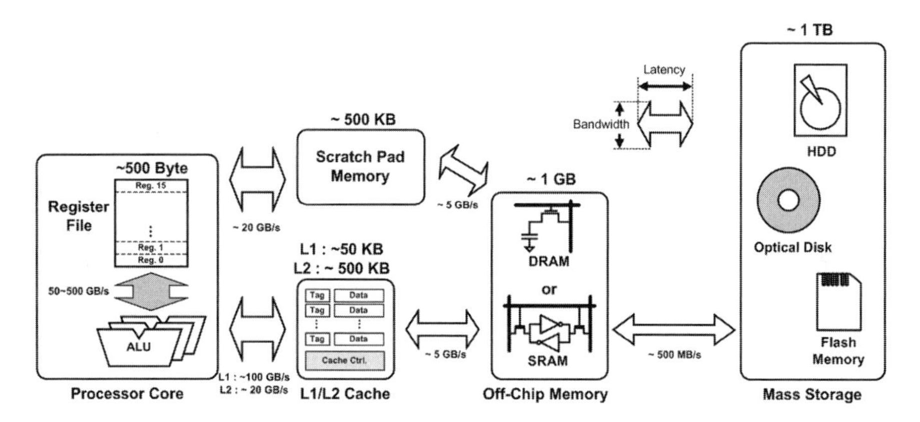

Fig. 2.1 Memory hierarchy

processor stalls to wait for the required data to be available from the memory. Such drawback is resolved by embedding a small memory that runs as fast as a processor core inside the chip. By maintaining an active working set inside the embedded memory, no processor stalls due to memory accesses occur as long as a program is executed within the working set. However, the processor could stall when a working set replacement is performed. This overhead can be reduced by pre-fetching the next working set in the additional in-between level of memory which is easier to access than the lowest level memory. In this way, the memory hierarchy builds up so that the number of levels and types of memories in the memory hierarchy are properly adjusted to minimize average wait cycles for memory accesses. However, finding the optimum configuration of the memory hierarchy requires sophisticated investigation of target application, careful consideration of processor core features, and exhaustive design space exploration. Therefore, design of the memory hierarchy has been one of the most active research fields from the emergence of the computer architecture.

The third advantage of the memory hierarchy is reducing power consumption of a memory system. Accessing an external memory consumes more power than accessing an on-chip memory because off-chip wires have larger parasitic capacitance due to their bigger dimensions. Charging and discharging such large parasitic capacitors result in significant power overhead of off-chip memory accesses. Adopting the memory hierarchy is advantageous to reduce the number of external memory transactions, thus also reducing the power overhead. In a program execution, dynamic data are divided into two categories. One of them is temporary data used to calculate and produce output data of a program execution, and the other is result data that are used by other programs or I/O devices. The result data need to be stored in an off-chip memory such as a main memory or hard disk drive for later reuse of the data. However, temporary data do not need to be stored outside of the chip. Embedding on-chip memories inside the processor enables keeping the temporary data inside the chip during program execution. This reduces the chance of reading or writing of the temporary data in the external memory, which is very costly in power consumption.

2.1.3 Components of the Memory Hierarchy

This section briefly describes different types of memories that construct typical memory hierarchy in conventional computer architectures.

2.1.3.1 Register File

A register file constructs the highest level of the memory hierarchy. A register file is an array of registers embedded in the processor core and is tightly coupled to datapath units to provide an immediate storage for the operands to be calculated. Each entry of the register file is directly accessible without address calculation in the arithmetic and logic unit (ALU) and is defined in an instruction set architecture

(ISA) of the processor. The register file is usually implemented using SRAM cells, and the I/O width is determined to match the datapath width of the processor core. The register file usually has larger number of read ports than conventional SRAMs to provide an ALU with required number of operands in a single cycle. In the case of superscalar processors or very long instruction word (VLIW) processors, the register file is equipped with more than two write ports to support multiple register writes resulting from parallel execution of multiple instructions. Typical number of entries in a register file is around a few tens, and the operation speed is the same as the processor core in most cases.

2.1.3.2 Cache

A cache is a special type of memory that autonomously pre-fetches a subset of temporary duplicated data from lower levels of the memory hierarchy. The caches are the principal part of the memory hierarchy in most computer architectures, and there is a hierarchy among the caches as well. Level 1 (L1) and level 2 (L2) caches are widely adopted and level 3 (L3) cache is usually optional. The L1 cache has the smallest capacity and the lowest access latency. On the other hand, the L3 cache has the largest capacity and the longest access latency. Because the caches maintain duplicated copy of data, cache controllers to manage consistency and coherency schemes are also required to prevent the processing core fetching outdated copies of the data. In addition, the cache includes a tag memory to look up which address regions are stored in the cache.

2.1.3.3 Scratch Pad Memory

A scratch pad memory is an on-chip memory under the management of a user program. The scratch pad memory is usually adopted as a storage of frequently and repeatedly accessed data to reduce the external memory transactions. The size of the scratch pad memory is in the range of tens or hundreds of kilobytes and its physical arrangements, such as number of ports, bank, and cell types, are application-specific. The scratch pad memory is generally adopted for real-time embedded systems to guarantee the predictability in program execution time. In cache-based systems, it is hard to guarantee worst execution time, because behaviors of caches are not under the control of a user program and vary dynamically depending on the dynamic status of the memory system.

2.1.3.4 Off-Chip RAMs

The random access memory (RAM) is a type of memory that allows a read/write access to any address in a constant time. The RAMs are mainly divided into dynamic RAM (DRAM) and static RAM (SRAM) according to their internal cell structures. The term RAM does not specify a certain level in the memory hierarchy, and most of the memories such as cache, scratch pad memory, and register files in the memory hierarchy are classified as RAMs. However, a RAM implemented in a separate package usually specifies a certain level in the memory hierarchy, which is lower than the caches or scratch pad memories. In the perspective of a processor, such RAMs are referred to as off-chip RAMs. The process technologies used to implement off-chip RAMs are optimized to increase memory cell density rather than fast logic operation. The off-chip SRAMs are used as L3 caches or main memory of handheld systems due to their fast operation speed and low-power consumption compared to the off-chip DRAMs. The off-chip DRAMs are used as main memory of a computer system because of their large capacity. In the DRAMs, the whole working set of a program that does not fit into the on-chip cache or scratch pad memory is stored. The DRAMS are usually sold in a single or dual in-line memory module (SIMM or DIMM) that is assembled with a number of DRAM packages on a single printed circuit board (PCB) to achieve large capacity up to few gigabytes.

2.1.3.5 Mass Storages

The lowest level of the memory hierarchy consists of mass storage devices such as hard disk drives, optical disk, and back-up tapes. The mass storage devices have the longest access latencies in the memory hierarchy but provide the largest capacity sufficient to store entire working set as well as other peripheral data such as operating system, device drivers, and result data of program executions for future use. The mass storage devices are usually non-volatile memories able to retain internal data without power supply.

2.2 Memory Access Pattern Related Techniques

In this and following sections, low-power techniques applicable to embedded memory system are described based on the background knowledge of the previous section. First, this section covers memory architecture design issues regarding memory access patterns.

If the system designers understand the memory access pattern of the system operation and it is possible to modify the memory interface, the system performance as well as power consumption can be enhanced by removing or reducing unnecessary memory operations. For some applications having predictable memory access patterns, it is possible to improve effective memory bandwidth with no cost overhead by understanding the intrinsic characteristics of the memory device. And sometimes the system performance is increased by modifying the memory interface. The following case studies show how the system performance and power efficiency are enhanced by understanding the memory access pattern.

2.2.1 Bank Interleaving

When accessing a DRAM, a decoded row address activates a word line and corresponding bit-line sense amplifiers so that the cells connected to the activated word line are ready to transfer or accept data. And then the column address decides which cell in the activated row is connected to the data-bit (DB) sense amplifier or write driver. After accessing data, data signals such as bit lines and DB lines are pre-charged for the next access. Thus, a DRAM basically needs "row activation," "read or write," and "pre-charge" operations to access data. The sum of their operation times decides the access time. The "row activation" and "pre-charge" operations occupy most of the access time and they are not linearly shrunk according to the process downscaling, whereas the operation time of "read or write" is sufficiently reduced to be completed in one clock cycle even with the faster clock frequency of smaller scale process technologies, as shown in Fig. 2.2. In the case of the cell array arranged in a single bank, these operations should be executed sequentially and cannot be overlapped. On the other hand, by dividing the cell array into two or more banks, it is possible to scatter sequential addresses into multiple banks by modulo *N* operations as shown in Fig. 2.3(b), where *N* is the number of memory

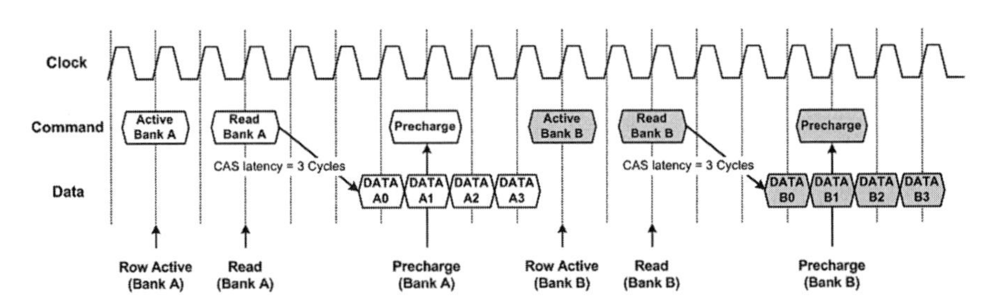

Fig. 2.2 Timing diagram of DRAM read operations without bank interleaving

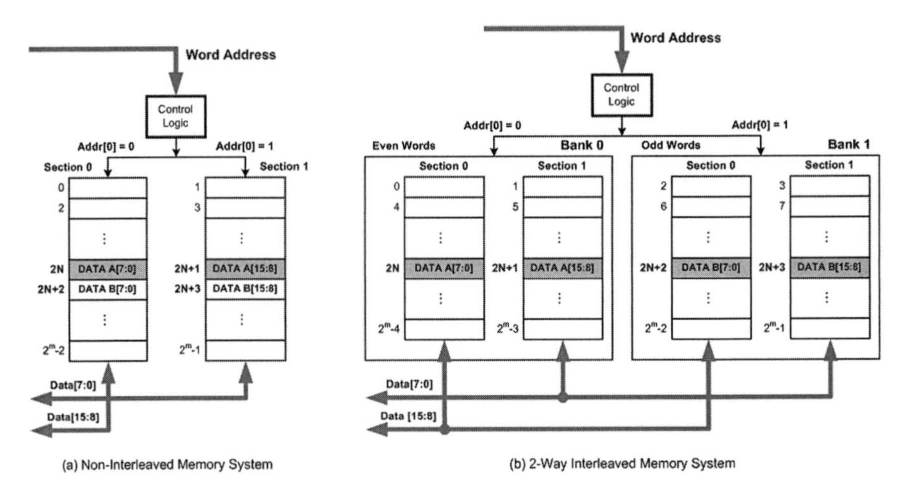

Fig. 2.3 Structures of non-interleaved and interleaved memory systems

banks. And this contributes to hiding the "row activation" or "pre-charge" time of the cell array.

Bank interleaving exploits the independency of the row activations in the different memory banks. Bank is the unit of the cell array which shares the same row and column addresses. By dividing the memory cells into multiple banks, we can obtain the following advantages [9, 10]:

- It hides the amount of time to pre-charge or activate the arrays by accessing one during pre-charging or activating the others, which means that high bandwidth is obtained with low-speed memory chip.
- It can save the power consumption by activating only a subset of cell array at a time.
- It keeps the size of each cell array smaller and limits the number of row and column address pins, and this results in cost reduction.

Figure 2.3 shows the memory configurations with and without bank interleaving. The configuration in Fig. 2.3(a) consists of two sections with each section covering 1-byte data. After it accesses one word, namely addresses $2N$ and $2N+1$, it needs pre-charge time to access the next word, addresses $2N+2$ and $2N+3$. And the row activation for the next access cannot be overlapped. The configuration in Fig. 2.3(b), however, consists of two banks and it can activate the row for the addresses $2N+2$ and $2N+3$ while the row for the addresses $2N$ and $2N+1$ is pre-charged.

Figure 2.4 shows the timing diagram of read operation with interleaved memory structure. Figures 2.2 and 2.4 both assume that column address strobe (CAS) latency is 3 and burst length is 4. On comparing with Fig. 2.2, interleaved memory structure generates 8 data in 12 clock cycles, while non-interleaved memory structure needs additional 5 clock cycles.

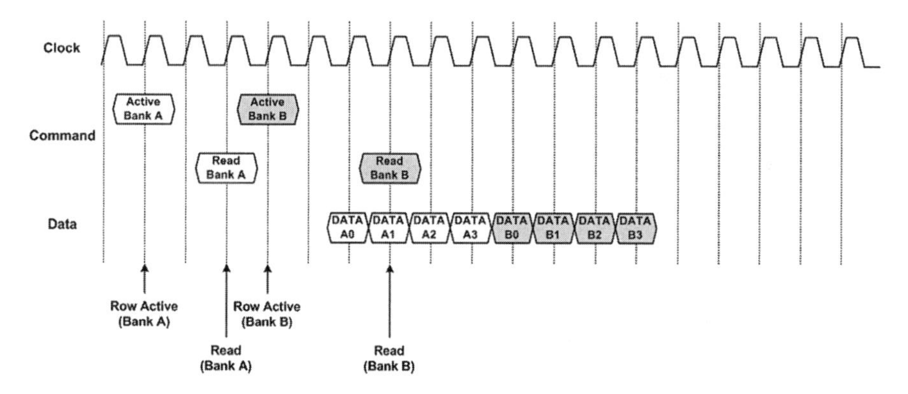

Fig. 2.4 Timing diagram of DRAM read operations with bank interleaving

2.2.2 Address Alignment Logic in KAIST RAMP-IV

In 3D graphics applications, rendering engine requires large memory bandwidth to render high-quality 3D images in real time. To obtain large memory bandwidth, the memory system needs wide data bus or fast clock frequency. Otherwise, we can virtually enlarge the bandwidth by reusing data which had been accessed.

The address alignment logic (AAL) [11] in the 3D rendering engine exploits the access pattern of the texture pixel (texel) from the texture memory. Generally a pixel is calculated using four texels as shown in Fig. 2.5. And corresponding four memory accesses are required. Observing the address of the four texels, they are normally neighbored to each other because of their spatial correlation. If the rendering engine is able to recognize which address it had accessed before, it does not need to access it again, because it has already fetched the data. Figure 2.6(a) shows the block diagram of the AAL. It checks the texel addresses spatially and temporally. If the address had been accessed before and is still available in the logic block, it does not generate the data request to texture memory. Although the additional check operation increases the cycle time, the average number of the texture memory accesses is reduced to less than 30% of the memory access count without the AAL. Figure 2.6(b) shows the energy reduction by the AAL; 68% of the total energy consumption is reduced by understanding and exploiting the memory access pattern.

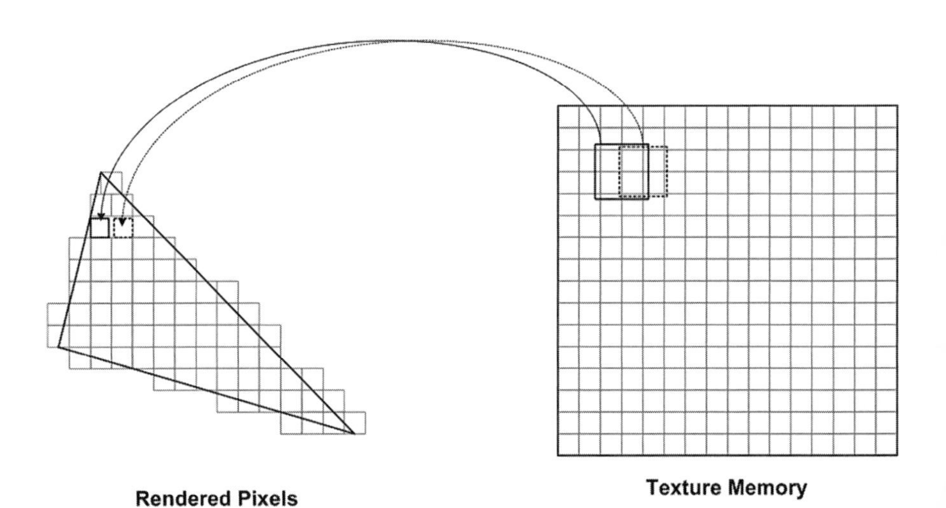

Fig. 2.5 Pixel rendering with texture mapping

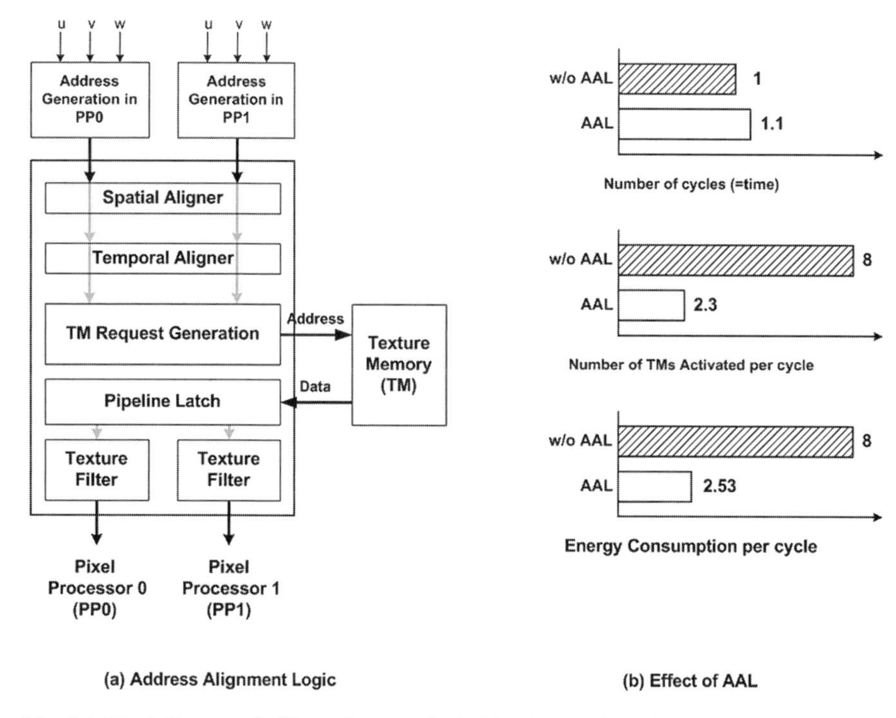

Fig. 2.6 Block diagram of address alignment logic (AAL) (**a**) and its effects (**b**)

2.2.3 Read–Modify–Write (RMW) DRAM

Read–modify–write (RMW) is a special case in which a memory location is first read and then re-written again. It is useful for 3D graphics rendering applications, especially for frame buffer and depth buffer. Frame buffer stores an image data to be displayed, and depth buffer stores the depth information of each pixel. Both of them are accessed by 3D graphics processor, and data are compared and modified. In the depth comparison operations, for example, depth buffer data are accessed and the depth information is modified. If the depth information of stored pixel is screened by newly generated pixel, it needs to be updated. And the frame buffer is refreshed every frame. Both memory devices require three commands: read, modify, and write. From the memory point of view, modify is just waiting. If the memory consists of DRAM cells, it needs to carry out "Row Activation–Read–Pre-charge–Nop (Wait)–Row Activation–Write–Pre-charge" sequences to the same address. If it supports RMW operations, the command sequence can be reduced to "Row Activation–Read–Wait–Write–Pre-charge" as shown in Fig. 2.7, which is compact with no redundant operations. The RMW operation shows that the data bandwidth and control complexity can be reduced by modifying the command sequences regarding the characteristics of memory accesses.

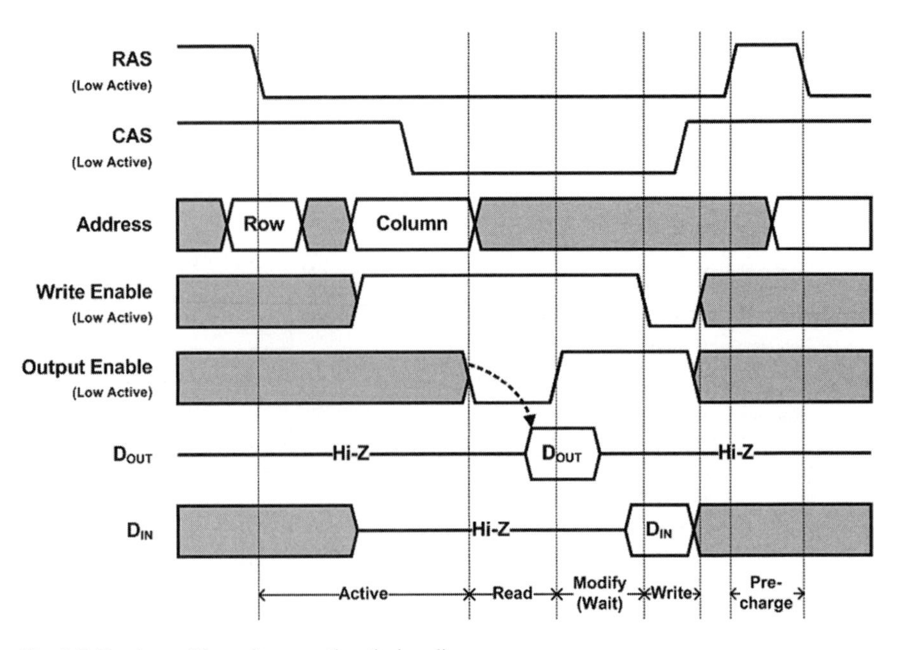

Fig. 2.7 Read–modify–write operation timing diagram

2.3 Embedded Memory Architecture Case Studies

In the design of low-power system-on-chip (SoC), architecture of the embedded memory system has significant impact on the power consumption and overall performance of the SoC. In this section, three embedded memory architectures are covered as case studies. The first example is a Marvell PXA 300 processor which represents a general-purpose application processor. The second example is an IMAGINE processor aimed at removing bandwidth bottleneck in stream processing applications. The last example is the memory-centric network-on-chip (NoC) which adopts co-design of memory architecture and NoC for efficient execution of pipelined tasks.

2.3.1 PXA300 Processor

The PXA series processors were first released by Intel in 2002. The PXA processor series were sold to Marvell technology group in 2006, and PXA3XX series processors are in mass production currently. The PXA300 processor is a general-purpose SoC which incorporates a processor core and other peripheral hardware blocks [12]. The processor core based on an ARM instruction set architecture (ISA) is integrated for general-purpose applications. The SoC is also featured with a 2D graphic processor, video/JPEG acceleration hardware, memory controllers, and an LCD controller. Figure 2.8 shows simplified block diagram of the PXA300 processor [13]. As

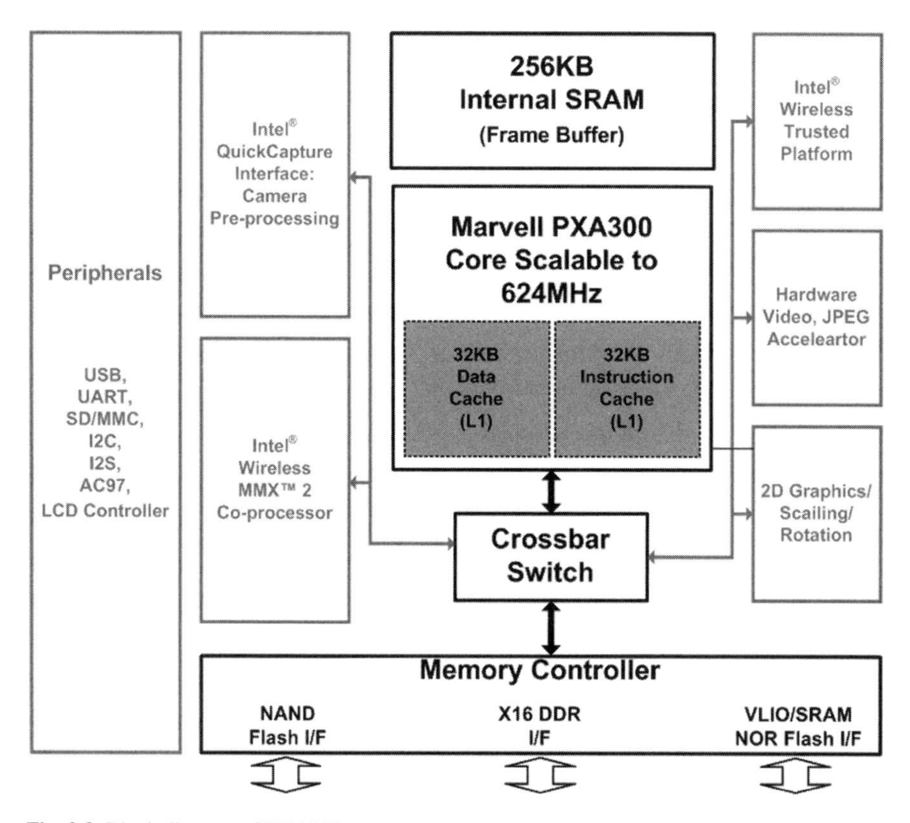

Fig. 2.8 Block diagram of PXA300 processor

shown in Fig. 2.8, the memory hierarchy of the PXA300 is rather simple. The processor core is equipped with L1 instruction/data caches, and both caches are sized to 32 KB. Considering relatively small difference in the operation speed of processor core and the main memory provided by an off-chip double data rate (DDR) SDRAM, absence of an L2 cache is a reasonable design choice. The operation speed of the DDR memory is in the range of 100–200 MHz, and the clock frequency of the processor core is designed to be just around 600 MHz for low-power consumption. Besides the L1 caches, a 256 KB on-chip SRAM is incorporated to provide frame buffer for video codec support. Because the frame buffer requires continuous update of its context and consumes large memory bandwidth, integrating the on-chip SRAM and LCD controller contributes to reducing the external memory transactions. The lowest level of the memory hierarchy consists of flash memories such as NAND/NOR flash memories and secure digital (SD) cards to adapt for handheld devices. Since the PXA300 processor is targeted for general-purpose applications, it is hard to tailor the memory system for low-power execution of a specific application. Therefore, the memory hierarchy of the PXA300 processor is designed similar to those of conventional computer systems with some modifications appropriate for handheld devices. Instead, low-power technique is applied for the entire processor so that operation frequency of the chip is varied according to the workload.

2.3.2 Imagine

In contrast to the general-purpose PXA300 processor, the IMAGINE is more focused on applications having streamed data flow [14, 15]. The IMAGINE processor has customized memory architecture to maximize the available bandwidth among on-chip processing units that consist of 48 ALUs. The memory architecture of the IMAGINE is tiered into three levels so that the memory hierarchy leverages the available bandwidth from the outside of the chip to the internal register files. In this section, the architecture of the IMAGINE processor is briefly described, and then the architectural benefits for efficient stream processing are discussed.

Figure 2.9 shows the overall architecture of the IMAGINE processor. The processor consists of a streaming memory system, a 128 KB streaming register file (SRF), and 48 ALUs divided into 8 ALU clusters. In each ALU cluster, 17 local register files (LRFs) are fully connected to each other through a crossbar switch and the LRFs provide operands for the 6 ALUs, a scratch pad memory, and a communication unit as shown in Fig. 2.10. The lowest level of the memory hierarchy in the IMAGINE is the streaming memory system, which manages four independent 32-bit wide SDRAMs operating at 167 MHz to achieve 2.67 GB/s bandwidth between external memories and the IMAGINE processor. The second level of the memory hierarchy consists of the SRF including a 128 KB SRAM divided into 1024 blocks. All accesses to the SRF are performed through 22 stream buffers and they

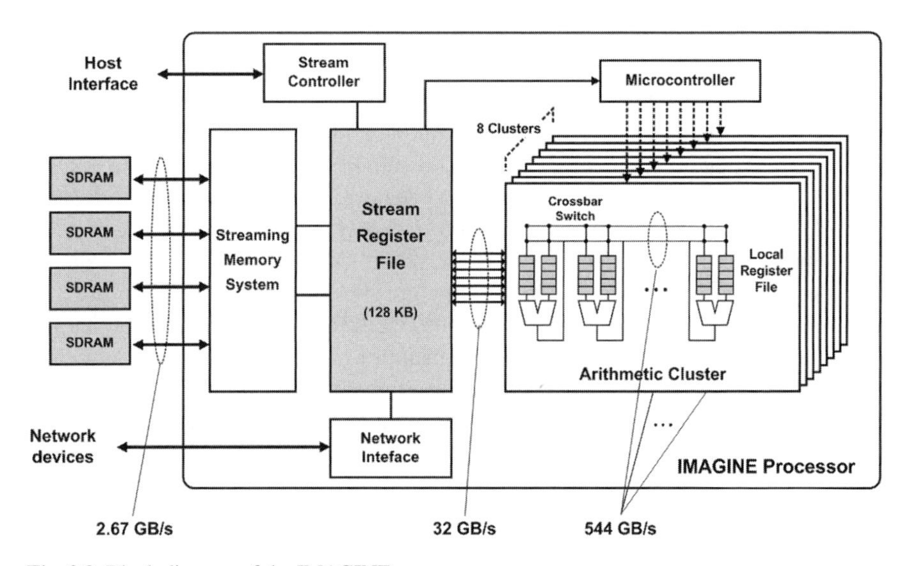

Fig. 2.9 Block diagram of the IMAGINE processor

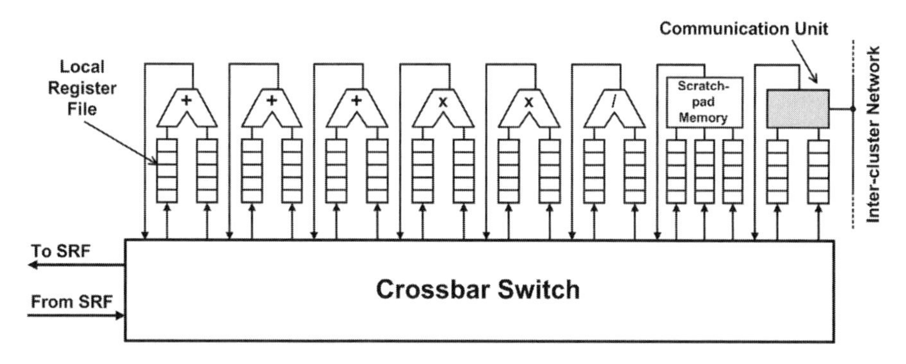

Fig. 2.10 Block diagram of an ALU cluster in the IMAGINE processor

are partitioned into 5 groups to interact with different modules of the processor. By pre-fetching the SRAM data into the stream buffers or utilizing the stream buffers as write buffers, the single-ported SRAM is virtualized as a 22-ported memory, and the peak bandwidth between the SRF and the LRF is 32 GB/s when the IMAGINE operates at 500 MHz. In this case, the 32 GB/s bandwidth is not a sustained bandwidth but a peak bandwidth because the stream buffers for the LRF accesses are managed in time-multiplexed fashion. Finally, the first level of the memory hierarchy is realized by the number of LRFs and crossbar switches. As shown in Fig. 2.10, the fully connected 17 LRFs in each ALU cluster provide a vast amount of bandwidth among the ALUs in each cluster. In addition, eight ALU clusters are able to communicate with each other throughout the SRF or inter-cluster network. The aggregated inter-ALU bandwidth among the 48 ALUs of the 8 ALU clusters reaches up to 544 GB/s.

The architectural benefits of the IMAGINE are found by observing characteristics of the stream processing applications. Stream processing refers to performing series of computation kernels repeatedly on a streamed data flow. In practical designs, the kernel has a set of instructions to be executed for a certain type of function. In stream processing applications such as video encoding/decoding, image processing, and object recognition, major portion of the input data is in the form of video streams. To process vast amount of pixels in a video stream with sufficiently high frame rate, stream processing usually requires intensive computation. Fortunately, in many applications, it is possible to process separate regions of the input data stream independently, and this allows exploiting data parallelism for stream processing. In addition, little reuse of input data and producer consumer locality are the other characteristics of stream processing.

The architecture of the IMAGINE is designed to take advantage of knowledge about the memory access patterns and to exploit intrinsic parallelism of stream processing. Since fixed set of kernels are repeatedly performed on an input data stream, memory access patterns of stream processing are predictable and scheduling of the memory accesses from the multiple ALU is also possible. Therefore, pre-fetching data from the lower level of memory hierarchy, i.e., the streaming memory system or SRF, is effective for hiding latencies of accessing the off-chip SDRAMS from the ALU clusters. In the IMAGINE, all data transfers are explicitly managed by the stream controller shown in Fig. 2.9. Once pre-fetched data are prepared in the SRF, the large 32 GB/s bandwidth between the SRF and the LRFs is efficiently utilized to provide the 48 ALUs with multiple data simultaneously. After that, background pre-fetch operation of the next data is scheduled while the ALU clusters are computing fetched data. However, in the case of general-purpose applications, large peak bandwidth of the IMAGINE is not always available because scheduling of data pre-fetching is impossible for some applications with non-predictable data access patterns.

Another aspect of the stream processing, data parallelism, is also considered in the architecture of the IMAGINE, hence the eight ALU clusters are integrated to exploit data parallelism. The eight clusters perform computations on a divided part of the working set in parallel, and the six ALUs in each cluster compute kernels in a VLIW fashion. Large bandwidth among the ALUs and LRFs is provided for efficient forwarding of the operands and reuse of partial data calculated in the process of computing the kernels. Finally, producer–consumer locality is the key characteristic of stream processing, which is practical for reducing external memory transactions. In stream processing, a series of computation kernels are executed on an input data stream and large amounts of intermediate data are transacted between the adjacent kernels. If these intermediate data are only produced by a specific kernel and only consumed by a consecutive kernel, there is a producer–consumer locality between the kernels. In this case, it is not necessary to share these intermediate data globally and to maintain them in the off-chip memory for later reuse. In the IMAGINE, the SRF provides temporary storage for such intermediate data, thus reducing external memory transactions. In addition the stream buffers facilitate parallel data transactions between the producer and the consumer kernels computed in parallel.

In summary, the IMAGINE is an implementation of the customized memory hierarchy based on the common characteristics of memory transactions in the stream applications. Regarding the predictability in the memory access patterns, the memory hierarchy is designed so that peak bandwidth is gradually increased from outside of the chip to the ALU clusters. The increased peak bandwidth is fully utilizable by explicit management of the data transactions and also practical for facilitating parallel executions of the eight ALU clusters. The SRF of the IMAGINE is designed to store intermediate data having producer–consumer locality, and this is useful for reducing power consumption because unnecessary off-chip data transactions can be reduced. The other feature helpful for low-power consumption is the LRFs in the ALU clusters which maintain frequently reused intermediate data close to the processing units.

2.3.3 Memory-Centric NoC

In this section, the memory-centric Network-on-Chip (NoC) [16, 17] is introduced as a more application-specific implementation of the memory hierarchy. A target application of the memory-centric NoC is the scale-invariant feature transform (SIFT)-based object recognition. The SIFT algorithm [18] is widely adopted for autonomous navigation of mobile intelligent robots [19–22]. Due to vast amount of computation and limited power supply of the mobile robots, power-efficient computing of object recognition is demanded. The memory-centric NoC was proposed to achieve power-efficient object recognition by reducing external memory transactions of temporary data and overhead of data sharing in the multi-processor architecture. In addition, special-purpose memory is also integrated into the memory-centric NoC to further reduce power consumption by replacing complex operation with simple memory read operation. In this section, target application of the memory-centric NoC is described first to discover characteristics of the memory transactions. After that, architecture, operation, and benefits of the memory-centric NoC to implement power-efficient object recognition processor are explained.

2.3.3.1 SIFT Algorithm

The scale-invariant feature transform (SIFT) object recognition [18] involves a number of image processing stages which repeatedly perform complex computations on the entire pixels of the input image. Based on the SIFT, points of human interest are extracted from the input image and converted into vectors that describe the distinctive features of the object. The vectors are then compared with the other vectors in the object database to find the matched object. The overall flow of the SIFT computation is divided into key-point localization and descriptor vector generation stages as shown in Fig. 2.11. For the key-point localization, Gaussian filtering with varying coefficients is performed repeatedly on the input image. Then, subtractions among the filtered images are executed to yield the difference of Gaussian (DoG) images. By performing the DoG operation, the edges of different scales are detected from the input image. After that, 3×3 search window is traversed over all DoG images to decide the locations of the key points by finding the local maximum pixels inside the window. The pixels having a local maximum value greater than a given threshold become the key points.

The next stage of the key-point localization is the descriptor vector generation. For each key-point location, $N \times N$ pixels of the input image are sampled first, and then the gradient of the sampled image is calculated. The sample size N is decided according to the DoG image where the key-point location is selected. Finally, a descriptor vector is generated by computing the orientation and magnitude histograms over $M \times M$ subregions of the sampled input image. The number of key points detected for each object is about a few hundreds.

As shown in Fig. 2.11, each task of the key-point localization consumes and produces a large amount of intermediate data, and the data should be transferred between the tasks. The data transaction between tasks has significant impact on the overall object recognition performance. Therefore, the memory hierarchy design should account for characteristics of the data transactions. Here, we note two important characteristics of the data transaction in the key-point localization stage of the SIFT calculation.

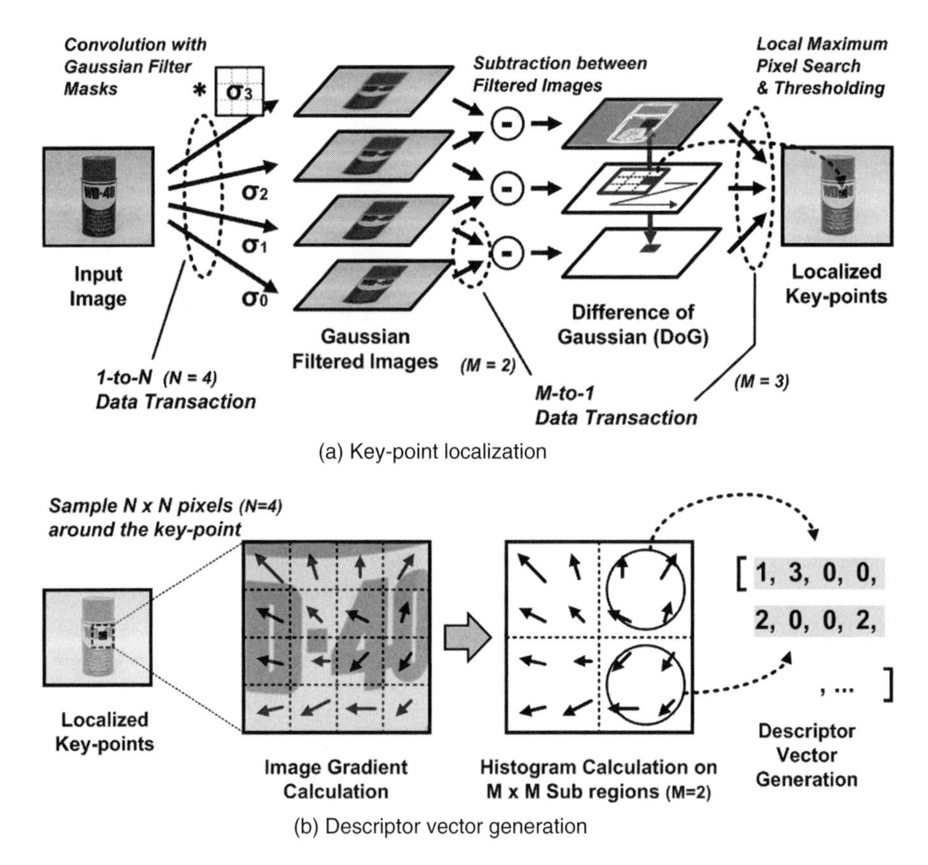

Fig. 2.11 Overall flow of the SIFT computation

The first point is regarding the data dependency between the tasks. As illustrated in Fig. 2.11(a), the processing flow is completely pipelined; thus data transactions only occur between two adjacent tasks. This implies that the data transaction of the SIFT object recognition has producer–consumer locality as well, and the memory hierarchy should be adjusted for tasks, organizing a task-level pipeline. The second point is concerning the number of initiators and targets in the data transaction. In the multi-processor architecture, each task such as Gaussian filtering or DoG will be mapped to a group of processors, and the number of processors involved in each task can be adjusted to balance the execution time. For example, the Gaussian filtering in Fig. 2.11(a), having the highest computational complexity due to the 2D convolution, could use four processors to filter operations with different filter coefficients, whereas all of the DoG calculation is executed on a single processor. Due to the flexibility in task mapping, the resulting data of one processor is transferred to multiple processors of subsequent task or the results from multiple processors are transferred to one processor. This implies that the data transaction will occur in the forms of not only 1-to-1 but also 1-to-N and M-to-1, as shown in Fig. 2.11(a).

By regarding the characteristics of the data transactions discussed, the memory-centric NoC realizes the memory hierarchy that supports efficient 1-to-N and M-to-1 data transactions between the pipelined tasks. Therefore the memory-centric NoC facilitates configuring of variable types of pipelines in the multi-processor architectures.

2.3.3.2 Architecture of the Memory-Centric NoC

The overall architecture of the object recognition processor incorporating the memory-centric NoC is shown in Fig. 2.12. The main components of the proposed processor are the processing elements (PEs), eight visual image processing (VIP) memories, and an ARM-based RISC processor. The RISC processor controls the overall operation of the processor by initiating the task execution of each PE. After initialization, each PE fetches and executes an independent program for parallel execution of multiple tasks. The eight VIP memories provide communication buffers between the PEs and accelerate the local maximum pixel search operation. The memory-centric NoC is integrated to facilitate inter-PE communications by dynamically managing the eight VIP memories. The memory-centric NoC is composed of five crossbar switches, four channel controllers, and a number of network interface modules (NIMs).

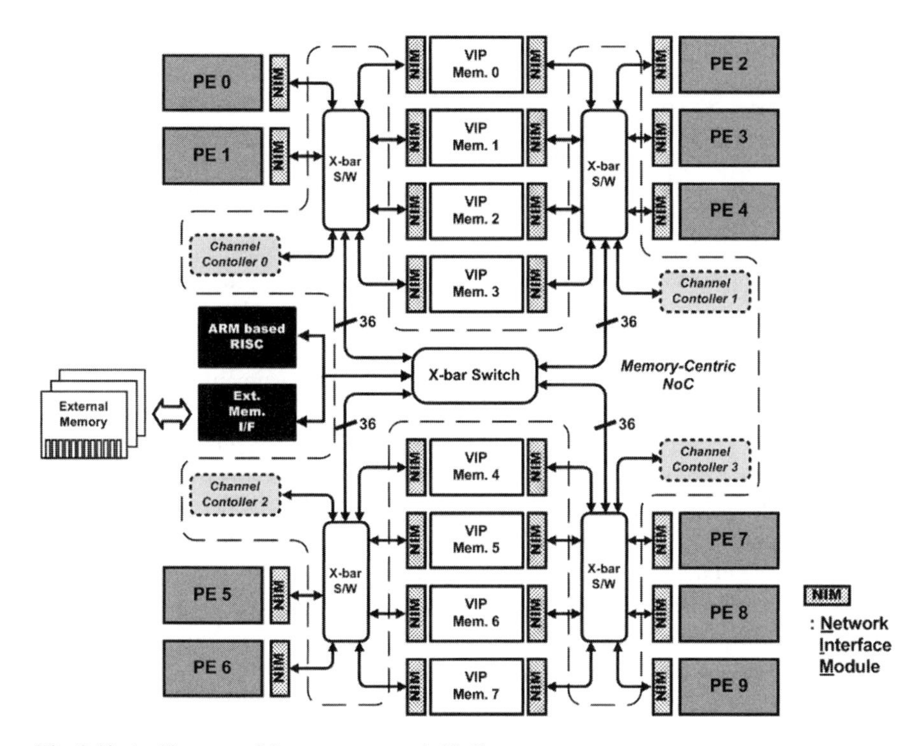

Fig. 2.12 Architecture of the memory-centric NoC

The topology of the memory-centric NoC is decided by considering the characteristics of the on-chip data transactions. For efficient support of the 1-to-N and M-to-1 data transactions shown in Fig. 2.11(a), using the VIP memory as a shared communication buffer is practical for removing the redundant data transfer when multiple PEs require the same data. Because the data flow through the pipelined tasks, each PE accesses only a subset of the VIP memories to receive the source data from its former PEs and send the resulting data to its following PEs. This results in localized data traffic, which allows tailoring of the NoC topology for low power and area reduction. There has been a research concerning power consumption and silicon area of the NoC in relation to NoC topologies [23], which concluded that a hierarchical star topology is the most efficient in case of interconnecting a few tens of on-chip modules with localized traffics. Therefore, the memory-centric NoC is configured in a hierarchical star topology instead of a regular mesh topology. By adopting a hierarchical star topology for the memory-centric NoC, the architecture of the proposed processor is able to be determined so that average hop counts between each PE and the VIP memories are reduced at the expense of a large direct PE-to-PE hop count, which is fixed to 3. This is also advantageous because most data transactions are performed between the PEs and the VIP memories, and direct PE-to-PE data transactions rarely occur. In addition, the VIP memory adopts dual read/write ports to facilitate short-distance interconnections between the ten PEs and the eight VIP memories. The NIMs are placed at each component of the processor to perform packet generation and parsing.

2.3.3.3 Memory-Centric NoC Operation

The operation of the memory-centric NoC is divided into two parts. The first part is to manage the utilization of the communication buffers, i.e., the VIP memories, between the producer and the consumer PEs. The other part is to support the memory transaction control after the VIP memory is assigned for the shared data transactions. The former operation removes the overhead of polling-available buffer spaces and the latter one reduces the overhead of waiting for valid data from the producer PE.

The overall procedure of the communication buffer management in the memory-centric NoC is shown in Fig. 2.13. Throughout the procedure, we assume that PE 1 is the producer PE and PEs 3 and 4 are consumer PEs. This is an example case of representing the 1-to-N (N=2) data transaction. The transaction is initiated by PE 1 writing an *open channel* command to the channel controller connected to the same crossbar switch (Fig. 2.13(a)). The *open channel* command is a simple memory-mapped write and transfers using a normal packet. In response to the *open channel* command, the channel controller reads the global status register of the VIP memories to check the utilization status. After selecting an available VIP memory, the channel controller updates the routing look-up tables (LUTs) in the NIMs of PEs 1, 3, and 4, so that the involved PEs read the same VIP memory for data transactions (Fig. 2.13(b)). The routing LUT update operation is performed by the channel controller sending the configuration (CFG) packets. At each PE, read/write accesses

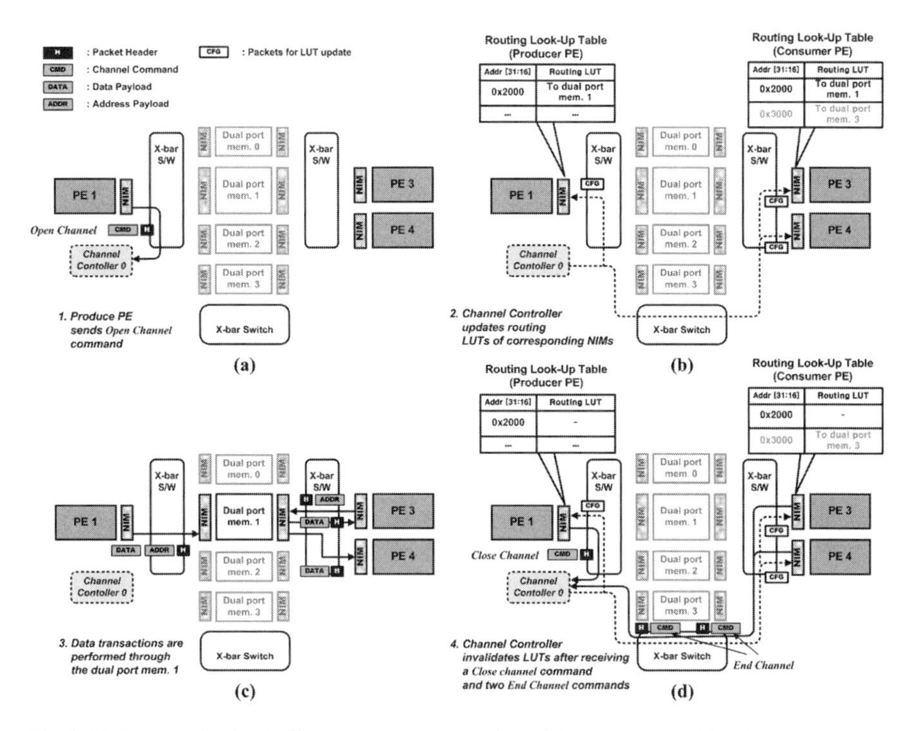

Fig. 2.13 Communication buffer management operation of the memory-centric NoC

for shared data transaction are blocked by the NIMs until the routing LUT update operation finishes. Once the VIP memory assignment is completed, a shared data transaction is executed using the VIP memory as a communication buffer. Read and write accesses to the VIP memory are performed using normal read/write packets that consist of an address and/or data fields (Fig. 2.13(c)). After the shared data transaction completes, PE 1 sends a *close channel* command, and PEs 2 and 3 send *end channel* commands to the channel controller. After that, the channel controller sends CFG packets to the NIMs of PEs 1, 3, and 4 to invalidate the corresponding routing LUT entries and to free up the used VIP memory (Fig. 2.13(d)).

From the operation of communication buffer management, efficient 1-to-N shared data transaction is clearly visible. Compared with the 1-to-1 shared data transaction, the required overhead is only sending additional (N–1) CFG packets at the start/end of the shared data transaction without making additional copy of shared data. In addition, an M-to-1 data transaction is also easily achieved by the consumer PE simply reading M VIP memories assigned to M producer PEs.

The previous paragraphs dealt with how the memory-centric NoC manages the utilization of VIP memories. In this paragraph, the memory transaction control scheme for efficient shared data transfer is explained. In the memory-centric NoC operation, no explicit loop is necessary to prevent consumer PEs reading the shared data too early before the producer PE writes valid data. To support the memory transaction control, the memory-centric NoC tracks every write access to the VIP memory from the producer PE after the VIP memory is assigned to shared data transactions. This is realized by integrating a valid bit array and valid check logic inside the VIP memory. In the VIP memory, every word has a 1-bit valid bit entry that is dynamically updated. The valid bit array is initialized when a processor resets or at every end of shared data transactions. By the write access from the producer PE, the valid bit of the corresponding address is set to HIGH. When an empty memory address with a LOW valid bit is accessed by the consumer PEs, the valid bit check logic asserts an INVALID signal to prevent reading false data. Figure 2.14 illustrates the overall procedure of the proposed memory transaction control. We assume again that PE 1 is the producer PE, and PEs 3 and 4 are consumer PEs. In the example data transaction, PE 3 reads the shared data at address 0×0 and PE 4 reads the shared data at address 0×8, whereas PE 1 writes the valid data only at address 0×0 of the VIP memory (Fig. 2.14(a)). Because the valid bit array has a HIGH bit for the address 0×0 only, the NIM of PE 4 obtains an INVALID packet instead of normal packets with valid data (Fig. 2.14(b)). Then, the NIM of PE 4 periodically retires reading valid data at address 0×8 until PE 1 also writes valid data at address 0×8 (Fig. 2.14(c)). Meanwhile, the operation of PE 4 is in a hold state. After reading the valid shared data from the VIP memory, the operation of the PE continues (Fig. 2.14(d)).

The advantages of the memory transaction control are reduced NoC traffic and PE activity, which contribute to a low-power operation. For consumer PE polls on the valid shared data, receiving INVALID notification rather than barrier value reduces the number of flits traversed through the NoC because the INVLAID notification does not have address/data fields. In addition, no polling loops are required for waiting valid data because the memory-centric NoC automatically blocks the

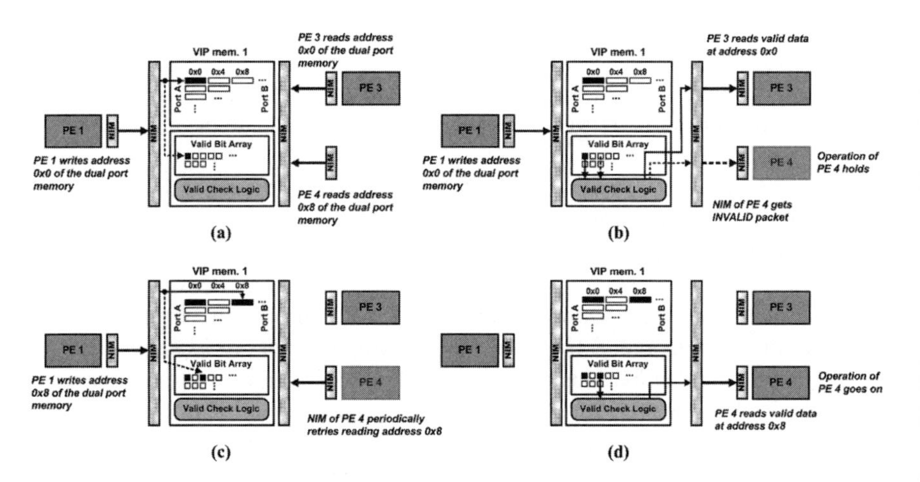

Fig. 2.14 Memory transaction control of the memory-centric NoC

access to the unwritten data. This results in reduced processor activity which is helpful for low-power consumption.

2.4 Low-Power Embedded Memory Design

At the start of this chapter, the concept of memory hierarchy was introduced first to draw a comprehensive map of memories in the computer architecture. After that, we discussed memory implementation techniques for low-power consumption regarding memory access patterns. Then, we discussed about the way of architecting the memory hierarchy considering the data flow of target applications for low-power consumption. As a wrap-up of this chapter, other low-power techniques applicable for memory design independent of data access pattern or data flow are introduced in this section. By using such techniques with application-specific optimizations, further reduction in power consumption can be achieved.

2.4.1 General Low-Power Techniques

For high-performance processors, providing data to be processed without bottleneck is as important as performing computation in high speed to achieve maximum performance. For that reason, there have been a number of researches for memory performance improvement and/or memory power reduction.

The common low-power techniques applicable to both DRAMs and SRAMs are summarized in [24]. This chapter reviews previously published low-power techniques such as reducing charge capacitance, operating voltage, and dc current, which focused on reducing power consumed by active memory operations. As the process technology has scaled down, however, static power consumption is becoming more and more important because the power dissipated due to leakage current of the on-chip memory starts to dominate the total power consumption in sub-micron process technology era. Even worse, the ITRS road map predicted that on-chip memory will occupy about 90% of chip area in 2013 [25] and this implies that the power issues in on-chip memories need be resolved. As a result, a number of low-power techniques for reducing leakage current in the memory cell have been proposed in recent decade. Koji Nii et al. suggested using lower NMOS gate voltage to reduce gate leakage current and peripheral circuits [26]. Based on the measured result that the largest portion of gate leakage current results from the turned on NMOS in the 6-transistor SRAM cell as shown in Fig. 2.15(a), controlling cell supply voltage is proposed. By lowering the supply voltage of the SRAM cells when the SRAM is in idle state, gate leakage current can be reduced without sacrificing the memory operation speed and this scheme is shown in Fig. 2.15(b). On the other hand, Rabiul Islam et al. proposed back-bias scheme to reduce sub-threshold leakage current of the SRAM cells [27]. This back-bias scheme is also applied when the SRAM is in idle state, and back-bias voltage is removed in normal operation.

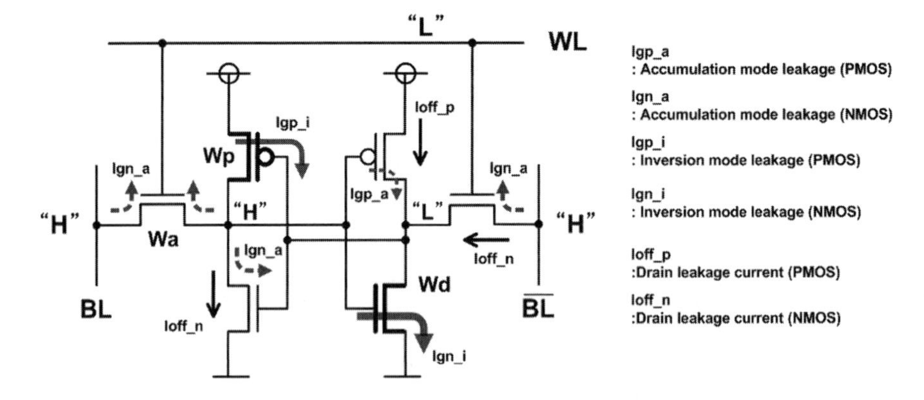

(a) Leakage model in six-transistor SRAM cell

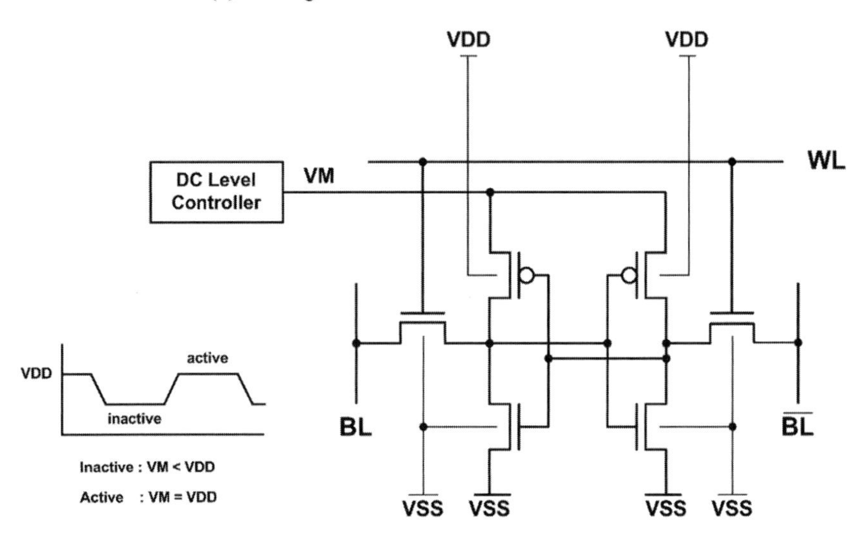

(b) Gate leakage current suppression in a cell

Fig. 2.15 Gate leakage model and suppression scheme [26]

More recent researches attempted to reduce leakage current more aggressively. Segmented virtual ground (SVGND) architecture was proposed to improve both static and dynamic power consumptions [28]. The SVGND architecture is shown in Fig. 2.16. The bit line of the SRAM is divided into M+1 segments, where each segment consists of a number of SRAM cells sharing the same segment virtual ground (SVG) and each SVG is switched between the real column virtual ground (CVG) and V_L voltage according to the corresponding segment select signals. In the SVGND architecture, only about 1/3–2/3 of power supply voltage is adaptively provided to the SRAM cells through the V_H and V_L signals instead of power and

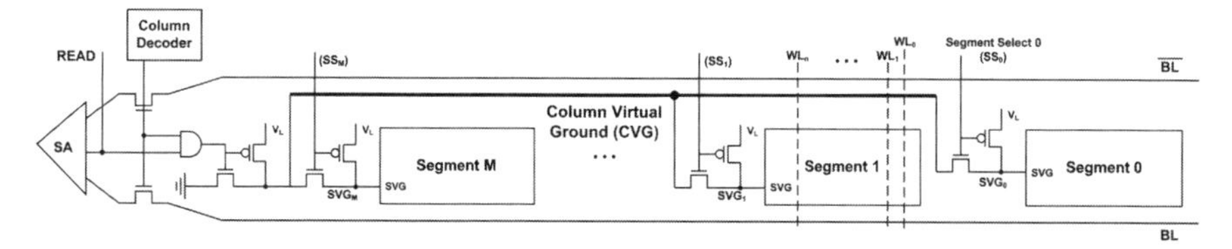

Fig. 2.16 Concept of SVGND SRAM [28]

ground signals, respectively. In this scheme the V_H is fixed and adjusted around two thirds of the supply voltage and the V_L is controlled between about one third of the supply voltage and the ground. At first, static power reduction is clearly visible. By reducing voltage across the SRAM cells, both gate and sub-threshold leakage currents can be kept in very low level. In addition, maintaining the source voltage of the NMOS (V_L) higher than its body bias voltage (Vss) has the effect of reverse biasing, and this results in further reduction of sub-threshold leakage current. The dynamic power consumption of the SRAM is also reduced by lower voltage across the SRAM cells. In the case of write operation, cross-coupled inverter chain in the SRAM cell can be driven to the desired value more easily. Compared to the SRAM cells with full supply voltages, the driving forces of SVGND SRAM cells have lower strength. When the read operation occurs, SVG line of each segment is pulled down to ground to facilitate sense amplifier operation. The power reduction in the read operation comes from selective discharge of SVG node, which prevents unnecessary discharge of internal capacitances of the neighboring cells in the same row.

As the process scales down to deep sub-micron, a more powerful leakage current reduction scheme is required. In the SRAM implementation using 65 nm process technology, Yih Wang et al. suggested using a series of leakage reduction techniques at the same time [29]. In addition to scaling of retention voltage in the SRAM cells, bit-line floating and PMOS back-gate biasing are also adopted. Lowering the retention voltage across the SRAM cell is the base for reducing gate and junction leakage current. However, there still remains junction leakage current from the bit line pre-charged to Vdd voltage which is higher than SRAM cell supply voltage. The bit-line floating scheme is applied to reduce the junction current through the gate NMOS. Finally, the PMOS back-gate biasing scheme suppresses leakage current through PMOS transistors in the SRAM cell which results from the lowered PMOS gate voltage due to retention voltage lowering. Figure 2.17 shows the concepts of leakage reduction schemes and their application to the SRAM architectures.

2.4.2 Embedded DRAM Design in RAMP-IV

Embedded memory design has advantages to system implementation. One of the biggest benefits is energy reduction in the memory interface and ease of memory utilization such as bus width. General system-on-board designs use off-the-shelf

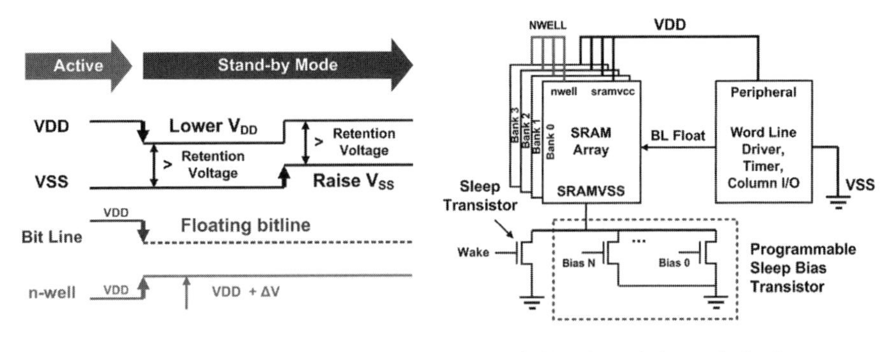

(a) SRAM cell leakage reduction techniques (b) SRAM array leakage reduction features

Fig. 2.17 SRAM cell leakage reduction techniques and their application to the SRAM [29]

memory devices and they have narrow bus width like 16 bit or 32 bit. For large data bandwidth, the system needs to increase the clock frequency or the bus width by using many memory devices in parallel. Figure 2.18 shows examples to realize 6.4 Gbps bandwidth. The first option increases the power consumption and the latter option occupies large system footprint. And both of them consume large power in pad drivers between the off chip memory and the processor. Embedded memory design, on the contrary, is free from the number of the bus width because interconnection between the memory and the processor inside the die occupies a little area. And the interconnection inside the die does not need large buffers like pad drivers. Another benefit is that the embedded memory does not need to follow conventional memory interface which is standard but somewhat redundant. Details will be shown through the 3D graphics rendering engine design with embedded DRAM memory.

Figure 2.19 shows the architecture of 3D graphics rendering processor [11]. Totally 29 Mb DRAMs are split into three memory modules: frame buffer, depth

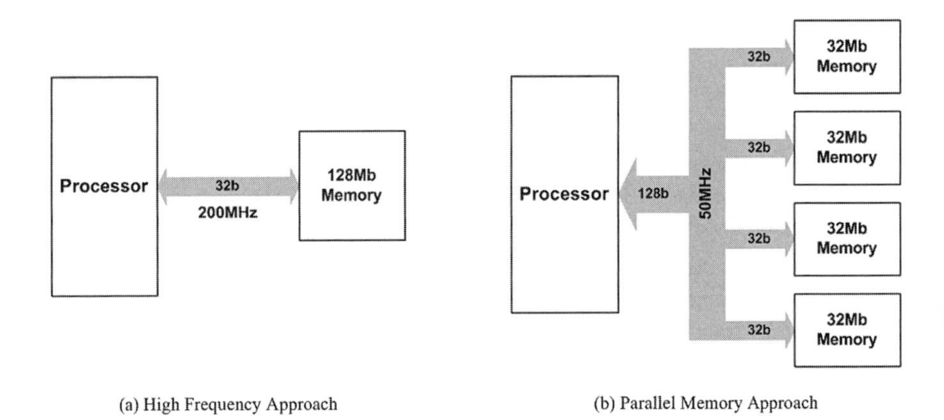

(a) High Frequency Approach (b) Parallel Memory Approach

Fig. 2.18 Large bandwidth approaches for system-on-board design

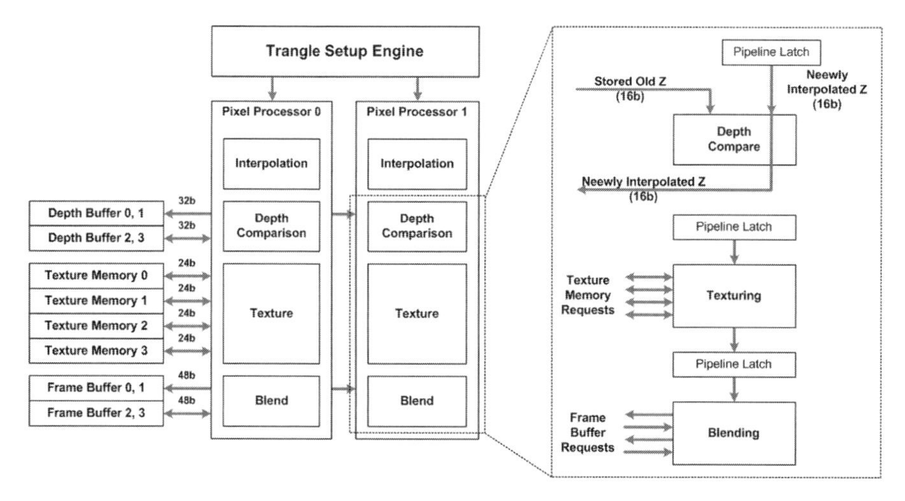

Fig. 2.19 Three-dimensional graphics rendering processor with DRAM-based EML approach

buffer, and texture memory. And each memory is physically divided into four memory modules so that totally 12 memory modules are used. In the pixel processors, scene or pixel is compared with the previous one. After that it is textured and blended in the pipeline stages. Each stage needs its own memory access to complete the rendering operations. And the memory access patterns are each different. For example, depth comparison and blending needs read–modify–write (RMW) operation while texturing needs just read operation. If the system is implemented by off-chip memories, the rendering processor needs 256 data pins and additional control signal pins, which cause increase in both package size and power consumption due to the pad driving. The processor shown in this example integrates the memories on a single chip and eliminates more than 256 pads for memory access.

For operation-optimized memory control, depth buffer and frame buffer are designed to support the single-cycle RMW operation with separate read and write buses, whereas the texture memory uses shared read/write bus. In order to provide the operation-optimized memory control for depth comparison and blending, the frame buffer and depth buffer support a single-cycle read–modify–write data transaction using separate read and write buses. It drastically simplifies the memory interface of the rendering engine and the pipeline, because the data required to process a pixel are read from the frame and depth buffers, calculated in the pixel processor, and written back to the buffers within a single clock period without any latency. Therefore, caching and pre-fetching, which may cause power and area overhead, are not necessary in the RMW-supporting architecture. The timing diagram of the RMW operation in the frame buffer is depicted in Fig. 2.20. To realize low-power RMW operation, command sequence is designed to use "PCG-ATV-READ-HOLD-WRITE" instead of "ATV-READ-HOLD-WRITE-PCG." In this case, the PCG sequence could be skipped in consecutive RMW operations. The write-mask signal, which is generated by the pixel processor, decides the activation of the write

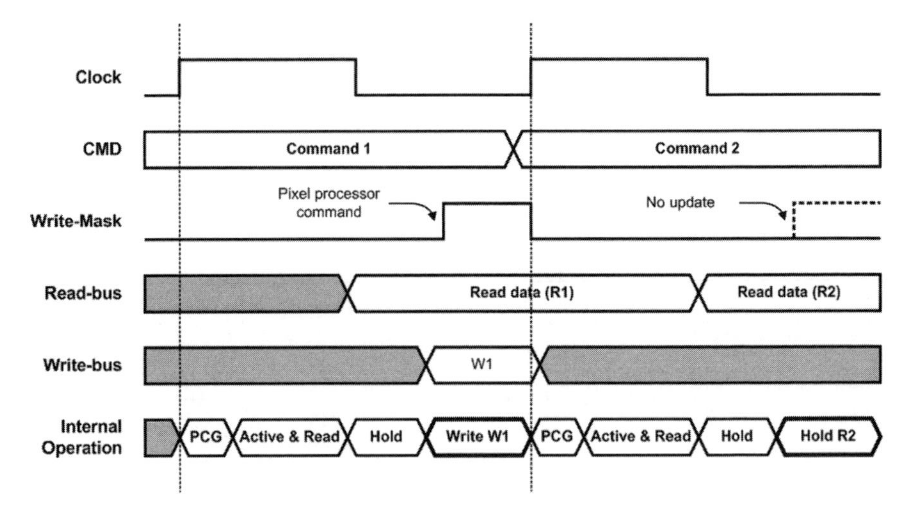

Fig. 2.20 Frame buffer access with read–modify–write scheme

operation. With single-cycle RMW operation, the processor does not need to use over-clocked frequency, and memory control is simple.

2.4.3 Combination of Processing Units and Memory – Visual Image Processing Memory

The other memory design technique for low-power consumption is to embed the processing units inside the memory. According to the target application domains, this technique is not always applicable in general. In the case of application-specific processor, however, the memory hierarchy is tailored for the target application and various types of memories are integrated together. Among them, some application-specific memories may incorporate the processing ability to improve overall performance of the processor. In case, large amount of data are loaded to a processor core from the memory and fixed operations are repeatedly performed on the loaded data, integrating the processing unit inside the memory is advantageous for removing the overhead of loading data into the processor core. A good implementation example of a memory with processing capability is a visual image processing (VIP) memory of KAIST [18, 30]. The VIP memory is briefly mentioned in Section 2.3 when describing the memory-centric NoC. Its function is to read out the address of local maximum pixel inside the 3×3 window in response to center pixel address of the $3\times$window.

The VIP memory has two behavioral modes: *normal* and *local-maximum* modes. In *normal* mode, VIP memory operates as a synchronous dual-port SRAM. It receives two addresses and control signals from the two ports and reads or writes two 32-bit data independently. While in *local-maximum* mode, the VIP memory finds

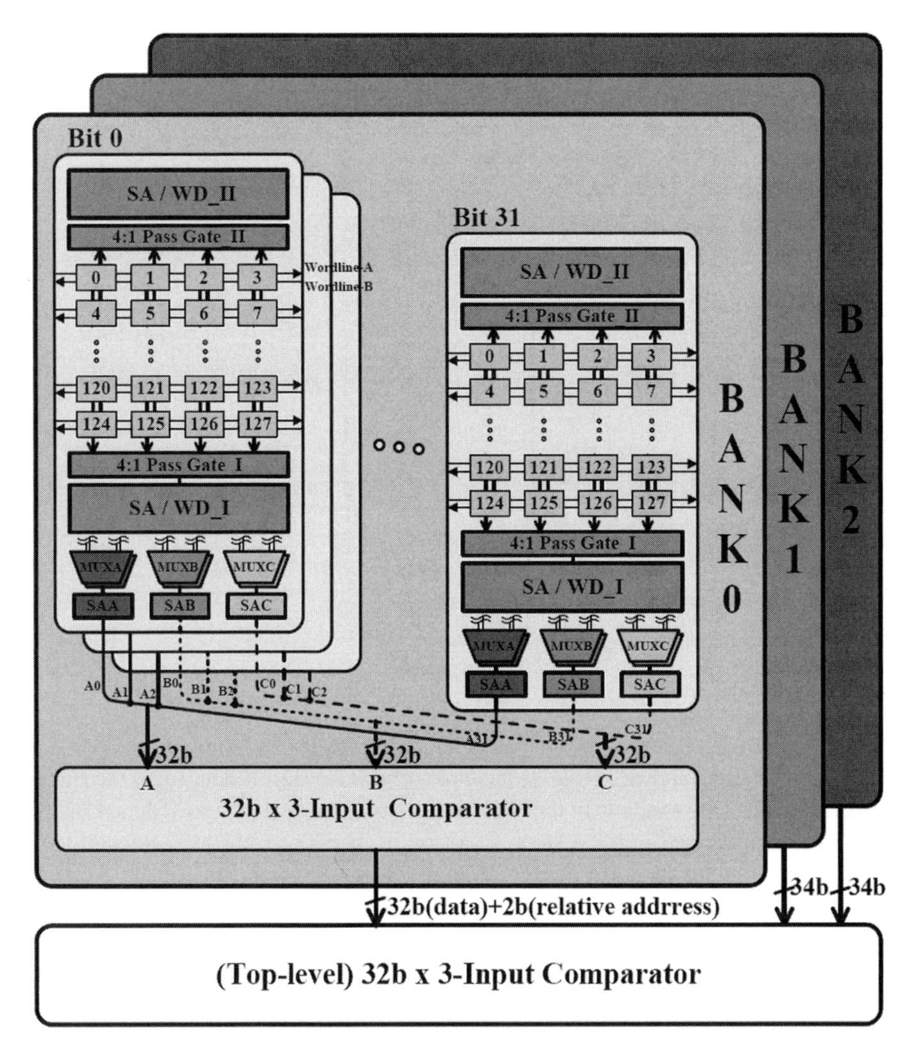

Fig. 2.21 Overall architecture of the VIP memory

the address of the local maximum out of the 3×data window when it receives the address of the center location of the window. Figure 2.21 shows the overall architecture of the VIP memory. It has a 1.5 KB capacity and consists of three banks. Each bank is composed of 32 rows and 4 columns and operates in a word unit. Each bit of four columns shares the same memory peripherals such as write driver and sense amplifier and the logic circuits for local maximum location search (LMLS). The LMLS logic is composed of multiplexers and tiny sense amplifiers. And a 3-input comparator for 32-bit number is embedded within the memory arrays.

Before the LMLS inside a 3×3 window, the pixel data of the image space should have been properly mapped into the VIP memory. First, the rows of the visual image

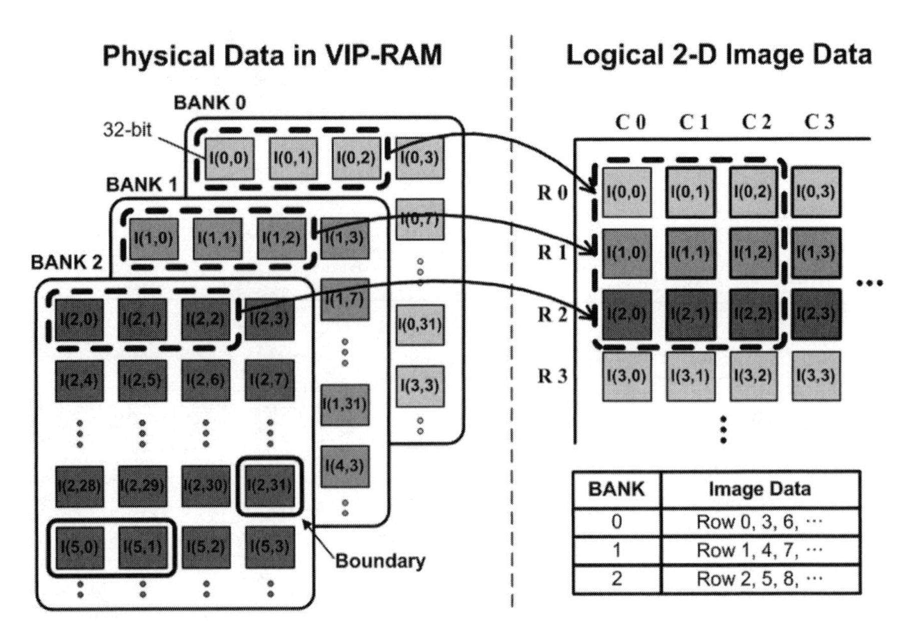

BANK	Image Data
0	Row 0, 3, 6, ···
1	Row 1, 4, 7, ···
2	Row 2, 5, 8, ···

Fig. 2.22 Data arrangement in the VIP memory

data are interleaved into different banks according to the modulo-3 operation on the row number as shown in Fig. 2.22. Then, the three 32-bit data from the three banks form the 3×3 window. In the VIP memory with properly mapped data, LMLS operation is processed in three steps. First, two successive rows are activated and three corresponding data are chosen by multiplexers. Second, the 32-bit 3-input comparators in three banks deduce the three intermediate maximum values among the respective three numbers of the respective bank. Finally, the top level 32-bit 3-input comparator finds the final maximum value of the 3×3 window from three bank-level intermediate results and outputs the corresponding address.

The VIP memory is composed of dual-ported storage cell which has eight transistors, as shown in Fig. 2.23. A word line and a pair of bit lines are added to the conventional 6-transistor cell. Pull-down NMOS transistors are larger than other minimum-sized transistors for stability in data retention. A single cell layout occupies 2.92 μm $\times$ 5.00 μm in a 0.18-μm process. Bitwise competition logic (BCL) is devised to implement a fast, low-power, and area-efficient 32-bit 3-input comparator. It just locates the first "1" from the MSB to the LSB of two 32-bit numbers and decides the larger number between the two 32-bit binary numbers without complex logics. The BCL enables the 32-bit 3-input comparator to be compactly embedded in the memory bank. Figure 2.24 describes its circuit diagram and operation. Before input to BCL comparator, each bit of two numbers are pre-encoded from A[i] and B[i] into (A[i] $\cdot$ ~B[i]) and (~A[i] $\cdot$ B[i]), respectively. Pre-encoding prevents the occurrence of logic failures in the BCL when both inputs have 1 at the same bit position. In the BCL, A line and B line are pre-charged to VDD initially. Then,

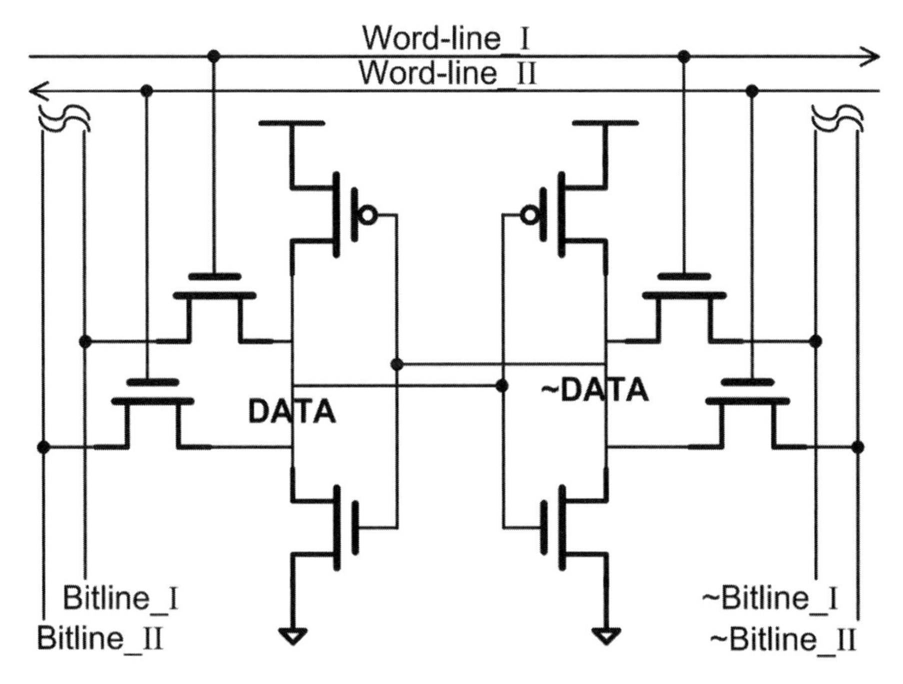

Fig. 2.23 A dual-ported memory cell of the VIP memory

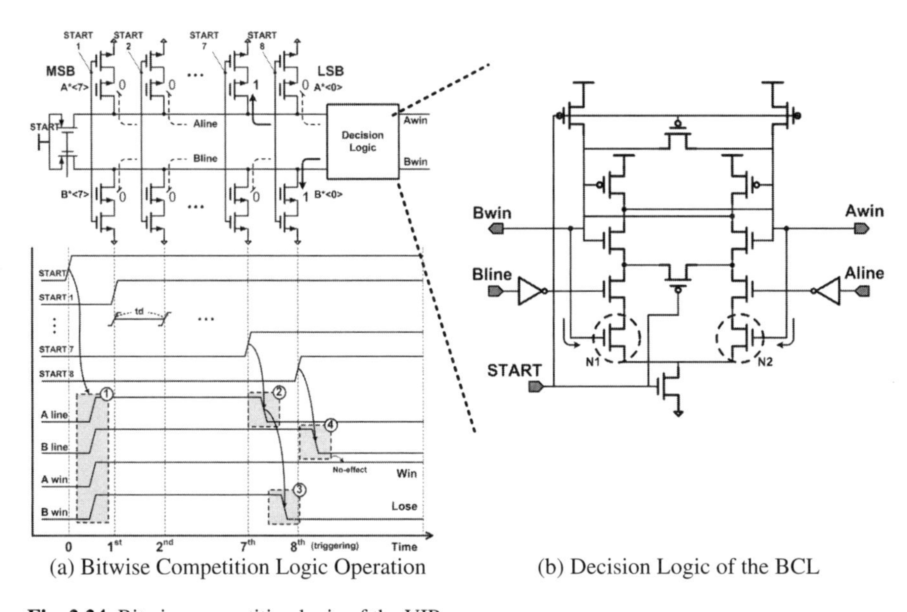

Fig. 2.24 Bitwise competition logic of the VIP memory

START signals are activated to trigger each bit of pre-encoded signals sequentially from MSB to LSB. If any triggered signal is 1, the path from the corresponding line to GND is opened and its voltage goes down immediately. Then, decision logic, at the right end of the lines, detects the line that first goes down and keeps the result until the bit comparisons end. As shown in Fig. 2.24, the circuit of decision logic is the same as the sense amplifier except transistors N1 and N2. N1 and N2 receive the feedback signals and disable input of the small number to preserve only the first decision or the large number. For example, the timing diagram of Fig. 2.24 illustrates its operation in case that the two pre-encoded inputs A∗ and B∗ are 00000010 and 00000001. The gray boxes represent the transitions of a few important events in BCL operation. At box (1), all lines are pre-charged when the START signal is low. Triggering starts but both lines stay in VDD by the seventh start signal. At the seventh triggering of box (2), A line is dropped to GND because the seventh bit of A is 1. The drop of A line forces the decision logic to turn down *Bwin* signal of box (3). Finally, *Awin* and *Bwin*, which represent the comparison results, are kept until the end of the cycle irrespective of B line voltage as shown in box (4). The 32-bit data comparator is composed of four parallel 8-bit BCLs. The 32-bit comparison results can be obtained from the four results of the four parallel BCLs by setting higher priority to the result of the MSB part BCL. As a result, the 32-bit 2-input comparator with BCL uses only 482 transistors, which are 38% less than the transistor count of the comparator reported in [31]. The 32-bit 3-input comparator is designed using three of 32-bit 2-input BCL comparators. Its worst case delay is 1.4 ns, which is sufficiently small considering the 5-ns timing budget of VIPRAM when operating at 200 MHz.

References

1. Lu Peng, et al., "Memory Performance and Scalability of Intel's and AMD's Dual-Core Processors: A Case Study," IEEE International Performance, Computing, and Communication Conference, pp.55–64, April, 2007.
2. Dac C. Pham, et al., "Overview of the Architecture, Circuit Design, and Physical Implementation of a First-Generation Cell Processor," IEEE Journal of Solid-State Circuits, Vol. 41, Issue 1, pp.179–196, Jan. 2006
3. Marc Tremblay and Shailender Chaudhry, "A Third-Generation 65 nm 16-Core 32-Thread Plus 32-Scout-Thread CMT SPARC Processor," Technical Digest of IEEE International Solid State Circuits Conference, pp.82–83, February, 2008
4. Seung-Jun Bae, et al., "A 60 nm 6 Gb/s/pin GDDR5 Graphics DRAM with Multifaceted Clocking and ISI/SSN-Reduced Techniques," Technical Digest of IEEE International Solid State Circuits Conference, pp.82–83, February, 2008
5. Kyungwoo Nam, et al., "A 512 Mb 2-Channel Mobile DRAM (oneDRAM™) with Shared Memory Array," IEEE Asian Solid-State Circuits Conference, pp.204–207, November, 2007.
6. Samsung High Speed SRAM product page, http://www.samsung.com/global/business/semiconductor/products/sram/Products_HighSpeedSRAM.html, 2008
7. International Technology Roadmap for Semiconductors, Interconnect, 2003 Edition, Semiconductors Industry Assoc. and SEMATECH.
8. John L. Hennessy and David A. Patterson, "Computer Architecture – A Quantitative Approach," third edition, pp.390–392, San Francisco, USA, Morgan Kaufmann Publishers, 2003.

9. Satoru Tanoi, et al., "A 32 Bank 256 Mb DRAM with Cache and TAG," Technical Digest of IEEE International Solid State Circuits Conference, Feb., 1994.
10. Barth J.E., Jr., et al., "A 500-MHz Multi-Banked Compliable DRAM Macro with Direct Write and Programmable Pipelining," IEEE Journal of Solid-States Circuits, Vol. 40, Jan. 2005.
11. Ramchan Woo, et al., "A 210-mW Graphics LSI Implementing Full 3-D Pipeline with 264 Mtexels/s Texturing for Mobile Multimedia Applications," IEEE Journal of Solid-States Circuits, Vol. 39, Feb., 2004
12. Marvell technology product page, http://www.marvell.com/products/cellular/applications.jsp, 2008.
13. Marvell technology, "PXA300 Product Brief," http://www.marvell.com/files//products/cellular/ application/PXA300_PB_R4.pdf, 2008
14. Brucek Khailany, et al., "IMAGINE: Media Processing with Streams," IEEE Micro, Volume 21, Issue 2, pp. 35–46, Mar.-Apr., 2001.
15. William J. Dally, et al., "Stream Processors: Programmability with Efficiency," ACM Queue, Vol. 2, Issues 1, pp. 52–62, 2004
16. Donghyun Kim, et al., "Solutions for Real Chip Implementation Issues of NoC and Their Application to Memory-Centric NoC," IEEE/ACM 1st International Symposium on Networks-on-Chip, pp. 30–39, May, 2007.
17. Donghyun Kim, et al., "An 81.6 GOPS Object Recognition Processor Based on NoC and Visual Image Processing Memory," IEEE Custom Integrated Circuits Conference, pp. 443–446, Sept. 2007.
18. David G. Lowe, "Distinctive Image Features from Scale-Invariant Key points," ACM Intl. Journal of Computer Vision, Vol. 60, Issue 2, pp. 91–110, 2004.
19. Sunghwan Ahn, et al., "Data Association Using Visual Object Recognition for EKF-SLAM in Home Environment," Proceedings of IEEE Intl. Conf. on Intelligent Robots and Systems, pp. 2760–2765, 2006.
20. Patric Jensfelt, et al., "Augmenting SLAM with Object Detection in a Service Robot Framework," IEEE Intl. Symposium on Robot and Human Interactive Communication, pp. 741–746, 2006.
21. Bertolli F., Jensfelt P., Christensen H.I., "SLAM using Visual Scan-Matching with Distinguishable 3D Points," IEEE/RSJ International Conference on Intelligent Robots and Systems, pp. 4042–4047, Oct., 2006.
22. Zhang Nan, Li Maohai, Hong Bingrong, "Active Mobile Robot Simultaneous Localization and Mapping," IEEE International Conference on Robotics and Biomimetics, pp. 1671–1681, Dec., 2006.
23. Kangmin Lee, Se-Joong Lee and Hoi-Jun Yoo, "Low-power network-on-chip for high-performance SoC design," IEEE Transactions on Very Large Scale Integration Systems, Vol. 14, Issue 2, pp. 148–160, Feb., 2006.
24. Kiyoo Itoh, Katsuro Sasaki, and Yoshinobu Nakagome, "Trends in Low-Power RAM Circuit Technologies," IEEE Digest of Technical Papers of Symposium on Low Power Electronics, pp. 84–87, Oct., 1994.
25. International Technology Roadmap for Semiconductors, 2001 Update, Semiconductors Industry Assoc. and SEMATECH.
26. Koji Nii, et al., "A 90-nm Low-Power 32-kB Embedded SRAM with Gate Leakage Suppression Circuit for Mobile applications," IEEE Journal of Solid-State Circuits, Vol. 39, Issue 4, pp. 684–693, Apr., 2004.
27. Rabiul Islam, Adam Brand, and Dave Lippincott, "Low Power SRAM Techniques for Handheld Products," IEEE Proceedings of the 2005 International Symposium on Low Power Electronics and Design (ISLPED), pp. 198–202, Aug., 2005.
28. Mohammad Sharifkhani, and Majog Sachdev, "Segmented Virtual Ground Architecture for Low-Power Embedded SRAM," IEEE Transactions on Very Large Scale Integration Systems (TVLSI), pp. 196–205, Feb., 2007.

29. Yih Wang, et al., “A 1.1 GHz 12 uA/Mb-Leakage SRAM Design in 65 nm Ultra-Low-Power CMOS Technology with Integrated Leakage Reduction for Mobile Applications,” IEEE Journal of Solid-State Circuits, Vol. 43, Issue 1, pp. 172–179, Jan., 2008.
30. Joo-Young Kim, et al., “Visual Image Processing RAM for Fast 2-D Data Location Search,” IEEE European Solid State Circuits Conference, pp. 324–327, Sept., 2007.
31. Shun-Wen Cheng, “A High-Speed Magnitude Comparator with Small Transistor Count,” IEEE Proceedings of International Conference on Electronics, Circuits and Systems, Vol. 3, pp. 1168–1171, Dec., 2003.

Chapter 3
Embedded SRAM Design in Nanometer-Scale Technologies

Hiroyuki Yamauchi

Abstract Static random access memory (SRAM) has been embedded in almost all of VLSI chips and has played a key role in the wide variety of applications required to enhance the performances of high speed, high density, low power, low voltage, low cost, time to market. Embedded SRAM has had a long reign in upper memory hierarchy than any other memories such as dynamic random access memory (DRAM). This is largely because SRAM is able to provide the highest random access speed performance among various embedded memory technologies. In addition, SRAM is fully compatible with CMOS logic process technology and operating voltage, enabling a seamless integration with logic circuits. Meanwhile, the device miniaturization driven by the technology scaling into nanometer regime has made it more challenging to maintain a sufficient SRAM cell stability margin while continuing to increase a random access speed as the transistor threshold voltage mismatching becomes significant. This also makes it more difficult to scale the operating voltage (V_{DD}) while keeping the compatibility with logic's. This chapter intends to provide an overview on the state-of-the-art SRAM circuit design technologies to address the key SRAM challenges in nanometer-scale technologies in terms of the read/write stability margins, cell current, and leakages.

3.1 Introduction

Primary reason for SRAM introduced in upper memory hierarchy in VLSI chip will be described in Section 3.1.1. Following the SRAM basics of circuit configuration and its control for operation in Section 3.1.2, scaling trend of SRAM cell size and operating voltage will be described in Sections 3.1.3 and 3.1.4, respectively.

H. Yamauchi (✉)
Fukuoka Institute of Technology, Fukuoka, Japan
e-mail: yamauchi@fit.ac.jp

K. Zhang (ed.), *Embedded Memories for Nano-Scale VLSIs*, Series on Integrated Circuits and Systems, DOI 10.1007/978-0-387-88497-4_3,

3.1.1 Embedded SRAMs in VLSI Chip

SRAM has had a long reign in upper memory hierarchy and has been used for the register file and cache memories of all levels from first to third in the embedded memory system as shown in Fig. 3.1. Primary reasons for this dominance are [27, 30] (1) SRAM is able to provide the highest random access speed and (2) SRAM enables a seamless integration with logic circuits due to its compatibility of process and operating voltage (V_{DD}). In addition, SRAM has been widely used for the buffer and register memories in application-specific integration circuit (ASIC), which requires a compiler to generate a wide range of SRAM configurations of the numbers of bit, access port, input/output, and banks. Compared to others, SRAM is able to provide the compiler with higher cell array area efficiency. For example, SRAM cell area efficiency is over 75%, but that of DRAM is less than 50% for 1-Mbit macro.

As multiple processing cores or central processing unit (CPU) cores are being integrated into one chip, the demand for integrated on-chip SRAM has become even higher to provide a sufficient data stream and to keep up with an increasing demand of memory capacity and bandwidth. For example, the total percentage of occupied SRAM area of overall chip size is ever increasing and has been reached over 70%. An over 128-Mbit SRAM has been implemented in an area of 100 mm^2 at a 45 nm process node [27, 30]. As a result, the silicon area and power consumptions from SRAM have become a major factor in the overall power budget and cost performance for advanced VLSI system designs. The SRAM power savings are particularly important for battery-operated mobile and hand-held applications where required battery size/weight for a certain lifetime depends on the power

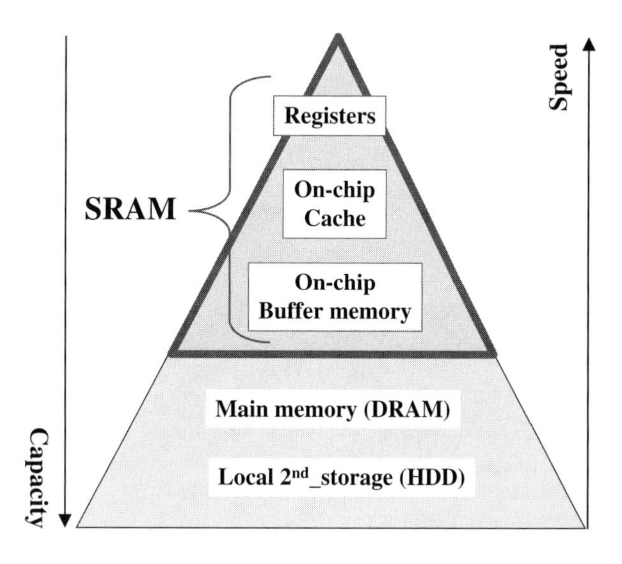

Fig. 3.1 Embedded SRAM positioning in memory system

consumption. The SRAM bit cell size scaling has also become important for not only saving silicon area but also increasing bit density.

On the other hand, the transistor and SRAM bit cell size reduction driven by the technology scaling has also made it even more challenging to maintain a sufficient cell stability margin while keeping the same scaling pace of access time and cell size as the mismatching of threshold voltage (V_T) between cross-coupled MOSFET pairs becomes larger and larger [27–30]. To maintain sufficient margins for read and write stability and read cell current has become challenging as V_{DD} has to be scaled down in order not only to meet the requirements for device scaling and power savings but also to keep the logic operating voltage compatibility. The V_{DD} scaling has quickly become one of the most critical challenges because of its strong dependency on the SRAM stability margins for read and write since the invention of 65 nm process node. In this chapter, the state-of-the-art SRAM circuit technologies including a bit cell design to address these challenges will be described.

3.1.2 SRAM Cells Array Configuration

SRAM elements are arranged in an array of rows and columns. Each row of bit cells shares a common word-line (WL), while each column of bit cells shares a common bit-line (BL) as shown in Fig. 3.2. The number of columns of the memory array is known as the bit width of each word. Depending on the relationship between the number of BLs (n) and the bit width (m), the BLs are multiplexed with ration of n to m. For example, 64-bit data (m) are output through 8 to 1 multiplexer from 512-BL pairs (n). This architecture is able to relax sense-amplifier layout pitch by

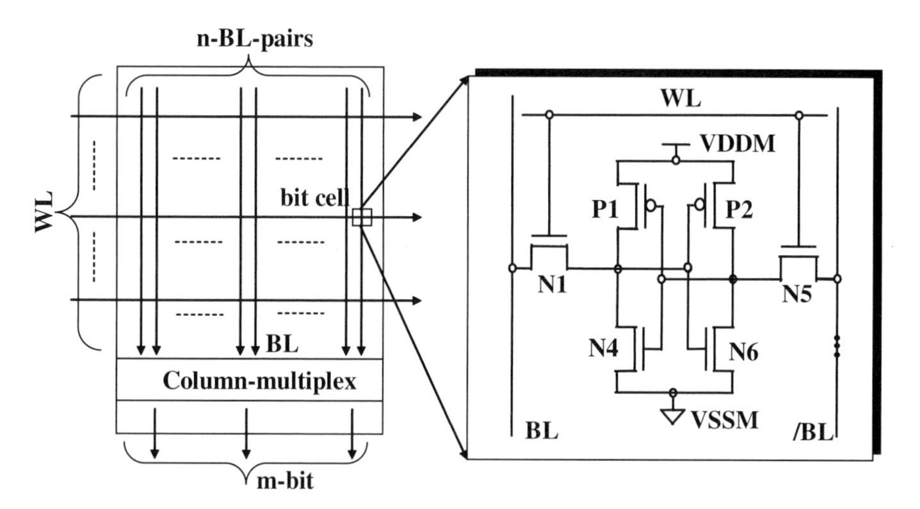

Fig. 3.2 SRAM basics

(*n*/*m*)-fold. In this section, following the SRAM data storage elements, the basic operations of reading, writing, and data holding are described.

3.1.2.1 Six Transistor SRAM Cell

The basic storage element of an SRAM consists of the pairs of inverter and pass-gate. A pair of inverters is cross-coupled such that it has the output of one inverter going into the input of the other and vice versa. The CMOS cross-coupled inverters can hold a "1" or "0" state as long as SRAM is powered up. As a result, SRAM does not need to refresh in order to hold its data, making its cycle time shorter than a DRAM, which needs data refreshing periodically. However, since SRAM memory cell consists of larger number of transistors, contacts, and wirings than that of a DRAM, it takes more area on the chip, making SRAMs more area consuming than DRAMs. The ratio of bit cell areas of SRAM to DRAM is about 4–5.

Cell layout of a 6T SRAM is shown in Fig. 3.3. To provide the best possible control over device channel length, all the poly-silicon gates for arrays are run in the same direction on a fixed pitch based on the restricted design rule (RDR). The cell layout that conforms to this arrangement is referred to as a thin-type cell as shown in Fig. 3.3. The thin-type cell configuration is narrow in the direction of bit-line, so the bit-line wiring length is minimized, enabling a shorter BL charging and discharging delay with lesser power consumption.

3.1.2.2 SRAM Read

The read cycle is started by pre-charging of the pair of BL to logical "1"-level V_{DD}, then asserting the WL, enabling to turn-on the pass-gate (PG) transistors. The second step occurs when the one side of the BL pair is connected to the "low" side of storage node pair through the PG transistor and the stored charge on the BL is transferred to the storage node. As a result, the BL level is lowered due to not only

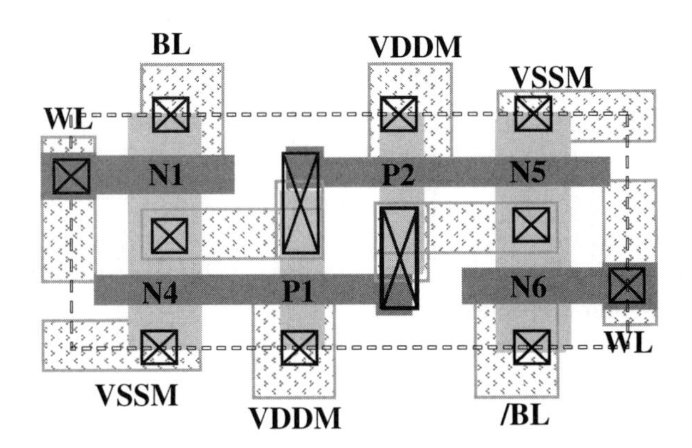

Fig. 3.3 SRAM cell layout

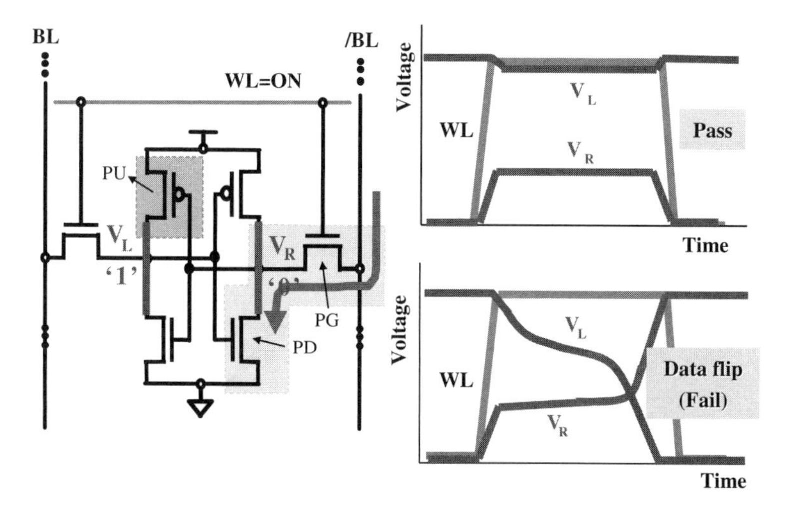

Fig. 3.4 SRAM read operation and its waveforms

the charge-sharing but also discharging to logical "0"-level V_{SS} through the PG and pull-down (PD) transistors connected in series as shown in Fig. 3.4. During the read, discharging current flows from BL to V_{SS}, the potential of the storage node of "0" side goes higher depending on the drivability/resistance ratio of the PG and PD transistors connected in series. If this level goes over the trip point of connected inverter, the stored data is flipped. To avoid such data flipping, the drivability of PG has to be weaker than that of PD. This ratio is one of the key SRAM design parameter which is referred to as β-ratio. Careful sizing of the transistors to realize the required β-ratio is needed to ensure stable read operation. On the other BL side, the PG transistor is not turned on as long as the potential difference from WL to BL or the storage node is smaller than V_T of PG. In that sense, BL pre-charging level has to be designed so that PG is not turned on [4]. If the content of the memory was inverted, the opposite would happen in BL discharging [27] in the same manner.

3.1.2.3 SRAM Write

The start of a write cycle begins by forcing the BL pair to the differential levels of "1" and "0" so as to be written to the corresponding storage nodes. If we wish to invert the storage data of the same cell, we would force the BL pair to the inverted differential levels compared to the previous one. The WL is then asserted and the PG connected to the BL is turned on and the potential of the corresponding storage node is lowered and it is dependent on the ratio of drivability of PG and pull-up (PU) transistors as shown in Fig. 3.5. To successfully write into the cell, the critical level has to be lowered than the trip point of the inverter in the storage element. This ratio is one of the key SRAM design parameter which is referred to as γ-ratio.

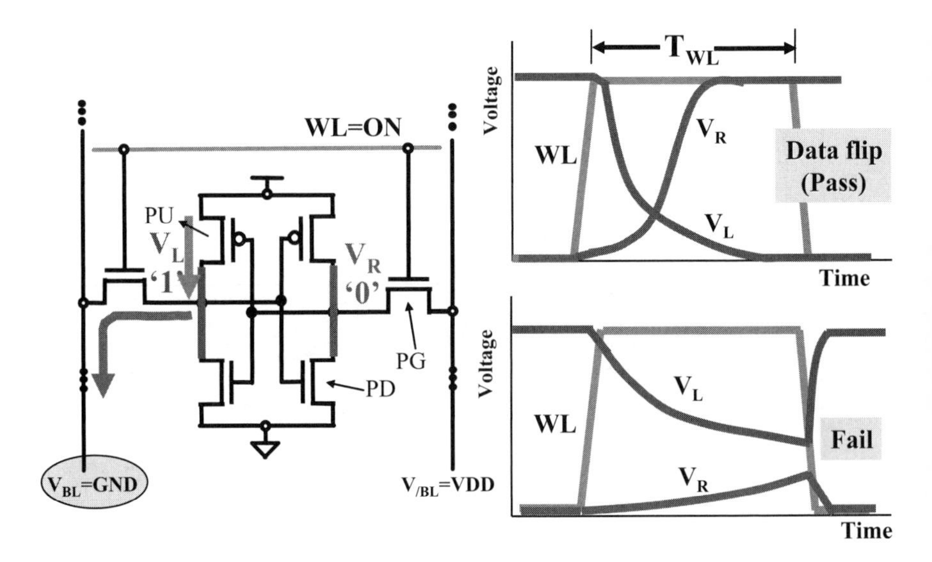

Fig. 3.5 SRAM write operation and its waveforms

Careful sizing of the transistors to realize the required γ-ratio is needed to ensure robust write operation [27].

3.1.2.4 SRAM Data Retention

When WL is not asserted, SRAM cell is in data retention mode. A sufficient level (VDDH) is required to turn-on the inverters as shown in Fig. 3.6. The cross-coupled inverters will reinforce each other without any disturbance from BL through PG. As a result, SRAM data can hold the full potential difference of (VDDH–V_{SS}). However, when VDDH gets lower than a certain point, which is referred to as SRAM data retention voltage (V_{HOLD}), no longer the inverters are able to hold the state, the storage nodes can latch into the wrong state as shown in Fig. 3.6 [27]. As explained later, nanometer-scale SRAM needs to lower the VDDH as low as possible to reduce the leakage while maintaining V_{HOLD}. However, the trend of V_{HOLD} gets higher as SRAM size is scaling.

3.1.3 SRAM Cell Scaling Trend

As Moore's law continues to drive the scaling of CMOS technology well into nanometer region, the advancement of the technology provides a greater opportunity for higher-level integration in VLSI systems. The recent trend of implementing homogeneous and heterogeneous multiple-cores in microprocessors and system LSIs is a perfect example where more on-die memory capacity is needed to meet its bandwidth requirement. The memory density is doubled every process generation

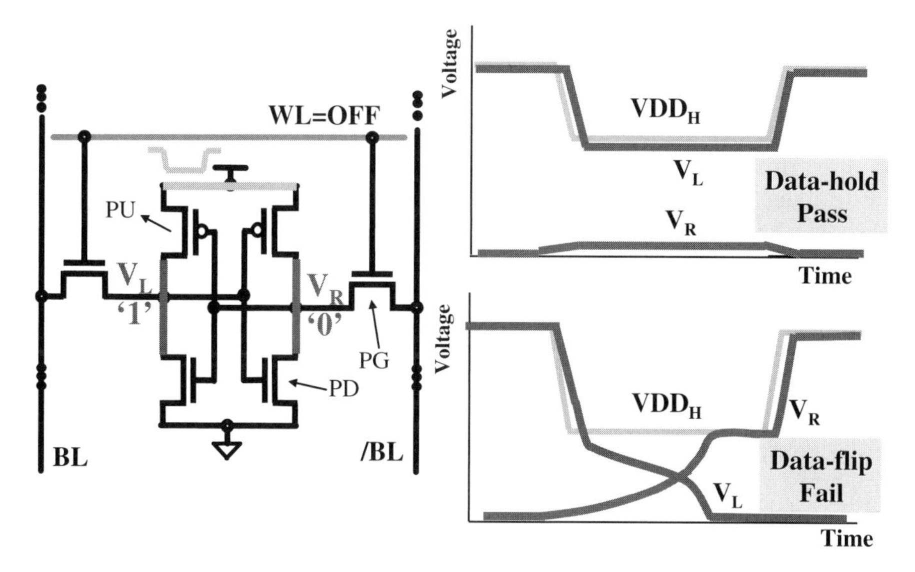

Fig. 3.6 SRAM data retention

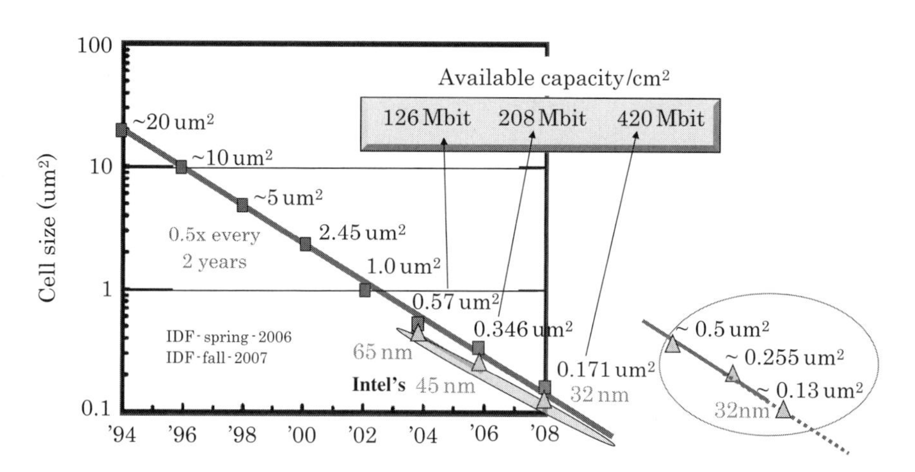

Fig. 3.7 SRAM memory cell size scaling trend

in order to meet the continuous demand as shown in Fig. 3.7. With rapid technology scaling trend, over 208 Mbit has been fabricated within 100 mm^2 at 45 nm process node. Although general scaling trends in the industries are similar, SRAM cell size itself is varied depending on the requirements for the minimum operation voltage (V_{DD_MIN}), access speed, and total bit density of embedded SRAM [28–30].

3.1.4 SRAM-Operating Voltage V_{DD} Scaling Trend

Unlike the cell size scaling trend, SRAM V_{DD} scaling trend has significantly slowed down recently and it saturates around 1.0 V as shown in Fig. 3.8. One primary reason why SRAM V_{DD} scaling has become so challenging was due to scaling limitation of transistor V_T limited exponentially increasing leakage. However, one more inevitable reason has been added since 65 nm process node. The reason stems from increasing transistor V_T random variations as the transistor feature size is scaled into nanometer region. The increasing V_T random variation degrades SRAM functional margins, resulting in an increase of the minimum SRAM V_{DD} [28–30].

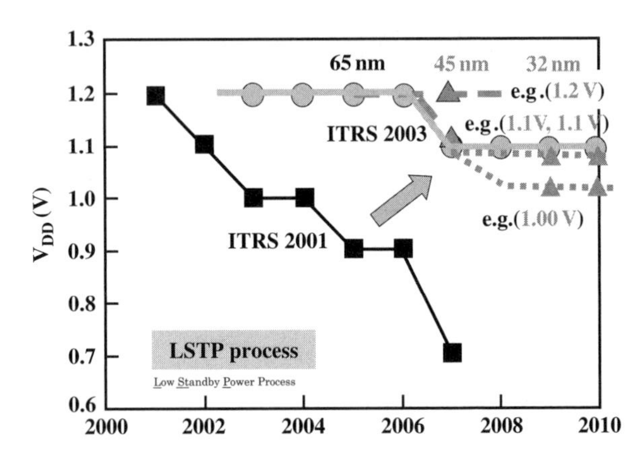

Fig. 3.8 SRAM-operating voltage scaling trend

3.2 Functional Margin Issues with Scaled Nanometer-Scale SRAM

SRAM functional margins are determined by three key SRAM design parameters of static noise margin (SNM), write margin (WRM), and cell current (Icell). Since all of them strongly depend on operating voltage (V_{DD}), transistor channel length (Lg), and width (Wg), scaling of all these key parameters will be examined in details in this section.

3.2.1 Static Noise Margin (SNM)

The static noise margin (SNM) is the maximum amount of noise voltage V_N that can be tolerated at the both inputs of the cross-coupled inverters in different directions while inverters still maintain bi-stable operating points and cell retains its data. In other words, the static noise margin (SNM) quantifies the amount of noise voltage V_N required at the storage nodes of SRAM to flip the cell data. The most common

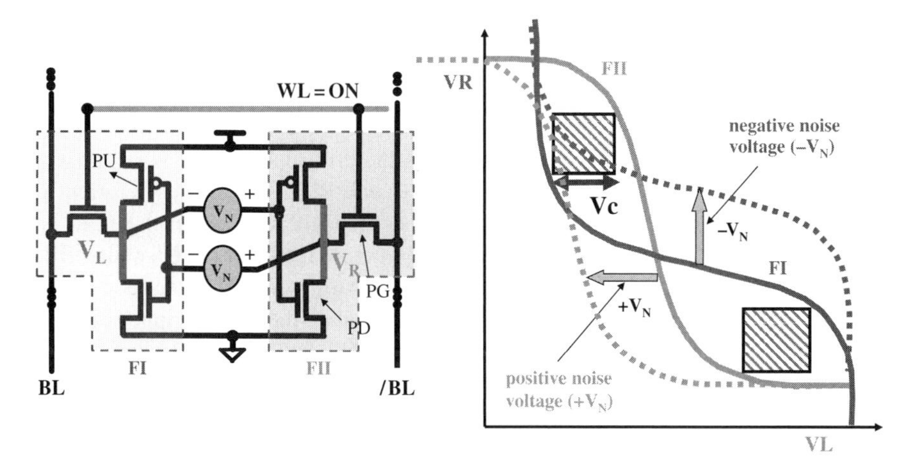

Fig. 3.9 SNM graphical definition

way of representing the SNM graphically is shown in Fig. 3.9 [27]. The two curves representing the voltage transfer characteristic (FI and FII) of the inverters that are inversed from each other are shown in Fig. 3.9. The resulting two-wing curve is called a "butterfly curve" and is used to determine the SNM. The SNM is defined as the length of the side of the largest nested square (Vc) that can be embedded inside smaller wing of the butterfly curve. Graphically, as seen in Fig. 3.9, the SNM equals the noise voltage necessary at each of the cell storage nodes to shift the static characteristics of the two cell inverters (FI and FII) vertically or horizontally along the side of the maximum nested square so that they intersect at only one point as shown in Fig. 3.9 [27].

The cell becomes more vulnerable to noise during a read access than data retention state since the "0" storage node rises to a voltage higher than ground (GND) due to a voltage division along the PG and inverter PD devices between the pre-charged BL and the GND terminal of the cell. The ratio of the transistor width of PD to PG, commonly referred to as the β-ratio determines how high the "0" storage node rises during a read access. Careful design of β-ratio is needed to maintain the required SNM. As the device size of NFETs is scaled down to nanometer regime, the variation of β-ratio is significantly increased. This is the primary reason for increasing SNM challenge in nanometer-scale SRAM. The ratio of inverter PD and PU also directly impacts the cell immunity to noise. Weaker PU due to the variations makes the cell easier to flip as lowering its trip point of inverter, making the cell more vulnerable to noise.

When the WL is off, the SNM becomes larger than that for read access because of no rising of "0" storage node from GND level. The two kinds of SNMs for data retention and read access are referred to as "hold SNM margin" and "read SNM margin" as shown in Figs. 3.4 and 3.6, respectively.

3.2.2 Write Margin (WRM)

The cell data is written by forcing the BL pair to the differential levels of "1" and "0" while WL is asserted to allow PG connected to the BL. The potential of the corresponding storage node is pulled down to the critical level that is dependent on the ratio of transistor strengths between PG and PU. This ratio is referred to as γ-ratio. In order to ensure robust write operation, the critical level has to be lowered than the trip point of connected inverter before the level of "0"-written BL is reached to the end-point (e.g., GND). The write margin (WRM) is defined as the rest of potential difference between the BL level at which the data is flipped and the end-point (e.g., GND) as shown in Fig. 3.10. If the cell data is flipped when the BL comes at X mV, where X mV is allowed to reach to the GND level, WRM is defined as *X* mV. Careful design of γ-ratio is needed to maintain the required WRM. As the device sizes of PG and PU are scaled down to nanometer regime, the variation of γ-ratio is significantly increased. This is the main reason why WRM has become just as difficult as read in nanometer-scale SRAM [27].

3.2.3 Cell Current (Icell) Distribution

The BL discharging time takes a large percentage of the total access time. The discharging time (T_{BL}) depends on the BL capacitance, the cell current, and the required BL discharging level (V_{SEN}). The amount of cell current (Icell) is determined by the strength of PG and PD connected in series between the BL and GND as shown in Fig. 3.11. Although the BL capacitance per one-bit cell is reduced so that the access time is reduced as the process is scaled, Icell is not simply increased

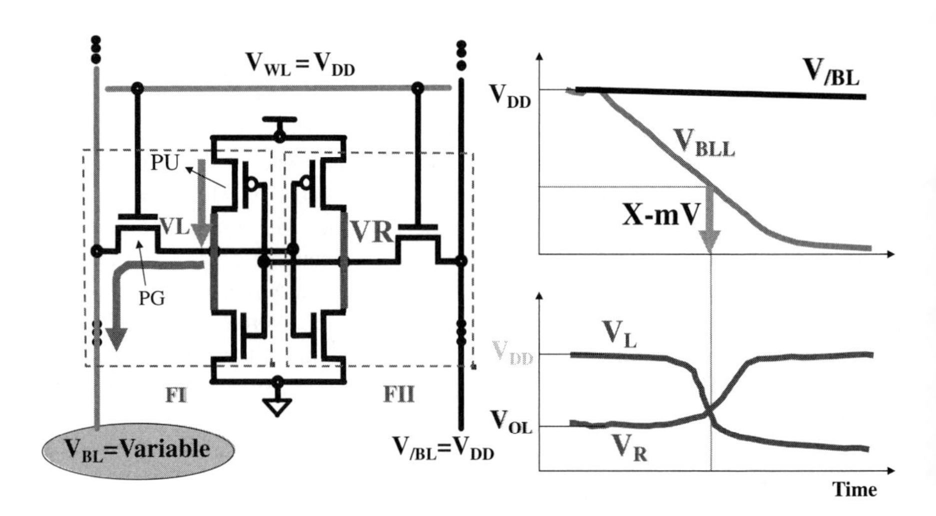

Fig. 3.10 WRM graphical definition

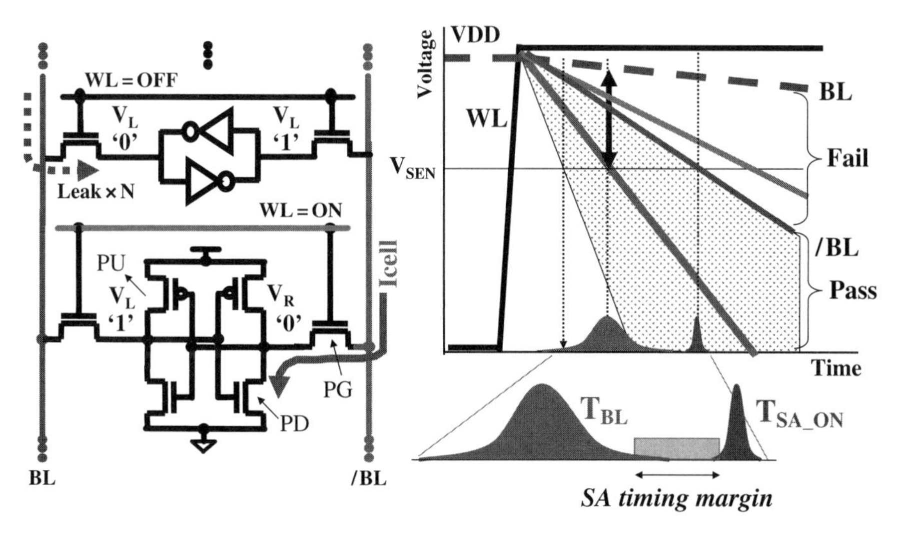

Fig. 3.11 Graphical definition of cell current margin

but it is significantly reduced at lower V_{DD}. This is due to the V_T variations of PG and PD, resulting in reduction of gate overdrive (V_{GS}−V_T) of PD and PG. As a result, T_{BL} has a larger variation as shown in Fig. 3.11. In addition, since the BL leakages (leak×N) flown through unselected cells cause to discharge the referenced BL, the BL differential voltage becomes smaller than V_{DD}−V_{SEN} as shown in Fig. 3.11. The higher V_T settings for PG, PD, and PU in SRAM can suppress the subthreshold leakage but it causes not only the reduction of Icell but also increases its variation [30]. The sensing margin is determined by the T_{BL} and T_{SA_ON} which depends on the input offset voltage and active timing of sense amplifier, as shown in Fig. 3.11. In most cases, since T_{BL} is much larger than T_{SA_ON}, the variation of Icell has larger impact on the sensing margins [27].

3.2.4 V_{DD} Scaling and SRAM Functionality

The dependencies of SNM, WRM, and the cell current (Icell) on V_{DD} are described. Figure 3.12 shows V_{DD} dependency of WRM. The two lines of top and bottom represent the WRMs at 0σ and 6σ points, respectively. It can be seen that WRM has strong V_{DD} dependency and WRM at 6σ is reduced to "0" at V_{DD}=0.95 V. If V_{DD} is reduced by 50 mV around V_{DD}=0.95 V, the failure rate is increased by 33 times. So, SNM has weaker V_{DD} dependency than WRM. SNM is down to "0" at higher voltage (V_{DD}=1.05 V) than WRM. If V_{DD} is reduced by 50 mV around V_{DD}=1.0 V, the failure rate is increased by three times. V_{DD} dependency of SNM is smaller than that for WRM in higher V_{DD} as shown in Fig. 3.13. The dependencies of σ_{WRM}/μ_{WRM} and σ_{SNM}/μ_{SNM} on V_{DD} are compared as shown in Fig. 3.14. It is also found that its difference becomes significant in higher V_{DD}. Figure 3.15 shows

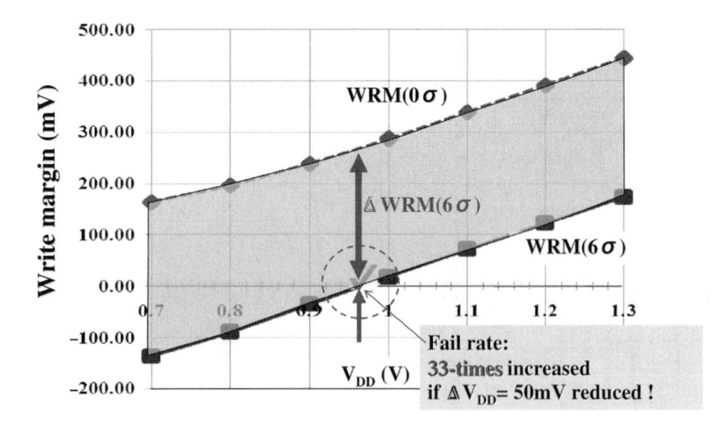

Fig. 3.12 V_{DD} dependency of WRM

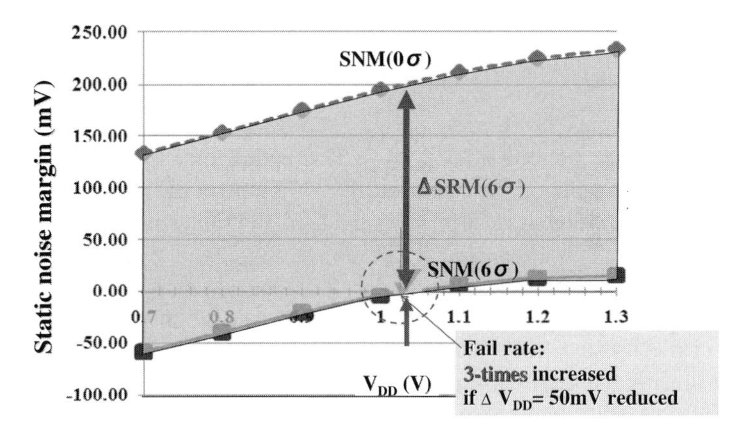

Fig. 3.13 V_{DD} dependency of SNM

the V_{DD} dependency of the cell current (Icell). As can be seen from a log-scaled Y-axis, Icell has strongest V_{DD} dependency. If V_{DD} is reduced by 50 mV around V_{DD}=1.0 V, the failure rate is increased by 400 times.

Figure 3.16 shows the dependencies of σ_{Icell} on V_{DD} and the number of σ (Z number), where $\sigma_{\text{Icell}}(Z{=}2)$ and $\sigma_{\text{Icell}}(Z{=}6)$ are given by (Icell(Z=2)–Icell(Z=0))/2 and (Icell(Z=6)–Icell(Z=0))/6, respectively. It can be seen that σ_{Icell} has larger Z number dependency in lower V_{DD} and the deviation from the Gaussian in the tails are increased as the Z number is increased.

3.2.5 Device Feature Size Dependency of Functional Margins

Since the functional margins of WRM, SNM, and Icell have dependency on the threshold voltage V_T's of NFET (PD/PG) and PFET (PU), the variations of them

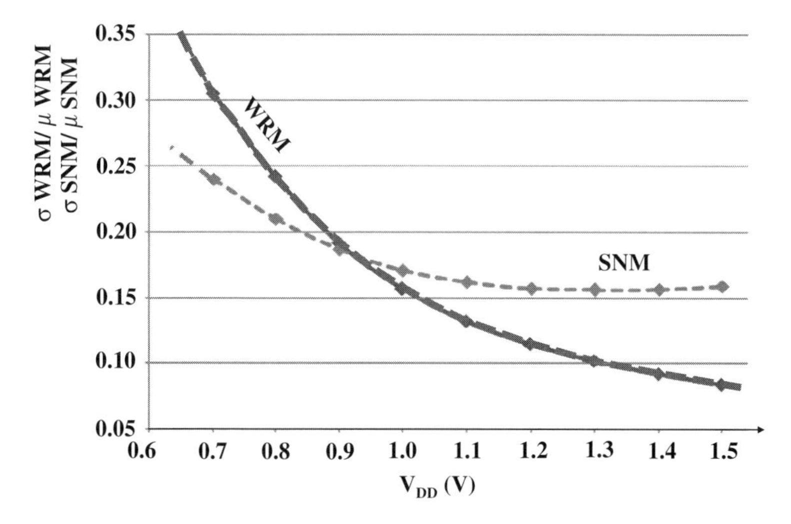

Fig. 3.14 Comparisons of between SNM and WRM V_{DD} dependency

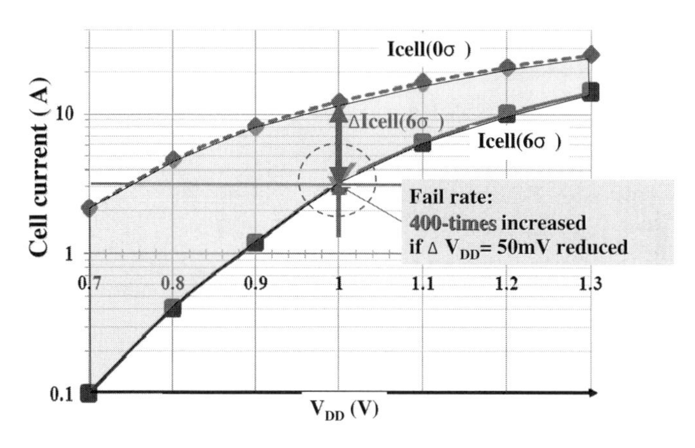

Fig. 3.15 Cell current dependency on V_{DD}

(σ_{WRM}, σ_{SNM}, and σ_{Icell}) also depend on the random variation σ_{VT} of V_T which is proportional to $EOT/\sqrt{L_g \times W_g}$, where EOT is electrical oxide thickness and Lg and Wg are channel gate length and gate width, respectively. As a result, if Lg and Wg are scaled by 30% each and EOT is not scaled, the V_T random variation σ_{VT} is increased by about 43%. If the distributions of WRM and SNM can be approximated by Gaussian distribution like V_T distribution, the variations of WRM (σ_{WRM}) and SNM (σ_{SNM}) is increased by the same percentage as the σ_{VT}. Figure 3.17 shows the device size dependencies of SNM including σ_{SNM}. The values of SNM(6σ) are approximated by (SNM(0σ)–$6\sigma_{SNM}$). It can be seen that size dependency becomes stronger as the number of σ (Z number) is increased. If the transistor channel areas defined by (Lg×Wg) are reduced by 10%, 30%, and 50%, the SNM failure rates

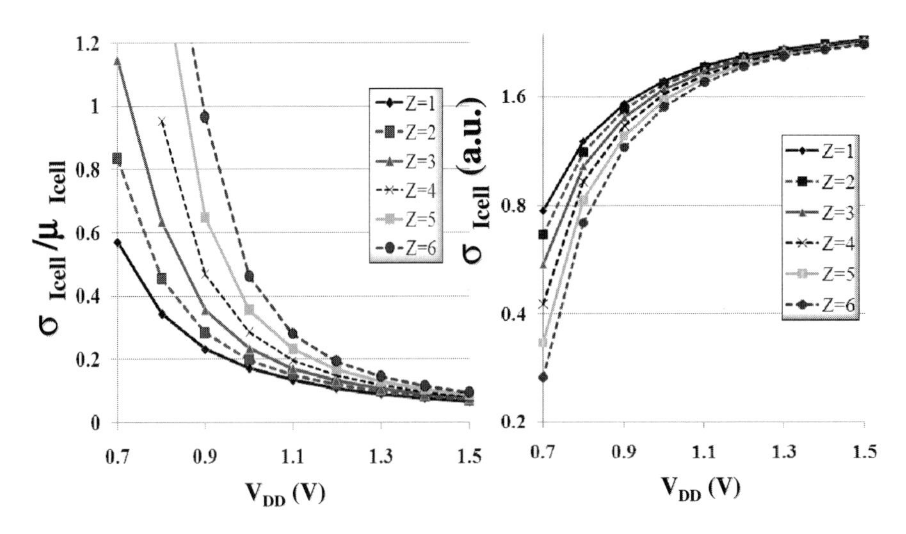

Fig. 3.16 V_{DD} and Z-position dependencies of cell current variation

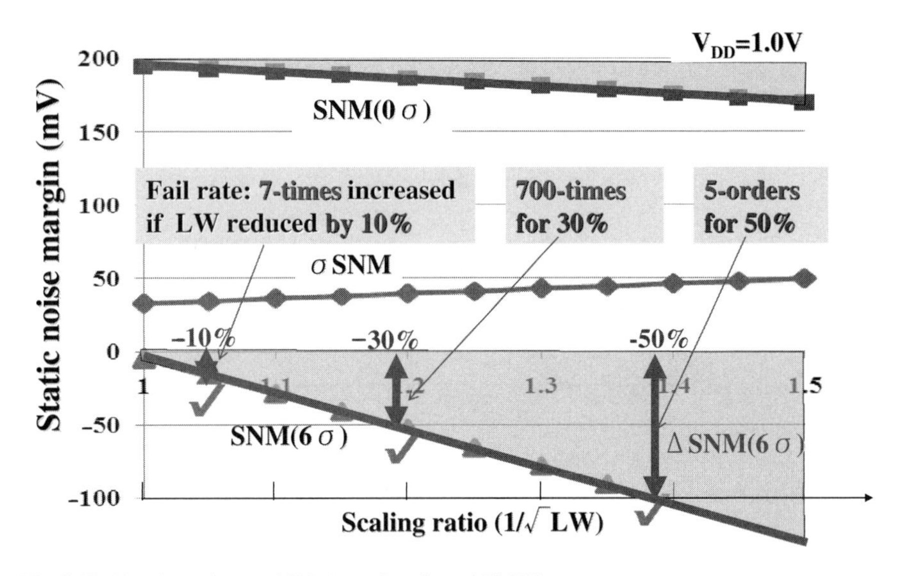

Fig. 3.17 Size dependency of SNM as a function of 1/•LW

are increased by 7-fold, 700-fold, and 5-orders of magnitude, respectively, as shown in Fig. 3.17. Figure 3.18 shows the comparisons of the minimum operating voltage (V_{DD_MIN}) of SNM, WRM, and Icell as a function of the size-scaling ratio ($1/\sqrt{L_g \times W_g}$). For example, it can be seen that SNM has the strongest area dependency among SNM, WRM, and Icell, and increase in the scaling ratio by 20% and 40% are needed to reduce the V_{DD_MIN} by 120 and 200 mV, respectively, as shown in Fig. 3.18.

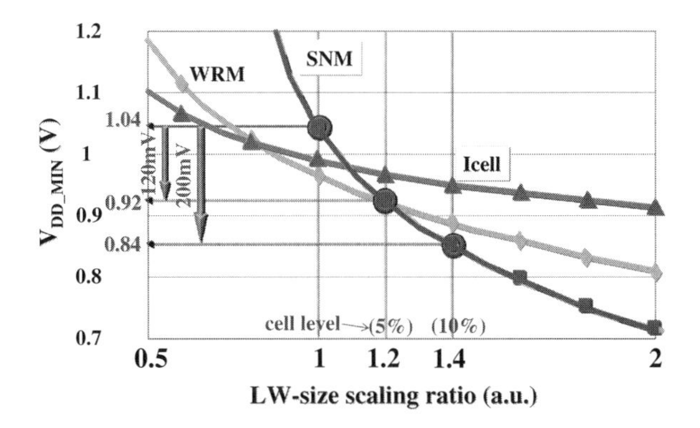

Fig. 3.18 Minimum V_{DD} scaling with SRAM cell area

3.3 Cell Stability Improvement

As explained in the Sections 3.2.3 and 3.2.4, the cell stability depends on the amount of V_T mismatch caused by the random variation (σ_{VT}) of threshold voltage V_T and operation voltage V_{DD} as well as cell ratios: γ-ratio for write and β-ratio for read. In this section, the design solutions to address the issues of the cell stability are discussed. Multiple power supply will be proved to provide the following advantages: (1) use offset voltage to compensate the cell ratios γ for write and β for read; (2) use designated power supply for SRAM to minimize V_{DD} scaling; and (3) use adaptive back biasing to the substrate for increasing and decreasing controls.

3.3.1 Read and Write Margin Assist Circuits

The primary reasons for reducing the read and write margins are due to the loss of balance of the transistors in memory cell. The variations in transistors lead to imbalance in the cell ratios γ and β.

3.3.1.1 Read Margin Assist and Its Limitation

To increase read margin, the potential of the gate electrode of the pull-down NFET (PD_R) can be made higher than that of the pass-gate NFET (PG_R) so that β-ratio becomes larger as shown in Fig. 3.19. Since the WL and cell power line CVDD are connected to the gate electrode of PG_R and PD_R, respectively, the WL voltage has to be lower than that of CVDD as shown in Fig. 3.19. In order to implement such relationship, there are two techniques: (1) suppress the WL level [13, 20, 22] and (2) boost the CVDD level [5]. Figure 3.19 shows the butterfly curves for the cases with and without V_T mismatch between PG_R and PD_R (PG_L and PD_L), respectively. It can be seen graphically that the butterfly curves become asymmetrical and the nested square becomes smaller than that for without V_T mismatch [20, 30]. When the WL level is suppressed by 100 mV so that WL level becomes

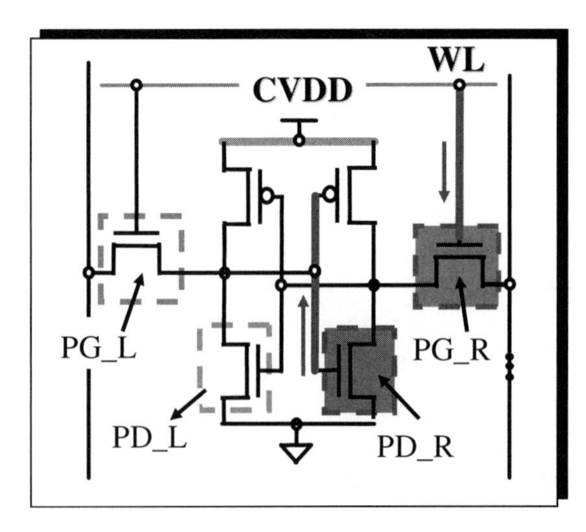

Δ Vt_PD_L=−100mV, Δ Vt_PD_R=+100mV
Δ Vt_PG_L=+100mV, Δ Vt_PG_R=−100mV

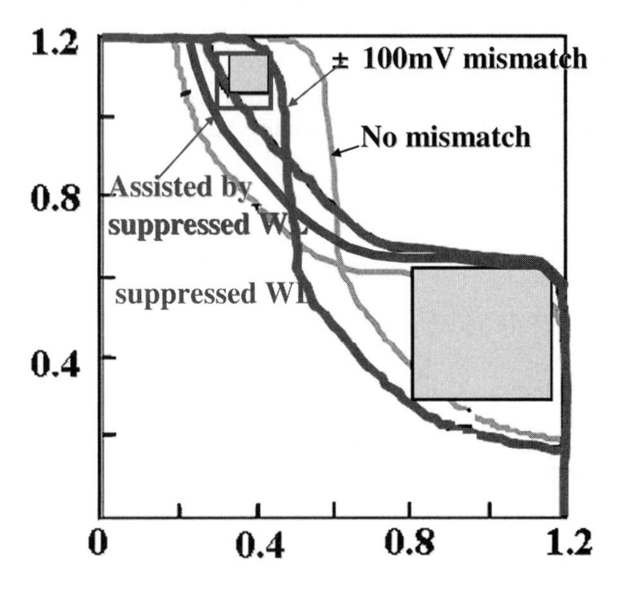

Fig. 3.19 SNM assist basics

lower than CVDD by 100 mV, it can be seen graphically that the nested squares are expanded, which leads to read margin improvement, as shown in Fig. 3.19. Figure 3.20 shows SNM improvements when using the suppressed WL by 100 mV as a function of V_{DD}. It can be seen that SNM is increased by about 50 mV for 100 mV-suppressed WL at V_{DD}=1.0 V. As described above, the intentional offset

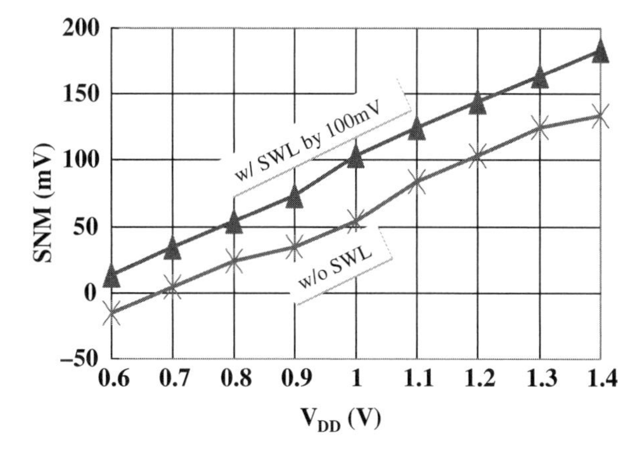

Fig. 3.20 SNM improvement by suppressed WL as a function of V_{DD}

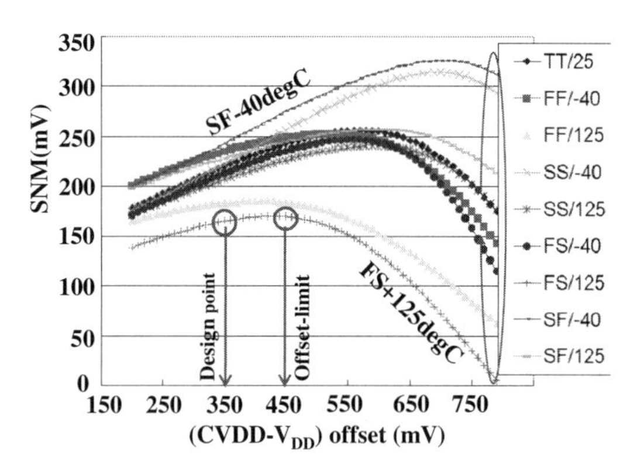

Fig. 3.21 SNM dependency on BL pre-charge level

biasing between WL and CVDD can compensate the lost read margin due to V_T mismatch. Instead of suppressing WL, boosting CVDD also enables such potential relationship between WL and CVDD. In addition to that, the BL pre-charged level also has an impact on the read stability due to the change in not only amount of the charge injection but also the raised equilibrium level of "0" from ground caused by the charge-sharing. As shown in Fig. 3.21, it can be seen that there is the optimal pre-charged level for SNM which is around the V_T of PG below CVDD [4]. However, the design solutions for read margin improvement have the limitation of the operating voltage window as shown in Fig. 3.22. In case of using suppressed WL scheme and boosted CVDD scheme as shown in the left and right sides of Fig. 3.22, it can be seen that the cell current becomes too low to read at the lower V_{DD} boundary due to the suppressed WL level. Higher power supply for the array can also become too excessive to ensure device reliability at higher V_{DD} boundary due to the boosted CVDD level [30].

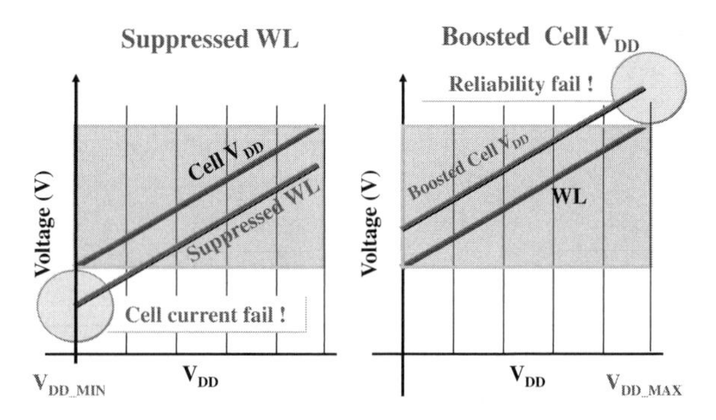

Fig. 3.22 V_{DD} window for suppressed WL and boosted cell V_{DD} scheme

3.3.1.2 Write Margin Assist and Its Limitation

To increase write margin WRM, the potential of the gate electrode of the pull-up PFET (PU_L) can be made lower than that of the pass-gate NFET (PG_L) so that γ-ratio becomes smaller as shown in Fig. 3.23. In order to implement such relationship, there are two different approaches: (1) pull-down the CVDD level and (2) boost the WL level. Figure 3.24 shows the write margin improvement trend as the offset biasing amount between WL and CVDD [37]. It can be seen that the WRM is proportional to the amount of offset biasing and WRM is increased by 50% for the offset biasing of 300 mV. To ensure the potential of the connected node of PU and PG to be lower than the trip point of the inverter, the BL negative overdriving scheme has been proposed, as shown in Figs. 3.23 and 3.25 [18, 21]. Figures 3.25

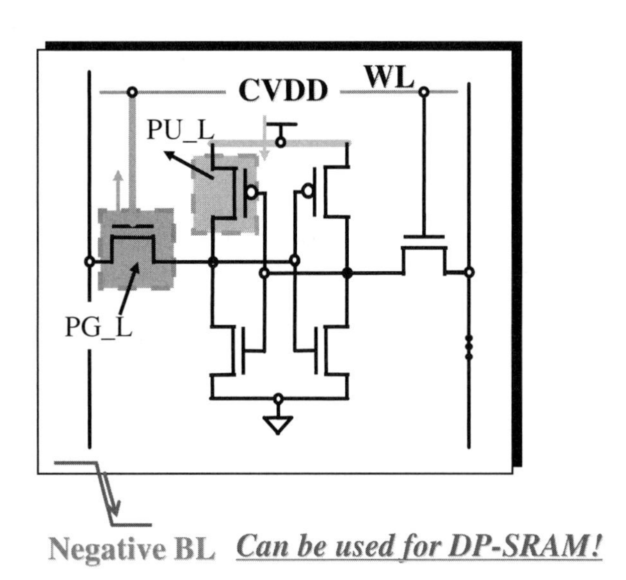

Fig. 3.23 WRM assist basics

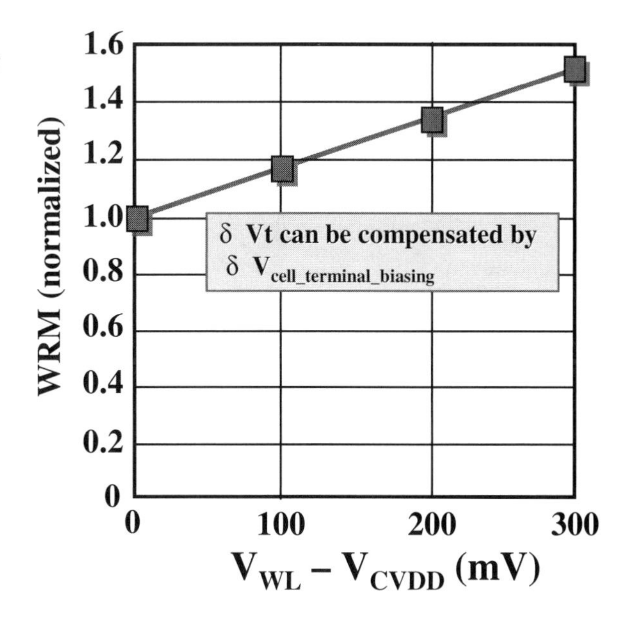

Fig. 3.24 WRM improvement as a function of offset voltage

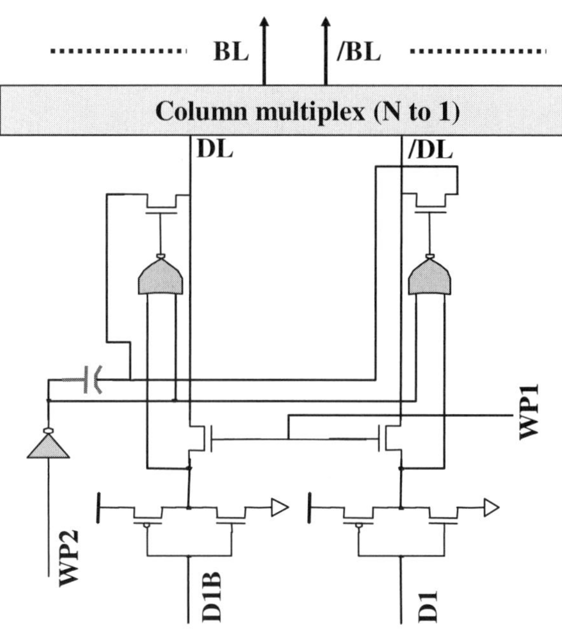

Fig. 3.25 Diagram of negative BL biasing scheme

and 3.26 show the circuit diagram and its control-timing diagram for BL negative overdriving, respectively. Since the required negative overdriving amount is <1/5 of V_{DD}, a area for kick-capacitor will not add a significant impact on macro size. The negative BL overdriving needs the two durations of pull-down from V_{DD} to

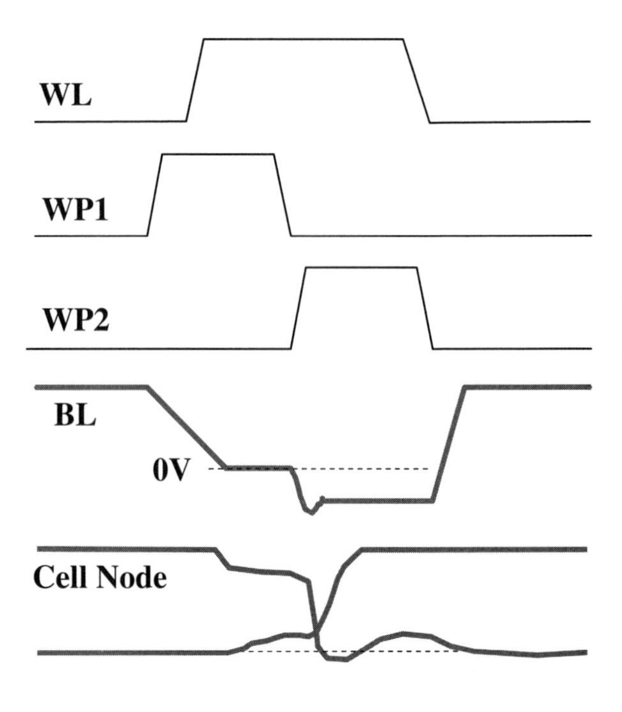

Fig. 3.26 Timing waveforms of negative BL biasing scheme

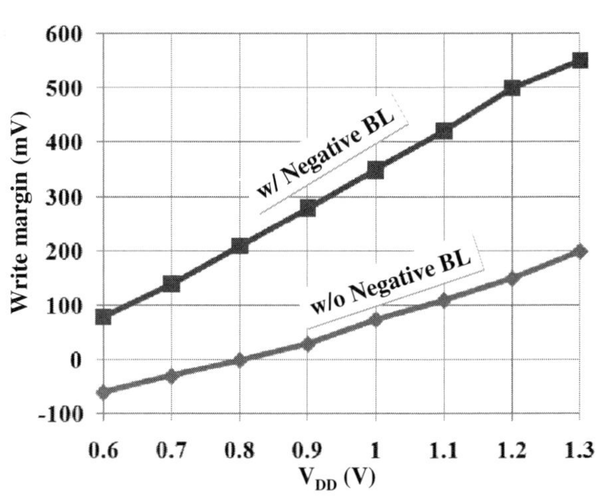

Fig. 3.27 WRM improvements by negative BL scheme as a function of V_{DD}

GND and overdrive from GND to a certain negative potential as shown in Fig. 3.26. Figure 3.27 shows the WRM improvements when using the negative BL scheme as a function of V_{DD}. It can be seen that WRM is increased by about 200 mV at V_{DD}=0.8 V. The WRM can be increased by the amount of negative BL overdriving (–200 mV at V_{DD}=0.8 V) as described in Section 3.2.2. Table 3.1 summarizes the advantage of negative BL scheme relative to lowering the CVDD. The negative BL

Table 3.1 Negative BL scheme advantages compared with CVDD scheme

	Area penalty	Noise tolerance (mesh power)	V_{DD} min for retention	SOC friendly
Negative BL	Small for each I/O mux block	Good for V_{DD} keep meshed	lower	Good for Metal4 as power mesh routing
CVDD-down	large for each column mux pitch	NG for no V_{DD} mesh	higher	Power routing from metal5

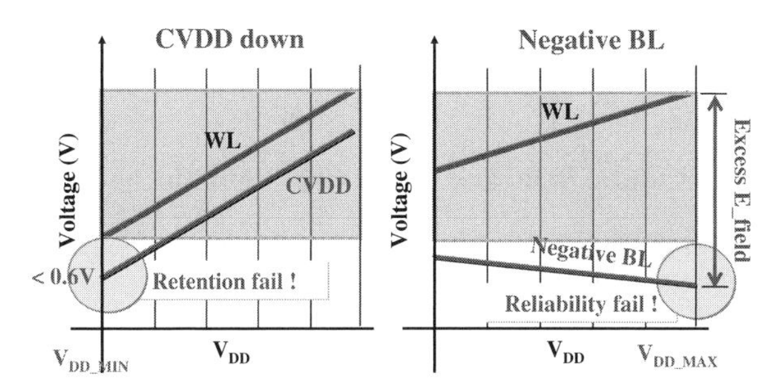

Fig. 3.28 V_{DD} window for CVDD-down and negative BL schemes

scheme can get the benefits for not only area saving but also reducing of CVDD line impedance. This is because CVDD-down scheme needs column-based separate CVDD line and power switch, but negative BL scheme does not need such column-based layout and control. As a result, meshed CVDD line layout becomes available and only one negative potential generator can be shared among the columns as shown in Fig. 3.25.

Boosting WL also can help to implement such potential relationship between CVDD and WL for write margin improvement. However, its design solution has the limitation of the operating voltage window as shown in Fig. 3.28. In case of using pull-down CVDD scheme and negative BL scheme as shown in the left and right sides of Fig. 3.28, it can be seen that the cell data retention voltage becomes too low to retain the data at the lower V_{DD} boundary due to the pull-down CVDD level and the applied electric field becomes too excessive to ensure device reliability at higher V_{DD} boundary due to the negative BL level, respectively [30].

3.3.1.3 Body Biasing Scheme for Read and Write Assists and Its Limitation

It is well known that forcing the reverse and forward back biases can increase and reduce the V_T of NFET and PFET, which can be used for reducing the subthreshold leakage and increasing the current drivability, respectively [6, 15, 23]. In addition, reverse back biasing for NFET can increase the read margin due to increase in the V_T of PG and PD and forward back biasing for PU also can increase the read margin due to increase the trip point of inverter so that the noise immunity can be increased. On the contrary, forward back biasing for NFET (PG/PD) can increase the write margin due to reduction in the V_T of PG and PD, and reverse back biasing for PFET also can increase the write margin due to reduction in the γ-ratio of the drivability of PU to PG. However, the design solutions using the reverse and forward back biases have the limitation of compensating the lost balance for read and write margins caused by random V_T variations. This is because V_T and β-ratio of NFET (PG/PD) cannot be separately changed due to the same substrate and back biasing, resulting in the same trend of V_T shift. In addition, the amount of V_T shift would be insufficient for compensating the lost balance for read and write margins compared with the significantly increased random V_T variations, which cause the increase in lost margins. Applicable back bias window for NFET and PFET is also limited by the junction leakage including gate-induced-drain-leakage (GIDL).

3.3.1.4 Dynamic vs Static Read and Write Cell Stabilities

Since the data flipping takes a certain time to be completed, it depends on the durations of on-state of the WL and high/low state of the BL. For example, to shorten WL pulse and BL length make the duration of flowing charge from BL into the storage node shorter and the amount of charge flowing from BL smaller, respectively [9, 12, 13]. As a result, not only potential voltage, but also duration of the raised "0"-level from GND can be reduced, resulting in improvement of the read stability. This is the reason why the dynamic cell stability looks more optimistic compared with DC read cell stability. In actual area-aware designs, this dynamic read cell stability analysis becomes more important because shortening the duration of WL pulse and/or the length of BL need to consume less area than that for using the multiple power supply. Since the shorter BL architecture can reduce the required discharging time, enabling shorter WL pulse, it becomes the key to improve the dynamic read cell stability. On the contrary, the dynamic write stability becomes worse compared with DC analysis where WL pulse is assumed as infinite. In practical design, the durations of on-state of WL and "0"-state of written BL are limited and as a result, the write margin looks pessimistic compared with the DC analysis.

3.3.1.5 SRAM-Designated Power Supply Scheme

Conventionally, since the power supply island for logic circuits and SRAM is not isolated, the higher minimum operating voltage $V_{DD\_SRAM\_MIN}$ of SRAM limits the overall V_{DD_MIN} performance as shown in Fig. 3.29. However, the trend of increasing $V_{DD\_SRAM\_MIN}$ and demand for lowering the V_{DD_LOGIC} for power-saving have driven the SRAM power supply island to be decoupled from the logic [12, 30]. Once the SRAM power supply island is separated from others, the designated SRAM power supply becomes possible to use for SRAM array only so that the operating voltage range can be optimized for SRAM as shown in Fig. 3.29. As explained above, since the functional margins of read and write and the cell current have strong V_{DD} dependencies, such designated SRAM power supply can improve the functional margins. In addition to that, the designated SRAM power supply can relax the lower

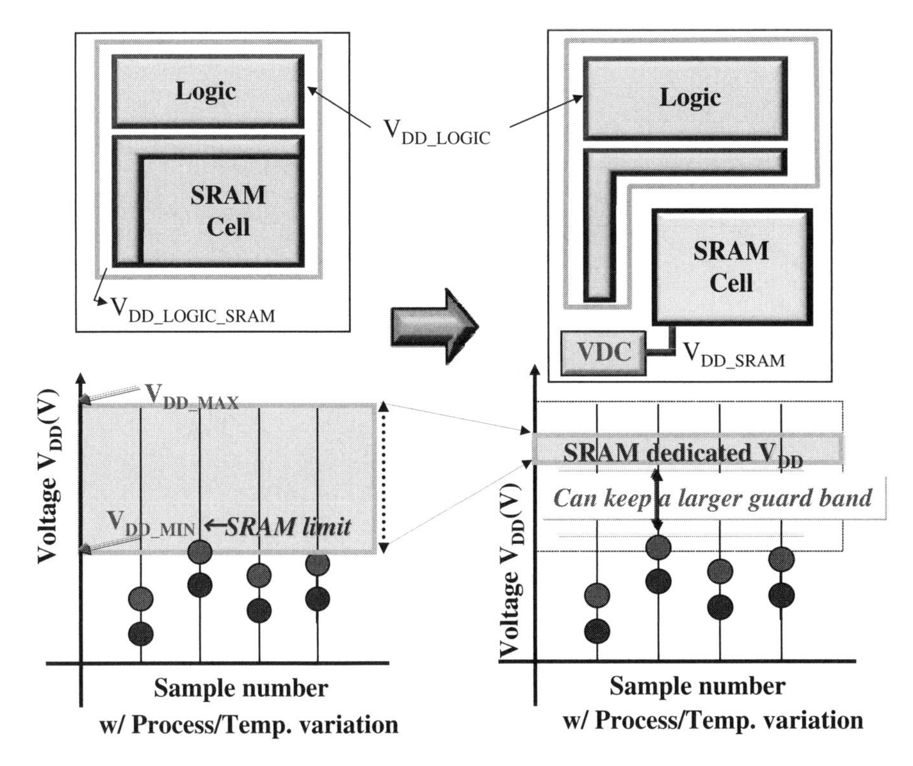

Fig. 3.29 Concept of using designated SRAM power supply scheme

and upper V_{DD} limitations, resulting in wider V_{DD} design windows, as shown in Figs. 3.30 and 3.31, respectively. Conventionally, the V_{DD} scaling limit is caused by the cell current fail due to suppressed WL or the retention fail due to the pull-down CVDD. V_{DD} scaling limit is also caused by the excessive applied electric field due to the negative-overdriven BL. These limitations can be relaxed by using the designated SRAM power supply as shown in Figs. 3.30 and 3.31, respectively.

3.3.2 SRAM with Multiple Power Supplies Against Single-Rail Scheme

Table 3.2 shows the comparisons of multiple power rails in SRAM including dual-rail, triple-rail, and quad-rail designs [4, 9, 12]. Both dynamic and static power switching are considered here. The single-rail scheme is used for the baseline of these comparisons which uses only the same V_{DD_LOGIC} for all terminals of CPL, WL, and BL. In the dual-rails scheme, CPL and WL is provided by CVDD whose potential is higher than V_{DD_LOGIC}. The dual-rails scheme is divided into the two cases, depending on which power of CVDD or V_{DD_LOGIC} is used for BL pre-charging. The triple-rails scheme uses dynamic power supplies to CPL and WL

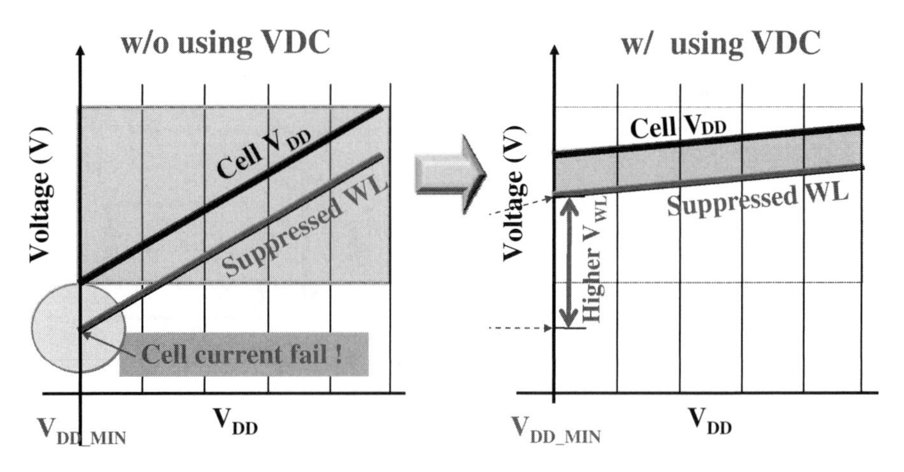

Fig. 3.30 Relaxation of $V_{DD\text{-}MIN}$ limits from suppressed WL

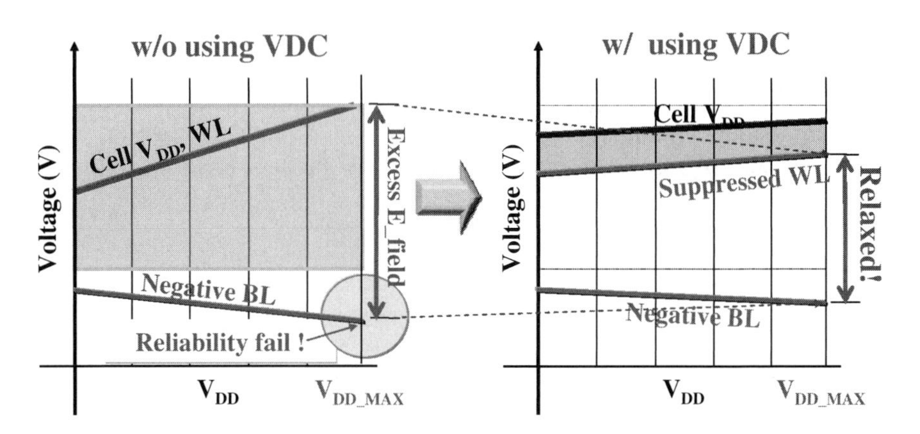

Fig. 3.31 Relaxation of $V_{DD\text{-}MAX}$ limits from negative BL

whose potentials are changed from CVDD_H to CVDD_L for CPL and from CVDD_L to CVDD_H for WL, depending on the read and write operations, respectively. The potentials of CVDD_H and CVDD_L are higher and lower than that of $V_{DD\text{-}LOGIC}$, respectively. The quad-rails scheme adds one more power supply of CVDD_WL whose potential is set in the middle of CVDD_H and CVDD_L for WL avoiding a cell current drop due to lower level of CVDD_L. Table 3.2 also shows the comparisons of the power supply impacts on SNM, WRM, the cell current Icell, $V_{DD\text{-}MIN}$, power, and area. It can be seen that quad rails can provide the best SNM and WRM. The triple-rails scheme is limited by the column-multiplexing (column-interleaving). This is because the triple-rails scheme cannot avoid the half-select issues for the unselected columns due to using higher WL voltage of CVDD_H for

Table 3.2 Comparisons of different power supply schemes

	Single	Dual		Triple	Quad
		BL (V_{DD_LOGIC})	BL (CVDD)		
Cell inverter	Static V_{DD_LOGIC}	Static CVDD	Static CVDD	Dynamic (R/W) CVDD_H/ CVDD_L	Dynamic (R/W) CVDD_H/ CVDD_L
WL	Static V_{DD_LOGIC}	Static CVDD	Static CVDD	Dynamic (R/W) CVDD_L/ CVDD_H	Static CVDD_WL
BL	Static V_{DD_LOGIC}	Static $V_{DD\_LOGIC} < CVDD$	Static CVDD	Static CVDD_H	Static CVDD_H (CVDD_WL)
SNM WRM	Worst	Better	Fair	Fair (Best w/o c-MUX)	Best
Icell	Fair	Best	Best	Worst	Better
V_{DD_MIN}	Worst	Better	Best	Better	Best
Power	Best	Better	Fair	Fair	Fair (Better)
Area	Best	Fair	Fair	Worst	Worst
Remarks	Baseline	Well balanced Suitable for compiler	Better for Lower-V_{DD_LOGIC}	Half-select limits the advantage for SNM when needed column multiplex.	Best performance at the cost of area and power

write operation. On the other hand, quad-rails scheme uses CVDD_WL for WL whose potential is lower than CVDD_H which is supplied to CPL, so that SNM can be improved. In addition, quad-rails scheme uses CVDD_L for CPL only for selected written column while forcing the CPL to CVDD_L for unselected columns, enabling the increase of WRM while avoiding the half-select issues. It can also be seen that the best cell current Icell is provided from the dual-rails scheme instead of quad-rails scheme. This is because the dual-rails scheme uses CVDD whose potential is higher than CVDD_WL for WL and CPL. A V_{DD_MIN} performance which is measured by how low V_{DD_LOGIC} can be reduced, is provided from both schemes of quad rails and one of dual rails. This is because they use the SRAM-designated power which does not depend on the potential of V_{DD_LOGIC} for all of WL, CPL, and BL unlike others. Since one of dual-rails schemes uses V_{DD_LOGIC} for BL pre-charging, the allowable potential difference between CVDD and V_{DD_LOGIC} is limited due to the SNM degradation [4] caused by the pull-down of "1" level of storage node toward the BL pre-charged level of V_{DD_LOGIC}. Meanwhile, the SRAM multiple-rail schemes need to separate the multiple power islands and to generate internally multiple power sources by using voltage down converter or to be supplied from additional external power pins. As a result, additional area and power consumptions are needed compared with the single-rail scheme. Overall, it can be said that one of dual-rails schemes where BL is pre-charged by V_{DD_LOGIC} is suitable for SRAM compiler application. This is because CVDD requires less power because of no need to supply for BL pre-charging which takes a large percentage of the total SRAM power consumption. As a result, it can relax the limitation of Place and Rout design flow overcoming IR-drop constraint. On the other hand, another dual-rails scheme can lower V_{DD_LOGIC} because it is independent from the SRAM-operating voltage. Quad rails can give the best SNM/WRM improvement at the most expensive cost of area and power consumptions. In practical design, we need to compare the overall cost performance depending on the requirements from the applications. Upsizing memory cell to suppress the loss of functional margins caused by the V_T random variation could be the best choice for some applications.

3.4 New Cell Topology to Improve Read and Write Stabilities

As discussed in Sections 3.3.1.5 and 3.3.2, using SRAM-designated multiple power supply schemes referred to as dual rail, triple rail, and quad rail allows to raise SRAM-operating voltage (V_{DD_SRAM}), to overcome the limitations caused by ever increasing V_T random variation σV_T. However, ever increasing σV_T will drive up V_{DD_SRAM}, running into V_{DD_MAX} limit determined by the maximum applicable voltage without device reliability issue as shown in Fig. 3.32. Once V_{DD_SRAM} reaches the ceiling, such kind of design solutions cannot help to address the SRAM scaling issues any more.

The pass-gate of 6T SRAM cell are shared for read and write data transfers between the BL and the storage nodes of SRAM and the strength requirement for PG between the read and write operations conflict each other. As a result, 6T SRAM cell

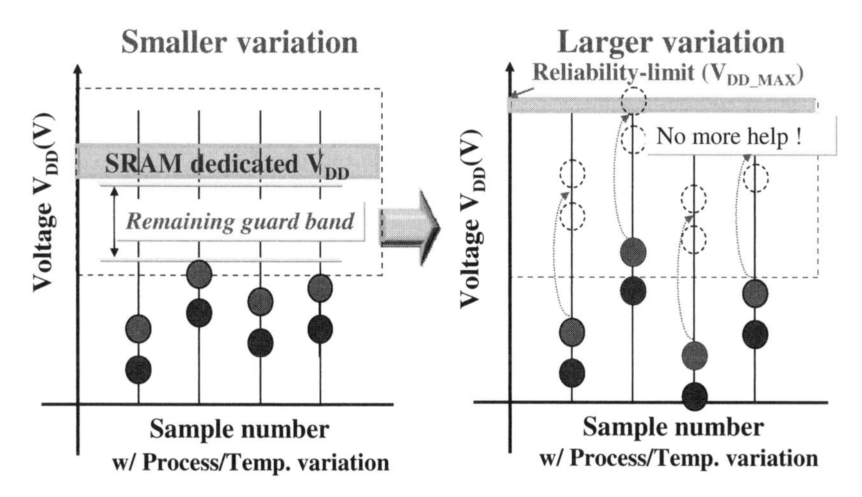

Fig. 3.32 8T SRAM benefits

design has faced a difficult trade-off between the read and write stabilities. In addition, another trade-off between the read stability and the cell current Icell has to be taken into account for the 6T cell design. For example, the suppressed WL scheme can improve the read stability but can reduce the Icell as a side effect. As a result, we need to take a difficult trade-off among the three functional margins of SNM, WRM, and Icell in designing 6T SRAM cell. In order to make 6T SRAM cell design free from the conflicting requirements between the read and write, two more nFETs for designated read port to 6T SRAM has been proposed and it is referred as 8T SRAM as shown in Fig. 3.33. However, 8T SRAM would still face the half-select issues during write operation if the column-multiplexing (column-interleaving) is needed. In order to solve this remaining issue, two more NFETs could be added to eliminate the column-interleaving challenge, which will be discussed later.

3.4.1 8T SRAM

In order to eliminate the conflicting requirements between the cell current and the read stability in 6T SRAM cell design, two NFETs along with extra read-word-line (RWL) and read-bit-line (RBL) are added to provide designated read port SRAM, as shown in Fig. 3.34, which is often referred to as 8T SRAM [2, 13, 30]. 8T SRAM can improve the following aspects of the SRAM cell: (1) the cell current Icell; (2) the dynamic and leakage power; and (3) read stability. The cell current can be increased by boosting the designated RWL and RBL without causing any read stability issues. In addition, the strength ratio of the two NFETs connected in series can be optimized so as to increase Icell without considering conventionally required β-ratio. The dynamic power-saving for read operation can be realized by selected

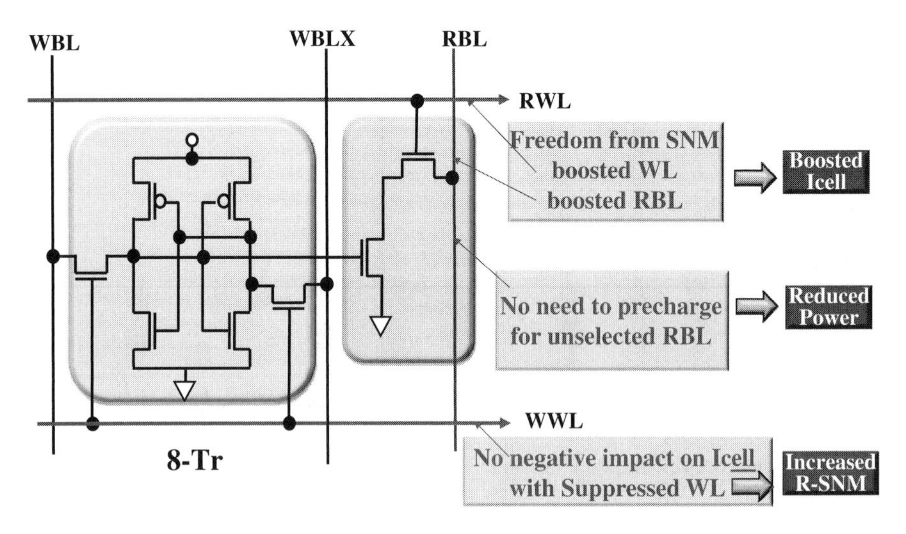

Fig. 3.33 8T SRAM cell

BL pre-charging, which can save the pre-charging power consumption for unselected BLs. The leakage power consumption of the two inverters of 8T SRAM can be reduced by using higher V_T in the inverter NFETs. This is because the amount of the cell current is determined by the strength of only two NFETs of read port. Meanwhile, lower V_T can be used for the two NFETs of read port, resulting in increasing Icell. The read stability against WL disturbing (i.e., half-select issue) can be increased by enhancing β-ratio without compromising Icell. However, 8T SRAM cannot address the half-select issue in write operation when requiring the column-interleaving operation. The trade-off between WRM and SNM has to be taken in account for the cell design, as shown in Fig. 3.34.

3.4.2 10T SRAM

In order to avoid the half-selected access issue, two more transistors can be added as shown in Fig. 3.35. The pass-gate for write operation consists of the two NFETs connected in series between BL and the storage nodes, which are selected by row and column decoding, respectively, as shown in Fig. 3.35. Meanwhile, the pass-gate for read operation bypasses the column-decoded NFET, which is connected to the storage node, so that the read disturbance can be eliminated. As a result, the performances of read stability and the cell current drivability becomes the same as the read port of 8T SRAM cell. On the other hand, since the pass-gate for write operation consists of the two NFETs connected in series, the equivalent γ-ratio might be reduced compared with 8 T. However, since the drivability of each of pass-gate NFETs can be increased without any concern for the β-ratio, the γ-ratio can be improved independently. This 10T SRAM cell can completely eliminate the

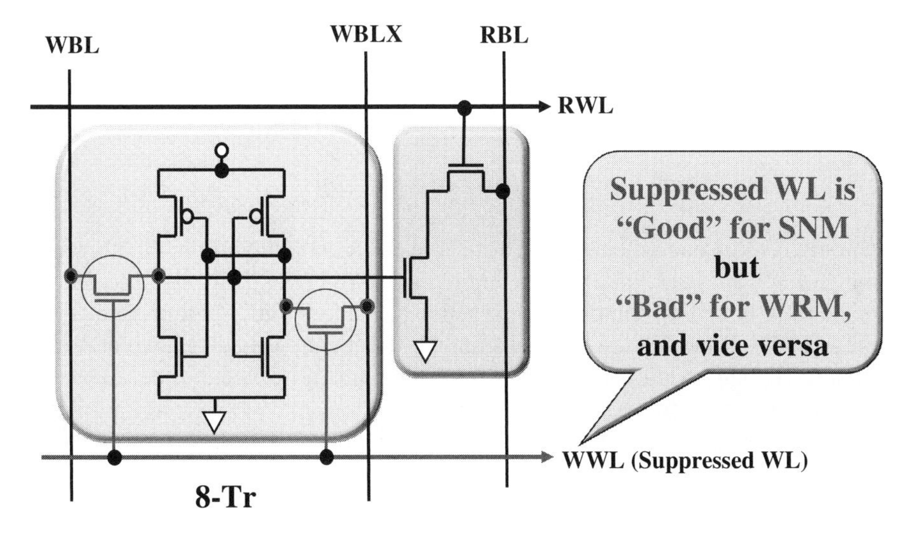

Fig. 3.34 8T SRAM column-interleaving

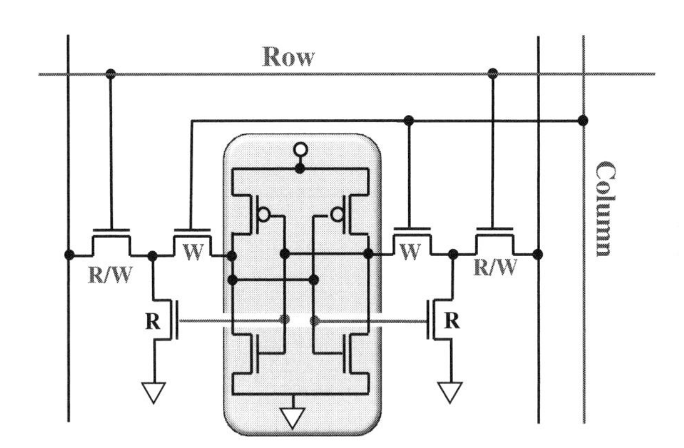

Fig. 3.35 10T SRAM

trade-off between the SNM, WRM, and the cell current at the expensive cost of cell area due to the use of more transistors.

3.5 Read and Write Multiplexing

As explained above, the SRAM read and write stability issues stem from the conflicting requirements for each other, resulting in a difficult trade-off in the design of the margins for read and write operations. To have the designated ports for read and write by adding two or four NFETs is one of the effective ways to avoid the conflicting requirements as explained in Section 3.4. In this section, another alternative way of using read and write multiplexing is described.

3.5.1 Pulsed Word-Line and Bit-Line Scheme

The pulsed word-line scheme [10, 16] features the controlling of the duration and timing of WL for read and write operations as shown in Fig. 3.36. For read operation, the WL turned-on duration is designed to be made sufficiently short so that the storage nodes of the cell are isolated from the BLs before the cell gets to the edge of faulty flipping, while ensuring that the minimum BL differential development required for correct sensing. Meanwhile, the write capability constraint on limited WL pulse width can be eliminated by combining it with the write back operation following the read operation, which is referred as "read modify write" (RMW). In the write back operation, the duration of WL-on pulse can be optimized only for write operation without any trade-offs. The challenge of this scheme is how to define the duration WL-on to meet the conflicting requirements for eliminating of cell data flipping and of BL sensing-margin [10, 16]. The pulsed bit-line scheme allows the voltage of pre-charged BL level to go down before WL is being turned on for read and write operations so that the amount of charge injection from BL to storage node can be reduced as shown in Fig. 3.37 [10, 16].

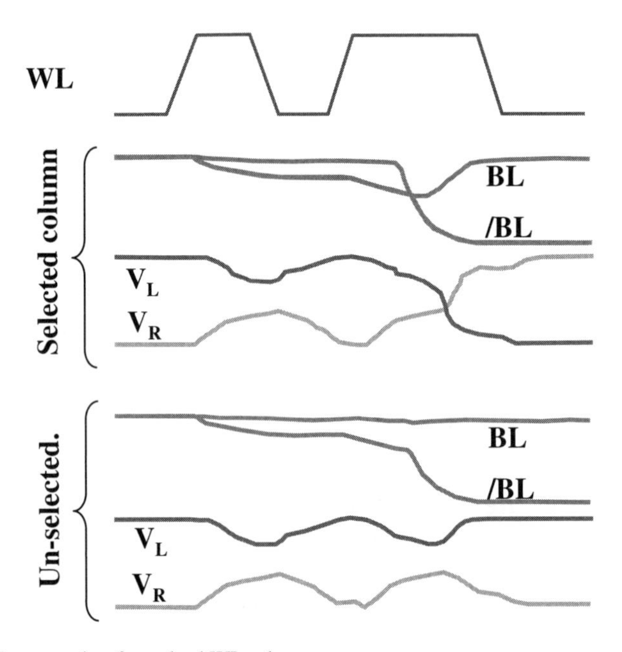

Fig. 3.36 Write operation for pulsed WL scheme

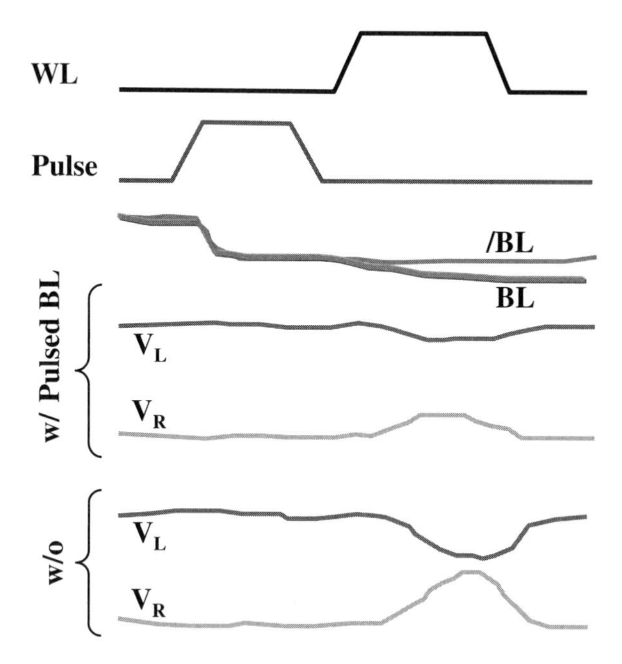

Fig. 3.37 Read operation for pulsed BL scheme

3.5.2 Time Division for Read with Suppressed WL and Write Operation

The limitations in using pulsed WL scheme stem from the difficulty in finding a good balance between read and write requirements. In order to relax such constraint, the potential level of WL can also be modulated so that sufficient read stability can be maintained without any difficulties in tuning of the WL pulse width [7, 30]. The potential levels of WL-on can be optimized separately for sequential read and write WL accesses so that its level becomes lower and higher for read and write operations, respectively, as shown in Fig. 3.38. The drawback in applying this scheme is the degradation of read access speed due to the reduction of the cell current by suppressed WL.

3.5.3 Time Division for Read with Decoupled Read Port and Write Operation

In order to solve the remaining issues discussed above, decoupled read port from the write port is introduced to eliminate the requirements of suppressed WL for read operation [13, 30]. The decoupled read port of 8T SRAM avoids the slowdown of the read access speed as shown in Fig. 3.39. This scheme can completely eliminate

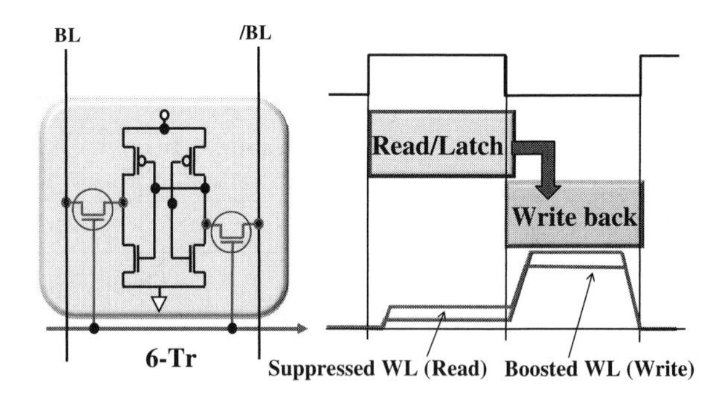

Fig. 3.38 Time division for read and write multiplexing for 6T SRAM

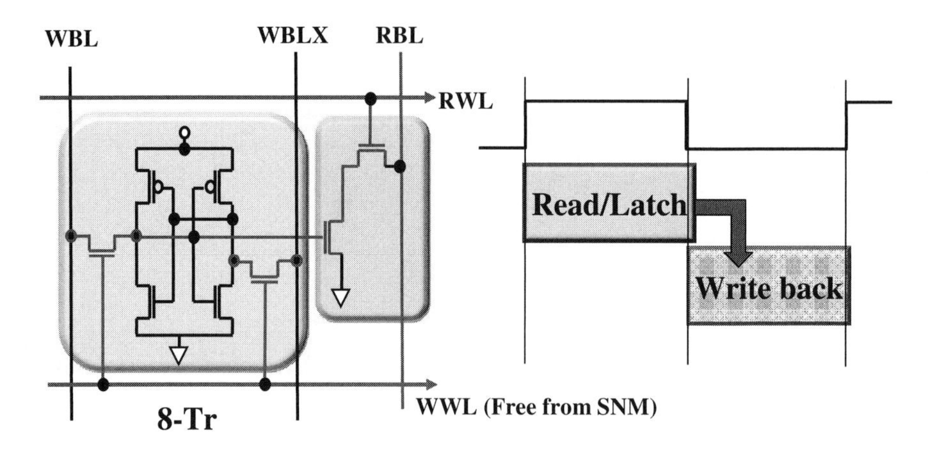

Fig. 3.39 Time division for read and write multiplexing for 8T SRAM

the trade-off between the SNM, WRM, and the cell current at the expensive cost of two more NFETs area and sequential timing slot of read-modify-write.

3.6 Analysis on SRAM Design Solutions with Scaling

This section describes the scaling trend of SRAMs with various design solutions for 6T, 8T, and 10T SRAM cells [13, 30]. SRAM is heavily modulated by V_T variation, and its scaling often hinges upon the future technology direction. The SRAM solutions will be analyzed in light of overall technology scaling. For example, it is well known that introducing high κ metal-gate can scale the electric oxide thickness (EOT) while eliminating the depletion layer in the gate, resulting in suppressing the random variation (σ_{VT}) of threshold voltage V_T. Conventionally, the EOT scaling has reached the limit due to the increasing of gate leakage. However, such kind of device innovation enables to extend the scaling limit of the MOSFET gate channel size defined by (Lg$\times$Wg). In order to predict the scaling trend for a 32 nm and beyond, the speed of increasing σ_{VT} is assumed in this section as shown in the following Section 3.6.1.

3.6.1 V_T Random Variation Trend

The V_T random variation amount (σ_{VT}) of MOSFET threshold voltage is proportional to EOT/$\sqrt{L_g \times W_g}$where EOT is electrical oxide thickness and Lg and Wg are channel gate length and gate width, respectively [11, 30]. When both Lg and Wg are scaled by 0.7, the channel area will be scaled by 0.49 without EOT scaling, the V_T random variation amount σ_{VT} will be increased by about 1.43 times as shown in Fig. 3.40. Figure 3.41 shows the gate leakage scaling for different EOT and gate materials (SiO_2, SiON, HfSiON, and other high-κ). The required EOT depends on the gate material. It can be found that SiON for 1.9 nm, HfSiON for 1.6 nm, and the new high-k for <1.4 nm are needed to suppress the maximum gate leakage Jg(V_{fb}–1 V) to < 2 E–2 A/cm^2 for low standby power (LSTP) process. Figure 3.42 shows the trend of EOT and σ_{VT} from 65 to 15 nm process generation. The σ_{VT} could be suppressed to <75 mV even in 15 nm process node if EOT could be successfully scaled down, otherwise, it could be increased to >135 mV as shown in Fig. 3.42 [30].

3.6.2 Limit of Design Solutions with Increasing σ_{VT}

As explained above, the increase of σ_{VT} has led to the degradation in cell stability and write margins. Figure 3.43 illustrates the design limits. It can be seen that the suppressed WL scheme for SNM assist could become nonpractical if σ_{VT} will become >50 mV due to the significant Icell reduction by >70%. If σ_{VT} is larger than 105 mV, the minimum data retention voltage would reach 1.1 V. The negative BL scheme and the CVDD-down scheme for WRM assists become unable to be

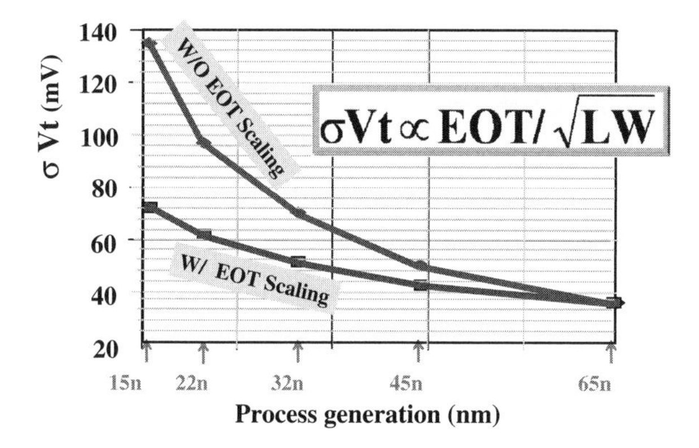

Fig. 3.40 σ_{VT} scaling trend

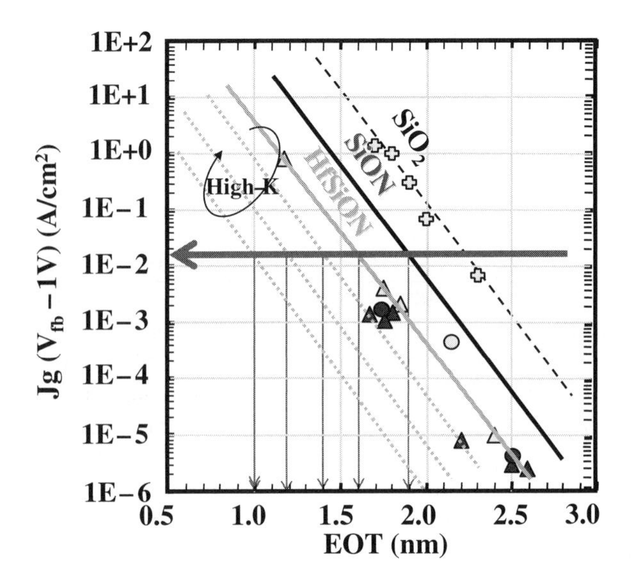

Fig. 3.41 Gate leakage scaling with different dielectrics and EOT

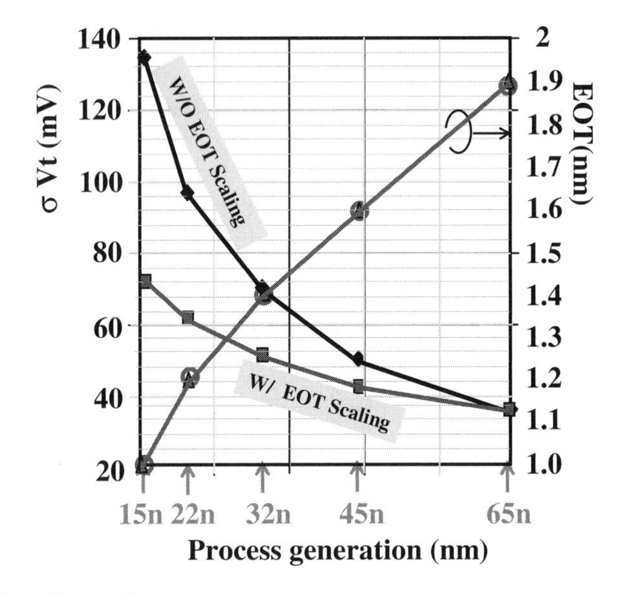

Fig. 3.42 σV_T scaling trend

adopted any more if σ_{VT} will become >75 and >60 mV, respectively. The most sensitive parameter to σ_{VT} is the cell current, Icell. If 20% reduction is the limit, often true, allowable σ_{VT} can only be <40 and <45 mV for 6T SRAM and 8T SRAM, respectively [30].

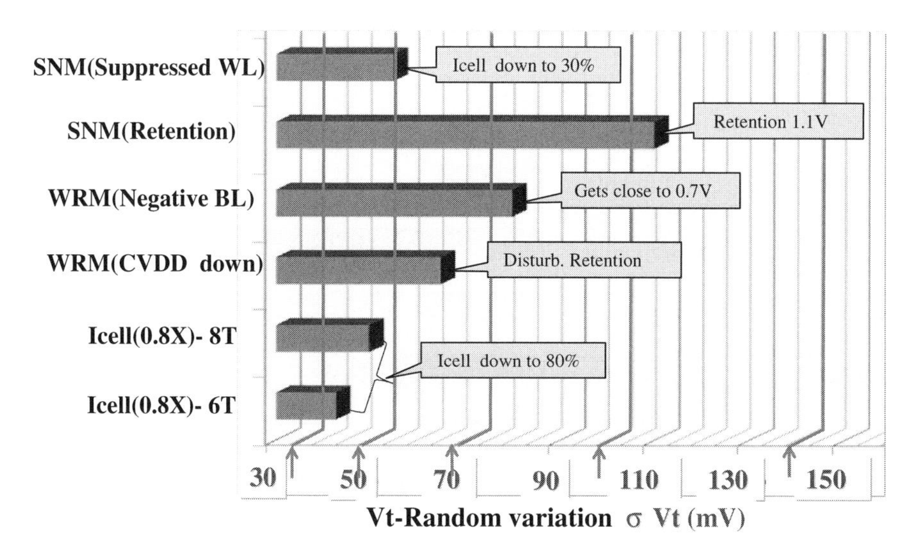

Fig. 3.43 Figure of merit for assist schemes

3.6.3 Extension of Limit of Design Solutions

As explained above, a increasing of σ_{VT} will place limits on the coverage of each design solution for the margin assists as SRAM is scaled into a deeper nanometer region. There are higher and lower bounds in the design windows for the design solutions and both of them give the impact on each other as a trade-off. For example, a deeper suppressed WL for SNM assist causes to require a deeper negative BL overdriving or CVDD-down for WRM. In this section, the three examples of (1) read modify write decoupled read port [13], (2) error-checking and correcting (ECC) circuit [8], and (3) V_{DD} regulation [3, 12], enabling to relax such kind of limitations are given.

3.6.3.1 RMW Operation with Decoupled Read Port

As explained above, it is the key to eliminate the requirements of suppressed WL scheme for SNM in order to relax the excessive requirements of CVDD-down and negative BL for WRM. Having a decoupled read port like 8T SRAM and read modify write operation becomes the key enabler to do so [13]. Figure 3.44 shows how each scheme plays the key role for extending the limit of design solution as a function of σ_{VT}. It can be seen that read modify write operation with decoupled read port referred to as 8 T+RMW can extend the timing of assist failure for SNM and WRM so that it cannot be limited even if σ_{VT} becomes >135 mV.

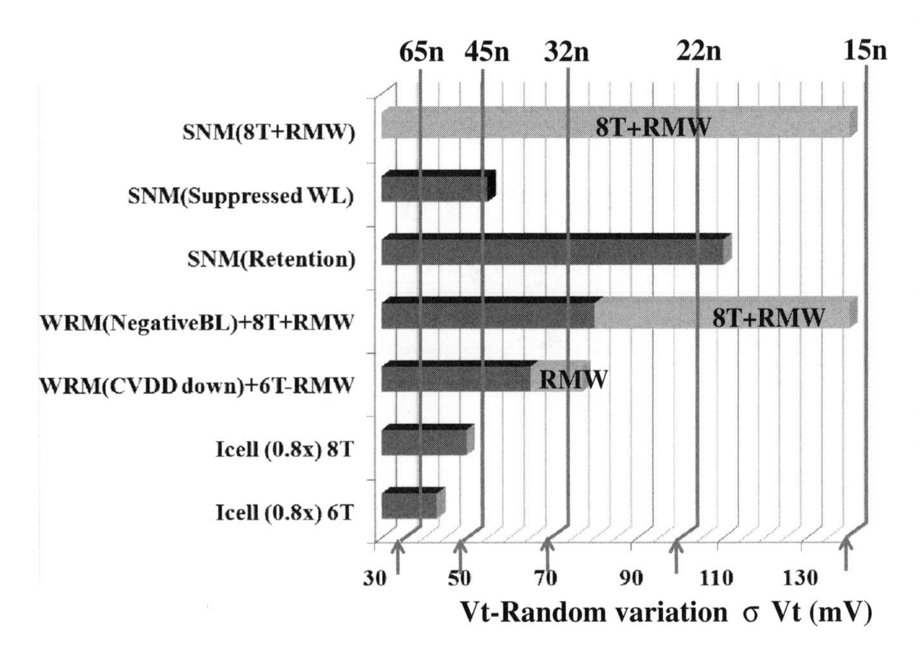

Fig. 3.44 Impacts on each design solution's availability given by read-modify write with decoupled R/W port

3.6.3.2 Error Correction Scheme for Redundancy

Error Check and Correct (ECC) scheme has been widely used in improving yield and design robustness in memory design. An ECC word includes N_0 data bits and N_P parity bits (e.g., N_0=128 and N_P=8), which can correct only one defect cell in the word [8, 30]. Meanwhile, ECC word with two or more defective cells can be replaced by adding a redundant ECC word. The yield improvement of SRAMs using both of such kind of ECC (N_0=128 and N_P=8) and ECC redundancy (ECC redundant words of R) can be estimated as follows.

If the bit-error rate is r, the probability (p) of an ECC word having two or more defective cells is expressed as:

$$p = 1 - (1 - r)^N - N \times r \times (1 - r)^{(N-1)}$$

where N= (N_0+N_P).

The probability that q redundant words are required is expressed as:

$$P_q = wC_q \times p^q \times (1 - p)^{(w-q)}$$

where W is the total number of ECC words excluding ECC redundant words of R.

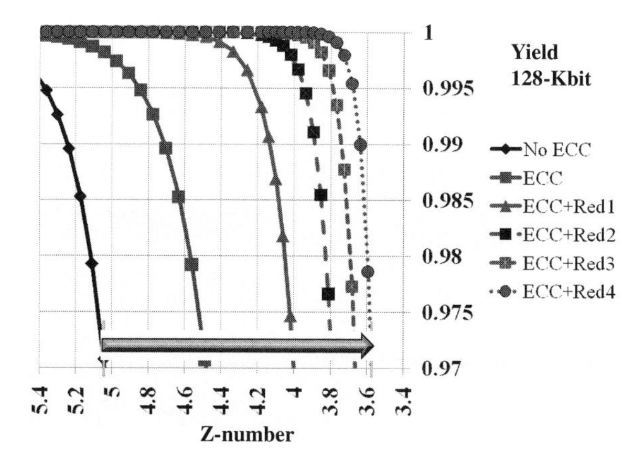

Fig. 3.45 ECC impacts on yield improvement

As a result, the yield can be expressed as: $Y = \sum_{q=0}^{R} P_q$, where the probability for defects of redundant ECC words of R is neglected for simplicity [8, 30].

Based on the above equation, ECC impact on yield improvement as a function of the bit-error rate r (which is represented by the number of Z) is calculated as shown in Fig. 3.45, where memory bits are assumed as N=136 consisting N_0=128 and N_P=8, W=1024, and R=0∼4. It can be seen that not only ECC but also ECC + its redundancy can significantly improve the bit-error tolerance. If R=4, the variation tolerance can be improved by 1.4 times of Z, which means that even if σ_{VT} is increased by 1.4 times due to device scaling to the half (which corresponds to about one process generation), the same yield can be maintained. This implies that ECC combined with its redundancy can extend the limit of SRAM scaling by one process generation without EOT scaling [30].

Figure 3.46 shows that how can each of (1) read-modify write with 8T SRAM, (2) V_{DD} regularity, and (3) ECC relax the limitations of each design solutions as a function of σ_{VT}. Figure 3.47 shows how EOT scaling can play a key role for SRAM scaling compared with the case of non-EOT scaling (shown in upper part). It can be seen that if EOT can be scaled down to 1.9 nm for 65 nm node → 1.6 nm for 45 nm node → 1.4 nm for 32 nm node → 1.2 nm for 25 nm node → 1.0 nm for 15 nm node, 6T SRAM can survive with V_{DD}=1.2 V or ECC V_{DD}=1.1 V without using a read-modify-write operation with 8T SRAM.

3.6.4 Area Scaling Trend Comparisons for Various Design Solutions

The design solutions for margin assists including 8T and 10T SRAM cells are realized at the cost of additional area penalties caused by adding two or four MOSFETs and required associated peripheral circuits. Meanwhile, upsizing 6T can

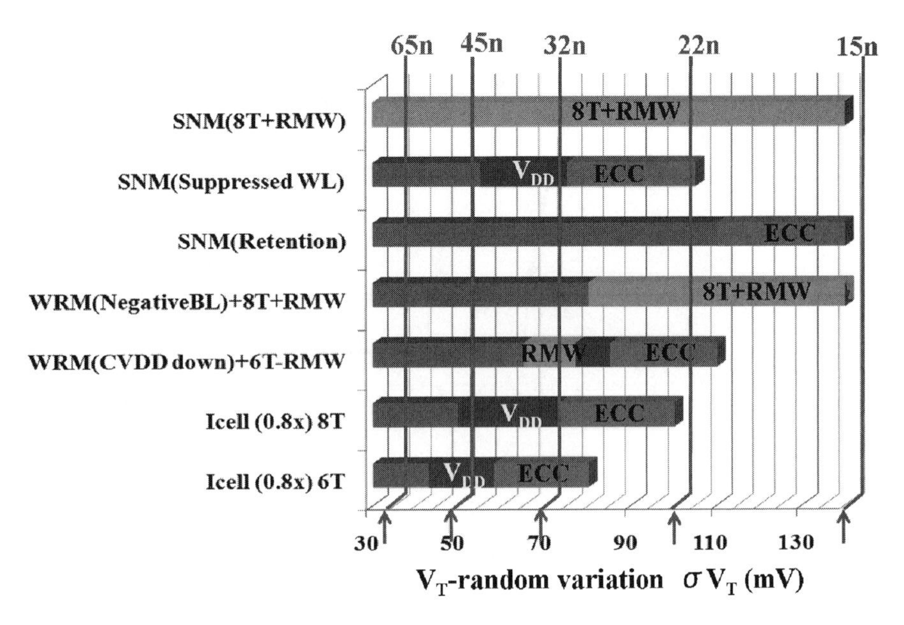

Fig. 3.46 Comparisons of SRAM scaling impacts given by each solution

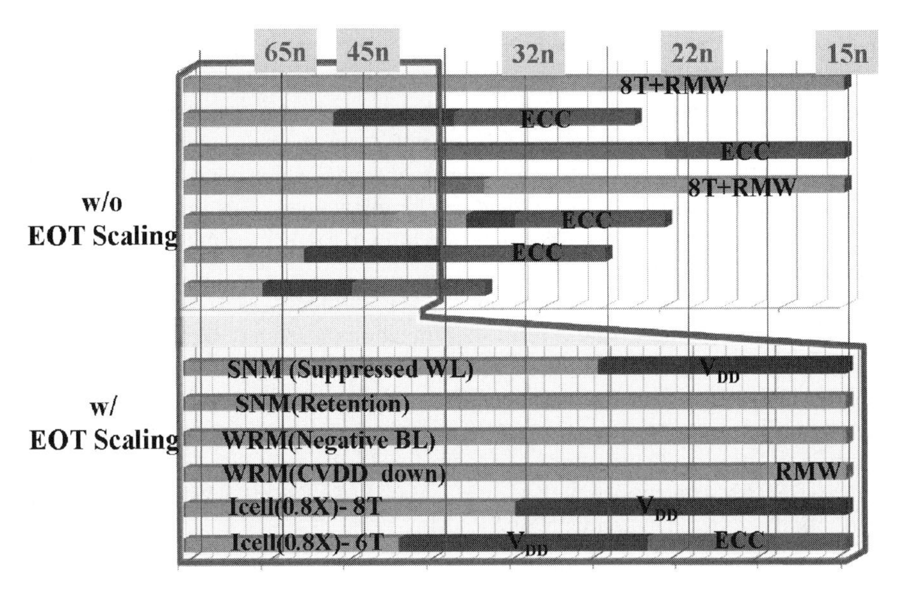

Fig. 3.47 Role of EOT in SRAM scaling

reduce σ_{VT} and improve the functional margins at the cost of cell area. In that sense, the area comparisons of SRAM macro as a function of σ_{VT} is needed to discuss which design solution including upsizing 6T becomes better solution for each request.

3.6.4.1 Area Comparisons of SNM and WRM Assists

As an example, upsizing 6T are compared with the suppressed WL depending on how much Icell down can be allowed as shown in Fig. 3.48, which shows the normalized required upsizing ratio comparisons as a function of σ_{VT} among (1) upsizing 6T, (2) suppressed WL (SWL) without care of Icell down, (3) SWL with 50% Icell down, and (4) SWL without Icell down at all. It can be seen that if 50% Icell down could be allowed, suppressed WL would have greater area advantage than just upsizing 6T when σ_{VT} becomes >60 mV. Meanwhile, when σ_{VT} is less than 50 mV, upsizing 6T is superior in terms of macro area saving than the suppressed WL due to no requirement of additional area overhead of peripheral circuits to generate suppressed WL level. In order to highlight which design solutions for SNM and WRM consumes more macro area, Fig. 3.49 shows how much the CVDD-down and negative BL schemes can save the macro area compared with the suppressed WL as a function of σ_{VT}. It can be seen that WRM assist design solutions of CVDD-down and negative BL consume smaller area than the suppressed WL for SNM and its areas becomes smaller than that of upsizing 6T when σ_{VT} becomes >50 mV. In that sense, SNM-enhanced cell combined with WRM assist can provide better option than simply upsizing 6T cell.

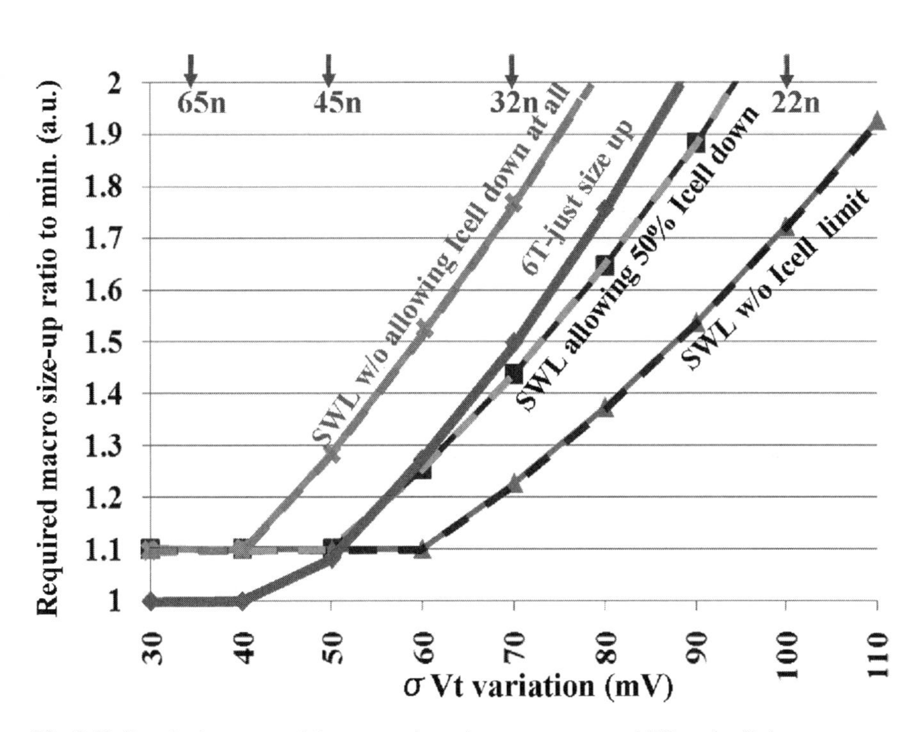

Fig. 3.48 Required macro upsizing comparisons between suppressed WL and cell size-up

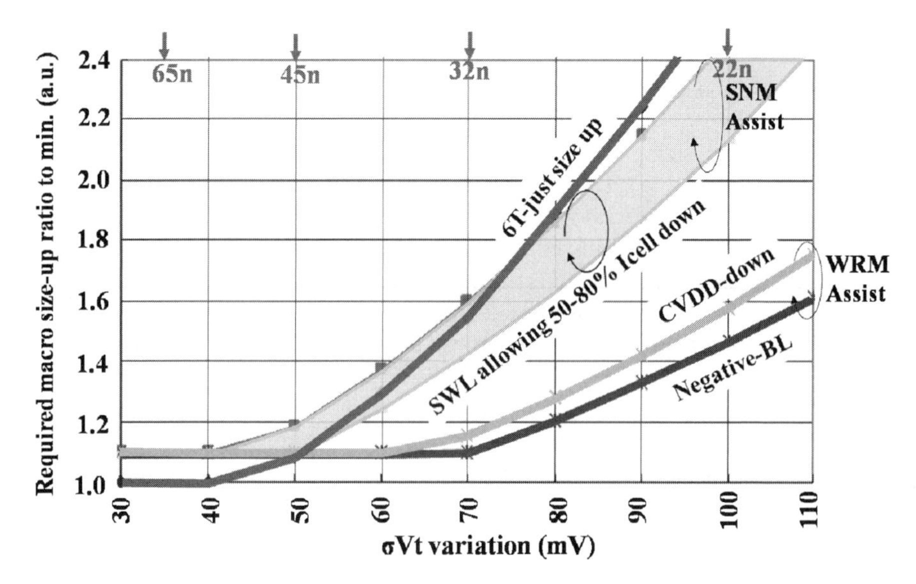

Fig. 3.49 Required macro upsizing comparisons between assist schemes for SNM and WRM

3.6.4.2 Area Comparisons of 8T and 10T SRAMs

Area benefits from the cells with more transistors depend on the following requirements and conditions: (1) increasing cell current; (2) lowering V_{DD} operation; and (3) increasing σ_{VT}. As a result, there are crossover points if areas of 6T, 8T, 8T with time-multiplexing, and 10T are compared as a function of the above conditions as shown in Fig. 3.50. At the initial point, the minimum macro sizes for 10T and 8T are about 1.9 and 1.4 times larger than 6T, respectively. However, the required conditions like Icell or σ_{VT} become severe, the required area for 6T crosses over 8T first and then 10T as shown in Fig. 3.46. It can be seen that 8T SRAM combined with time-multiplexing can be smallest at the end of the day if the time-multiplexing (read-modify-write operations) can be allowed. This is due to no need of trade-off among SNM, WRM, and Icell.

In order to highlight the trend of area advantage as a function of σ_{VT}, Figure 3.51 shows the normalized required upsizing ratio comparisons as a function of σ_{VT} among (1) upsizing 6T, (2) 6T combined with both negative BL and suppressed WL, (3) 8T combined with negative BL and read-modify-write, (4) 8T combined with negative BL and suppressed WL, and (5) cross-point 10T. It can be seen that simply upsizing 6T crosses over the two 8Ts combined with negative BL for WRM and read-modify-write and suppressed WL for SNM when σ_{VT} becomes >70 and >80 mV, respectively. 6T combined with negative BL and suppressed WL crosses over 8T combined with negative BL and suppressed WL when σ_{VT} becomes >90 mV. 10T crosses over 8T combined with negative BL and suppressed WL when σ_{VT} becomes >110 mV.

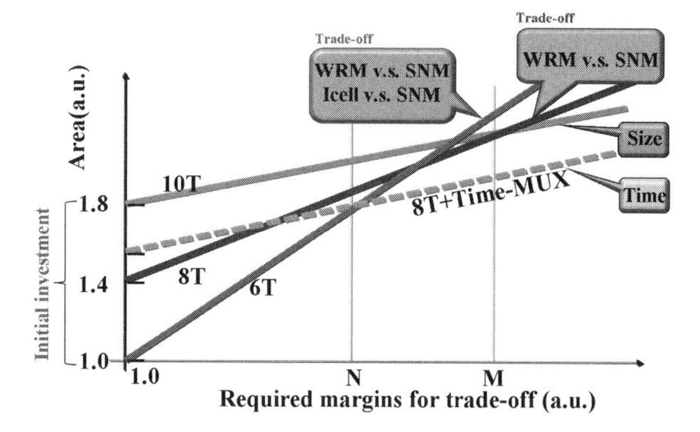

Fig. 3.50 Area comparisons of various SRAM cells

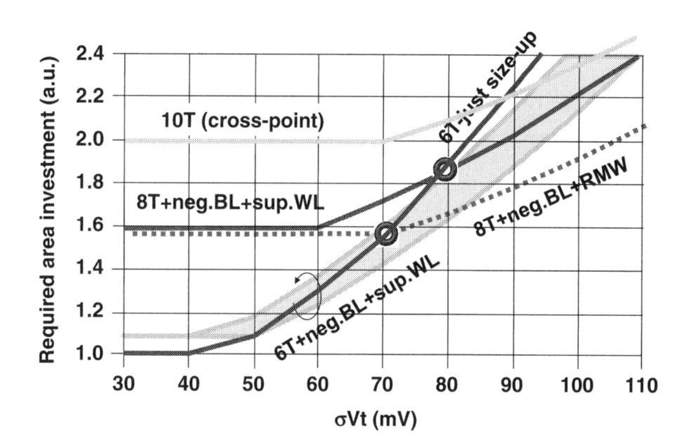

Fig. 3.51 Area comparisons of various SRAM cell options

If σ_{VT} could be suppressed to less than 70 mV, thanks to EOT scaling even at a 15 nm CMOS generation as shown in Fig. 3.42, 6T cells would be allowed long reign as shown in Fig. 3.51.

3.6.5 Comparisons of Design Options

As discussed above, many design options has been proposed. However, it depends on the requirements and conditions that can provide the best performance. Figure 3.52 shows the concept for which option can become majority in each application. In Fig. 3.52, each application zone is partitioned based on the required cell current Icell as *Y*-axis and V_{DD} range or the amount of σ_{VT} as *X*-axis. If V_{DD} is excessively lowered and σ_{VT} becomes extremely larger for some applications, cross-point 10T SRAM or 8T SRAM combined with read-modify write would be more needed than upsizing 6T SRAM as shown in Fig. 3.45 and they would become majority in such region, as shown in Fig. 3.52. Meanwhile, if the required Icell is extremely large for higher-speed application for other applications, 8T SRAM would become majority even if σ_{VT} is not extremely large and V_{DD} is not excessively lowered, as

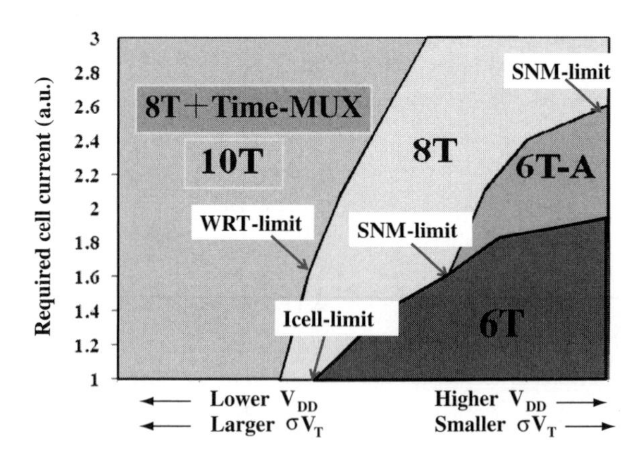

Fig. 3.52 Comparisons of design options

shown in Fig. 3.52. 6 T-A (Asymmetrical 6T SRAM) also has been proposed [9, 30] targeting for this region, as shown in Fig. 3.52.

3.7 SRAM Leakage Issues

SRAM leakage has become more significant with technology scaling for two main reasons: (1) the geometry of the transistor keeps shrinking, which leads to higher leakage current in channel, gate, and junction; (2) higher-performance VLSIs demands more and more on-die SRAMs. For example, in today's high-end CPUs there are more than 30 MB total of SRAMs that occupied over 70% of the total chip area. In this section, the SRAM leakage breakdown will be first analyzed in light of technology scaling. Various circuit mitigation schemes will then be summarized.

3.7.1 SRAM Leakage Breakdown

Table 3.3 shows the comparisons of the different transistor leakage components: (1) the subthreshold (sub-V_T) leakage; (2) gate leakage; and (3) gate-induced-drain-leakage (GIDL) for NFET and PFET including the mechanisms of leakage and their dependencies on temperature, process, and size scaling [27]. The sub-V_T leakage has strong dependencies on V_T and temperature. As well known, V_T depends on (1) the device feature size Lg and Wg as referred as "short and narrow" or "inverse-short and narrow" channel effects; (2) temperature; (3) biasings of back to source and drain to source. As a result, V_T can be increased by 1.5–1.7 orders of magnitude if temperature is changed by 100°C. The gate leakage, which is often attributed to quantum-mechanical gate-tunneling, depends on physical gate oxide thickness T_{OX}

Table 3.3 SRAM leakage break down

Junction leakage is included

	Sub-threshold leakage	Gate leakage	Gate-induced- -drain leakage
NMOS	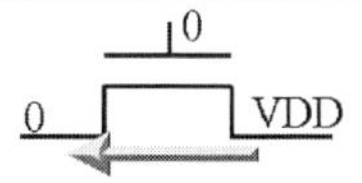 0 0 VDD	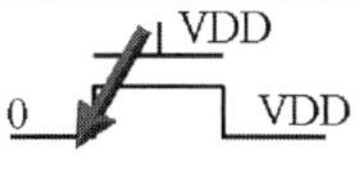 VDD 0 VDD	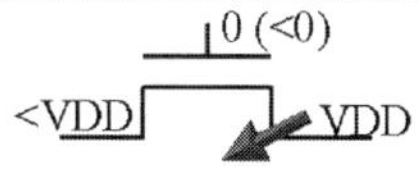 0 (<0) <VDD VDD
PMOS	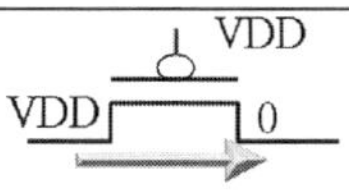 VDD VDD 0	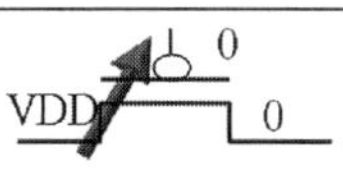 0 VDD 0	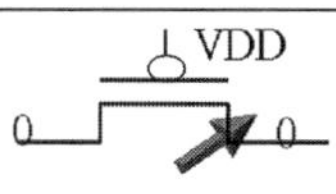 VDD 0 0
Mechanism	Sub_Vt_leak	Quantum-mechanical Gate-tunneling	Band to band(BTB) Band to Trap(BTT) i/f state assist
Dependency	Vt, Lg, Wg scaling	Tox Oxide material SiO2,SiON, High-K	Higher doping Shallow junction Sidewall space,...
Temp. dependency	Strong, 1.5-1.7 orders/100°C	Weak	Weak (BTB) Strong (BTT)

and it has relatively weak temperature dependence. As conventional T_{OX} cannot be scaled any more due to gate leakage, high-κ gate materials whose dielectric constant is higher than SiO_2 has been introduced at 45 nm process node, helping to continue scale the EOX without introducing excessive gate leakage. GIDL and junction leakages are based on three different types of tunneling mechanisms: (1) band-to-band (BTB); (2) band-to-trap (BTT); and (3) interface-state assist trap, which are modulated by the process conditions of doping, junction, side-wall space, and so on. The temperature dependencies of the tunneling of BTB and BTT are different, such as BTT is much stronger than BTB.

3.7.2 SRAM Leakage Scaling Trend

Figure 3.53 shows the leakage scaling trend for different components, including GIDL, subthreshold, and gate leakage at the temperature of 25°C from 90 to 45 nm for a typical low standby power (LSTP) process. It can be seen that the gate leakage and GIDL have become more dominant at room temperature. The gate leakage has become dominant at a 45 nm process generation with conventional gate oxide, as shown in Fig. 3.53. Based on the trend of increasing gate leakage as the process node is advanced, it is essential to introduce high-κ gate at 32 nm for further scaling [6, 11].

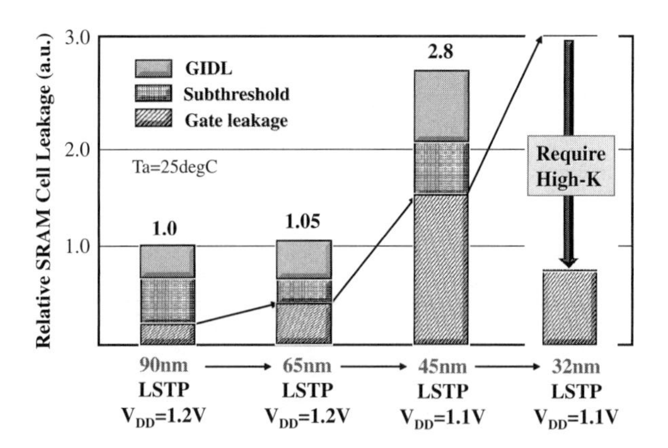

Fig. 3.53 SRAM leakage scaling trend for LSTP processes

3.7.3 High-κ Benefit for Leakage Reduction

Figure 3.54 shows the significant leakage reduction from 65 nm to 45 nm SRAM with high-κ gate [6, 11]. The gate leakage has been reduced by two and three orders of magnitudes for NFET and PFET, respectively. In addition, the leakages of GIDL and sub-V_T can be reduced further due to the use of the high-κ, which enabled the better scaling in channel doping concentration and reduction in short-channel effect. GIDL has become the most dominant leakage for high-κ SRAM at room temperature, as shown in Fig. 3.55. However, since sub-V_T leakage has strong temperature dependency, it still dominates overall SRAM leakage at 125°C, which results in limiting the V_T scaling.

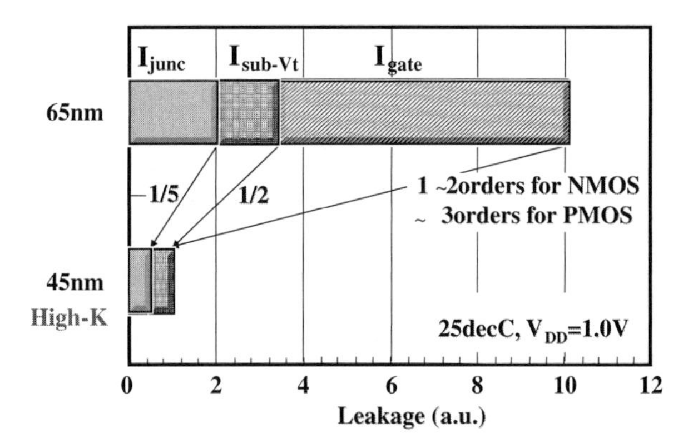

Fig. 3.54 High-κ benefit for leakage reduction

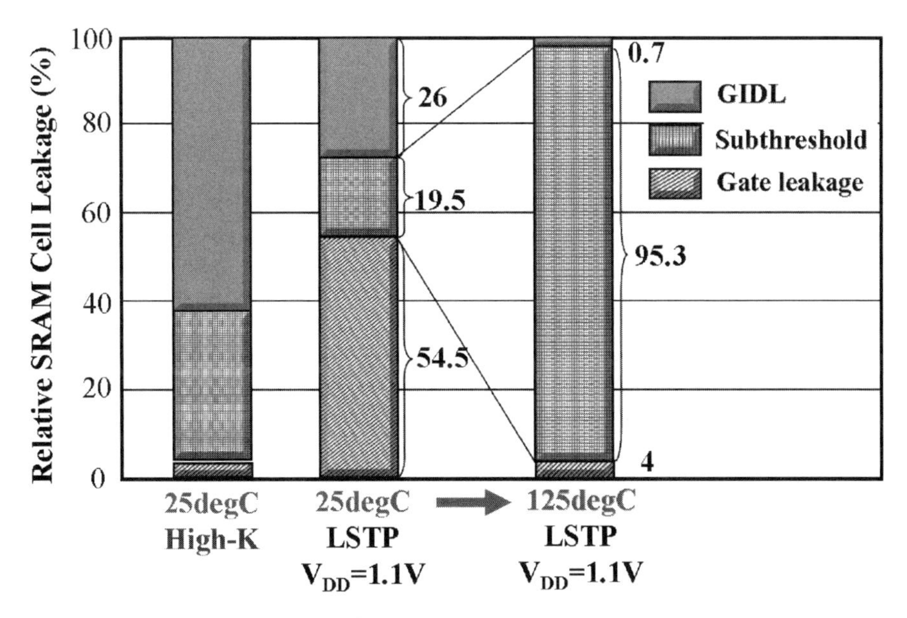

Fig. 3.55 Temperature dependency of leakage

3.7.4 Leakage Reduction

As discussed the leakage mechanism above, the leakages depend on V_T and electric field across gate and junction [15]. Therefore, the available choices are limited to three ways: (1) the increasing V_T; (2) biasing each electrode so as to relax the applied electric field; and (3) forcing negative potential difference of gate to source (V_{GS}) to suppress the leakage besides changing the gate material like high-κ and device structures like double-gate MOSFET as shown in Fig. 3.56(a) and (b). Sub-V_T leakages in SRAM happen through three MOSFETs of PU, PD, and PG. In order to reduce sub-V_T leakage of PG, the source of SRAM (VSSM) has to be raised by α mV, which results in negative V_{GS} biasing by α mV of PG. Floating BL also can eliminate the leakage path through single off-state PG, which can reduce the leakage by one order of magnitude. Meanwhile, the available choices in order to reduce sub-V_T leakages of PD and PU are limited to (1) back biasing to each substrate and (2) raising source potential of VSSM than P-well-bias (ground) or lowering source potential VDDM than N-well-bias (V_{DD}). In order to relax the forced electric field to reduce GIDL and gate leakage, VSSM and/or VDDM have to be changed so as to reduce the potential difference between the two electrodes [15].

3.7.4.1 Power Gating Switches

Power gating transistors can be inserted between V_{DD} and SRAM array or V_{SS} and SRAM array. Figure 3.57 shows the comparisons of power gating switches which consist of large and small MOSFETs for active and sleep modes, respectively [15,

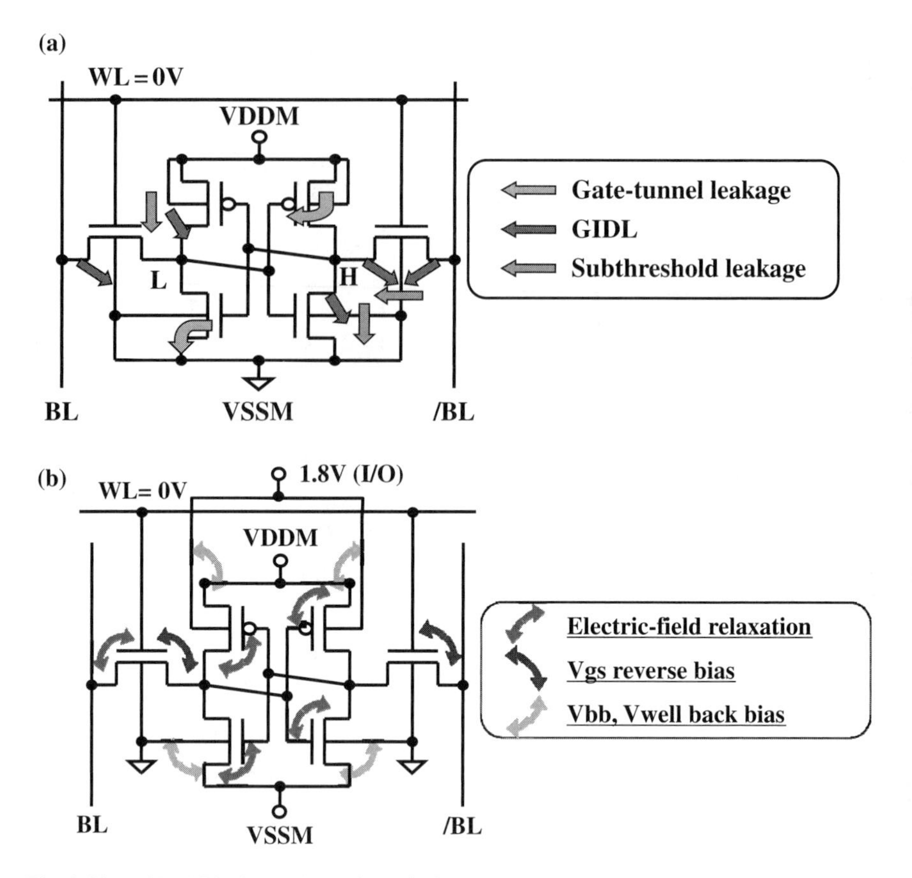

Fig. 3.56 (a) SRAM leakages (b) Various biasing means to reduce SRAM leakages

37]: (1) PFETs power switch inserted between the power line and VDDM which is referred to as "header" combined with its level clamping circuit provides a certain potential of VDDM which is lower than V_{DD}, and (2) NFETs ground switch inserted between the ground line and VSSM which is referred to as "footer" combined with its level clamping circuit provides a certain potential of VSSM which is higher than ground level. The leakages from V_{DD} power line to ground make VDDM level lower than V_{DD} (VSSM level higher than ground), but the clamping circuits avoid an excessive pull-down of VDDM (pull-up of VSSM) so as to retain data. The header and footer can help to reduce not only the sub-V_T leakages and gate leakages but also the junction leakages for NFET whose body tied to ground and for PFET whose body tied to V_{DD} due to relaxation of the electric filed between the body and drain, respectively. The footer can help to raise VSSM to significantly reduce the sub-V_T leakage of pass-gate NFET due to negatively biasing of V_{GS}. The power switch for active mode needs to be sized properly to meet a wake-up

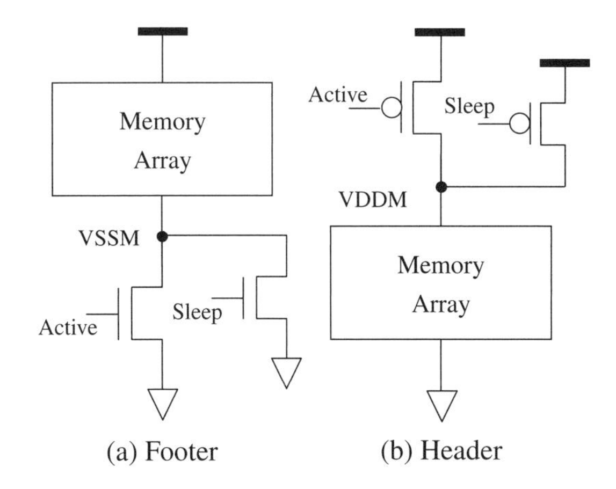

Fig. 3.57 Comparison of power switching gates: (**a**) footer and (**b**) header

timing requirement and to maintain small enough IR drop of the virtual power line VDDM. Since SRAM needs much larger current drivability of footer than header due to BL discharging for read operation, the header consumes less area than the footer for SRAM. As a result, header with smaller gate width of MOSFET not only consumes less dynamic power consumption caused by gate switching but also provides faster gate-switching than footer. The more frequent power gate switching can further reduce the DC leakages but the dynamic gate-switching power consumption increases depending on the frequency of its switching. Such kind of trade-off has to be taken in account for overall power savings.

3.7.4.2 Virtual Power Line Level Control

VDDM and VSSM are virtual power line and ground line of SRAM, respectively. The levels of VDDM and VSSM are determined by the ratios of the leakages path between the power switch transistor and SRAM leakage. The challenge with this control is that the level needs to be lowered as much as possible to meet the leakage requirement while preserving the data. Figure 3.58 shows the various VDDM level control schemes for sleep mode in which the large switch is off. The simple NMOS-diode can clamp the VDDM to one V_T drop from V_{DD}, preventing the VDDM from going too low to corrupt the retained data as shown in Fig. 3.55(a). This scheme has the following shortcomings: (1) NMOS-diode value can be too large compared with the minimum V_{DD} in the data retention mode in which V_{DD} is 0.6–0.7 V and (2) V_T fluctuation will directly give an impact on VDDM. Adding the programmable bias transistors is practical way to stabilize against the process fluctuation as shown in Fig. 3.58(b). Two primary benefits can be expected by programming a number of parallel switches: (1) the bias level can be tuned based on actual silicon results and (2) the wide range of operating voltage and temperature windows can be allowed by dynamically changing bias settings depending on the conditions of V_{DD} and Ta. Figure 3.55(c) shows the active feedbackcontrol based

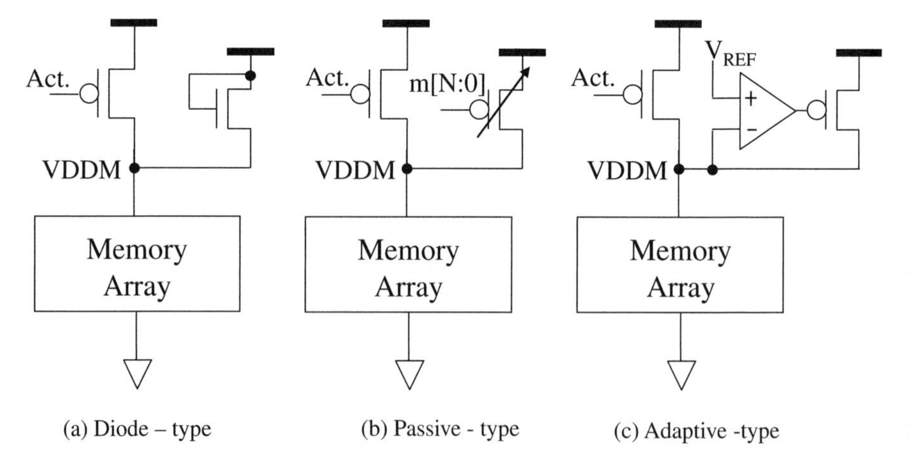

Fig. 3.58 Comparison of virtual power line level control schemes: (**a**) diode-type, (**b**) passive type, (**c**) adaptive type

on Op-amp which can be more adaptively tuned including within-die, die-to-die, temperature-induced, and stress-induced variations like NBTI (positive biased temperature instability) and PBTI (positive biased temperature instability). The reference voltage (V_{REF}) can be either generated internally or supplied externally. The major drawback of this scheme is the DC current consumption in Op-amps [6].

3.8 Conclusion

Embedded SRAM will continue to have a long reign in memory for embedded applications. It plays a key role in wide range of VLSI system designs from high-performance to low-power/low-voltage-aware applications as SRAM scaling continues to be driven by Moore's law. The SRAM scaling challenges in nanometer CMOS technologies will require simultaneous optimization in terms of area, speed, and power consumption based on the application requirements in order to achieve the product goals. In order to address these key challenges due to the increasing leakages and variations, both device and circuit innovations are required for future scaling as described in this chapter.

References

1. I. Chang, J-J. Kim, S. Park, K. Roy, West Lafayette, 21.7 "A 32 kb 10 T Subthreshold SRAM Array with Bit-Interleaving and Differential Read Scheme in 90 nm CMOS" IEEE Solid States Circuits Conference 2008
2. L. Chang, D. M. Fried, J. Hergenrother, J. W. Sleight, R. H. Dennard, R. K. Montoye, L. Sekaric, S. J. McNab, A. W. Topol, C. D. Adams, K. W. Guarini, and W. Haensch, "Stable SRAM Cell Design for the 32 nm Node and Beyond" IEEE 2005 Symposium on VLSI Technology

3. L. Chang, Y. Nakamura, R. K. Montoye, J. Sawada, A. K. Martin, K. Kinoshita, F. H. Gebara, K. B. Agarwal, D. J. Acharyya, W. Haensch, K. Hosokawa, and D. Jamsek, "A 5.3 GHz 8T-SRAM with Operation Down to 0.3.1 V in 65 nm CMOS", IEEE 2007 Symposium on VLSI Circuits
4. Y. H. Chen, W. M. Chan, S. Y. Chou, H. J. Liao, H. Y. Pan, J. J. Wu, C. H. Lee, S. M. Yang, Y. C. Liu, H. Yamauchi, "A 0.6 V 45 nm Adaptive Dual-rail SRAM Compiler Circuit Design for Lower VDD_min VLSIs 21.3" IEEE 2008 Symposium on VLSI Circuits
5. Y. Chung and S.-W. Shim, "An Experimental 0.8 V 256-kbit SRAM Macro with Boosted Cell Array Scheme", ETRI Journal, Volume 29, Number 4, 2007
6. H. F. Hamzaoglu, K. Zhang, Y. Wang, H.-J. Ahn, U. Bhattacharya, Z. Chen, Y.-G. Ng, A. Pavlov, K. Smits*, M. Bohr, 21.1, "A 153 Mb SRAM Design with Dynamic Stability Enhancement and Leakage Reduction in 45 nm Hi-K Metal Gate CMOS Technology" IEEE Solid States Circuits Conference 2008
7. K. Ishibashi, K.-I. Takasugi, T. Yamanaka, Member, T. Hashimoto, and K. Sasaki, "A 1-V TFT-Load SRAM Using a Two-step Word-Voltage Method", IEEE Journal of Solid-state Circuits. Volume 21, Number 2 1992
8. K. Itoh, M. Horiguchi, M. Yamaoka, "Low-Voltage Limitations of Memory-Rich Nano-Scale CMOS LSIs" 33rd European Solid State Circuits Conference, 2007. ESSCIRC 11–13 Sept. 2007 pp. 68–75
9. A. Kawasumi, N. Otsuka, T. Yabe, Y. Takeyama, O. Hirabayashi, K. Kushida, A. Tohata, T. Sasaki, A. Katayama, G. Fukano, Y. Fujimura, 21.4 "A Single-Power-Supply 0.7 V 1 GHz 45 nm SRAM with a Asymmetrical Unit-β-Ratio Memory Cell" IEEE Solid States Circuits Conference 2008
10. M. Khellah et al. "Wordline & Bitline Pulsing Schemes for Improving SRAM Cell Stability in Low-Vcc 65 nm CMOS Designs", IEEE Symposium on VLSI Circuits 2006
11. K. J. Kuhn, "Reducing Variation in Advanced Logic Technologies: Approaches to Process and Design for Manufacturability of Nanoscale CMOS" pp. 471–474, IEEE IEDM 2007
12. H. Mair, A. Wang, G. Gammie, D. Scott, P. Royannez, S. Gururajarao, M. Chau, R. Lagerquist, L. Ho, M. Basude, N. Culp, A. Sadate, D. Wilson, F. Dahan, J. Song, B. Carlson, U. Ko, "A 65 nm Mobile Multimedia Applications Processor with an Adaptive Power Management Scheme to Compensate for Variations", IEEE 2007 Symposium on VLSI Circuits
13. Y. Morita, H. Fujiwara, H. Noguchi, Y. Iguchi, K. Nii, H. Kawaguchi, and M. Yoshimoto, "An Area-Conscious Low-Voltage-Oriented 8T-SRAM Design under DVS Environment" 2007 Symposium on VLSI Circuits
14. S.Ohbayashi, M. Yabuuchi, K. Nii, Y. Tsukamoto, S. Imaoka, Y. Oda, M. Igarashi, M. Takeuchi, H. Kawashima, H. Makino, Y. Yamaguchi, K. Tsukamoto, M. Inuishi, K. Ishibashi, H. Shinohara, "A 65 nm Soc Embedded 6T-SRAM Design for Manufacturing with Read and Write Cell Stabilizing Circuits", IEEE 2006 Symposium on VLSI Circuits
15. K. Osada, Y. Saitoh, E. Ibe, K. Ishibashi, "16.7fF/Cell Tunnel-Leakage-Suppressed 16-Mb SRAM for Handing Consmic-Ray-Induced Multi-errors", IEEE Solid-States-Circuits Conference, Vol. 1 pp. 302–494, 2003
16. H. Pilo, J. Barwin, G. Braceras, C. Browning, S. Burns, J. Gabric, S. Lamphier, M. Miller, A. Roberts, F. Towler, "An SRAM Design in 65 nm and 45 nm Technology Nodes Featuring Read and Write-Assist Circuits to Expand Operating Voltage", IEEE 2006 Symposium on VLSI Circuits
17. H. Pilo, V. Ramadurai, G. Braceras, J. Gabric, S. Lamphier, Y. Tan, "A 450 ps Access-Time SRAM Macro in 45 nm SOI Featuring a Two-Stage Sensing-Scheme and Dynamic Power Management" 21.7, IEEE Solid States Circuits Conference 2008
18. N. Shibata, et al., "A 0.5-V 25-MHz 1-mW 256-kb MTCMOS/SOI SRAM for solar-power-operated portable personal digital equipment – sure write operation by using step-down negatively overdriven bitline scheme", IEEE Journal of Solid-State-Circuits, Volume 3.1, Number 3, pp.728–73.2. 2006

19. T. Suzuki, H. Yamauchi et al., "A Stable SRAM Cell Design Against Simultaneously R/W Disturbed Accesses", IEEE Symposium on VLSI Circuits, 2.2, 2006
20. Y. Tsukamoto et al., 5A.2 "Worst-Case Analysis to Obtain Stable Read/Write DC Margin of High Density 6T-SRAM-Array with Local Vth Variability", International Conference on Computer-Aided Design, 2005
21. D.-P. Wang, H.-J. Liao, H. Yamauchi, W. Hwang, Y. L. Lin, Y. H. Chen, H. C. Chang, "A 45 nm Dual-Port SRAM with Write and Read Capability Enhancement at Low Voltage", IEEE SOCC 2007
22. M. Yabuuchi, K. Nii, Y. Tsukamoto, et al., "A 45nm Low-Standby-Power Embedded SRAM with Improved Immunity Against Process and Temperature Variations", Solid-State Circuits Conference, 2007. ISSCC 2007. Digest of Technical Papers. IEEE International, pp. 326–606, 11–15 Feb. 2007
23. M. Yamaoka, N. Maeda, Y. Shimazaki, K. Osada, 21.5 "A 65 nm Low-Power High-Density SRAM Operable at 1.0 V Under 3σ Systematic Variation Using Separate Vth Monitoring and Body Bias for NMOS and PMOS" IEEE Solid States Circuits Conference 2008
24. M. Yamaoka, K. Osada, R. Tsuchiya, M. Horiuchi, S. Kimura, T. Kawahara, "Low Power SRAM Menu for SoC Application using Yin-Yang-Feedback Memory Cell Technology", 2004 Symposium on VLSI Circuits
25. H. Yamauchi, USP 5680356, Oct. 21, 1997
26. H. Yamauchi, USP 6898111, May.23.2005
27. H. Yamauchi "Embedded SRAM Design", tutorial in IEEE ASSCC 2006
28. H. Yamauchi "Embedded SRAM Circuit Design Technologies for a 45 nm and Beyond" (Invited Paper) 5.2-I, The 7th International Conference on ASIC (ASICON) 2007, Oct. 2007
29. H. Yamauchi, "Embedded SRAM Trend in Nano-Scale CMOS" (Invited Paper), IEEE International Workshop on Memory Technology, Design, and Testing (MTDT), Dec. 2007
30. H. Yamauchi, "Embedded SRAM Design and Trend" IEEE ISSCC 2008, Memory forum "Embedded Memory Design for Nano-scale VLSI system", 2008
31. H. Yamauchi et al., "A Circuit Design to Suppress Asymmetrical Characteristics in High-Density DRAM Sense Amplifiers", IEEE Journal of Solid-State Circuits, Volume 25, 1990
32. H. Yamauchi, et al., "A 0.8 V/100 MHz/sub-5 mW-operated mega-bit SRAM cell architecture with charge-recycle offset-source driving (OSD) scheme" IEEE Symposium on VLSI Circuits 1996, 126–127, June, 1996
33. H. Yamauchi et al., "A 0.5 V/100 MHz over-VCC grounded data storage (OVGS) SRAM cell architecture with boosted bit-line and offset source over-driving schemes", IEEE Low Power Electronics and design., pp.49–54, 1996
34. H. Yamauchi et al., "Gate-over-driving CMOS architecture for 0.5 V single-power-supply-operated devices", IEEE Solid States Circuits Conference, pp.290–291, 1997
35. H. Yamauchi et al., "A 0.5 V single power supply operated high-speed boosted and offset-grounded data storage (BOGS) SRAM cell architecture", IEEE Transaction on Very Large Scale Integration (VLSI) Systems Volume 5, Number 4, pp. 377–387, 1997
36. H. Yamauchi et al., "A Differential Cell Terminal Biasing Scheme Enabling A Stable Write Operation Against A Large Random Threshold Voltage (Vth) Variation", IEICE EC , Volume 11, 2006
37. K. Zhang, U. Bhattacharya, Z. Chen, F. Hamzaoglu, D. Murray, N. Vallepalli, Y. Wang, B. Zheng, M. Bohr, "A 3 GHz 70 Mb SRAM in 65 nm CMOS Technology with Integrated column-Based Dynamic Power Supply" IEEE Journal of Solid-State Circuits, Volume 3.1, number. 1, 2006

Chapter 4
Ultra Low Voltage SRAM Design

Naveen Verma and Anantha P. Chandrakasan

Abstract Aggressive scaling of the supply voltage to SRAMs greatly minimizes their active and leakage power, a dominating portion of the total power in an increasing number of applications. Hence, highly energy-constrained systems, where performance requirements are secondary, benefit greatly from SRAMs that provide read and write functionality at the lowest possible voltage, particularly down to 0.3 V. However, conventional bit-cells and architectures, designed to operate at nominal supply voltages, come far short of achieving the voltage scalability required. This chapter investigates the basic degradation mechanisms in the underlying MOSFET devices, and the resulting failures modes plaguing low-voltage SRAMs. Specific solutions to manage all of these are analyzed with respect to the associated density, performance, and power trade-offs. Actual design examples are cited that achieve full read and write functionality down to 0.3 V, where the leakage-power savings can exceed a factor of 50 compared to nominal supplies.

4.1 Introduction

Power and energy consumption are the critical limitation in an ever increasing number of applications. As a result, in virtually every digital design today, the power management strategy, often employing a multi-pronged suite involving device, circuit, and architecture techniques, has become a paramount concern. In this regime, voltage scaling has emerged as a vital enabler industry-wide, thanks to its strong impact on energy across all digital circuits [6].

But while power and energy benefit tremendously, low-voltage operation places severe duress on SRAM functionality; reduced noise margins and increased sensitivity to threshold voltage variation vehemently oppose our ability to achieve stable bit-cells and acceptable performance while maintaining density.

N. Verma (✉)
Massachusetts Institute of Technology, Cambridge, MA, USA
e-mail: nverma@mit.edu

K. Zhang (ed.), *Embedded Memories for Nano-Scale VLSIs*, Series on Integrated Circuits and Systems, DOI 10.1007/978-0-387-88497-4_4,

Nonetheless, device–circuit codesign, specialized circuits and architectures, and prudent application of density-stability trade-offs demonstrate the ability to achieve wide voltage scaling, down to 0.7 V, in a 65 nm low-power CMOS technology, with a 0.667 μm^2 bit-cell [35]. While this level of voltage scaling, achieving over a factor of three reduction in leakage power and nearly a factor three reduction in active energy, is viable for high-performance and medium-performance desktop computing applications, the class of highly energy constrained applications, encompassing portable, implantable, and highly remote devices, requires far more aggressive voltage scaling. In particular, operating voltages down to 0.3 V are required to enable many such systems, as their complete power budget is often much less than 1 mW. Of course, the ability of the SRAMs to scale accordingly is essential as they occupy a dominating portion of the total power, energy, and area.

As one might expect, at these ultra low voltages, fundamental device characteristics and circuit topologies critical to SRAMs are degraded by several orders of magnitude. Achieving functional high-density SRAMs, then, requires highly specialized techniques to manage these severe degradations. Figure 4.1 shows the minimum operating voltage achieved by ultra low voltage logic and SRAM designs targeting highly energy constrained applications [33]. The primary limitation posed by SRAMs, particularly at advanced technology nodes, is highlighted, pointing to the major challenges at hand. This chapter starts by motivating the need for aggressive voltage scaling in digital circuits generally and, even more critically, in SRAMs specifically. Subsequently, the severe associated challenges are identified and characterized, and then solutions at both the bit-cell and array level are presented and analyzed.

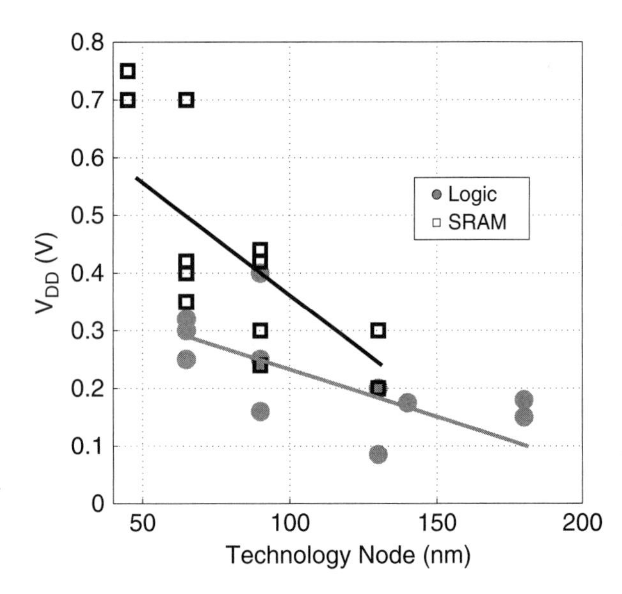

Fig. 4.1. Minimum operating voltage of reported designs with respect to the implementation technology node (© [2008] IEEE)

4.2 Minimum Energy and Leakage-Power Operation

In modern digital systems, active switching and leakage are the dominant sources of energy consumption. The total energy for an operation is given by Eq. (4.1)

$$E_{total} = E_{active} + E_{leakage} = C_{eff} V_{DD}^2 + \int_{op.time} I_{leakage} V_{DD} dt. \tag{4.1}$$

A circuit's leakage energy, E_{leakage}, is the integral of its leakage power over the time it requires to complete an operation. Once the operation is complete, the assumption, for non-state-retaining logic circuits, is that the circuit can be power-gated using high-threshold devices to suppress the idle leakage currents [22]. It should be noted that power-gating itself increases E_{active} due to the overhead of controlling the gating device and restoring the power/ground node voltage before subsequent circuit operation.

Although voltage scaling reduces the active $CV_{DD}{}^2$ energy, it also reduces circuit speed and, therefore, results in a longer leakage power integration time during which the circuit cannot be power-gated. Hence, the leakage energy increases [36]. These opposing trends are shown in Fig. 4.2 for the case of a logic circuit (32-b carry look-ahead adder) in 90 nm CMOS. The minimum total energy occurs at a very low voltage of approximately 0.25 V, where the energy saving, with respect to the nominal 1.1 V supply voltage, is over a factor of 15, but the leakage energy increases rapidly below this. This result is highly typical, as most practical digital circuits have a minimum energy voltage that occurs at or below the threshold voltage of the devices [5, 37].

SRAMs that can operate down to these voltages are essential for compatibility with minimum energy logic. However, the energy and power considerations for the SRAMs themselves make these low voltages even more compelling. In particular,

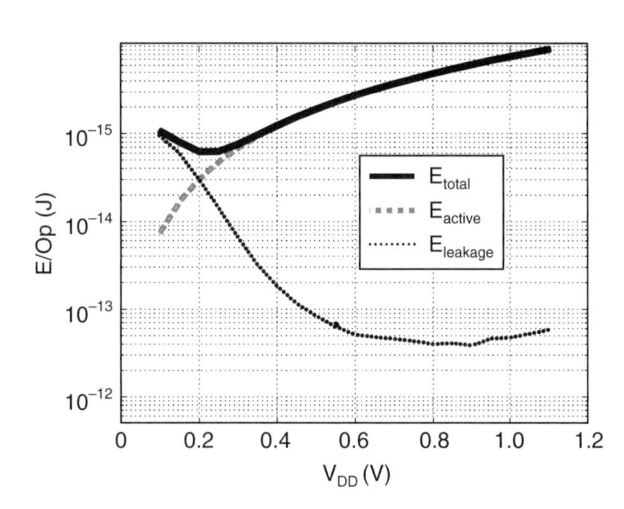

Fig. 4.2 Energy profiles (total, active, and leakage) with respect to supply voltage for a 90 nm 32-b carry look-ahead adder (© [2008] IEEE)

the argument above is slightly modified for SRAMs, where data buffering requirements imply a need for long-term retention capability. In this scenario, the SRAMs must stay on for an arbitrary length of time unrelated to their operation or access delay. Without the ability to power-gate after an access, leakage power becomes more important than leakage energy, and the two critical metrics, namely leakage power and active energy, both benefit from supply voltage scaling. In fact, the reduction in leakage power can be quite large, since, in nano-scale CMOS technologies, a reduced V_{DD} implies a reduced V_{DS}, and therefore significantly alleviates drain-induced barrier lowering (DIBL). As a result, the leakage current is greatly reduced. Figure 4.3 shows the normalized leakage current for minimum-sized devices in low-power 90, 65, and 45 nm technologies, highlighting the dramatic benefit of voltage scaling. In actual SRAM implementations, for instance a 65 nm low-power design, a supply voltage reduction from 1 to 0.3 V actually reduced the leakage current by over a factor of 5, resulting in a reduction of the leakage power by over a factor of 15 [30].

Of course, the primary cost associated with such aggressive voltage scaling is reduced performance both for the digital logic and the SRAMs. However, in many highly energy constrained applications, including, for instance, remote sensing devices and biomedical implants, performance constraints are significantly relaxed and often very modest. Specifically, in these cases, the circuit speeds are typically limited to less than 1 MHz. Accordingly, the voltage can be reduced with primary consideration to leakage power and active energy until these performance constraints become relevant. It should be noted, however, as mentioned earlier, the energy of non-state-retaining digital logic increases rapidly below the minimum energy voltage. Since introducing separate supply voltages for the logic and SRAMs

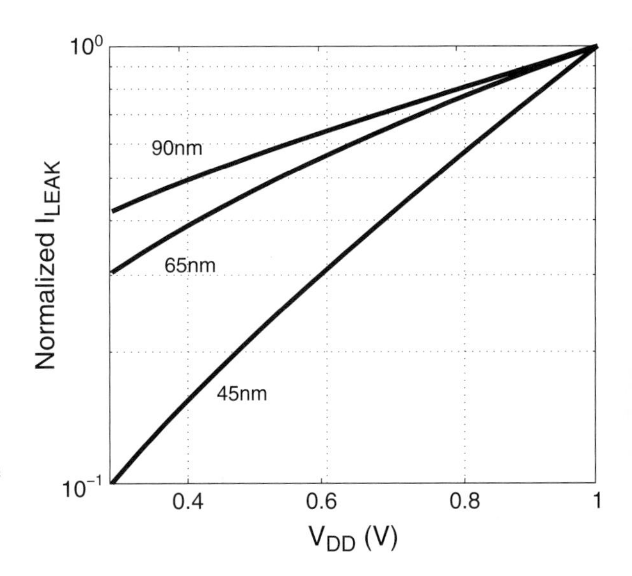

Fig. 4.3 Normalized leakage currents with respect to supply voltage for 90, 65, 45 nm devices

can be prohibitive from a system perspective, the energy, or average power, of the logic must be balanced with the leakage power of the SRAMs in choosing the optimal supply voltage.

In addition to ultra low voltage operation, standby modes, where the array supply is further reduced between active access, are a widely leveraged technique to exploit the leakage-power reductions [11]. These have the benefit that during active operation, where the operating margins are most severely stressed, the supply voltage is raised, improving both the noise margins and sensitivity to variation. However, there is an active energy overhead associated with transitioning between active and standby modes which involves supply voltage recovery and standby circuitry control. This leads to a standby breakeven time, below which the integrated leakage-power savings are not sufficient to justify the overhead.

Nonetheless, during sufficiently long periods, when the standby mode is viable, it is desirable to reduce the supply voltage to the data retention limit, known as the data retention voltage (DRV), so that the leakage-power savings are maximized. Since the DRV typically occurs in subthreshold, its value, among cells in the array, is highly affected by variation and can be difficult to track over operating conditions. To minimize the guard-band required for robust data retention, closed loop techniques can be employed, involving replica cells, to track the operating conditions, and statistical tail measurements or distribution models, to account for local variation [34].

4.3 Ultra Low Voltage SRAM Challenges

The effect of voltage scaling on MOSFET device performance and standard SRAM circuit topologies is dramatic. Moreover, these continue to become more pronounced with technology scaling. This suggests that one path toward easing the design challenges is to target larger geometry technologies, where the variation effects are much less severe and the leakage currents are typically much lower. However, density, due to its strong impact on cost, is a critical metric for most SRAMs and retains its importance in highly energy constrained, ultra low voltage applications. Of course, the trade-off between density and V_{MIN} does require renewed consideration. However, Fig. 4.4 illustrates the urgent need to maximize SRAM density in such applications, as the 128 kb cache dominates the area of this practical MSP430 DSP SoC, which is implemented in a 65 nm technology and designed to operate down to 0.3 V [17]. More generally, there exists three classes of highly energy constrained applications, all of which benefit from technology scaling:

1. *Low-speed requirement.* Applications such as environment and biological signal monitoring require circuit speeds of ten to hundreds of kilohertz. Voltage scaling can be applied aggressively to operate statically at a low voltage, as in the example from Fig. 4.4.

Fig. 4.4 Die photo of ultra low voltage MSP430 DSP in 65 nm CMOS. 128-kb on-chip SRAM cache dominates area

2. *Dynamic-speed requirement.* Applications such as cellular multimedia handsets have relaxed workloads for the vast majority time, but can provide bursts of high performance. Dynamic voltage scaling and ultra-dynamic voltage scaling [3] allow these systems to operate momentarily at a higher voltage and advanced technologies afford significant speed-ups. An important challenge is managing the trade-off between low-voltage functionality and the accompanying overheads to performance and area, which limit higher-voltage operation.
3. *Constant high-speed requirement.* Applications such as baseband radio processors must meet system throughput specifications. These can leverage the benefits of extreme parallelism in scaled technologies to assign unit operations to many separate hardware blocks, each operating efficiently at a reduced voltage and rate [28].

An important benefit of advanced technologies for highly energy constrained systems is the reduction in CV_{DD}^2 active energy, thanks to the reduced switch capacitance. A major, disadvantage, however, is the increase in leakage power due to the scaling of threshold voltages and oxide thicknesses. Importantly, though, these can be specifically managed even in highly scaled technologies. As energy-constrained applications gain importance, device parameters can be optimized accordingly: a direction that is already emerging, where low-power (LP) technologies at 90 nm and beyond are specifically designed to address leakage-current density.

The following sections examine the challenges relevant to high-density, ultra low voltage SRAMs. These start with an examination of the MOSFET device characteristics and then consider their impact on standard SRAM topologies. Specifically, they point to the need for entirely new topologies and highly specialized circuit assists to enable ultra low voltage SRAM operation.

4.3.1 MOSFET Degradations at Ultra Low Voltages

Figure 4.5 shows the normalized I_D versus V_{GS} behavior of an LP 65 nm NMOS in two different lights. More specifically, two effects are shown that are of critical importance to SRAMs at low voltages: (1) Fig. 4.5a shows the severe effect of threshold voltage variation, and (2) Fig. 4.5b shows the degradation in the on-to-off ratio of the current.

In Fig. 4.5a, the on-current initially increases exponentially in subthreshold, and then far more slowly in strong inversion. Threshold voltage shifts are, of course, a prominent result of processing variation and random dopant fluctuation (RDF) [20, 26], and they essentially cause sideways offsets. The $\pm 4\sigma$ case for local variation, which occurs commonly in large SRAM arrays, is shown in the I_D versus V_{GS} characteristic. Although the resulting variability is relatively small at high voltages, at ultra low voltages (e.g., 0.3 V), the resulting change in drain current can easily exceed three orders of magnitude. This suggests, very emphatically, that relative device strengths cannot reliably be set using conventional techniques like *W/L* sizing. As described in Section 4.3.2, standard high-density SRAM topologies rely heavily on ratiometric sizing and are, consequently, extremely sensitive to this failure mechanism.

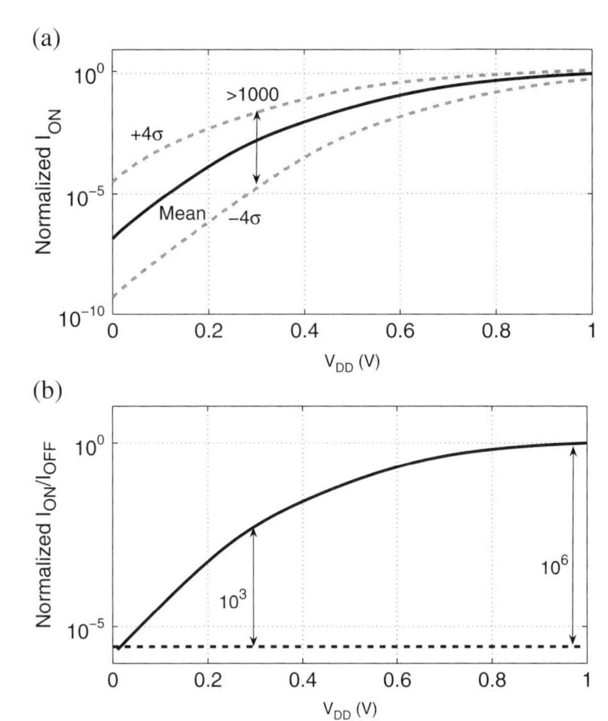

Fig. 4.5 MOSFET characteristics with respect to voltage shown (**a**) severe effect of variation and (**b**) degraded I_{ON}/I_{OFF} at low voltages

Figure 4.5b plots the ratio of the on-current to the off-current for the same LP 65 nm NMOS. As shown, the nominal ratio of I_{ON}/I_{OFF} degrades from over 10^6, at nominal voltages, to less than 10^3 at 0.3 V. This effect is even more severely amplified when variation is considered; for instance, with 4σ degradation to I_{ON}, the ratio, at low voltages, is reduced to less than 10^2. This implies that there is now a strong interaction between both the "on" and the "off" devices when it comes to setting the static voltages of critical nodes. This, once again, is highly problematic for SRAMs, where high-density requirements call for the integration of many devices on shared nodes.

4.3.2 Conventional SRAM Degradations at Ultra Low Voltages

In this section, the fundamental degradations to MOSFET operation, described previously, are more specifically related to the challenges of high-density ultra low voltage SRAM design. First, the precise effect on the standard symmetric-6T bit-cell is considered, and then, array-level effects associated with cell read-current and the ability to sense read-data are discussed.

4.3.2.1 Symmetric-6T Failures

The 6T bit-cell, which is shown in Fig. 4.6, fails to operate at ultra low voltages because of reduced signal levels and increased sensitivity to RDF [2]. In this configuration, both read and write accesses are ratioed making it highly difficult to overcome the severe effects of variation and manufacturing defects.

Specifically, during read accesses the cell must remain bi-stable to ensure that both data logic values can be held and read without being upset by transients that occur at the internal nodes, *NC/NT*, when the access devices, *M5/6*, are enabled. The read static noise margin (SNM) [25] quantifies the margin against loss of

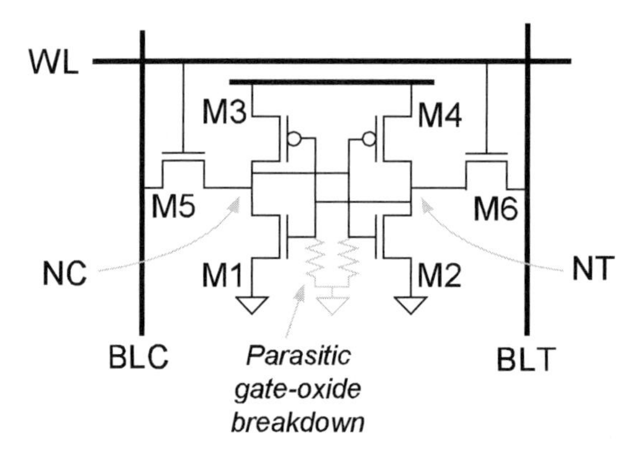

Device	Function
M1-2	Driver
M3-4	Load
M5-6	Access

Fig. 4.6 Standard symmetric 6T bit-cell

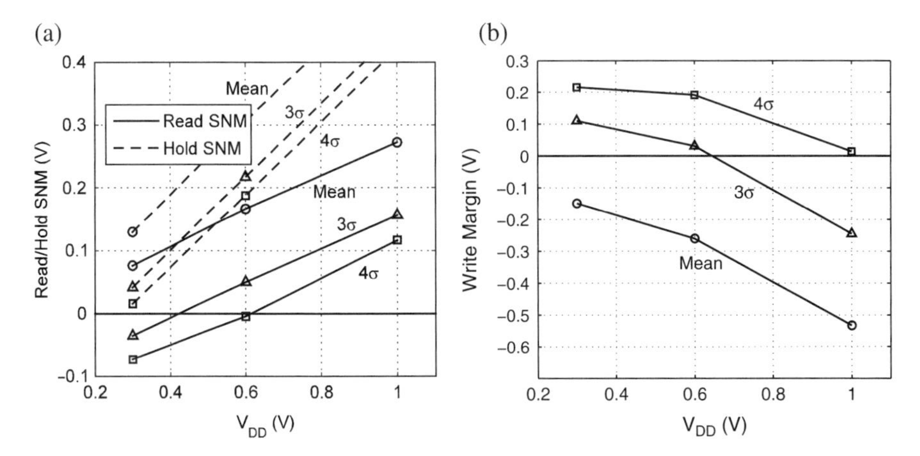

Fig. 4.7 Standard 6T bit-cell functionality metrics with respect to supply voltage showing (**a**) read and hold SNM and (**b**) write margin (© [2008] IEEE)

bi-stability by considering the worst-case condition, where the bit-lines, *BLT/BLC*, remain at their pre-charge voltage, V_{DD}. It should be noted that in many low-voltage designs, the cell read-current is very small as a ratio of the bit-line capacitance, and, as a result, *BLT/BLC* do in fact remain near the pre-charge voltage for an extended period. Figure 4.7a shows Monte Carlo simulations of a bit-cell in LP 65 nm CMOS considering the effect of RDF and gate-length variation. As shown, at ultra low voltages the read SNM vanishes and becomes negative, indicating read failures. Similarly, the write-margin measures the extent to which the cell can be made mono-stable to write the desired data, and as shown in Fig. 4.7b, it is also lost at low voltages, where a positive value, in this case, indicates write failures. Finally, the hold SNM measures the ability of the cell to remain bi-stable while the access devices are disabled. As shown, it is preserved to very low voltages and will form the basis for several of the ultra low voltage bit-cell designs described in Section 4.4.1.

Generally, read and write failures arise due to the reduced signal levels at low voltages and threshold-voltage variation, which increases exponentially in sub-threshold. The electrical-β ratio, shown in Fig. 4.8, characterizes the effective strength of the driver devices, *M1/2*, relative to the access devices, *M5/6*, and is critical to the cell immunity against problematic transients on *NT/NC* during read accesses. As a result, it serves to isolate the severe contribution from variation toward read failures. Figure 4.8 shows the electrical-β ratio with respect to supply voltage, where it can be seen that it is degraded by almost four orders of magnitude at 0.3 V.

An additional effect limiting minimum supply voltage in the 6T bit-cell is gate-oxide soft breakdown, resulting in extremely high gate leakage in the driver devices, *M1/2* [1]. In 65 nm and beyond, even with very high quality oxide, soft breakdown

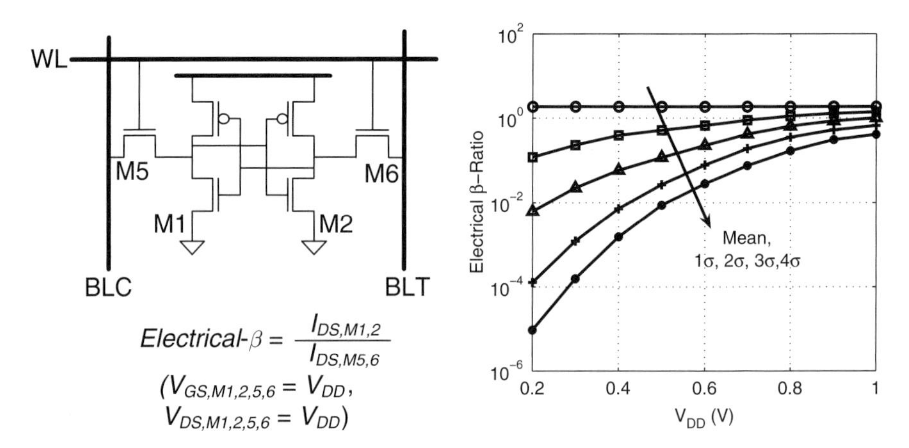

Fig. 4.8 Definition of electrical-β in 6T bit-cell and its degradation, due to variation, with respect to supply voltage (© [2008] IEEE)

unfavorably distorts the read butterfly curves. This can be seen by considering the additional current path shown in Fig. 4.6, which at low voltages can be comparable to the PMOS load-device currents. As a result, it degrades the strength of the driver device which is required to hold *NT* or *NC* low during a read access [24]. In this manner, soft-oxide breakdown limits the minimum voltage for read-stability similar to RDF.

Study of the design trade-offs between electrical-β ratio and the cell read-current suggest that the symmetric-6T topology imposes inherent restrictions to ultra low voltage operation. Specifically, the electrical-β ratio can be increased by reducing the strength of the access devices; however, this degrades the cell read-current, which fatally limits our ability to sense read-data (as discussed further in Section 4.3.2.3). Alternatively, the electrical-β ratio can be increased by upsizing the driver devices. However, the upsizing required to overcome the degradation shown in Fig. 4.8 is far too drastic to achieve the stability needs, particularly since the negative effect on density would be too costly. Additionally, a large increase in gate area for the driver devices can exacerbate the limiting effect of gate-oxide soft breakdown [24], opposing the read SNM improvement.

4.3.2.2 Cell Read-Current Degradation

The cell read-current, I_{READ}, is the current sunk from the pre-charged bit-lines during a read access when the access devices are enabled. At ultra low voltages, we expect a significantly reduced read-current because of the lower gate-drive voltage. However, the increased effect of threshold-voltage variation severely degrades the weak cell read-current even further. Figure 4.9 normalizes the read-current distribution by the mean read-current to highlight just the further degradation due to

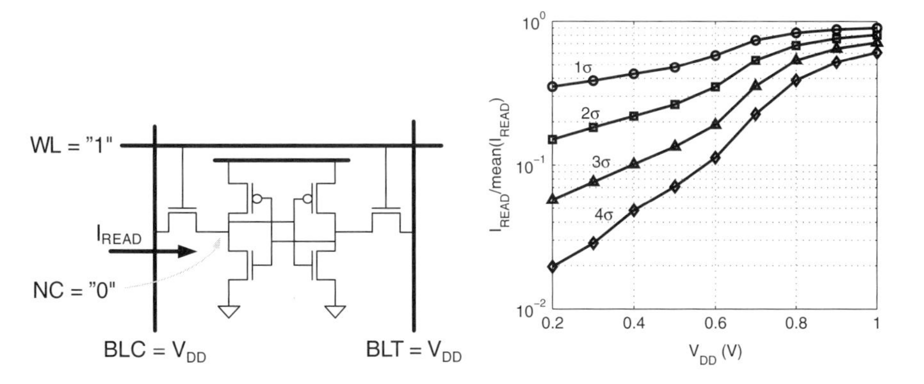

Fig. 4.9 Read-current distribution normalized to mean read-current to illustrate the severe effect of variation with respect to supply voltage (© [2008] IEEE)

variation [31]. At ultra low voltages, where the mean read-current is already greatly reduced, the effect is particularly pronounced, and the weak cell read-current can easily be further degraded by a couple of orders of magnitude. This combination of drastic variation on top of already reduced mean read-current implies that the read access time can extend almost arbitrarily. Of course, this is undesirable from a performance point of view, but, even more importantly it affects the ability to correctly sense data, as described in the following section.

4.3.2.3 Bit-Line Leakage

An implied consequence of the reduced read-current is that the aggregate leakage currents from the unaccessed cells on the same bit-lines can make conventional data sensing impossible. Typically, during read-data sensing, we detect a droop on one of the bit-lines, *BLT* or *BLC*, differentially with respect to the other bit-line. During this time, we expect the other bit-line to dynamically remain high. However, the aggregate leakage currents on this other bit-line depend on the data stored in the unaccessed cells, and, because of the reduced I_{ON}-to-I_{OFF} ratio and severe degradation from read-current variation, these can exceed the actual read-current of the accessed cell. Figure 4.10a shows the worst-case bit-line leakage scenario where the data in all of the unaccessed cells is such that the leaking devices on the bit-line, which should nominally remain high, face a large V_{DS} voltage drop. As a result, the aggregate leakage current, $I_{\mathrm{LEAKAGE,tot}}$, is maximized. Figure 4.10b plots the statistical read-current normalized to this worst-case bit-line leakage current assuming 256 cells per column. As shown, at ultra low voltages the bit-line leakage current exceeds the weak cell read-current, making the droop on the two bit-lines indistinguishable.

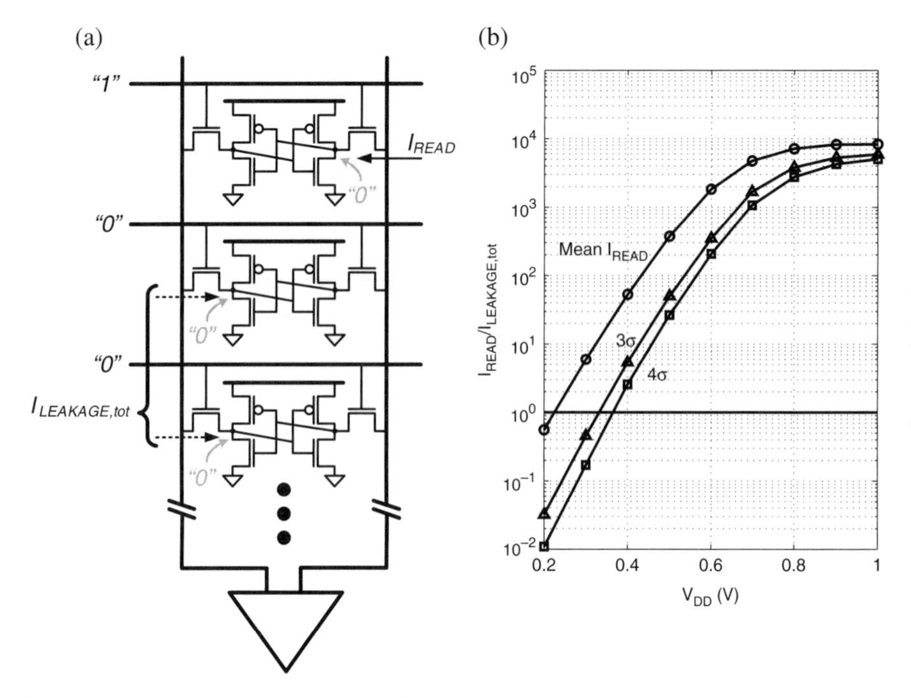

Fig. 4.10 Bit-line leakage from unaccessed cells sharing *BLT/BLC*, leading to (**a**) parasitic droop and (**b**) exceeding the actual read-current (© [2008] IEEE)

4.4 Ultra Low Voltage Bit-Cell Design

The challenges faced by ultra low voltage SRAMs point to the need for new bit-cell and array topologies to manage the reduced signal levels and increased sensitivity to variation. Specifically, failures arise because the conventional SRAM topology is optimized for density at the cost of V_{MIN}, and it is not longer viable for severely energy-constrained applications. As a result, the ratioed bit-cells and long bit-lines need to be replaced with alternatives. Of course, even in energy-constrained applications, density remains a vital design constraint. Accordingly, the objective here is to achieve robust ultra low voltage operation with a minimum density overhead. It should be noted before proceeding, however, that with the appropriate design discipline [33], static non-ratioed logic structures can achieve nearly fundamental levels of low-voltage operation [27], as also indicated in Fig. 4.1. Although, their associated data retention circuits, like flip-flops and latches, typically employ many more devices than SRAM storage structures, as a boundary case, they ensure the possibility of ultra low voltage memory. In fact, ultra low voltage designs have used these as the basis of their memory cache. Both [37, 10], for instance, employ a register-file memory that is fully static and has storage cells with a structure approaching data retention standard logic. In both cases, regularity is exploited to improve the density relative to general logic by using specialized layout techniques and access circuits.

Nonetheless, the overall density can be improved substantially by utilizing high-density storage structures that leverage peripheral circuit-assists as much as possible to achieve ultra low voltage operation. The following sections start by describing buffered-read bit-cells, which are able to achieve operation below 0.35 V, with a high level of integration. Following this, alternate low-voltage bit-cells are discussed for improving the density metric by trading off voltage scalability.

4.4.1 Buffered-Read Bit-Cells

In current 90, 65, and 45 nm technologies, the 6T bit-cell provides a good balance between density, performance, and stability if the supply voltage is not reduced below 0.8 V. However, as discussed in Section 4.3.2.1, read accesses cause problematic transients on the internal storage nodes of the bit-cell and can upset the stored data. Specifically, this effect depends on the relative strength of the access devices (*M5/6* in Fig. 4.6), which attempt to pull the internal storage nodes up to the bit-line pre-charge voltage, typically at V_{DD}, and the strength of driver devices (*M1/2* in Fig. 4.6), which attempt to hold the storage nodes down depending on the stored data. Unfortunately, however, as the supply voltage is reduced, sensitivity to variation increases rapidly, making it impossible to guarantee the relative strengths with the given area constraints.

A very effective solution is to separate the storage-element, where data is written and held, from the read-element, where data is only read. This can be done with the addition of at least two devices, leading to the 8T topology shown in Fig. 4.11. The evolution from a 6T cell to the 8T cell is shown, and the additional devices, *M7/8*, form a read buffer which isolates the storage nodes from the read bit-line,

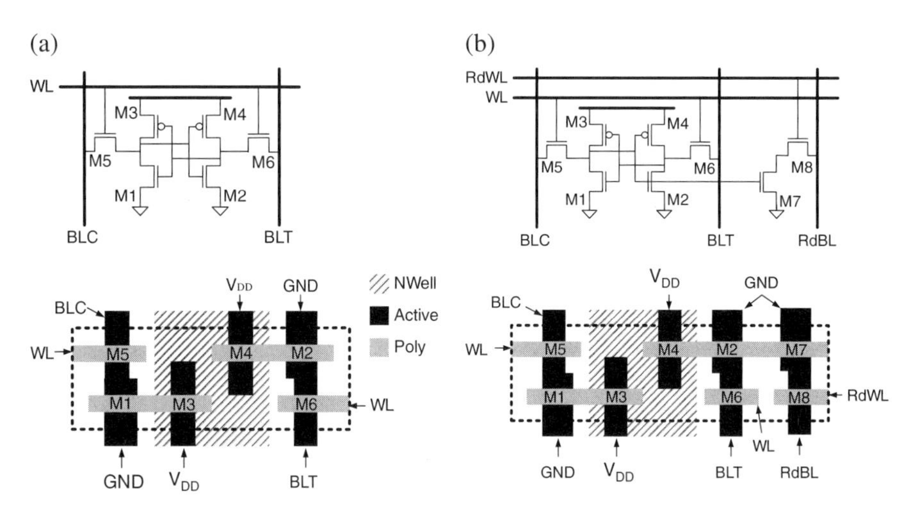

Fig. 4.11 Efficient bit-cell layouts for (**a**) 6T cell and (**b**) 8T cell allowing sharing of most abutting source–drain junctions and poly between adjacent cells

RdBL. A comparison of the layout for the two cells is also shown; both adhere to the "thin cell" topology [13], alleviating lithography stresses and device mismatch sources by minimizing jogs in the poly. The 6T layout, in Fig. 4.11a, is limited by the pitch of four devices, while the 8T layout [8], in Fig. 4.11b, is limited by the pitch of five devices. As shown, this layout is highly efficient in the sense that it amenably allows sharing of all source–drain junctions and poly wires with abutting cells. Nonetheless, even though the number of adjacent devices has only increased by 25% relative to the 6T layout, practically the *baseline* 8T layout area increases by approximately 30% [8].

8T area versus 6T area. Of course, this baseline picture changes dramatically as we start to optimize for low-voltage operation in the presence of variation. Section 4.3.2.1 describes the read SNM limitation faced by 6T cells, which, for the 65 nm technology bit-cell considered, limits the minimum voltage to above 0.7 V. However, with regards to data retention stability, an 8T cell faces only the hold SNM constraint, which is imposed by the storage element. As shown in Fig. 4.7a, this is preserved to very low voltages, below 0.3 V. The required analysis, then, must compare the 6T cell size necessary to maintain the stringent *read* SNM to the 8T cell size necessary to maintain the more relaxed *hold* SNM, both with respect to supply voltage. The analysis is complicated by the importance of simultaneously comparing the read-current for the two cases, which, in the 6T cell, is affected when selecting the electrical-β ratio. Nonetheless, a very simple analysis shows that, for the process considered, even at 0.8 V, the 8T cell size continues to scale amenably well below 0.4 μm^2, whereas the 6T cell scaling saturates much before 0.5 μm^2 [21]. Appropriate process optimization will enable much better scaling for the 6T cell; however, its increased sensitivity to variation cannot be avoided, and the situation worsens severely as the supply voltage is reduced further.

Read-buffer sizing. Although the 8T cell overcomes the interdependence between stability and read-current, the dependence between density and read-current remains; however, the associated trade-offs change drastically at ultra low voltages. Specifically, at nominal voltages, increasing the read-current through device upsizing is unattractive due to the resulting increase in cell size. Importantly though, the standard deviation of threshold voltage is inversely related to the square root of device areas [23]. Accordingly, this approach greatly reduces the variation-induced degradation at ultra low voltages, where the dependence on threshold voltage approaches exponential levels. Consequently, upsizing, at least slightly, has enhanced appeal. Figure 4.12a shows the gain in read-current if the read-buffer device widths are increased by 25% and 50%, respectively. Although we expect the mean read-currents to increase by factors of only 1.25 and 1.5, the weak cell read-currents, which limit the functionality and performance of the entire array, increase by factors of 1.8 and 2.8, respectively, at 0.3 V. Similarly, as shown in Fig. 4.12b, increasing the device lengths by 40% and 80% increases the weak cell read-currents by factors of 3.3 and 5, respectively. It is important to note that, at ultra low voltages, increasing device lengths can have a significant impact on increasing even the mean read-current since the reverse short-channel effect [19] essentially causes a decrease in the effective threshold voltage.

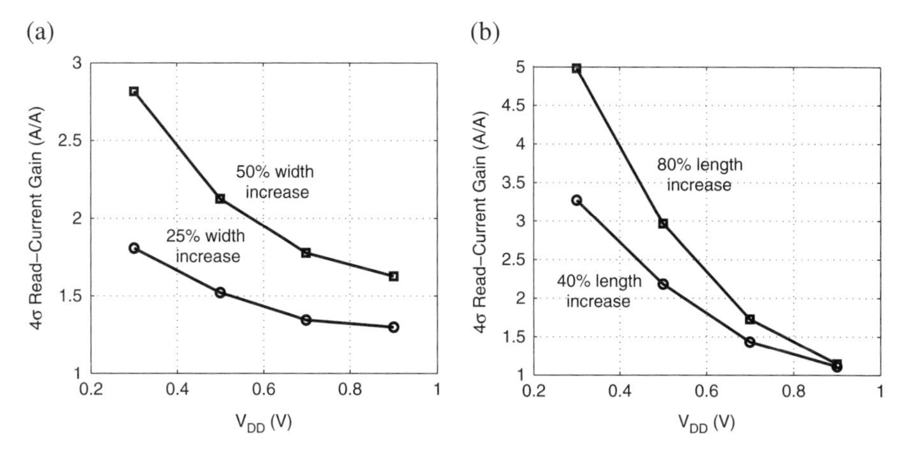

Fig. 4.12 Read-current gain with respect to supply voltage as a result of (**a**) increasing device widths and (**b**) increasing device lengths (© [2008] IEEE)

8T leakage power. Leakage power is a critical SRAM metric for highly energy constrained applications. Although the 8T cell introduces an additional leakage path through the read-buffer devices, this does not necessarily increase leak-power during the active operation. Specifically, with regards to the read access considered in Fig. 4.13, during the pre-charge portion, both branches of the cross-coupled inverters and one of the access devices pose leakage paths in the 6T cell. However, in the 8T cell, the write bit-lines, *BLT/BLC*, can be allowed to float, potentially reducing the leakage current through the access device by over an order of magnitude. Additionally, however, leakage current flows through the devices of the read-buffer with half probability (due to the dependence on the stored data). The other half of the

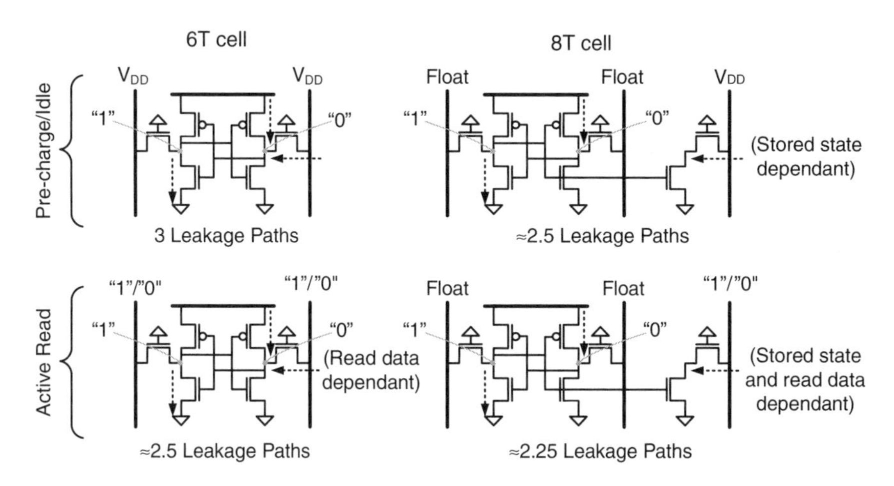

Fig. 4.13 Leakage paths of 6T and 8T bit-cells during read-access phases

time, the leakage is reduced somewhat owing to the stacked effect of the two "off" NMOS devices in series [38]. During the bit-line discharge portion of the read cycle, the access-device leakage in the 6T cell can actually contribute to the read-current depending on the accessed data. In that case, it can be disregarded. Alternatively, the leakage currents in the storage-element of the 8T cell remain unchanged, but similarly, the possibility that the read-buffer contributes a leakage current is further reduced depending on both the stored and accessed data.

The remainder of this section considers various forms of buffered-read bit-cells that can achieve ultra low voltage operation. The precise attributes of each cell topology depends heavily on its associated array structure. Accordingly, the array configuration and biasing requirements are also discussed.

4.4.1.1 10T Bit-Cells

Although the baseline 8T bit-cell eliminates the read SNM limitation, which is critical to ultra low voltage operation, it does not address the problem of bit-line leakage, which degrades our ability to sense read-data correctly. Specifically, the problem with the single-ended 8T cell is analogous to that with the differential 6T cell discussed in Section 4.3.2.3. The only difference here is that the leakage currents from the unaccessed cells sharing the read bit-line, *RdBL*, affect the same node as the read-current from the accessed cell. Consequently, the aggregated leakage current, which depends on the data stored in all of the unaccessed cells, can pull-down *RdBL* even if the accessed-cell, based on its stored value, does not mean to. Once again, if the aggregated leakage current exceeds the actual read-current, correct data sensing becomes impossible.

A solution to manage the bit-line leakage from unaccessed cells is shown in the 10T bit-cell of Fig. 4.14 [4]. Here, *M7-10* form a read-buffer that aims to cut-off the leakage on *RdBL* when *RdWL* is low, independent of the stored data at node *NC*. In particular, if *NC*=0, both *M7* and *M9* are "off," but *M10* is "on," and, therefore, node *NCB* is pulled up to V_{DD}. Accordingly the read-access device, *M8*, does not impose subthreshold leakage pulling *RdBL* low. Alternatively, if *NC*=1, *M7* is "on," but *M8-10* are all "off." As a result, two leakage reduction effects are simultaneously achieved. First, the leakage-path through *M8* and *M9* is weakened due to the stacked effect of two series "off" devices [38]. Second, the leakage current through *M10* actually contributes to pulling NCB up, and, importantly, in the technology used, PMOS devices have much higher leakage than NMOS devices, so this effect is quite strong. As a result, the V_{GS} of *M8* is further reduced below zero, further reducing its leakage.

The reduced leakage from unaccessed read-buffers, the bit-cell of Fig. 4.14 has allowed the integration of 256 cell per column in a 256 kb SRAM implemented in LP 65 nm CMOS [4]. It is fully functional down to 400 mV with an operating frequency of 475 kHz, and the leakage power is reduced by a factor of over 30 compared to a 1.2 V supply.

Although bit-line leakage is greatly reduced by the cell of Fig. 4.14, the precise amount of the reduction can depend on the process skew. Specifically, as the PMOS

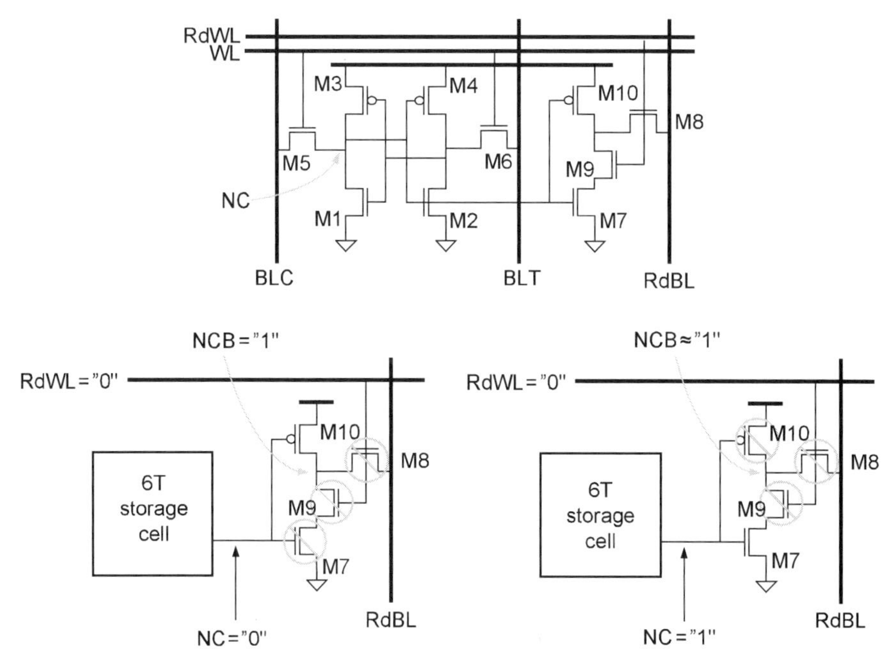

Fig. 4.14 10T bit-cell (in 65 nm CMOS) to manage bit-line leakage and allow 256 cell per column [4]

strength is reduced relative to the NMOS strength, the leakage current ratio for *M10* to *M9*, which is critical to determining the voltage of *NCB* when *NC*=1, is reduced. In this case, the bit-leakage has an increasing dependence on the data stored in the unaccessed cells. An alternate 10T bit-cell, shown in Fig. 4.15, eliminates the data dependence [14]. Here, when the cell is accessed, *RdWL* is asserted, and *RdBL* is conditionally pulled low through *M7-9* depending on the value of *NC*. For unaccessed cells, however, *RdWL* is low, and *M9* gates the read-buffer pull-down path, while *M10* actively pulls *NCB* high, unconditionally. Interestingly, bit-line leakage now flows in the opposite direction, from *NCB* to *RdBL*, if *RdBL* is pulled low by the accessed cell. This can be highly problematic, particularly in higher-density bit-cells where the read-current can be degraded by variation, since the accessed read-buffer must now overcome the aggregated bit-line leakage in order to ensure any sensing margin. Nonetheless, the bit-line leakage is data independent, implying that replica techniques (discussed further in Section 4.5.3.1) can be used to reduce the amount of margin required. The design in Ref. [14] is implemented in 130 nm CMOS, where variation is considerably less severe. As a result, a 480 kb prototype integrates 1024 cells per bit-line while achieving full operation down to 0.2 V.

The buffered bit-cells discussed thus far have the limitation that all cells sharing *WL* must be written to simultaneously. Specifically, during a write operation, *WL* is asserted to access a desired cell of the associated row. Data is then written to that cell by driving the appropriate signals on *BLT/BLC*. Alternatively, if unaccessed 6T

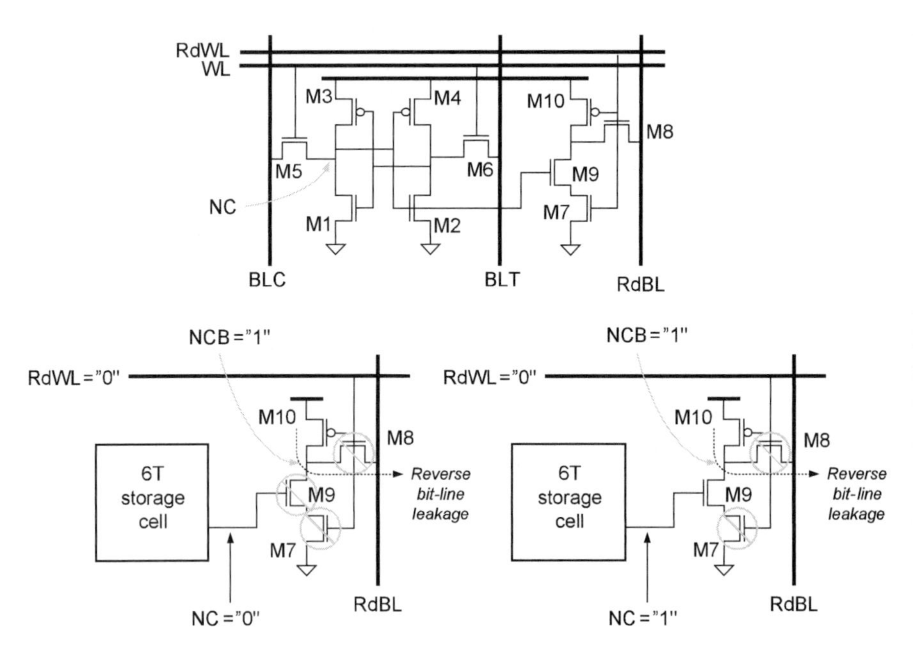

Fig. 4.15 10T bit-cell (in 130 nm CMOS) resulting in reverse bit-line leakage, but its data independence enables replica techniques to reduce the required sensing margin, thereby allowing the integration of 1024 cells per column [14]

storage elements in the same row, to which we do not want to write data, have sufficient read SNM, their data can be retained by maintaining V_{DD} on *BLT/BLC*. However, ultra low voltage designs aim to scale the supply voltage below the read SNM limit. Consequently, adjacent cells sharing *WL* must be written at once, implying that bit-cells of a logic word must occur contiguously and cannot be interleaved. As discussed in Section 4.5.1.1, this requires special considerations and array layouts to deal with the possibility of soft errors from ionizing particles in the atmosphere. When these interact with the SRAM array, they exhibit spatial locality upsetting the data in several bit-cells, and, therefore, when bit-cells of a logical word occur adjacently, they are susceptible to multi-bit errors, which require advanced correction techniques.

The buffered 10T bit-cell in Fig. 4.16, however, aims to achieve bit-cell interleaving while also providing a differential read port [7]. Here, during read accesses, only the row-wise *WL* signal is asserted, but the *VGND* signal, which is shared among multiple columns, is also pulled low. As a result, *RdBLT* (*BLC*) or *RdBLC* (*BLT*) is pulled low by the buffer devices, *M7/8*, depending on the stored data. Since *VGND* is shared by the column, however, this bit-cell still faces the bit-line leakage problem from the unaccessed cells sharing *RdBLT/RdBLC*. During write accesses, *VGND* remains high (as shown in Fig. 4.16), and both the row-wise *WL* signal *and* the column-wise *WrWL* signals are asserted. This provides access to the *NT/NC* storage nodes for only the cells in the selected row *and* columns, leaving the unaccessed

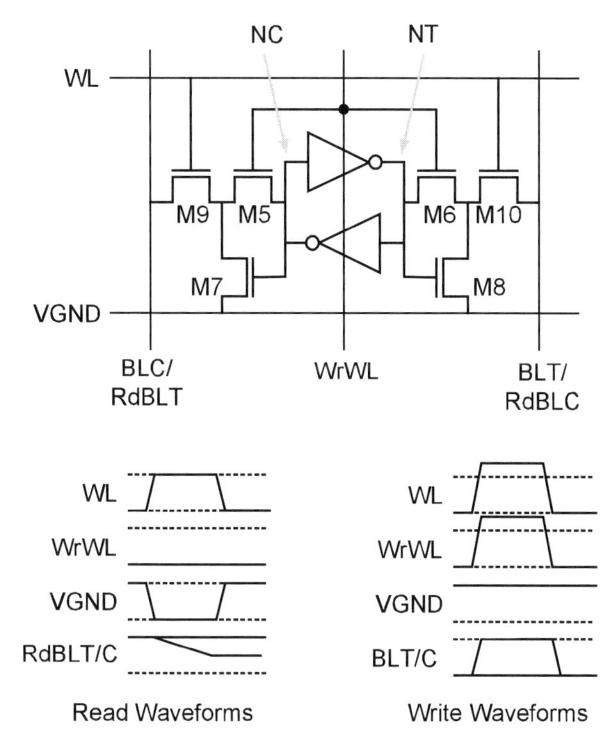

Fig. 4.16 Differential buffered-read 10T cell to allow column interleaving for soft-error immunity [7]

cells in the associated row undisturbed. Importantly, however, since data must be written through two series devices, *M9/5* or *M10/6*, the write margin is significantly degraded. In this design [7], both *WL* and *WrWL*, which have significant capacitance, must be boosted beyond the supply voltage to ensure write-ability. The 32-kb prototype in 90 nm CMOS operates down to 190 mV.

Another buffered bit-cell design that achieves a differential read port, using only nine transistors, is shown in Fig. 4.17 [18]. Here, the single-access device, *M9*, also serves as the driver. When enabled, it conditionally pulls down either *BLT* or *BLC*, depending on the state of the gating buffer devices *M7/8*. Like the 10T cell of Fig. 4.16, this cell does not address the bit-line leakage problem. Additionally, although it employs one transistor less, its use of an odd number of NMOS devices precludes an efficient layout structure similar to that in Fig. 4.11b, making it challenging to achieve a density improvement over the 10T cell.

4.4.1.2 8T Bit-Cells

Previously discussed 10T designs solve the issue of bit-line leakage by adding devices within the bit-cell to eliminate data-dependant droop induced by the unaccessed cells. Their associated area overhead, due to extra devices and layout com-

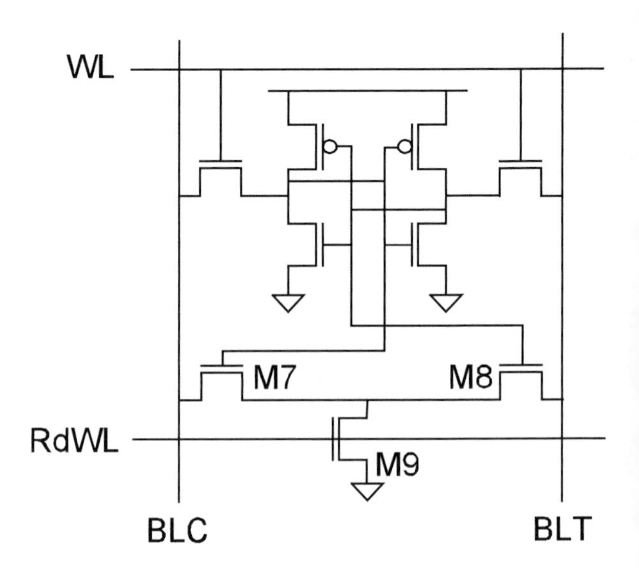

Fig. 4.17 9T bit-cell with buffered differential read port [18]

plexity, however, can be quite severe. Although, this is amortized considerably by the improvement in array area efficiency that comes with being able to integrate many more bit-cells on the shared bit-lines, since the bit-line leakage problem is not fundamental to an isolated bit-cell, other solutions exist to address bit-line leakage while employing the high-density 8T cell topology.

One solution is to greatly reduce the number of cells sharing the read bit-line. The design in Ref. [9], for instance, integrates just eight cells per read bit-line. As a result, the aggregated leakage current, from the seven unaccessed cells, is small enough that, regardless of their stored data, it never exceeds the accessed cell's read-current, ensuring sufficient sensing margin. Of course, as shown in Fig. 4.18a, this implies that local read-data evaluation circuitry is required in the column every eight cells, introducing additional area overhead; importantly, however, the use of short read bit-lines also greatly reduces the capacitance that must be discharged during the read access. As a result, the bit-cell can discharge the local read bit-line completely, precluding the need for complex small-signal sensing circuitry. Hence, the area overhead for local read-data evaluation can be minimized. Additionally, Fig. 4.18a shows that the write-port associated with the six-transistor storage element can be optimized independently of the read port. As a result, *WrBL* can be shared by many more cells in the column (512 in this case) to avoid further degradation in array area efficiency.

Figure 4.18b shows the full structure of the array. After local read-data evaluation, longer global read lines, driven by larger devices which are less sensitive to variation, can be sensed using more complex circuitry whose area overhead is amortized over the 256 rows. With this approach, the design in Ref. [9], implemented in a general-purpose (GP) 65 nm CMOS process, achieves voltage scaling from 1.2 V down to 0.41 V, covering a frequency range from 5.3 GHz to 295 MHz.

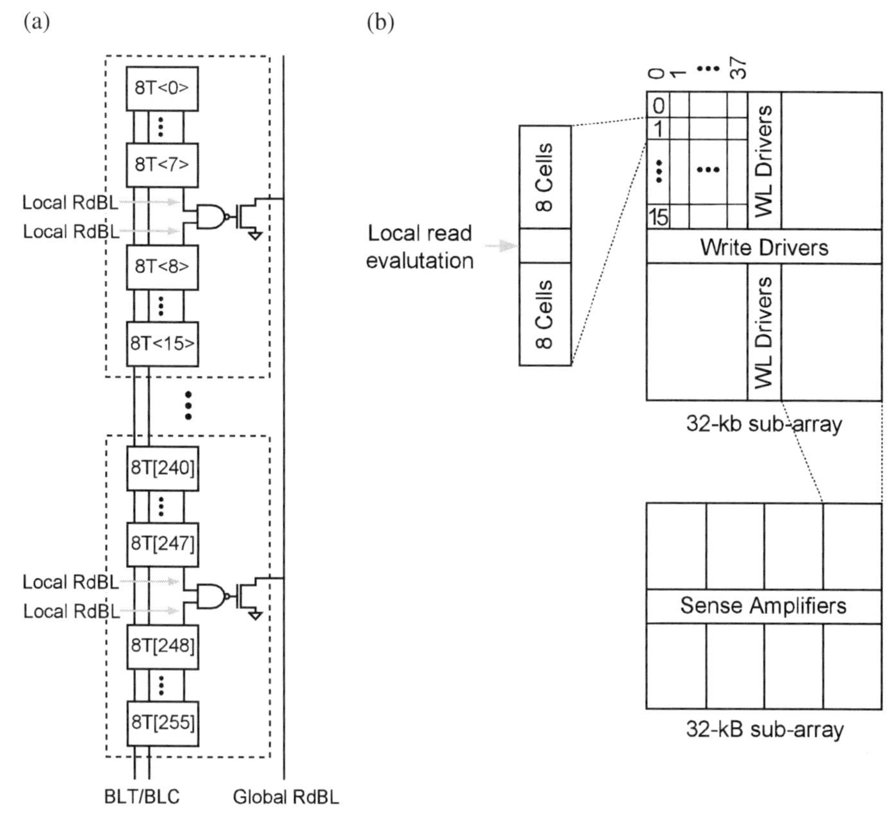

Fig. 4.18 8T SRAM employing hierarchy to manage bit-line leakage by using (**a**) short local read bit-lines but (**b**) much longer write bit-lines [9]

Increasing the number of cells per read bit-line, however, can lead to a much better performance–density trade-off at the array level. Specifically, the sensing circuitry overhead can be amortized over more rows, permitting the use of more sophisticated sense-amplifiers to support reduced read bit-line signal swings. The design in Ref. [30] employs an 8T bit-cell but mitigates the bit-line leakage from the unaccessed cells using a peripheral assist in order to maintain a high array density.

The standard 8T bit-cell is modified as shown in Fig. 4.19, where the feet of all read-buffers in a particular row are connected to the row-wise *BffrFt* control signal. A peripheral driver pulls this node low only in the case of the accessed row, and pulls it high for all other rows.

The resulting benefit with regards to the bit-line leakage can be seen in Fig. 4.19. With the feet of the unaccessed read-buffers statically connected to ground, the transient simulation shows that the logic "1" level for the read bit-line can be much lower than the logic "0" level for high levels of bit-line integration. In particular, the number of bit-cells sharing the read bit-line must be drastically reduced to ensure

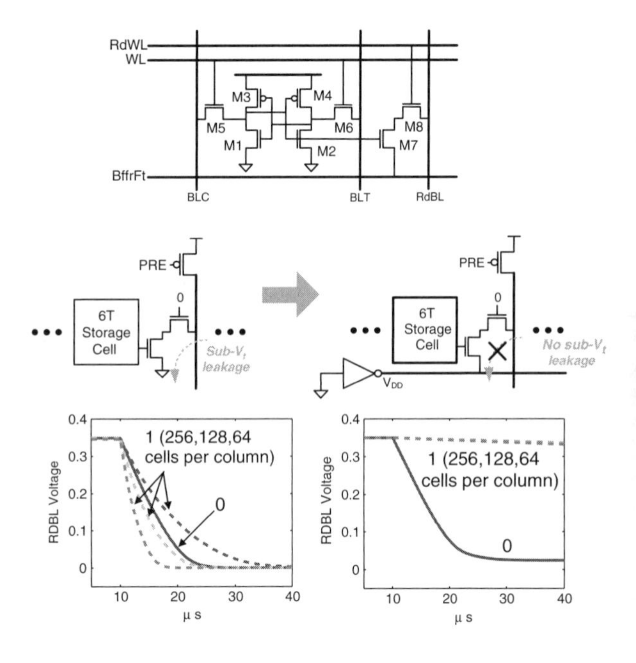

Fig. 4.19 8T bit-cell with peripheral control of *BffrFt* to eliminate subthreshold bit-line leakage [30] (© [2008] IEEE)

any sensing margin between the two logic states. However, the use of a peripheral driver to selectively control the *BffrFt* node ensures that the feet of all unaccessed read-buffers are pulled up to V_{DD}. Consequently, after the *RdBL* is pre-charged, the unaccessed read-buffer devices have no voltage drop across them, and they impose no subthreshold leakage current pulling it low. As a result, the *RdBL* remains high in the logic "1" state, ensuring ample sensing margin even when over 256 cells are integrated. Some very small residual droop does remain; however, this comes about primarily due to gate leakage from the read-buffer access devices, and junction leakage from their drains.

Although bit-line leakage in the unaccessed rows is effectively managed, an important concern with this approach does arise in the case of the accessed row. As shown in Fig. 4.20a, here the peripheral *Bffrft* driver must sink the read-current from potentially all accessed cells. In the case of the design in Ref. [30], for instance, 128 cells are integrated in each row, making its current requirement impractically large. Unfortunately, this driver also faces a two-sided constraint, and, as a result cannot simply be upsized to meet that drive requirement. In particular, as shown in Fig. 4.20a, upsizing would impose too much leakage current in the case of all the unaccessed rows, and, further, the associated area overhead would negate the density advantage of using a peripheral assist. The solution in Fig. 4.20b leads to much more favorable trade-offs. Here, a simple charge-pump circuit is used to boost the gate voltage of the *BffrFt* driver to nearly twice V_{DD}. Accordingly, since the NMOS current is related exponentially to its gate voltage in the subthreshold operating regime, its current increases by a factor of over 500, as shown. Consequently, no

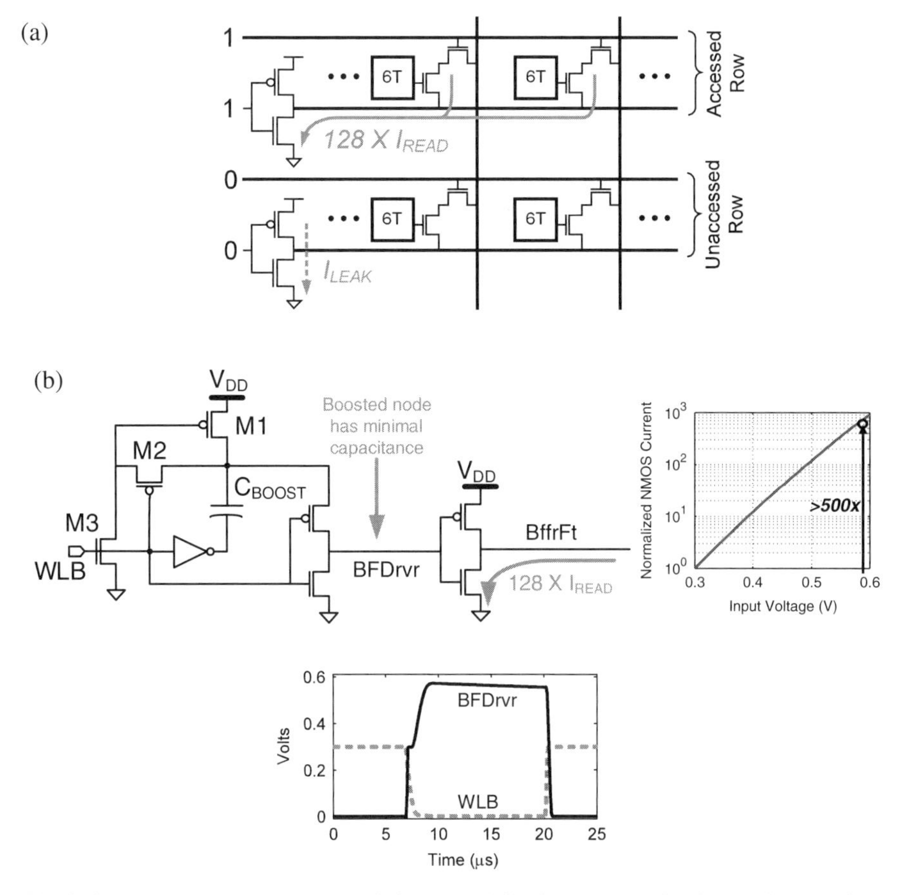

Fig. 4.20 *BffrFt* driver (**a**) read-current limitation and (**b**) charge-pump circuit to achieve required drive strength with minimal area and leakage-power overhead [30] (© [2008] IEEE)

device upsizing is required, minimizing both the area and leakage power overhead of the driver. Further, the use of nearly minimum-sized driver devices implies that the charge-pump must drive minimal load capacitance, allowing its boost capacitor to be small; the entire charge-pump layout is slightly larger than just two bit-cells. Lastly, it is effective at providing a nearly $2V_{DD}$ boost, since it employs the PMOS device *M1* to pre-charge the boost capacitor. As a result, the design in Ref. [30] integrates 256 bit-cells per column and 128 bit-cells per row.

4.4.2 Non-buffered-Read Bit-Cells

Buffered-read bit-cells, described in the previous section, overcome the read SNM limitation to achieve much lower operating voltages than the symmetric 6T bit-cell.

However, non-buffered-read bit-cell topologies also exist to overcome the read SNM limitation or mitigate it considerably.

4.4.2.1 7T Bit-Cell

During a read access with a standard 6T bit-cell, the voltage of either *NT* or *NC* tends to get raised by transients that are caused by the pull-up effect of the access devices *M5/6* that fight with the driver devices *M1/2* (Fig. 4.6). Such transients lead to data disturbances because high-gain cross-coupled paths exist from *NT* to *NC*, through *M1/3/5*, and from *NC* to *NT*, through *M2/4/6*. Consequently, such transient increases on either *NT* or *NC* can get amplified through the cross-coupled path resulting in erroneous regeneration to the opposite state.

The bit-cell in Fig. 4.21a [29], however, breaks the problematic cross-coupled path using the additional device *M7*. Here, like many of the buffered-read bit-cells,

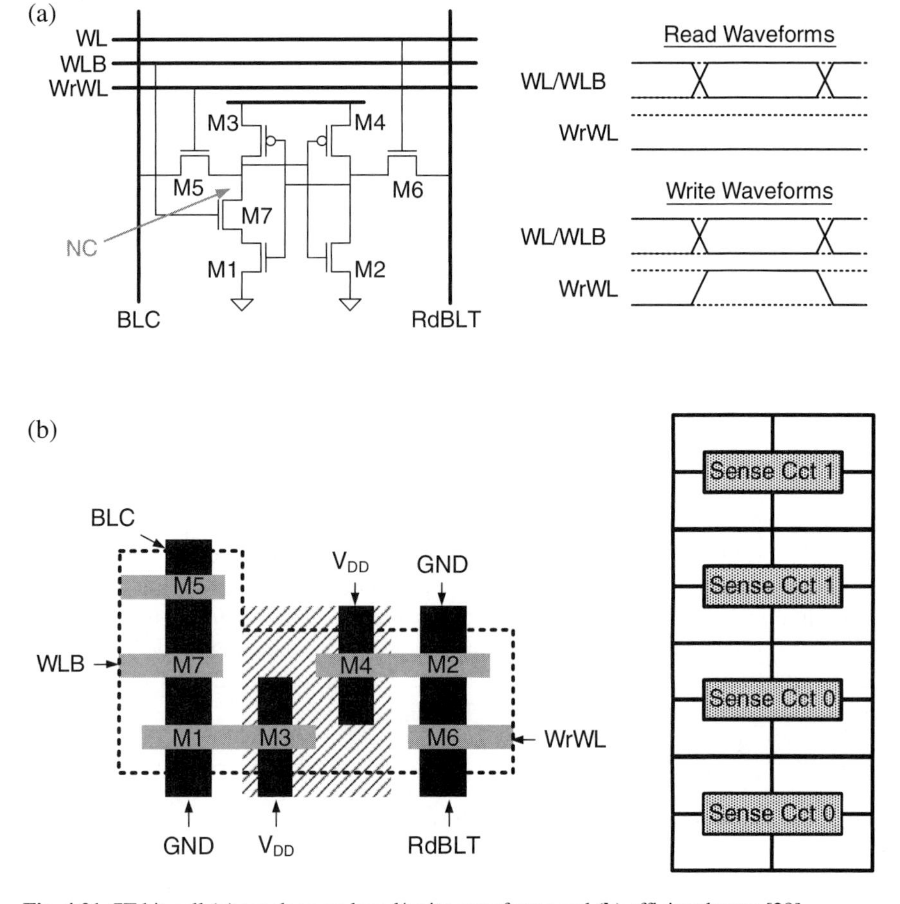

Fig. 4.21 7T bit-cell (**a**) topology and read/write waveforms and (**b**) efficient layout [29]

data can only be read from the single-ended port, *RdBLT*; however, writes are performed as in the 6T case, though both *RdBLT* and *BLC*. Importantly, this implies the need for three row-wise control signals. As shown in the waveforms, during a read, *WL* is asserted (*WLB* is de-asserted) and *WrWL* is de-asserted, and, during a write, *WL* is asserted (*WLB* is de-asserted) and *WrWL* is also asserted. Importantly, however, this means that, during read assesses, voltage raising transients on *NT* cannot activate the cross-coupled path and flip the data on *NC*, since the pull-down path through *M5* is broken. As a result, *NC* dynamically remains high, and, after the read operation is complete, the cross-coupled regenerating paths are restored and the data is retained.

Of course, this bit-cell does not explicitly address the bit-line leakage problem. As a result, the number of cells sharing the local *RdBLT/BLC* must be minimized as in the case of the 8T design of Ref. [9]. In particular, eight bit-cells are used per local *RdBLT/BLC*. However, an efficient layout for the array results from the bit-cell layout is shown in Fig. 4.21b. Here, the additional device leads to an irregular "L" shape, yielding gaps throughout the array layout. Nonetheless, the full-swing sensing circuitry required for the two columns shown can be distributed in these gaps, and then this entire structure, comprising eight rows and two columns, can be further arrayed in a regular manner.

It is important to note that during the read access, the voltage at *NC* may only be held dynamically. As a result, during very long access times, which can occur at ultra low voltages, leakage paths can compromise this voltage level, leading to a data disruption hazard. Accordingly, with this bit-cell, the access-delay cannot exceed the associated leakage time, posing a primary limitation to the achievable voltage scaling. Nonetheless, the design in Ref. [29] achieves operation to below 0.5 V in a 90 nm process, where the access delay is 20 ns.

4.4.2.2 Asymmetric 6T Bit-Cell

It was mentioned in Section 4.3.2.1 that the static-noise-margin analysis conservatively assumes that *BLT/BLC* remain statically at the pre-charge voltage of V_{DD}, emulating the worst-case conditions for data disturbing transients on the storage nodes *NT/NC*. Although this sort of analysis is typically practical for ultra low voltage designs, where the very low read-currents imply that *BLT/BLC* can in fact remain near V_{DD} for very long periods, some improvement in voltage scaling can be afforded by specifically designing the bit-cell to rapidly discharge the bit-line so that this worst-case static condition is alleviated.

The bit-cell in Fig. 4.22 [12], for instance, uses a 6T topology but an asymmetric layout. As a result, the cell read-current on one side is greatly increased, thanks to the larger mean strength, and the much lower effect of variation (as described in Section 4.4.1). The 3.7*x* increase in only one of the NMOS pull-down paths leads to 22% area savings compared to symmetric sizing, while increasing the associated read-current by over a factor of 9. As a result, read accesses are performed through this single-ended port to ensure rapid full-swing bit-line discharge and, therefore, alleviate the hazardous static condition. Additionally, as in Refs. [9, 29], the number

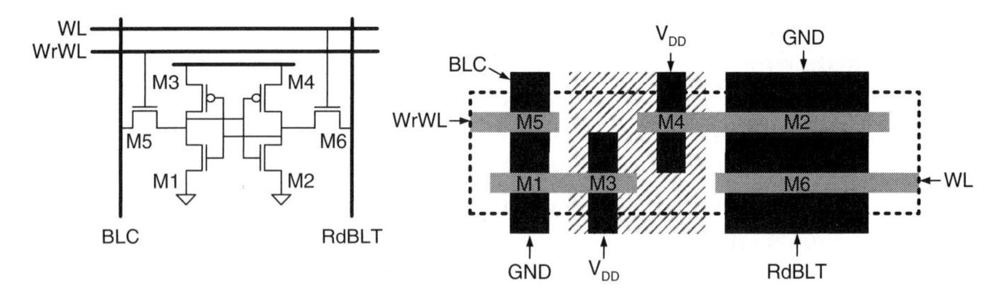

Fig. 4.22 Asymmetric 6T bit-cell and layout to alleviate static read upset condition [12]

of cells sharing the bit-lines is kept to a minimum, in this case, 16; the primary purpose here, however, is to reduce the bit-line capacitance and further support rapid discharge to enhance read stability.

Using this asymmetric bit-cell, the array in Ref. [12] achieves operation down to 0.7 V at 1 GHz in a 45 nm technology. Although, the high-performance operation and the low-voltage read-stability enhancement strategy are consistent, both require the use of low-threshold devices for the NMOS pull-down paths, which increases the leakage power of the array considerably.

4.4.2.3 10T Schmitt Trigger Bit-Cell

The Schmitt trigger bit-cell of Fig. 4.23 [16] increases robustness to pull-up transients on *NT/NC*, by adjusting the switching threshold of the cross-coupled paths using hysteresis. Specifically, devices *M7-9*, adjust the switching threshold of *M1* by selectively raising its source voltage based on the stored data; *M8-10* perform the same function with regards to *M2*. As a result, a larger voltage increase on *NT* or *NC* is required in order to switch the alternate node's logic state.

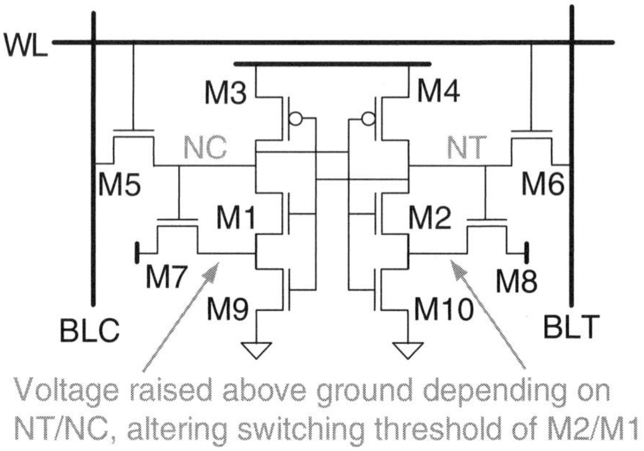

Fig. 4.23 10T Schmitt trigger based bit-cell that adjust the switching thresholds of the cross-coupled paths depending on the initial stored data [16]

Since three series devices are required for this bit-cell, its layout height, assuming a "thin cell" structure [13], is increased by approximately 50%, and the additional NMOS devices, *M7,8*, require a width increase of approximately 40%. The three-device pull-down path also leads to additional degradation of the cell read-current. Nonetheless, the design in Ref. [16] achieves operation down to 0.175 V in a 0.13 μm technology for a 4 kb array.

4.5 Ultra Low Voltage Periphery Design

The problem of read instability is highly coupled with degraded cell read-current and, correspondingly, introduces an unavoidable trade-off constrained by the bit-cell density (Section 4.3.2.1). As a result, alternate bit-cell topologies, yielding a more favorable trade-off, are necessary at ultra low voltages. However, where possible, it is desirable to implement solutions to the low-voltage failures in the array periphery so that their overhead can be amortized more effectively.

Fortunately several of the challenges arising at low voltages are not inherent to the bit-cell, so this approach can be applied in many cases. This section discusses critical failures related to the periphery, including robust layout against soft errors, write failures, and sensing failures, and presents low-area-overhead solutions to address these, leading to practical ultra low voltage high-density SRAMs.

4.5.1 Column-Interleaved Layout

In a standard SRAM array, bit-cells corresponding to a logical data access are interleaved as shown in Fig. 4.24. For instance, in a 4:1 interleaved layout, each bit of words *A*, *B*, *C*, and *D* are layed-out contiguously in a row. Then, during a read operation, the row-wise word-line corresponding to the accessed word, which is word *C*in the case shown, is asserted. However, since all bit-cells in the row share the word-line, the cells from all four words are enabled simultaneously. Accordingly,

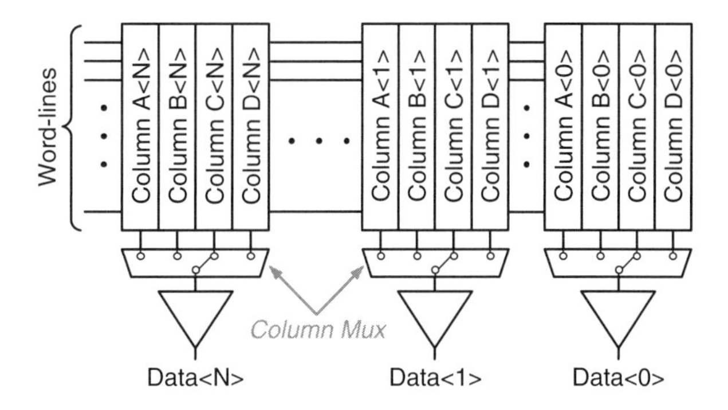

Fig. 4.24 Column-interleaved array layout

the bit-lines for all words in all columns are driven. The bit-lines corresponding to the accessed word, word *C*, are then selected by the column multiplexer to drive the sense-amplifier, so that the accessed data can be resolved. In this case, words *A*, *B*, and *D*, are referred to as *half-selected*, since they are activated by the row-wise word-line, but not the column multiplexer. Similarly, during a write operation, only the bit-lines corresponding to the accessed word are driven (differentially) to the data value that must be written. The bit-lines of the half-selected cells are both pre-charged, so that when these cells are enabled by word-line assertion, they simply drive their bit-lines as they would during a read access.

Half-selection, however, poses a critical limitation to the low-voltage operation of buffered-read bit-cells. The intention behind these bit-cells is to eliminate the read SNM condition. However, since their storage portion must share its word-line with adjacent cells in the row, in an interleaved layout, this implies sharing with different words. Consequently, during a *write operation*, when the word-line is asserted to write to the target cells, the storage portion of all of the remaining cells is half-selected and, therefore, must be read SNM stable. As a result, in this configuration, buffered-read cells overcome the read SNM condition during the read access but are still limited by it during the write access.

It is, however, worth noting that the storage portion of a buffered-read cell can be highly optimized for read SNM without regard for read-current, since the actual read operation involves separate read-buffer devices. Consequently, the constraints facing a conventional 6T cell are alleviated somewhat.

Nonetheless, much more aggressive low-voltage operation can be achieved by eliminating the read SNM condition entirely. Unfortunately, the efficient and highly manufacturable "Thin Cell" layout (Fig. 4.11) does not amenably support multiple row-wise word-lines that would allow only the accessed cells to be selected. Topologies such as the one in Fig. 4.16 provide a viable alternative; but they require additional devices beyond the 8T cell that lead to more complex layouts. As a result, the most straightforward array layout compatible with the 8T bit-cell is shown in Fig. 4.25 and involves eliminating column interleaving so that bit-cells of each word exist contiguously. As a result, each word has its own word-line driver, and any particular word-line is shared only by bit-cells from the same logical word; so the half-select condition is eliminated.

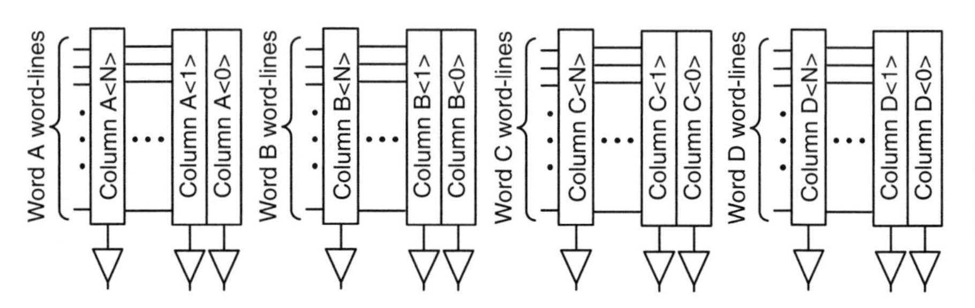

Fig. 4.25 Non-interleaved array layout

Such a layout, however, leads to several practical challenges. The following sections describe these in detail.

4.5.1.1 Soft-Error Correction Coding Complexity

Soft errors, caused by radiation of energetic particles, exhibit spatial locality. As a result, several contiguous bits can be corrupted during a single event. An advantage of the interleaved layout is that the affected bits will likely be associated with different logical words. As a result, under the assumption that only single-bit errors occur in each logical word, the error correction coding (ECC) required to recover from these errors is greatly simplified. However, with a non-interleaved layout, the assumption of single-bit errors is impractical, and more sophisticated ECC schemes are required.

A solution with minimal additional overhead is to maintain a physically adjacent layout for bit-cells of the same word, but, within this, use different error correction codes for interleaved bits. Specifically, a different code is used for the odd/even bits. For the 128-b per word configuration in Ref. [9], the extra ECC penalty to support interleaved coding is approximately 5%.

4.5.1.2 Sense-Amplifier Sharing

A further challenge associated with a non-interleaved layout is that, since adjacent cells are no longer just half-selected, the read-data from adjacent columns must be resolved simultaneously. Consequently, sense-amplifier sharing, via a column multiplexer, is no longer possible. This leads to the scenario in Fig. 4.25, where each column must have its own sense-amplifier, and each of these must fit within a column pitch or be stacked above and below each other in the layout.

This result significantly stresses the sense-amplifier design constraints. Specifically, small-signal sensing requires that the sense-amplifiers have minimal input offset, which implies the use of large devices to overcome variation and leads to larger layouts [23, 39]. Further, because the total number of sense-amplifiers in the entire array increases, the need to reduce this statistical offset is more critical. As a result, sense-amplifiers can pose a critical limitation that must be overcome with specialized techniques to address their area-offset trade-off. These are described further in Section 4.5.3.2.

4.5.2 Write-Assists

In a 6T SRAM cell, the write margin depends on the NMOS access devices (*M5/6*) being stronger than the PMOS load devices (*M3/4*) so that the existing data can be reliably overwritten. A critical advantage of the buffered-read bit-cells, such as the 8T topology, is that they use separate devices for data writing/storage and data reading. As a result, if the half-select condition is eliminated, the 6T storage element in these topologies can be highly optimized for improved write margin without concern for read-stability, by increasing the strength of the access devices.

An efficient strategy for improving the strength of the access devices at ultra low voltages is to utilize long channel lengths, taking advantage of the reverse short-channel effect. Source–drain halo pocket implants, which are employed in modern devices to counteract the effects of drain-induced barrier lowering, also tend to raise the threshold voltage of the device. However, longer device lengths imply that less of the channel is effected by this increased doping, yielding a lower net threshold voltage. For instance, the design in Ref. [15] uses $3X$ minimum length access devices to reduce the threshold voltage by approximately 20%, leading to significantly improved write-ability.

Nonetheless, in the presence of extreme variation, the relative strengths required for write operations are very difficult to guarantee with sizing alone; they can, however, be enforced electrically through explicit biasing. In particular, this can be achieved by selectively increasing the gate-drive (V_{GS}) of the NMOS access devices or reducing the gate-drive of the PMOS load-devices during the write operation. Such an approach involves very little additional area overhead since no topology change in the 6T storage element is required. It is worth noting that, while electrically enforcing relative strengths in this manner can be effective for write-ability, it cannot be applied as readily for read-ability. For instance, reducing the word-line bias, to weaken the NMOS access devices, improves read-stability but adversely affects the cell read-current, which is also critical for proper functionality. Consequently, topology changes (discussed in Section 4.4) are required to ensure read-ability.

With regards to improving write-ability though electrical biasing, increasing the gate-drive of the NMOS devices can be accomplished by either boosting the word-line voltage beyond V_{DD} or boosting the appropriate bit-line voltages below ground. Both of these, however, require boosting a large capacitance, either the word-line or the bit-lines, beyond one of the rails. Reducing the PMOS gate-drive, on the other hand, requires lowering the accessed cell's supply below V_{DD} and, therefore, avoids the overhead of generating an explicit bias voltage. Figure 4.26 shows that as the bit-cell supply voltage is reduced, the strength required of the access devices is eased, which is reflected by the decrease in the minimum word-line voltage that results in a successful write.

Accordingly, to support dynamic load biasing in the bit-cells to be written, the virtual supply node, VV_{DD}, can be introduced, as shown in Fig. 4.27 [30]. In a non-interleaved layout, adjacent cells are written simultaneously, and, as result, they can all share VV_{DD}. During the first half of the write-cycle, VV_{DD} get actively pulled low by the shared peripheral supply driver. However, as shown in the waveforms, it does not go all the way to ground because all of the bit-cells actually contribute to pulling it back up. Specifically, one of the bit-lines of each cell gets pulled low, causing the corresponding storage node, *NC* in the case shown, to also get pulled low. Accordingly, the alternate PMOS load device, *M4*, tends to turn on, introducing a weak current path from *NT* to VV_{DD}. In this manner, each accessed bit-cell contributes to pulling VV_{DD} back up through one of its PMOS load devices and one

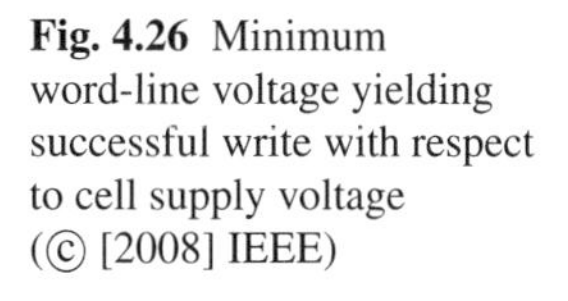
Fig. 4.26 Minimum word-line voltage yielding successful write with respect to cell supply voltage (© [2008] IEEE)

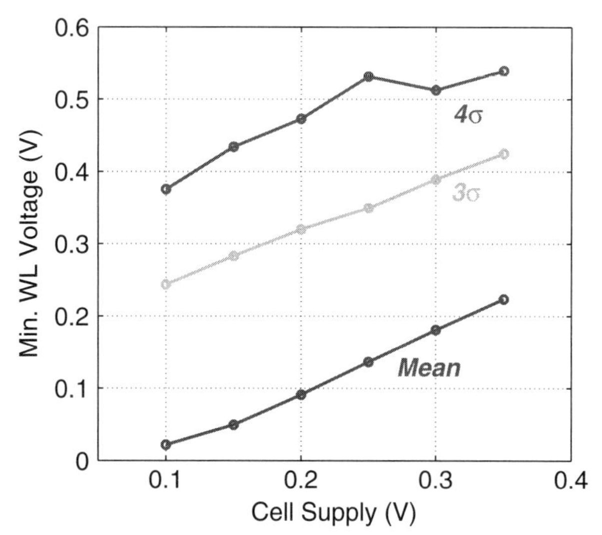

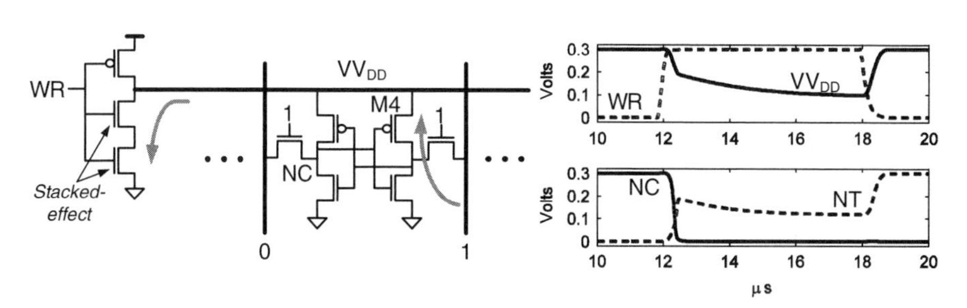

Fig. 4.27 Write-assist to enforce device strengths required for only the accessed cells (© [2008] IEEE)

of its NMOS access devices. Fortunately, however, this interaction is actually quite accurately controllable, since the pull-down devices of the supply driver are large enough that they are subject to minimal local variation, and, similarly the pull-up path through all of the accessed bit-cells tends to average. It is important to note that the peripheral supply driver does introduce an additional leakage path in all of the unaccessed rows. As a result, to minimize this leakage current, series NMOS pull-down devices are used, taking advantage of the stacked effect [38], where the V_{GS} of the top NMOS is reduced and the V_{DS} of the bottom NMOS is reduced to set the net series leakage current.

It is important to note that similar to the word-line sharing constraint for buffered-read bit-cells, which precludes column interleaving, VV_{DD} sharing in the manner described also complicates column interleaving. The scheme of selective bit-cell supply biasing, however, can also be applied in a column-wise manner to improve the write margin [40]. Here, the supply voltage of selected columns in the interleaved array can be reduced to weaken the associated PMOS loads. Unfortunately, this scheme is not compatible with very aggressive low-voltage operation, where the operating supply has been scaled to the data retention limit. In this case, further reduction to the supply voltage of bit-cells in the unaccessed rows will lead to erroneous loss of data.

4.5.3 Sensing Circuits

Sensing circuitry is becoming a limiting factor in SRAM scaling [39]. Fundamentally, this is because of the trade-off associated with its parasitic input offset voltage, arising from variation in the constituent devices, and its physical size, which must be increased in order to manage this variation. Of course, as discussed earlier, cell read-currents face severe degradation, especially at ultra low voltages, due to bit-cell variation. Further, to enhance density, it is desirable to maximize the number of cells per bit-line, leading to large bit-line capacitance. Consequently, the ratio of read-current to bit-line capacitance, which is critical to developing sense-amplifier input swing, is diminishing, pointing to the urgent need for highly sensitive sense-amplifiers with small input offset.

At ultra low voltages, the bit-line leakage further stresses the sense-ability constraints. As mentioned in Section 4.3.2.3, when the cell read-current is exceeded by the bit-line leakage, sensing can become impossible since margin between the logic "1" and "0" states cannot be guaranteed. However, if the effect of bit-line leakage, and its dependence on the data stored in the bit-cells of a column is managed, the margin required on the part of the sense-amplifier can be reduced through circuit techniques.

4.5.3.1 Leakage Replica Scheme

The bit-cell of Fig. 4.15 eliminates the data dependence of bit-line leakage. It does not, however, eliminate bit-line leakage and the resulting degradation to sensing margin. To accommodate the reduced, but data independent, sensing margin, the design in Ref. [15] utilizes a replica column to emulate the fairly consistent logic levels. In particular, as shown in Fig. 4.28, the logic "0" level is degraded due to reverse subthreshold leakage from the unaccessed bit-cells, which tends to pull *RdBL* up (see Section 4.4.1.1). The replica column, however, generates a logic "0" reference, which serves as the virtual ground for the sense-amplifier inverters, thereby centering their trip-point between the expected *RdBL* logic levels.

4.5.3.2 Sense-Amplifier Redundancy

Although replica techniques can reduce the required sensing margin, they do not address the problem of local variation obscuring the sense-amplifier trip-point, which is particularly important in advanced technologies. Sense-amplifier

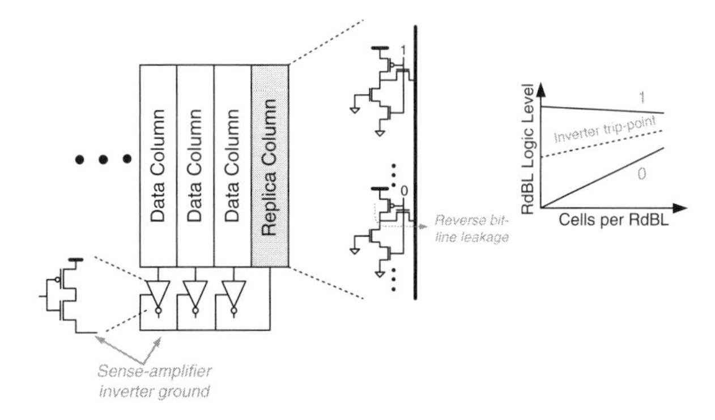

Fig. 4.28 Adjustment of sense-amplifier inverter trip-points using replica column scheme to reduce sensing margin [15]

redundancy [31], however, can significantly improve the area-offset trade-off originating from variation.

This concept requires that the *RdBL* from each column be connected to *N* different sense-amplifiers. Importantly, however, as shown in Fig. 4.29, the total area is constrained, so each of the individual sense-amplifiers must be proportionally smaller. Unfortunately, this implies that their individual error probabilities are correspondingly worsened; the offset distributions for an individual sense-amplifier as *N* is increased from 1 to 8 are shown to get wider. More specifically, an individual error probability $P_{\mathrm{ERR},N}$can be defined for an individual sense-amplifier, as the area under its offset distribution, where the magnitude of the offset exceeds the expected input voltage swing. As shown in Fig. 4.29, these individual error probabilities also worsen as *N* increases. However, if 1 out of the *N* structures with sufficiently small offset can be selected, then the overall error probability, $P_{\mathrm{ERR,tot}}$ for the entire sensing network is the joint probability that all *N* of the individual sense-amplifiers yield and error. As a result, the total error probability is given by the following:

$$P_{ERR,tot} = (P_{ERR,N})^N \tag{4.2}$$

The resulting total error probabilities are plotted in Fig. 4.30, normalized to the case of a single full-sized sense-amplifier (i.e., *N*=1. As shown, for the input voltage

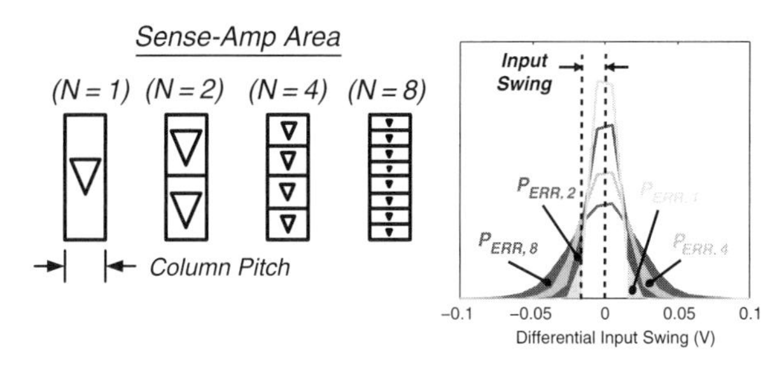

Fig. 4.29 Sense-amplifier redundancy concept and individual error probabilities for different values of *N* (© [2008] IEEE)

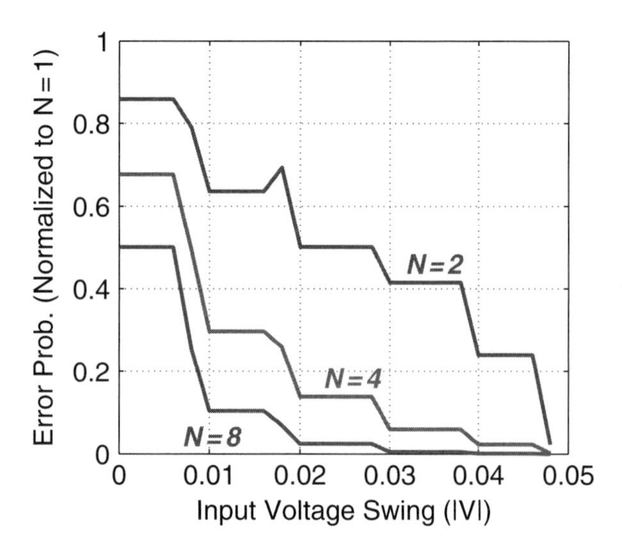

Fig. 4.30 Overall error probabilities (normalized to nominal case of a single full-sized sense-amplifier) with respect to the input voltage swing (© [2008] IEEE)

swings of interest, which are approximately 50 mV, increased levels of redundancy significantly improve the overall error probability.

Of course, actual implementation of sense-amplifier redundancy raises practical design considerations [31]. First, the analysis above assumes that the errors in the individual sense-amplifiers are uncorrelated. As a result, their actual implementation requires a pseudo-differential structure, which cancels correlations via symmetry. Second, the analysis did not consider the overhead of selecting between the N different instances. However, if N is limited to 2, the selection logic can be minimal, specifically, just a couple of latches and a couple of logic gates. Accordingly, sense-amplifier selection for all columns is performed simultaneously on start-up by using dummy bit-cells to drive reference logic "0" and "1" levels on each *RdBL*; then, one sense-amplifier that can correctly resolve both logic states is enabled.

4.5.3.3 Small-Signal Single-Ended Sensing

Many of the buffered-read bit-cells, which achieve the lowest-voltage operation, provide only single-ended read ports in order to maximize density. However, single-ended small-signal sensing introduces several challenges. As a result, the approach that is most commonly used with single-ended bit-cells is full-swing sensing where the *RdBL* is allowed to discharge completely [39], and the sense-amplifier can be reduced to just a logic gate. This approach, however, leads to a severe performance penalty in high-density arrays, where the read-current is small and the bit-lines are loaded by as many cells as possible; therefore, the bit-line discharge time is very long. Alternatively, the small-signal discipline can be maintained with a pseudo-differential sense-amplifier; however, here, generation of a complimentary reference

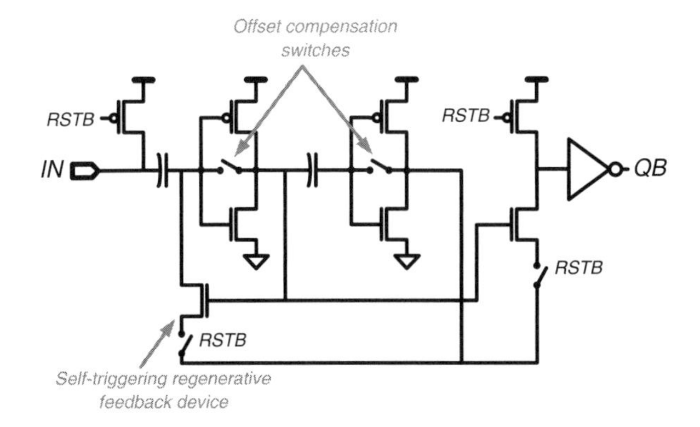

Fig. 4.31 Offset compensating small-signal sense-amplifier employing self-triggering regeneration [32]

that can accurately track the worst-case bit-cell behavior over all operating conditions is highly challenging.

Accordingly, the design in Fig. 4.31 [32] uses an offset-compensating small-signal sense-amplifier that is single-ended. Offset compensation improves the sense-amplifier sensitivity in the presence of variation, and efficient small-signal sensing, which requires very large gain, is achieved through regenerative positive feedback. Importantly, however, triggering regeneration typically requires an external strobe control signal. Unfortunately, in the presence of extreme variation, the timing of such a control cannot track the delay through the SRAM array, and consequently requires very excessive timing margin, which greatly limits the overall array performance. In this design [32], however, the stable voltage references, generated implicitly by offset compensation, are leveraged as an internal threshold to self-trigger regeneration without an external control path. Consequently, the associated tracking uncertainty and required timing margin are overcome.

4.6 Summary and Conclusions

Voltage scaling is an effective strategy for minimizing the power consumption of SRAMs. Further, as SRAMs continue to occupy a dominating portion of the overall area and power in modern ICs, the resulting total power savings are significant. Unfortunately, however, in modern technologies, conventional SRAMs, based on the 6T bit-cell, fail to operate at voltages below approximately 0.7 V both because of reduced signal levels and increased effects of variation. In subthreshold voltage regime, threshold-voltage variation has an exponential effect on the drive current, resulting in increased cell instability and a severely degraded read-current. To address these limitations alternate bit-cell topologies must be employed. Among these, buffered-read cells can achieve the greatest voltage scaling by isolating their internal storage nodes to overcome disruptive transients that limit read functionality. Such cells minimally require two additional devices, leading to an 8T topology.

Further enhancements, to address the critical problem of bit-line leakage, can be applied within the cell, or in the periphery, to maintain maximum array density. Alternatively, significant voltage scalability can also be achieved using non-buffered-read cells if the associated low-voltage degradations are targeted. Accordingly, a number of options exist for the wide range, and ever increasing number, of energy-constrained applications requiring embedded SRAMs.

Beyond just the bit-cells, however, array periphery is also stressed by low-voltage operation. In particular, the issues of write-ability, soft-error correction, and sense-ability all require specialized considerations. In particular, sense-amplifier area and offset are emerging as a significant limiting in high-density SRAMs, and especially in ultra low voltage designs. To address these, redundancy techniques already used to overcome 5 or 6σ failures in the array bit-cells can be used to overcome the 2 or 3σ failures in the periphery, when the sigma is amplified due to the extreme voltage scaling desired.

Ultra low voltage SRAM techniques enable practical arrays that form the basis for extremely low-power computing devices that are urgently required in many emerging portable and implantable electronic systems. However, their functionality is severely threatened by device-level challenges brought on by technology scaling. As a result, the design constraints are constantly and rapidly changing. Accordingly, much more than previous SRAM design, our ability to continue to achieve low-voltage SRAMs implies that the design decisions be regularly updated, requiring a very deep understanding of the low-voltage trade-offs and limitations.

References

1. M. Agostinelli, J. Hicks, J. Xu, B. Woolery, K. Mistry, K. Zhang, S. Jacobs, J. Jopling, W. Yang, B. Lee, T. Raz, M. Mehalel, P. Kolar, Y. W. J. Sandford, D. Pivin, C. Peterson, M. DiBattista, S. Pae, M. Jones, S. Johnson, and G. Subramanian, "Erratic fluctuations of SRAM cache Vmin at the 90 nm process technology node," in *IEDM Dig. Tech. Papers*, Dec. 2005, pp. 671–674.
2. A. Bhavnagarwala, X. Tang, and J. Meindl, "The impact of intrinsic device fluctuations on CMOS SRAM cell stability," *IEEE Journal of Solid-State Circuits*, vol. 36, no. 4, April 2001, pp. 658–665.
3. B. Calhoun and A. Chandrakasan, "Ultra-dynamic voltage scaling using sub-threshold operation and local voltage dithering in 90 nm CMOS," in *IEEE Int. Solid-State Circuits Conf. Dig. Tech. Papers*, Feb. 2005, pp. 300–301.
4. B. Calhoun and A. Chandrakasan, "A 256 kb sub-threshold SRAM in 65 nm CMOS," in *IEEE Int. Solid-State Circuits Conf. Dig. Tech. Papers*, Feb. 2006, pp. 480–481.
5. B. H. Calhoun, A. Wang, and A. Chandrakasan, "Modeling and sizing for minimum energy operation in subthreshold circuits," *IEEE Journal of Solid-State Circuits*, vol. 40, no. 5, Sept. 2005, pp. 1778–1786.
6. A. P. Chandrakasan and R. Brodersen, "Minimizing power consumption in digital CMOS circuits," in *Proc. IEEE*, vol. 83, no. 4, April 1995, pp. 498–523.
7. I. J. Chang, J.-J. Kim, S. P. Park, and K. Roy, "A 32 kb 10T subthreshold SRAM array with bit-interleaving and differential read-scheme in 90 nm CMOS," in *IEEE Int. Solid-State Circuits Conf. Dig. Tech. Papers*, Feb. 2008, pp. 388–389.

8. L. Chang, D. M. Fried, J. Hergenrother, J. W. Sleight, R. H. Dennard, R. Montoye, L. Sekaric, S. J. McNab, A. W. Topol, C. D. Adams, K. W. Guarini, and W. Haensch, "Stable SRAM cell design for the 32 nm node and beyond," in *Proc. IEEE Symp. VLSI Circuits*, June 2005, pp. 128–129.
9. L. Chang, Y. Nakamura, R. Montoye, J. Sawada, A. K. Martin, K. Kinoshita, F. H. Gebara, K. B. Agarwal, D. J. Acharyya, W. Haensch, K. Hosokawa, and D. Jamsek, "A 5.3 GHz 8T-SRAM with operation down to 0.41 V in 65 nm CMOS," in *Proc. IEEE Symp. VLSI Circuits*, June 2007, pp. 252–253.
10. J. Chen, L. Clark, and T.-H. Chen, "An ultra-low-power memory with a sub-threshold power supply voltage," *IEEE Journal of Solid-State Circuits*, vol. 41, no. 10, Oct. 2006, pp. 2344–2353.
11. F. Jumel, P. Royannez, H. Mair, D. Scott, A. Er Rachidi, R. Lagerquist, M. Chau, S. Gururajarao, S. Thiruvengadam, M. Clinton, V. Menezes, R. Hollingsworth, J. Vaccani, F. Piacibello, N. Culp, J. Rosal, M. Ball, F. Ben-Amar, L. Bouetel, O. Domerego, J. L. Lachese, C. Fournet-Fayard, J. Ciroux, C. Raibaut, U. Ko, "A leakage management system based on clock gating infrastructure for a 65-nm digital base-band modem chip," in *Proc. IEEE Symp. VLSI Circuits*, June 2006, pp. 214–215.
12. A. Kawasumi, T. Yabe, Y. Takeyama, O. Hirabayashi, K. Kushida, A. Tohata, T. Sasaki, A. Katayama, G. Fukano, Y. Fujimura, N. Otsuka, "A single-power-supply 0.7 V 1 GHz 45 nm SRAM with an asymmetrical unit-β-ratio memory cell," in *IEEE Int. Solid-State Circuits Conf. Dig. Tech. Papers*, Feb. 2008, pp. 382–383.
13. M. Khare, S. H. Ku, R. A. Donaton, S. Greco, C. Brodsky, X. Chen, A. Chou, R. DellaGuardia, S. Deshpande, B. Doris, S. K. H. Fung, A. Gabor, M. Gribelyuk, S. Holmes, F. F. Jamin, W. L. Lai, W. H. Lee, Y. Li, P. McFarland, R. Mo, S. Mittl, S. Narasimha, D. Nielsen, R. Purtell, W. Rausch, S. Sankaran, J. Snare, L. Tsou, A. Vayshenker, T. Wagner, D. Wehella-Gamage, E. Wu, S. Wu, W. Yan, E. Barth, R. Ferguson, P. Gilbert, D. Schepis, A. Sekiguchi, R. Goldblatt, J. Welser, K. P. Muller, P. Agnello, "A high performance 90 nm SOI technology with 0.992 μm^2 6T-SRAM cell," in *IEEE IEDM Dig. Tech. Papers*, Dec. 2002, pp. 8–11.
14. T.-H. Kim, J. Liu, J. Kean, and C. H. Kim, "A high-density subthreshold SRAM with data-independent bitline leakage and virtual ground replica scheme," in *IEEE Int. Solid-State Circuits Conf. Dig. Tech. Papers*, Feb. 2007, pp. 330–331.
15. T.-H. Kim, J. Liu, J. Kean, and C. H. Kim, "A 0.2 V, 480 kb subthreshold SRAM with 1 k cells per bitline for ultra-low –voltage computing," *IEEE Journal of Solid-State Circuits*, vol. 43, no. 2, Feb. 2008, pp. 518–529.
16. J. P. Kulkarni, K. Kim, and K. Roy, "A 160 mV robust schmitt trigger based subthreshold SRAM," *IEEE Journal of Solid-State Circuits*, vol. 42, no. 10, Oct. 2007, pp. 2303–2313.
17. J. Kwong, Y. Ramadass, N. Verma, M. Koesler, K. Huber, H. Moormann, and A. Chandrakasan, "A 65 nm Sub-V_t Microcontroller with Integrated SRAM and Switch Capacitor DC-DC Converter," in *IEEE ISSCC Dig. Tech. Papers*, Feb. 2008, pp. 318–319.
18. Z. Liu and V. Kursun, "High read stability and low leakage SRAM cell based on data/bitline decoupling," in *Proc. IEEE Int. Syst Chip Conf.*,Sept. 2006, pp. 115–116.
19. C.-Y. Lu and J. M. Sung, "Reverse short-channel effect on threshold voltage of submicrometer salicide devices," *IEEE Electron Device Letters*, vol. 10, no. 10, Oct. 1989, pp. 446–448.
20. T. Mizumo, J.-I. Okamura, and A. Toriumi, "Experimental study of threshold voltage fluctuations using an 8 k MOSFET's array," in *Proc. IEEE Symp. VLSI Tech.*, May 1993, pp. 41–42.
21. Y. Morita, H. Fujiwara, H. Noguchi, Y. Iguchi and H. Kawaguchi, and M. Yoshimoto, "An Area-Conscious Low-Voltage-Oriented 8T-SRAM Design Under DVS Environment," in *Symp. VLSI Circuits*, June 2007, pp. 256–257.
22. S. Mutoh, T. Douseki, Y. Matsuya, T. Aoki, S. Shigematsu, and J. Yamada, "1-V power supply high-speed digital circuit technology with multithreshold-voltage CMOS," *IEEE Journal of Solid-State Circuits*, vol. 30, no. 8, Aug. 1995, pp. 847–854.

23. M. J. M. Pelgrom, A. C. J. Duinmaijer, and A. P. G. Welbers, "Matching properties of MOS transistors," *IEEE Journal of Solid-State Circuits*, vol. 24, no. 5, Oct. 1989, pp. 1433–1439.
24. R. Rodriguez, J. H. Stathis, B. P. Linder, S. Kowalczyk, C. T. Chuang, R. V. Joshi, G. Northrop, K. Bernstein, A. J. Bhavnagarwala, and S. Lombardo, "The impact of gate-oxide breakdown on SRAM stability," *IEEE Electron Device Letters*, vol. 23, no. 9, Sept. 2002, pp. 559–561.
25. E. Seevinck, F. J. List, and J. Lohstroh, "Static-noise margin analysis of MOS SRAM cells," *IEEE Journal of Solid-State Circuits*, vol. SC-22, no. 5, Oct. 1987, pp. 748–754.
26. S.-W. Sun and P. G. Y. Tsui, "Limitations of CMOS supply-voltage scaling by MOSFET threshold-voltage variation," *IEEE Journal of Solid-State Circuits*, vol. 30, no. 8, Aug. 1995, pp. 947–949.
27. R. M. Swanson and J. D. Meindl, "Ion-implanted complementary MOS transistor in low-voltage circuits," *IEEE Journal of Solid-State Circuits*, vol. sc-7, no. 2, Apr. 1972, pp. 146–153.
28. V. Sze, R. Blazquez, M. Bhardwaj, and A. Chandrakasan, "An energy efficient subthreshold baseband processor architecture for pulsed ultra-wideband communications," in *IEEE Int. Conf Acoust., Speech and Signal Process.*, May 2006, pp. 908–911.
29. K. Takeda, Y. Hagihara, Y. Aimoto, M. Nomura, Y. Nakazawa, T. Ishii, and H. Kobatake, "A read-static-noise-margin-free SRAM cell for low-*VDD*and high-speed applications," in *IEEE Int. Solid-State Circuits Conf. Dig. Tech. Papers*, Feb. 2005, pp. 478–479.
30. N. Verma and A. Chandrakasan, "A 256 kb 65 nm 8T sub-*Vt*SRAM employing sense-amplifier redundancy," in *IEEE Int. Solid-State Circuits Conf. Dig. Tech. Papers*, Feb. 2007, pp. 328–329.
31. N. Verma, and A. P. Chandrakasan, "A 256 kb 65 nm 8T subthreshold SRAM employing Sense-amplifier Redundancy," *IEEE Journal of Solid-State Circuits*, vol. 43, no. 1, Jan. 2008, pp. 141–149.
32. N. Verma and A. P. Chandrakasan, "A high-density 45 nm SRAM using small-signal non-strobed regenerative sensing," *ISSCC Digest of Technical Papers*, Feb. 2008, pp. 380–381.
33. N. Verma, J. Kwong, and A. P. Chandrakasan, "Nanometer MOSFET variation in minimum energy subthreshold circuits," *IEEE Transaction Electron Devices*, vol. 55, no. 1, Jan. 2008, pp. 163–174.
34. J. Wang and B. H. Calhoun, "Canary replica feedback for near-DRV stand-by VDD scaling in a 90 nm SRAM," in *Proc. IEEECustom Integr. Circuits Conf.,*Sept. 2007, pp. 29–32.
35. Y. Wang, H. Ahn, U. Bhattacharya, et al., "A 1.1 GHz 12AμA/Mb-leakage SRAM design in 65 nm ultra-low-power CMOS with integrated leakage reduction for mobile applications," *ISSCC Digest of Technical Papers*, Feb. 2007, pp. 324–325.
36. A. Wang, A. Chandrakasan, and S. Kosonocky, " Optimal supply and threshold scaling for sub-threshold CMOS circuits," in *Proc. IEEE Comp Soc. Annu. Symp. VLSI*, April 2002, pp. 5–9.
37. A. Wang and A. Chandrakasan, "A 180 mV FFT processor using sub-threshold circuit techniques," in *IEEE Int. Solid-State Circuits Conf. Dig. Tech. Papers*, Feb. 2004, pp. 292–293.
38. Y. Ye, S. Borkar, and V. De, "A new technique for standby leakage reduction in high performance circuits," in *Proc. IEEE Symp. VLSI Circuits*, June 1998, pp. 40–41.
39. K. Zhang, K. Hose, V. De, and B. Senyk, "The scaling of data sensing schemes for high speed cache design in sub-0.18μm technologies," in *Proc. IEEE Symp. VLSI Circuits*, June 2000, pp. 226–227.
40. K. Zhang, U. Bhattacharya, Z. Chen, F. Hamzaoglu, D. Murray, N. Vallepath, Y. Wang, B. Zheng, M. Bohr, "A 3-GHz 70 Mb SRAM in 65 nm CMOS technology with integrated column-based dynamic power supply," in *IEEE Int. Solid-State Circuits Conf. Dig. Tech. Papers*, Feb. 2005, pp. 474–475.

Chapter 5
Embedded DRAM in Nano-scale Technologies

John Barth

5.1 Introduction

Dynamic random access memory (DRAM) is a type of random access memory that uses charge stored on individual capacitors to hold data within an integrated circuit. Since these capacitors are non-ideal and suffer from parasitic leakages, the information eventually fades and the charge stored requires periodic refresh. Because of this refresh requirement, this memory type is classified as dynamic, in contrast to static random access memory (Fig. 5.1a) where a cross-coupled pair maintains the data state.

The advantage of DRAM is its structural simplicity: only one transistor and one capacitor (1T1C, Fig. 5.1b) are required per bit, compared to six transistors in a conventional SRAM. This allows DRAM to reach very high density. Like SRAM, it is in the class of volatile memory devices, since both lose the stored data state when the power supply is removed. To achieve ultimate densities, low-leakage access transistors and high-capacitance cells require technology features tailored specifically to these devices. Through the early 1990 s these process technologies have been optimized for low cost, low power, and high-density commodity stand-alone memory devices.

As application-specific integrated circuit (ASIC) technologies expanded into new markets, the need for denser embedded memory grew. To accommodate this increased demand, embedded DRAM macros have been offered in state-of-the-art ASIC library portfolios [1, 2]. The introduction of embedded DRAM extended the on-chip capacity to more than 40 MB, allowing historically off-chip memory to be integrated on chip and enabling system-on-a-chip (SoC) designs. With memory on the chip, applications can take advantage of the high bandwidth naturally offered by a wide I/O DRAM and achieve data rates greater than those previously limited by pin count and off-chip pin rates. Applications for this memory include network processors, digital signal processors, and cache chips for microprocessors.

J. Barth (✉)
IBM, Essex Junction, Vermont
e-mail: jbarth@us.ibm.com

K. Zhang (ed.), *Embedded Memories for Nano-Scale VLSIs*, Series on Integrated Circuits and Systems, DOI 10.1007/978-0-387-88497-4_5,

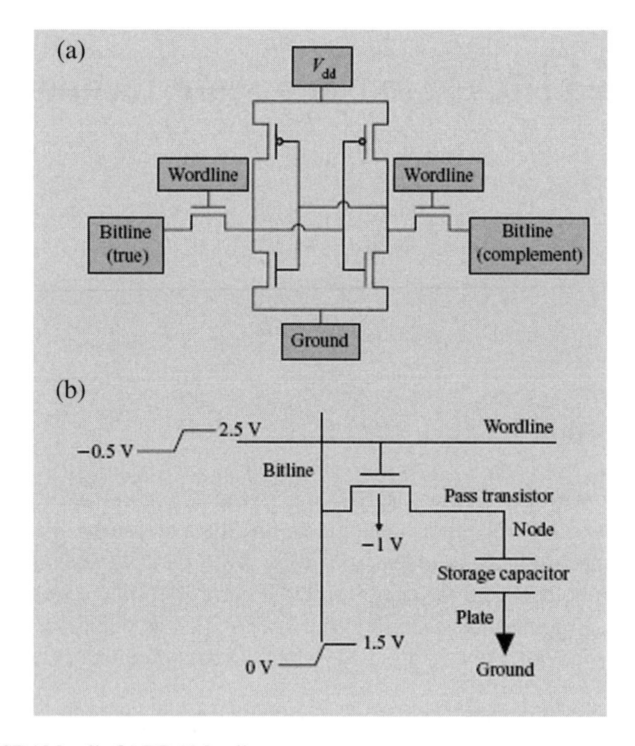

Fig. 5.1 (**a**) SRAM cell, (**b**) DRAM cell

The integration of embedded DRAM into ASIC designs has intensified the focus on how best to architect, design, and test a high-performance, high-density macro as complex as dynamic RAM in an ASIC logic environment. The ASIC environment itself presents many difficult elements that have historically challenged DRAMs—specifically wide voltage and temperature operating ranges and uncertainties in surrounding noise conditions. These challenges dictate a robust architecture that is noise-tolerant and can operate at high voltage for performance and at low voltage for reduced power.

5.1.1 Migrating from the Commodity DRAM Base to the Logic Base

Initially embedded DRAM developed from a DRAM base, this work at IBM and elsewhere focused on leveraging DRAM technology primarily to fabricate custom DRAMs for use in off-chip cache and graphic DRAM applications [3, 4]. The main rationale for this approach was the belief that the complexity of DRAM process architecture should not be modified, and any logic functions should be contained within the available menu of devices offered by the DRAM technology. The yield and line learning of this part would have to depend on the presence of a commodity part with the same subarray running in volume in the semiconductor line. At the 0.5–0.25 lm nodes, this meant relatively poor performance for the logic devices and, consequently, limited logic function. In 1998, IBM introduced a DRAM-based embedded DRAM [5] at the 0.25 μm node called 7LD, with additional performance-oriented changes in the devices. These included a shorter channel logic device for higher performance and the introduction of logic, such as multilevel back-end technology, with up to six aluminum wiring levels and a variety of logic-like packaging options. In addition, a limited custom library was also offered in this technology. It is worth summarizing some important conclusions from this exercise:

- The embedded DRAM macro performance was modest. In fact, it was comparable to commodity memory performance. This resulted from the fact that the subarray was designed to meet the standardized performance metrics required of the commodity part.
- The relatively poor performance of the peripheral logic devices resulted in a somewhat marginal library. This limited the widespread use of the technology.
- Density suffered on two counts: first, DRAM processes were optimized to print array features at very tight pitches, but random patterns were significantly looser than logic technology at the time; second, the low drive of the devices required them to be sized large enough to drive the on-chip load. As a result, the logic density lagged by more than a generation.
- The assumption of DRAM line learning did not apply as well as initially envisioned. The on-chip voltage variations, junction temperatures, and retention characteristics were different enough from standard DRAM that additional process-sensitivity learning was required. Furthermore, a strategic redirection of focus to logic rather than commodity DRAM meant that there was no commodity DRAM learning effect on the embedded DRAM parts in any case.

On the basis of these findings, subsequent embedded DRAM technologies were developed to be integrable with high-performance logic [6]. Furthermore, the embedded DRAM design intellectual property was integrated into the IBM Blue Logic∗ libraries and followed the same integration and test methodologies for ease of use. Both of these factors have contributed immensely to the success of embedded DRAM at IBM and have facilitated realization of the Blue Gene/L SoC, including the integration of the PowerPC 440 cores. Nevertheless, it should be emphasized that the commodity DRAM pedigree of IBM embedded DRAM has been a crucial element of its success. By incorporating the learning of several generations of commodity DRAMs and retaining crucial elements of the process architecture, our embedded DRAM macros have avoided the pitfalls of the well-known complex DRAM pattern sensitivities and retained a robust character over a wide range of operating conditions.

With the advent of embedded DRAM offerings in a logic-based ASIC technology [6], the performance of embedded DRAM macros has improved significantly over that of DRAM-based technologies. Users are increasingly replacing SRAM implementations with embedded DRAM, placing additional pressure on macro performance and random cycle time. This pressure extends into testing, where use of traditional direct memory access (DMA) is costly in silicon area and wiring complexity, and introduces uncertainty in performance-critical tests. A more attractive solution to this test problem is the use of a built-in self-test (BIST) system that is adapted to provide all of the necessary elements required for high fault coverage on DRAM, including the calculation of a two-dimensional redundancy solution, pattern programming flexibility, at-speed testing, and test-mode application for margin testing [7, 8].

When microprocessors enter the highly multi-core/multi-threaded era, higher-density, lower-latency embedded memory will be required to meet their cache design needs. System-level processor simulations show that doubling cache capacity results in respectable double-digit percentage gains for cache-constrained commercial applications [9]. Improving cache latency also has an impact on system performance, and integrating the cache on-chip eliminates delay, power, and area penalties associated with high-frequency off-chip I/O channels. Moving forward, trends in virtual machine technology, multi-threading and multi-core processors will continue to stress cache subsystems [10, 11]. The availability of a high-performance embedded DRAM macro, offering a 3× density advantage over SRAM, would allow cache-dominated chips to double their cache capacity, while actually fitting in a smaller footprint. Smaller cache size not only reduces chip cost, but offers latency reduction from shorter wire run lengths to retrieve data. For power-constrained applications, eDRAM consumes one-fifth the keep-alive power of SRAM, while its high capacitance and small collector area offer a soft error rate three orders of magnitude lower than SRAM.

5.2 Fundamental DRAM Operation

DRAM memory arrays are composed of wordlines (or rows) and bitlines (columns) (see Fig. 5.2). At the crosspoint of every row and column is a storage cell consisting of a transistor and capacitor [12, 14]. The data state of the cell is stored as charge on the capacitor, with the transistor acting as a switch controlling access to the capacitor. With the switch on (wordline activated), charge can be read from or written to the storage cell. The rest of the DRAM support circuits are dedicated to controlling the wordlines and bitlines to read and write the memory array.

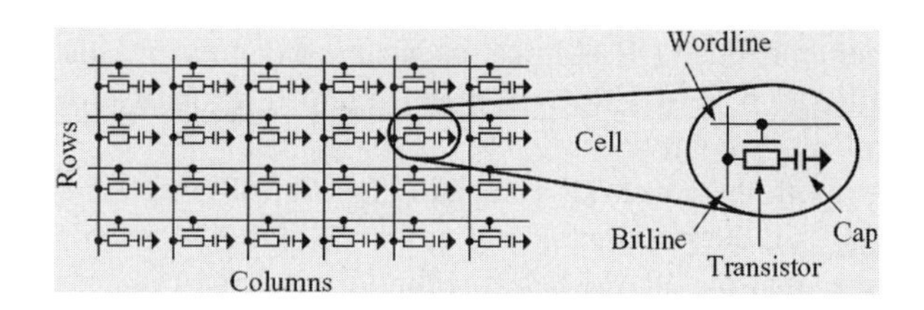

Fig. 5.2 DRAM cell array

5.2.1 The Capacitor

The heart of a DRAM cell is its capacitor. Whether the capacitor is formed as a deep trench or a stacked type, most of the challenge in scaling a DRAM to the next technology node involves achieving the same high capacitance, but in a smaller space. This leads to aggressive scaling of the aspect ratio of the fabricated capacitor structures and to reducing the equivalent thickness of the capacitor insulator through physical thinning or increasing the dielectric constant of the material.

The basic operation of reading a DRAM cell involves turning on the array device and allowing the charge stored on the capacitor on one side of the device to pass onto a metal wire connected to the other side of the device. This metal line, or bitline, is shared with as many as 500 other nonselected cells. The voltage generated on the bitline is detected or sensed by circuitry at the end of the bitline. A key metric in DRAM design is the amount of voltage or signal generated when a data cell is read by dumping its stored charge onto its bitline. Since the charge initially stored in the cell capacitor equilibrates with the bitline capacitance, the voltage ultimately created is given by

$$\text{Signal} = V_{\text{cell}} \times C_{\text{cell}}/(C_{\text{bitline}} + C_{\text{cell}})$$

where V_{cell} is the voltage initially stored in the cell capacitor, C_{cell}, the capacitance of the cell capacitor, and C_{bitline}, the capacitance of the bitline. The quotient of the capacitance terms in the preceding equation is known as the *transfer ratio* and is typically of the order of 0.2.

As technology shrinks, the capacitance of the bitline wire typically remains relatively constant, since the wire thickness is not scaled and the capacitance reduction due to the shrinkage in the length of the line is offset by the capacitance increase due to the reduced spacing between the lines. As long as the number of cells per bitline remains constant, the capacitance of the cell capacitor must remain constant to maintain the transfer ratio and the signal available for sensing cell data correctly. Unfortunately, as technology scales, the devices that detect the bitline signal suffer from the same matching degradation as was mentioned for the SRAM cell, requiring high levels of signal to detect the correct data state of the cell.

Since simple dimensional scaling causes the capacitance of a cell to decrease, maintaining a constant cell capacitance from generation to generation is possible only with continuous improvements in fabrication technology or increasing permittivity of the dielectric materials. Since new invention is not always assured, it is desirable to explore options that allow capacitance to be reduced. The obvious option is to reduce the bitline capacitance by reducing the number of bits per bitline. This reduces the capacitance on the bitline and allows a reduction of the cell capacitance while maintaining a high signal transfer ratio. This solution, unfortunately, degrades the density of the memory system by causing the fixed area overhead of the sensing circuitry at the end of the bitline to be amortized over fewer memory cells. The key to enabling fewer cells per bitline is to have a very efficient design

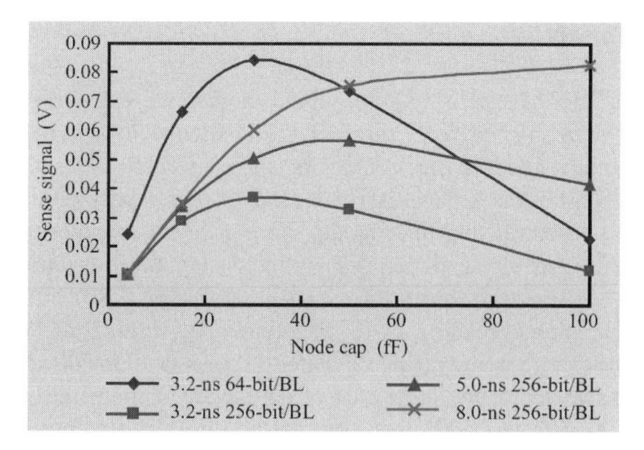

Fig. 5.3 Signal versus node capacitance

of the array sensing and control circuitry. With proper design area, efficient designs using 64 or fewer cells per bitline are feasible.

Reducing cell and bitline capacitance has additional performance benefits. A lower-capacitance bitline is faster to switch, improving array speed. Similarly, a lower-capacitance cell charges more quickly, enabling faster write time when switching the cell to its opposite data polarity state as shown in Fig. 5.3. It should be noted that reducing cell capacitance too far will limit the data retention time and increase the susceptibility of the cell to SER caused by energetic background radiation.

5.2.2 The Transistor

Because CMOS logic device technology has been aggressively scaled to create the fastest possible transistor switching, we have now arrived at a point at which the channel length of the conventional DRAM array transistor is significantly longer than the minimum used in the logic devices, and is limiting the continued area scaling of the array cell.

DRAM array transistors must maintain off currents in the range of 10^{-14}A, which allows data to be retained for the several-millisecond interval required between cell refreshes. (In comparison, logic device leakage ranges from 10^{-12}A to 10^{-7}A per device.) A gate voltage swing of approximately 3 V is necessary to achieve both the low off current and high-gate overdrive that enables writing a high voltage to the storage node. This high operating voltage requires the use of a gate oxide of approximately 5–6 nm compared with the 1–1.5 nm seen in the state-of-the-art logic devices of today.

In addition, the low-leakage requirement necessitates careful grading of the doping profiles of the junctions used in the source and drain of the device in order to limit the electric fields that increase leakages in the junctions. As a result, such

DRAM cells have reached their minimum channel limit. Further scaling of the cell area causes the width and available drive current of these devices to be significantly reduced. Such reduced drive current becomes a limitation in the read and write performance for the cell.

If it were possible to relax the leakage constraints for the embedded DRAM by significantly decreasing the interval between cell refreshes, the techniques of gate oxide reduction and device halo implantation used for logic device scaling could be employed to drive continued reduction of the embedded DRAM array device area. During a cell refresh cycle, the DRAM memory is unavailable to the system, and just increasing the rate of refresh for the DRAM would reduce this availability unacceptably. However, innovative circuit designs have enabled reduction of the retention interval to less than 100 μs while maintaining 99% memory availability to the system [13]. With data retention in the tens of seconds range, the off-current requirements can be relaxed to several pA per cell. Such off currents are achieved today by the low-power logic FET designs using a gate oxide of 2 nm to 2.5 nm and a gate voltage swing of 1.5 V (Fig. 5.4). The use of aggressive oxide thickness scaling can enable shrinking of the array device length and width while still maintaining drive current. The higher doping levels of the junctions in the scaled transistors also improve the resistance of the connection to the trench. The switching speeds of the circuits driving the DRAM gate also improve as their oxide thickness and voltage swing decrease. In a DRAM memory system, the DRAM gate voltage is typically generated internally with a system of charge pumps. Reducing the level of the generated voltage significantly improves the efficiency and the area required for these pump circuits.

The leakage current through the gate of the device will, however, ultimately limit the data retention for the array cell. For the 64-ls retention, this limit will be approximately 1.8 nm to 2.0 nm for conventional oxide gate dielectrics. These same device considerations can be applied to adapt the silicon-on-insulator devices used in the highest-performance logic technologies for use as an array device [14].

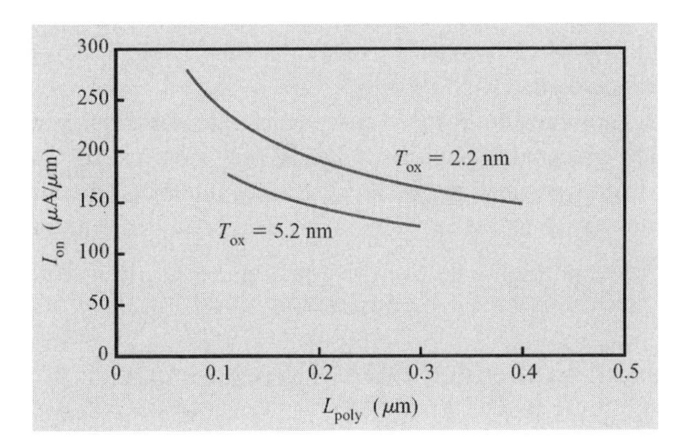

Fig. 5.4 Drain current

5.2.3 Cost and Process Complexity

A major misconception in the industry is that embedded DRAM is more complex and, therefore, more expensive. While the additional processing associated with the fabrication of the capacitor and DRAM access device does mean that an embedded DRAM wafer is more expensive to process, this does not mean that an embedded-DRAM-based die is more expensive than an embedded-SRAM-based die. The discussion must be focused on cost effectiveness. Consider the following argument:

> S is fraction of the die dedicated to SRAM
> D is relative DRAM area (typically 0.3333, DRAM being 3× more dense)
> C is relative DRAM cost (1.15 = 15% DRAM cost adder)
>
> SRAM area = S
> Logic area = 1—S
> DRAM area = $S \times D$
> DRAM chip area = Logic area + DRAM area = $(1–S) + S \times D$
> DRAM chip cost = DRAM cost × DRAM chip area = $C(1–S(1–D))$
>
> For DRAM to be cost effective, relative DRAM chip cost < 1
> $C(1–S(1–D)) < 1$
> $S(1–D) > 11/C$
> $S > (1 - 1/C) / (1 - D)$

Assuming DRAM is 3× more dense than SRAM (D=0.333) and DRAM cost adder is 15% (C=1.15), then the cost would cross over for dies with greater than 19.6% SRAM.

If the Blue Gene/L chip was designed with embedded SRAM, approximately two-thirds of the die would be composed of replaceable SRAM, so that the embedded DRAM solution is, in fact, 40% more cost effective, a significant cost saving. Interestingly, this cost savings has been achieved with no significant increase in latency. This is because the smaller footprint of the DRAM cache compared with the hypothetical SRAM cache results in a shorter time-of-flight delay from the farthest bits in the macro.

The logic technologies build more complexity into the base process – such as additional device types, additional metal levels, precision resistors and capacitors, and silicon-on-insulator substrates – the cost and complexity adder for embedded DRAM continues to fall and is expected to be well below 10% at the 65 nm high-performance node.

5.3 Overview of Embedded DRAM Architecture

The CU-11 embedded DRAM macro was developed around the idea of user simplicity while including a high degree of flexibility, function, and performance. For

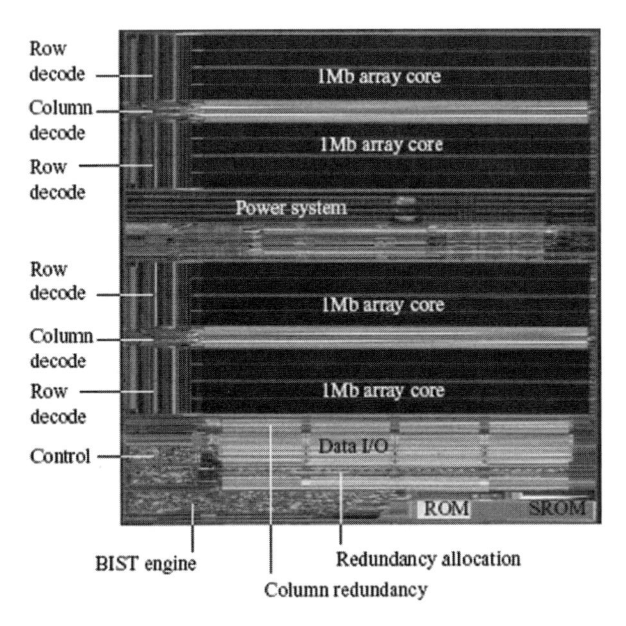

Fig. 5.5 A floor plan

application flexibility, the embedded DRAM is growable in 1 Mb increments to provide macro sizes from a 1 Mb minimum to a 16 Mb maximum and offers a 256-I/O width and a 292-I/O width for applications requiring parity. The wide I/O was chosen to provide maximum bandwidth; for applications that do not require the full width, bit-write control was included to facilitate masking. Multiple embedded DRAM macros can be instantiated on an ASIC die, enabling customers to make a performance/die-area trade-off specific to their application. Figure 5.5 shows a high-level floor plan of the embedded DRAM. This architecture lends itself well to providing two modes of macro operation: single-bank and multi-bank interleave modes. The single-bank operation provides a simple SRAM replacement function, while the multi-bank mode extends the macro performance by allowing concurrent operations to independent banks.

5.3.1 Single-Bank Operation

Single-bank operation was intended to resemble an embedded SRAM, supporting simple broadside addressing with read/write control. To improve bandwidth, the user can optionally use page mode, which was carried over from conventional DRAM. The addressing is broken down as follows:

- A0–A2 decodes one of eight-page (or column) addresses.
- A3–A11 decodes one of 512 row addresses within a 1 Mb block.
- A12–A15 decodes which 1 Mb block is to be accessed.

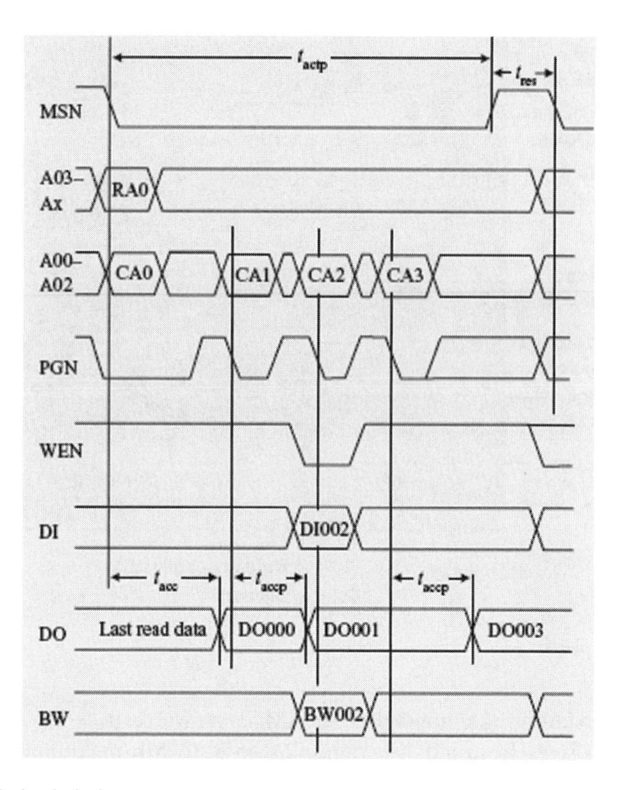

Fig. 5.6 Single-bank timing

The number of high-order addresses (A12–A15) is determined by the macro size. The diagram in Fig. 5.6 shows an example of timing for a macro in single-bank-mode operation. The horizontal lines indicate the logical data state of the corresponding signals as they change with time. The macro select signal (MSN) controls the active (t_{actp}) and restore timing (t_{res}) of the macro and latches the row (A3–A11), column (A0–A2), and block addresses (A12–A15) on every falling edge of MSN (indicated by the vertical lines). For a read cycle, write enable (WEN) is held high; data-out (DO) is latched and transmitted off the macro within time t_{acc}. For a write cycle, WEN is held low; data-in (DI) and bit write (BW) are received and latched. Page mode, controlled by the page signal (PGN), allows access to the additional bits along a wordline not selected at the MSN falling edge. For successive page cycles, falling PGN latches only WEN and a new column address (A0–A2); the row and bank addresses latched during the MSN falling edge are reused. In page mode, DO is latched and transmitted off the macro within time t_{accp}.

5.3.2 Multi-bank Operation

For the multi-bank-mode configuration, each 1 Mb block of the macro acts as an independent bank that shares a common address and data bus with all other 1 Mb

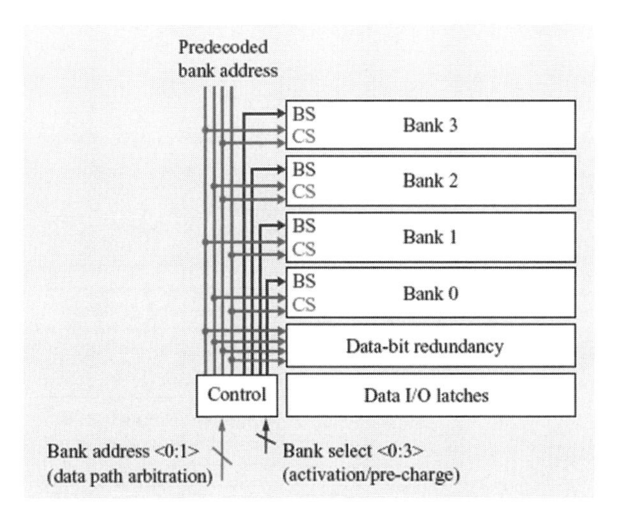

Fig. 5.7 Multi-bank architecture

blocks within the macro. The number of banks within a macro is determined by the macro size. Figure 5.7 shows a 4 Mb macro with four banks. A bank select (BS) pin is associated with each bank (1 Mb block) and controls activation and precharge of that bank. The bank address (BA) is decoded by control logic and arbitrates which bank has control of the data path.

In multi-bank configuration, the macro does not employ broadside addressing. Rather, the embedded DRAM macro operates similarly to a synchronous DRAM (SDRAM), in which addressing is performed in a row address strobe/column address strobe (RAS/CAS) manner and the macro select input (MSN) is treated like a master input clock, latching the state of all other input pins with each falling MSN edge. Figure 5.8 shows three cycles: Cycle 0 activates Bank 0, Cycle 1 activates Bank 1 and reads or writes Bank 0, and finally Cycle 2 activates Bank 2, reads or writes Bank 1, and precharges Bank 0. The MSN input can be cycled at a maximum rate of 250 MHz (4 ns assuming a nominal 50/50 clock duty cycle). All bank select (BS) inputs must be defined at every MSN falling edge to indicate whether each bank is to remain open or closed or whether a bank is to become active/open (from the precharge state) or become precharged/closed (from the active state). This protocol supports simultaneous activate, read/write, and precharge to three different banks.

Any combination of banks within a macro can be active simultaneously as long as each 1 Mb bank is opened in a sequential fashion. To avoid power-supply design limits, multiple banks cannot be activated or precharged on a single MSN clock cycle. Maximizing the number of banks in a macro improves the probability of avoiding an open (or busy) bank and maintaining the pseudorandom peak bandwidth of 8 GB/s.

To activate a bank, the corresponding BS signal must be low, and the row address for that bank must be supplied on address pins A3–A11 when MSN is clocked low. The row address for each active bank remains latched until the bank is precharged.

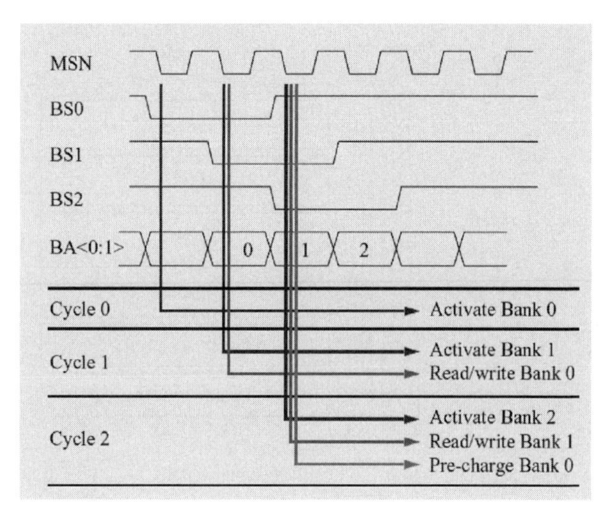

Fig. 5.8 Multi-bank timing

This frees the row address bus to allow other banks to be activated at a different row address on subsequent MSN clock cycles. A bank is precharged by placing a high level on the corresponding BS input when the MSN is clocked low. No address information is required to precharge a bank. Once a bank has been activated, a read or write operation can be performed to that bank by bringing PGN low, selecting the state of the WEN pin (read – high, write – low), specifying the column address on A0–A2, and specifying the bank address on A12–A15 as MSN is clocked low.

5.3.3 Macro Organization

The embedded DRAM is constructed from building blocks, shown in Fig. 5.5 floor plan: a 1 Mb array core, a power system for generating boosted voltage levels used by the array core, a control system for buffering and generating the array core timing signals, column redundancy for replacing defect data bits, data I/O for receiving and transmitting off-macro data, and BIST for testing the embedded DRAM macro. BIST is composed of a microprocessor-based engine, instruction memory (ROM/SROM), a data comparator, and a redundancy allocation unit. The 1 Mb array and its support circuitry are replicated to construct the desired macro size. Each embedded DRAM macro contains a single control system, a common power system, and a BIST.

The power system, for the scope of this book, simply supplies the necessary voltage network levels required for biasing the DRAM cell matrix. The power system is located at the midpoint of the 1 Mb arrays (or 1 Mb offset in the case of odd-sized macros) to provide optimal power distribution. The row decoder selects one of 512 words, while the column decoder selects 256 of 2048 bits in the 1 Mb array. The control system is the primary unit controlling signals to the array(s). It receives input signals from either primary macro inputs or the BIST. Signals to the array are selected by the system function mode: normal operation mode or test (BIST) mode. The control path also determines whether the embedded DRAM is operated as a single bank or as a multi-bank part.

The final block in the embedded DRAM macro is the BIST. The design goal of the BIST is to provide a test engine, operable in the logic test environment on low-cost, low-pin-count testers, that stimulates the control, data paths, and array of the embedded DRAM and provides fault coverage equivalent to that traditionally supplied by high-cost memory testers to discrete DRAM. The flexibility of the BIST system enables the test development engineers to alter the instruction memory to create new or modified test patterns or to change the sequence or number of patterns applied at each manufacturing test gate. Ultimately, the BIST locates all faults in each 1 Mb array segment, calculates the two-dimensional redundancy solution required to repair these faults, and reports this solution via standard scan string methods [15]. The redundancy solution is permanently stored in a remote fuse memory (nonvolatile) programmed with a laser after testing.

5.3.4 Array Core Organization

The array core, shown in Fig. 5.9, is the fundamental building block of the embedded DRAM macro; it includes the following:

1. DRAM cell array,
2. Row decoder for selecting one of many wordlines,
3. Level shift and driver for elevating the voltage level of the selected wordline,
4. Wordlines for coupling storage cells to bitlines,
5. Wordline stitch regions for reducing the wordline propagation delay with metal connections,
6. Bitlines for connecting storage cells to sense amplifiers (sense amps),
7. Restore devices and sense amps for reading the small signal levels from bitlines,

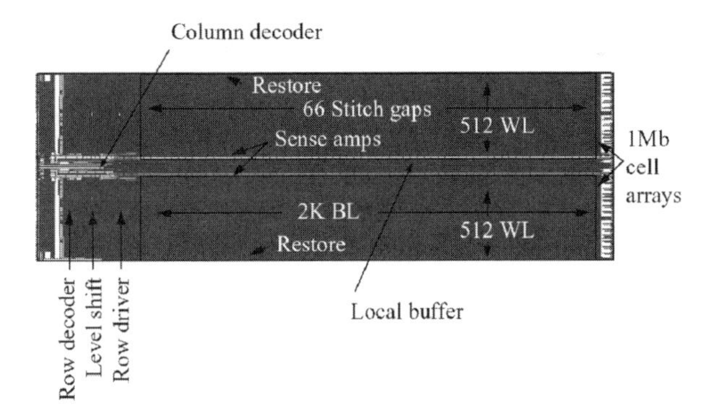

Fig. 5.9 An array core

8. Column decoder for selecting one of many sense amps,
9. Local buffers for driving the select sense amps to the edge of the macro on a dataline.

Each 1 Mb array core is organized as 512 wordlines (512 WL) by 2048 bitlines (2 K BL). The 2048 bitlines are decoded from 8 to 1, producing a 256-bit data word with an 8-bit page depth (additional bits available in page mode). The 292-parity option is implemented by adding 288 bitlines (or 36 data bits), extending the length of the wordline. The cell array and sense amplifiers are mirrored around a shared local buffer that drives the selected sense amps onto a global dataline running over the cell array, parallel to the bitlines on third-level metal (M3). Each of these 1 Mb array segments (or banks) contains eight redundant wordlines and eight redundant data bits that can be used to repair faults in that 1 Mb array. The redundancy repair region is limited to the 1 Mb bank to allow independent repair, simplifying test and redundancy allocation for macros constructed from multiple array cores.

This organization provides high array utilization while maintaining performance typically lost by creating longer wordlines or bitlines. Faster core performance can be achieved with shorter wordlines and bitlines, but at a cost: every wordline requires a decoder/driver, and every bitline requires a sense amp. Cutting wordlines and bitlines in half requires four times as many word drivers and four times as many sense amps, resulting in a larger core area for an equivalent memory size. Although smaller arrays can run faster, there is a diminishing return in performance due to increased total area and resulting propagation delays, segment decoding, and data multiplexing. A metric for measuring efficiency is array utilization, which is calculated by multiplying the cell area by the number of bits and dividing by the total macro area. The large array block achieves a higher efficiency because it can amortize the support circuit overhead (word decoders/sense amps) across more memory cells.

5.3.5 Array Core Operation

The following sections describe in more detail the operation of the array core components including the row system for activating wordlines, the sense system for reading and writing data from the cell, and the data path for delivering data to and from the array core.

5.3.6 Row System

The logical function of a DRAM word system is simple: select one wordline out of 512. However, the electrical and physical implementations are far from simple. Most DRAMs boost the wordline voltage above the bitline high level to increase the voltage written to the one transistor cell. The wordline high level is chosen to provide full write-back while staying within device reliability constraints. Overdriving the array cell allows the device designer to increase the array FET threshold voltage to achieve the low off-current required for data retention while maintaining a reasonable device performance. The physical implementation of the wordline driver is challenged by the aggressive wordline pitch (or periodicity). Each wordline must be decoded, level-shifted, and transmitted on minimum-pitch second-level wiring (M2). To minimize DRAM cell leakage, salicide used by the logic process to lower diffusion and polysilicon resistance must be blocked from the array. The resulting DRAM polysilicon wordlines are highly resistive and must be periodically connected to a parallel low-resistance M2 (or stitched) to meet the performance objectives. In the row decode system (see Fig. 5.5, floor plan), addresses (A3–11) are received, and true complement pairs are generated (Addr T/C) and simultaneously sent to the predecode system and redundancy compare circuits. The control block is activated with the bank select signal and enables address predecode and row latching. The redundancy circuits compare the current address with stored addresses of defective wordlines. If the incoming addresses match a stored defective address, the defective wordline is held inactive and a spare wordline is activated in its place. Redundancy enables repair of defective elements and improves yield.

5.3.7 Charge Sensing

There are a variety of mechanisms for sensing charge stored on a capacitor. Conventional VDD/2 sensing is reviewed for background, followed by the high-speed GND sensing scheme utilized by this work. The two schemes are contrasted, highlighting the benefits of GND sensing.

DRAMs operate on the principle of charge sharing, as shown in Fig. 5.10. By activating a wordline (WL), charge from the cell storage capacitor (NODE) is transferred to the bitline (BL), altering the potential of the bitline. The change in bitline potential is limited by the transfer ratio of the cell capacitance (C_{cell}) to the sum of the bitline capacitance (C_{BL}) and cell capacitance:

$$\mathrm{d}V = (V_{\mathrm{BL}} - V_{\mathrm{cell}}) * (C_{\mathrm{cell}}/(C_{\mathrm{BL}} + C_{\mathrm{cell}}))$$

For example, a transfer ratio of 1/5 and a bitline-to-cell voltage difference ($V_{\mathrm{BL}} - V_{\mathrm{cell}}$) of 600 mV would ideally result in an active bitline voltage change ($\mathrm{d}V$) of 120 mV. The most reliable, low-power means to amplify this small signal is with a cross-coupled differential sense amplifier. Operation of the amplifier involves precharging the true and complement bitlines (BT/BC) to an equivalent voltage (precharge level) with an equalize pulse (EQP), activating the select wordline (WL) to transfer charge from the cell to the bitline, creating a small signal difference between BT and BC (reading the cell), then driving the common source lines of the p-FETs (SETP) to V_{DD} and common source lines of the n FETs (SETN) to ground (GND). A multiplexor device, controlled by the MUX input signal,

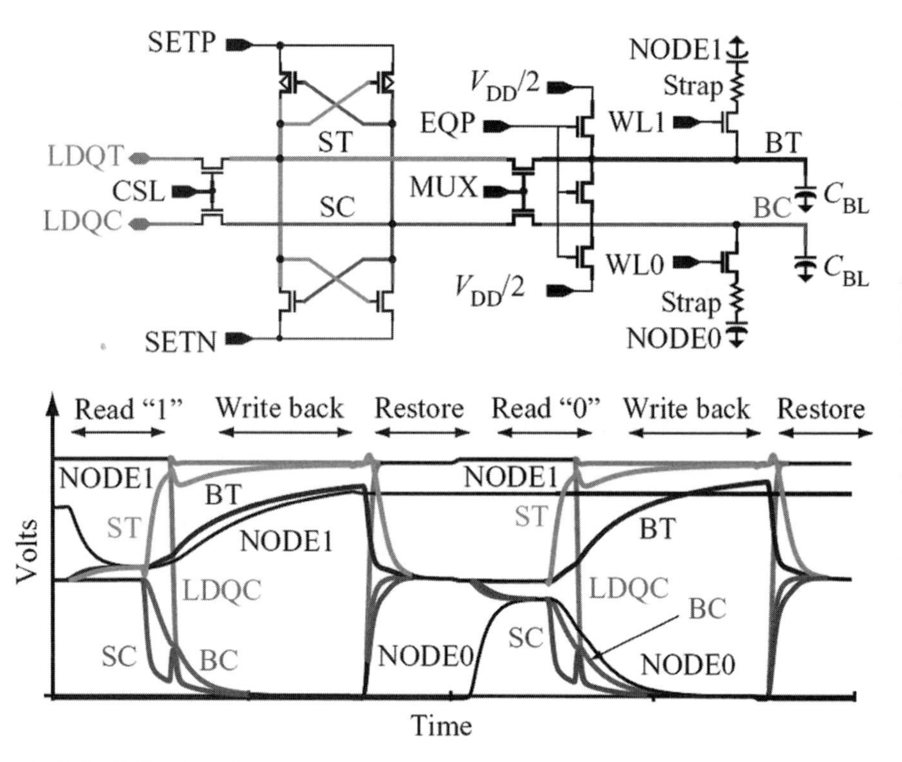

Fig. 5.10 Mid-level sensing

isolates the sense amp nodes (ST/SC) from the bitlines, allowing fast amplification. The bitline with the higher potential is driven to V_{DD} and the bitline with the lower potential is driven to GND. Once the sense amp has stabilized, the data can be transmitted off the macro through the local data lines (LDQT/LDQC) selected by the column-select line (CSL). Bitlines are then returned to their precharge level by reactivating the equalize signal (EQP).

With differential sense, the active bitline (connected to a selected cell) is compared to a reference bitline (not connected to the selected cell). Positive signal, as measured by the voltage difference between active bitline and reference bitline, is amplified to a logical "1." Negative signal is amplified to a logical "0." The reference bitline must be conditioned to allow the sense amp to reliably distinguish a low level stored in the cell from a high level stored in the cell. There are a variety of methods for preconditioning the reference bitline. Conventional DRAMs precondition both the active bitline and the reference bitline at $V_{DD}/2$. Reading a low level from the cell couples the active bitline below the reference, and a logical "0" is sensed. Reading a high level from the cell couples the active bitline above the reference bitline, and a logical "1" is sensed.

5.3.8 Precharge Level

For $V_{DD}/2$ precharge, the precharge level itself provides an excellent reference. Reading a high level from the selected cell moves the active bitline above $V_{DD}/2$, creating positive signal. Reading a low level from the selected cell moves the active bitline below $V_{DD}/2$, creating negative signal. An alternative to $V_{DD}/2$ preconditioning, shown in Fig. 5.11 is to precharge both the active and the reference bitline to GND. This scheme, however, cannot use the precharge level alone to provide a reference, because reading a low level from a cell does not move the active bitline ($V_{BL} = V_{cell} = 0$), resulting in zero signal and unpredictable amplifier operation. The most reliable means of generating a reference level for GND precharge is to condition the reference bitline with half charge from a reference cell (RN) activated by a reference wordline (RWL). This can easily be accomplished by writing $V_{DD}/2$ into a reference cell with an additional access device controlled by a reference equalize signal (REQP). When the selected wordline is activated, an associated reference wordline is activated, placing the half charge on the reference bitline and resulting in a level exactly between reading a high level and reading a low level.

GND precharge deviates from conventional $V_{DD}/2$ precharge, but offers a wider operating range, and the improved read and precharge performance required by the

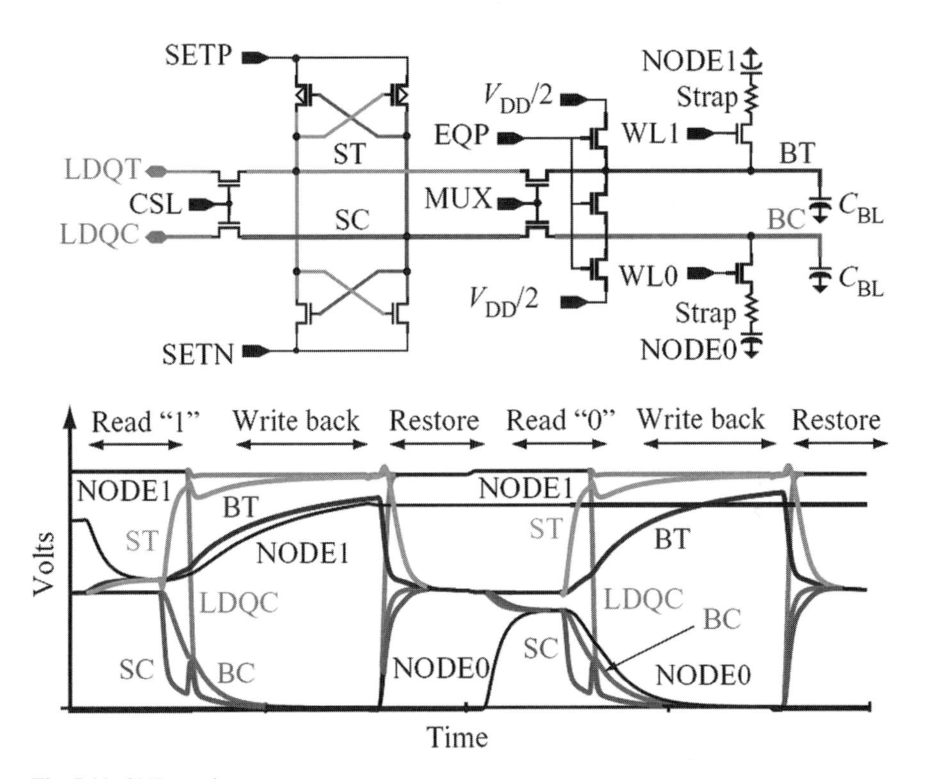

Fig. 5.11 GND sensing

ASIC environment. A difference to note is that GND precharge transfers charge to the active bitline only while reading a high level from the selected cell. Because there is no charge transfer when reading a low level from the active cell (cell and bitline are at the same potential), GND precharge relies on half charge transfer from the reference cell for reading a low level. In contrast, $V_{DD}/2$ sense schemes require charge transfer from the active cell for both a low and a high level for a read operation.

5.3.9 Cell Read Performance

When transferring a high level stored in a cell to the bitline, charge transfer does not start until the wordline is an array device threshold above the bitline. For $V_{DD}/2$, it takes more time for the wordline to reach $V_{DD}/2$ than the GND precharged scheme in which charge transfer begins when the wordline is a threshold above GND. This is critical for a longer wordline that will have a significant slew rate: a 1–V/ns slew introduces an extra 750 ps delay for $V_{DD}/2$ precharge (VDD = 1.5 V). Additionally, as a high level is read out of the cell in the GND precharge case, device overdrive (defined as $V_{gs} - V_t$) increases and is always $V_{DD}/2$ greater than the $V_{DD}/2$ precharge case, resulting in faster charge transfer. Although reading a zero in the $V_{DD}/2$ case begins when the wordline reaches a threshold above the cell (GND), it loses overdrive as the cell charges to the bitline $V_{DD}/2$ potential. As a result of wordline slew and overdrive, GND precharge develops signal faster than the $V_{DD}/2$ precharge.

5.3.10 Reference Cells

The reference cells and reference wordlines required by GND precharge increase core area; however, they provide many valuable features, including the following:

1. Bitline balance,
2. Equivalent WL-to-BL coupling for reference BL,
3. Lower sense amp operating point and more overdrive, avoiding stall at low voltage/low temperature,
4. Interlock signal to mimic circuit performance and generate sense amp timings,
5. SETN tied to GND, eliminating control and drive.

Implementation of reference cells provides both static and dynamic bitline balancing. Static bitline balancing is achieved by placing the equivalent capacitance of a reference cell on the reference bitline. Without a reference cell, the active bitline would see the extra capacitance of the storage capacitor of the selected cell, creating a 20% capacitance mismatch between bitline and reference bitline. Transient bitline balancing is accomplished by switching of the reference wordline, providing dynamic coupling to the reference bitline that is equivalent to the coupling from the selected wordline to the active bitline. The reference cell and reference wordline minimize the mismatch and coupling, allowing the sense system to operate on less stored charge, which enables increased performance and improved retention characteristics.

GND precharge simplifies sense amp control and provides faster amplification. With bitlines precharged to GND, gating the sense amp n-type cross-coupled pair source (SETN) is not required; consequently, SETN can be tied directly to GND, as shown in Fig. 5.11, GND sensing. In contrast, $V_{DD}/2$ precharge requires controlling both SETP and SETN. GND sense achieves faster amplification by applying full overdrive to the p-cross-coupled pair during sense amp set (SETP driven to V_{DD}). In the $V_{DD}/2$ precharge case, SETN is driven from $V_{DD}/2$ to GND and SETP is driven from $V_{DD}/2$ to V_{DD}, resulting in less overdrive, essentially splitting the overdrive between the n-cross-coupled pair and the p-cross-coupled pair. At low voltage, this may not be enough overdrive to turn on either the n or p devices, resulting in sense amp stall. GND sensing was chosen for providing increased overdrive to enable low-voltage operation and improved performance at nominal voltage operation.

5.3.11 Bitline Twisting

A key technology feature in achieving ASIC density and performance is back-end wiring characteristics, specifically metal pitch, resistance, and capacitance. Although a given design point may be good for logic, it does not necessary meet the ultralow-capacitance needs of the M1 bitline. DRAMs, which can tolerate higher-resistance bitlines, typically opt for metal half the thickness of the same generation logic process, resulting in less line-to-line capacitance. The higher line-to-line capacitance not only increases total bitline capacitance (and signal; see the section on charge sensing) but also increases the noise created by a neighboring bitline pair, introducing data-pattern dependence. This data-pattern sensitivity must then be included when testing for signal margin. To counteract line-to-line noise, bitline twisting was implemented. Bitline twisting uses a series of twists to distribute the coupling effect of neighboring noise equally into both the active bitline and reference bitline, converting the noise into common mode (see Fig. 5.12). Bitline twisting has been shown to decrease the minimum raw signal required from 100 to 30 mV, allowing the sense amp to operate at a faster cycle time and lower latency without suffering manufacturing yield loss.

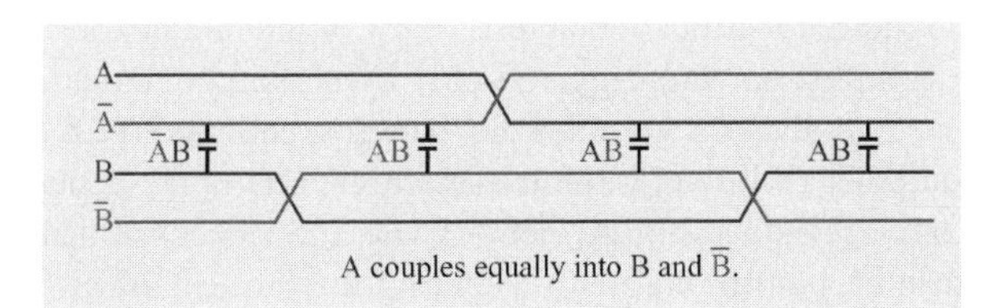

Fig. 5.12 Bitline twist

5.3.12 Data Path

Once the selected cell data has been captured by the sense amplifiers, it is ready to be shipped to the edge of the macro. The data path is responsible for selecting a subset of sense amplifiers, sending data to the edge of the macro and latching and then transmitting data to customer logic. A wide I/O data path is desirable in order to maximize core bandwidth. For the area efficiency described in the section on array core organization, DRAM cores typically access 2048 bits simultaneously and naturally offer very wide I/O. Because of the minimum pitch of the bitlines, it is unrealistic to drive all of the bits on a wordline off the macro. A column decoder is placed adjacent to the sense amps, selecting 256 of the 2048 bitlines and reducing the I/O by 8:1. This reduction provides one bit of data on approximately the same physical pitch as a 12-track logic cell. This feature reduces wiring congestion and facilitates bit-slice design within and outside the DRAM macro. Reducing the width from 2048 bits to 256 bits provides for enough wiring room to accommodate independent read and write data buses with ample tracks for power and noise shielding. Because of noise uncertainties in the surrounding ASIC environment and wide operating conditions (voltage and temperature), a robust, full-rail static data bus design was chosen. Dynamic data-line precharge schemes can offer higher performance, but typically burn more power and are less noise-tolerant. Separating the read and write paths eliminates design issues associated with bidirectional bus control and enables higher-frequency read-to-write performance. The timing specification in Fig. 5.6 shows a read cycle followed by a write cycle in which write data is latched at the same time read data from the previous cycle is being transmitted. With a bidirectional data bus, this would typically require one dead cycle to allow for bus turnaround.

Now that we have presented the DRAM macro operation and core, we can turn our attention to methods for guaranteeing the embedded DRAM macro functional specifications in an ASIC environment.

5.4 Test Strategy

As previously described, the cell arrays used in the embedded DRAM macro are very similar to those of their DRAM ancestors. We must assume that the cell arrays will have the same sensitivities as those well known from the development of DRAM, and require identification at test. Many of the interactions in the DRAM cell matrix are complex and are triggered only by certain combinations of defects and test patterns. To deliver the complex test patterns, commodity DRAMs use specialized test equipment with algorithmic pattern capability for generating the test sequences and large/fast-data-capture memory with redundancy allocation hardware for identifying and repairing faults. Considering how to test a DRAM embedded in logic creates a dilemma: logic tester or memory tester [16–18]. The logic test platform that has been developed for past generations of ASICs without embedded

DRAM can be characterized as a low-cost, reduced-pin-count tester with no algorithmic pattern generation or redundancy allocation hardware; it is therefore unable to test DRAM without assistance. The logic test patterns implemented are automatically generated with software based on the customer's netlist. The test strategy comes down to either a two-tester solution (memory tester and logic tester) or comprehensive BIST.

The two-tester approach suffers from the following issues: (1) multiple test gates are required, with an associated increase in wafer handling; (2) cumbersome requirements are placed on the customer to multiplex the macro I/O to package pins; and (3) the required part-number-specific test-pattern development is typically difficult to automate. In the high-part-number ASIC environment, it is essential to implement a single tester platform utilizing BIST for memory test and automated test generation for logic test.

5.4.1 BIST Engine Design Point

The use of BIST to test embedded SRAM in ASIC designs is not new. Much work since the late 1980s [19, 20] has proven this technique, which capitalizes on the similarities between the six-transistor SRAM memory cell and standard logic. The engine used was a state machine that simply ran through a predetermined sequence of standard patterns to uncover faults in either the memory array or the activation or decode support logic and determine pass/fail. Only in the most sophisticated cases was redundancy even considered, and then only in the row dimension.

Since the same set of six to eight SRAM test patterns have remained constant for a number of generations, little, if any, capability to alter the SRAM test sequence was required. However, DRAM is sensitive to far more subtle systematic and random process variations and defects, requiring a more complicated and dynamic test sequence to identify and repair these faults. It is generally not possible to predict all data patterns or write/read patterns that will ultimately be needed to reduce the test escape rate to less than 10 parts per million (10 ppm). The complicated dynamic pattern set and the multidimensional redundancy allocation (row and column) required by DRAM testing rendered earlier generations of BIST technology inadequate for DRAM testing. The new fail mechanism and pattern sensitivities specific to DRAM required reconsideration of the BIST engine dexterity. To address unique requirements of embedded DRAM test, a processor-based BIST engine with a high degree of program flexibility was developed [1]. A programmable BIST engine design provides a very flexible test pattern development system as well as a standard set of patterns for manufacturing test. This enables the test and characterization engineers to define the necessary patterns used at each test gate and the order in which they are applied. Test instructions that communicate with tester pin stimulation allow for external control of pauses used during retention-time testing and provide extended capability for exercises such as burn-in.

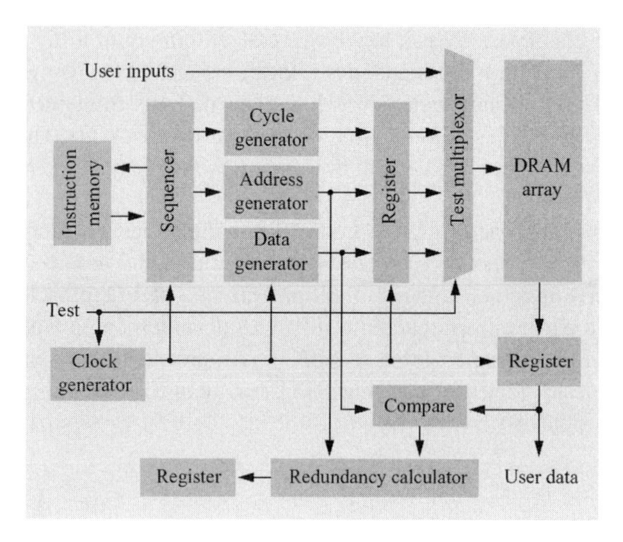

Fig. 5.13 BIST block diagram

The BIST can be subdivided into eight major components: test multiplexor, instruction memory, sequencer, address generator, data generator, cycle generator, redundancy calculator, and clock generator, as shown in Fig. 5.13.

5.4.1.1 Test Multiplexor

The test multiplexor, controlled by the test input, enables the BIST to take control of the DRAM macro. With the test signal deactivated, the DRAM operates normally, responding to the user inputs. These two modes are referred to, respectively, as test mode and mission mode.

5.4.1.2 Instruction Memory

The instruction memory is composed of a ROM and scannable read-only memory (SROM) and consists of 226 addressable words by 34 bits, with the SROM contributing 34 words. The ROM is programmed in the back end of the chip fabrication and contains instructions that represent predetermined patterns expected to be used during test of the embedded DRAM. The purpose of the SROM is to dynamically extend the programmability of the BIST. It allows new and unique patterns to be loaded into the program space in addition to the patterns hard-coded into the ROM. If a pattern is continuously used within the register array, it can be programmed into the ROM on a new metal mask release. There are also various utility instructions programmed into the ROM that allow the processor to reorder, reconfigure, eliminate, or add test patterns. The index addresses of the instruction memory are partitioned such that the processor can operate out of hard-coded ROM, SROM initialized at the start of test, or a combination of both.

The recent improvement made to this scheme consists of the addition of the capability to allow the SROM to be reloaded an unlimited number of times in any given test pass without upsetting the state of the rest of the BIST processor. This is especially important for the generation and preservation of cumulative redundancy calculation results over all patterns tested. The use of this feature eliminates concern over the choice of the size of ROM/SROM included in the design being adequate to contain all patterns necessary for a particular test gate. The SROM need now be only large enough to store the longest pattern, and the ROM can be sized to improve test time and efficiency.

5.4.1.3 Sequencer

The sequencer is the backbone of the BIST engine. It is responsible for the execution flow of the programs stored in the BIST instruction memory. It contains a program counter used to fetch an instruction from the instruction memory. Once an instruction has been fetched, the sequencer calculates the next address for the program counter based on the instruction's branch and branch operation code (opcode) fields.

Two types of instructions are used in the BIST. The first is a normal instruction indicated when bit 31 of the instruction word is set to a logical "0." The second type of instruction is a test-mode instruction, indicated when bit 31 of the instruction word is set to a logical "1." It is used to set various latches in the BIST. These latches control modes of operation, set digital-to-analog converts (DACs) to offset internal voltages and timings for margin testing, program the data scramble, and set the utility counter.

5.4.1.4 Address Generator

The address generator is responsible for generating the addressing for the array during self-test. This is done with the aid of three counters: row, column, and block, which are controlled by the appropriate fields of the BIST instruction. The counters can be incremented or decremented as defined by the increment/decrement bit of the instruction.

5.4.1.5 Data Generator

Because of the difficulty in predicting what actual data patterns will be necessary in the embedded DRAM array matrix to activate and expose subtle faults, a flexible data generator was introduced. The previous versions allowed only a short list of precoded patterns that provided blanket lows or highs, checkerboard lows/highs, row stripes, and column stripes to the matrix. This approach was found to be inadequate for proper test-pattern development and not directly reusable for varying array architectures. Thus, a design that accounts for current row and column address, adjacent data bit, and physical or logical data state stipulated in the pattern was added [7, 21]. The algorithm to be used by the data generator for a given test is specified with a mode-set instruction prior to executing any writes or reads to the DRAM. This instruction alters the states of latches in the design that govern the operation of the generator. Odd and even data bits are generated and replicated across the 256-bit data word. Odd and even data bits are a function of the program data scramble (PDS), the current address, and the data state specified in the read or write instruction. The PDS contains latches that are programmed with a test-mode instruction. These latches allow the test pattern to generate stripes of various sizes, checkerboards, or other desired patterns in the 1 Mb array(s).

The term logical data refers to the state of the data at the macro inputs, logical "1" being a high voltage level and logical "0" being a low voltage level. Logical data is simply written as is, regardless of the bitline connection to the sense amp and its location in the array. The term physical data refers to the voltage level stored on the DRAM cell capacitor. DRAM architectures routinely invert logical data when writing to the cell; that is, a logical "1" may be stored as a physical "0." Inversion is address dependent and is acceptable as long the architecture reinverts the data when it is read. Controlling physical data is required because many defect mechanisms are sensitive to the physical data state stored. When physical data is desired, the data is modified according to the bitline connection to either the true or complement side of the sense amp. Also taken into consideration is the location of the bit with respect to array topography. During a read or write operation, data is generated at the time of instruction execution. In a write operation, data is sent directly to the data-path latches. During a read operation, the data generator produces expected data for the comparator.

5.4.2 Test and Diagnostic Capability

In the ASIC environment, where the BIST has all of the capability and flexibility to test the DRAM with minimal tester interaction, the tester must still obtain the results from the BIST. Upon completion of a functional pattern, several types of data must be acquired from the BIST. Each of these data types has been considered in designing the BIST and the rules for integration of the DRAM/BIST core into an ASIC design system. The most frequently needed type of information is pass/fail data, which is supplied by three bits: one bit to flag successful completion of the BIST sequence; one bit to separate "perfect" from "not perfect;" one bit to separate "fixable" from "not fixable." Since the design system integration rules are set up in such a way that these bits are immediately accessible, overhead for the results-unload operation is minimized. Another data type to be considered is the repair data needed for fusing. The three pass/fail bits are the first out during a serial unload operation. If the part is fixable, the redundancy solution is unloaded from the BIST and passed on to the fusing tool for permanent redundancy implementation. If the part is either perfect or not fixable, no further BIST unloading is necessary. Skipping the redundancy unload step for perfect and non-fixable parts further reduces overhead. Another data type that deserves design-for-test attention is bit-map results.

The essential elements for BIST bitmapping are the ability to identify failing cycles and acquire the data-out states for the failing cycles. Enhancements to this minimum requirement can significantly reduce the effort required to create bit-maps. Specifically, the failing address is acquired directly, eliminating the need to create and query a cycle count to address cross-reference. The failing data-out state is flagged and positioned for easy acquisition from the BIST comparator, so the need to create and query a cycle count to expected data cross-reference is eliminated. Providing easy access to a synchronized set of address and compare states makes it possible to create bit-maps for any BIST pattern with a minimal amount of offline processing. In this scheme, BIST design-for-test efforts have had a significant positive impact on both test overhead costs and diagnostic capabilities.

In the early stages of embedded DRAM development, painful steps involving an iterative process of trial and error were taken to ensure the validity of the BIST patterns for manufacturing test. This process required the interactions of several individuals and many hours because of the lack of a methodology and an environment for test-pattern developers to use when coding and verifying patterns. Patterns were initially developed as binary vectors or strings of 1 s and 0 s that were loaded into the BIST scan chains. It was virtually impossible to ensure that the correct bits were set without sitting at a tester and debugging the vectors. This led to the development of a microcode for constructing ROM patterns and a software package to ensure the correctness of the binary vector and of the pattern developed.

The test-pattern development methodology involves three steps centered around logic simulation. These steps form a cyclical path of verification performed by a pattern developer. The first step involves writing the test-pattern file (TPF). Once the TPF is written, it is translated into a vector that can then be applied to the macro using a simulator. When the simulator has successfully completed, the final step involves verifying the results and extracting the initialization vector and expect vector required by manufacturing test. The final step required to produce a valid manufacturing test vector is to extract the data from the simulation graphics file. The data extractor takes as input the same scan-chain files used by the compiler and the graphics file generated from simulation. The extractor parses the graphics file and extracts the initialization vector and expect vector from a desired trigger provided as an argument. Once the data has been extracted, the vectors can be written to a file format appropriate for the tester platform to be used during manufacturing test.

5.5 Cycle Time Advances

As integration of DRAM into ASIC and foundry applications took hold, further pressure was being applied to close the performance gap with SRAM. Although DRAM page mode access and cycle times approach that of SRAM, page mode is not well suited for applications with random address patterns. Multi-bank interleaving offers a higher availability over page mode when presented with a random address sequence. To further improve bandwidth and availability the multi-bank interleave cycle and random bank (single bank) cycle time must be minimized. This work describes a 500 Mhz Compiled DRAM macro fabricated in 90 nm 1.2 V logic-based process and builds on material presented at ISSCC 2002 [22]. The random bank cycle was reduced 50% over the previous generation through segmentation and a direct write scheme. A 500 MHz operation was achieved with a configurable four-stage pipeline, reducing the multi-bank interleave cycle to 2 ns, which meets the random cycle of SRAMs offered in the same ASIC library.

5.5.1 Fast Cycle Architecture

Figures 5.14 and 5.15 show macro construction from either a 512 or 584 Kb local array kernel. The local array kernels, organized in 512 rows by 8 columns by 128 or 146 data bits, contain redundant row and data-bit elements, local 2.5 V row drivers, sense amps, and control logic. The 0.1848 μm^2 DRAM cell utilizes a 5.2 nm gate oxide array device coupled to a deep trench capacitor. Global kernels enable compilation in both row and data bit dimensions in 512 rows and 128/146 data bit increments. The global row and data bit kernels contain circuits for the hierarchical row, column, and data-path systems and include redundancy compare and steering circuits. Clock trees are dynamically grown to optimize each macro's performance. The largest macro (72 Mb – 33.72 mm^2) is configured as 8 K rows × 8 columns × 1168 data bits. The high degree of segmentation allows for higher single and multi-bank cycle times. An embedded DLL controls the read pipeline and write-back timings across voltage, temperature, and process variation. Built-in self test circuits were improved to support multiple built-in self repair (BISR) opportunities with eFuse and extensive bit fail mapping and diagnostic capabilities.

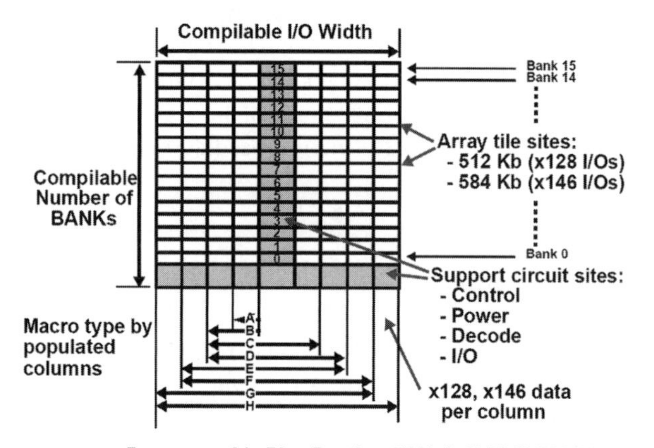

Fig. 5.14 Compilation

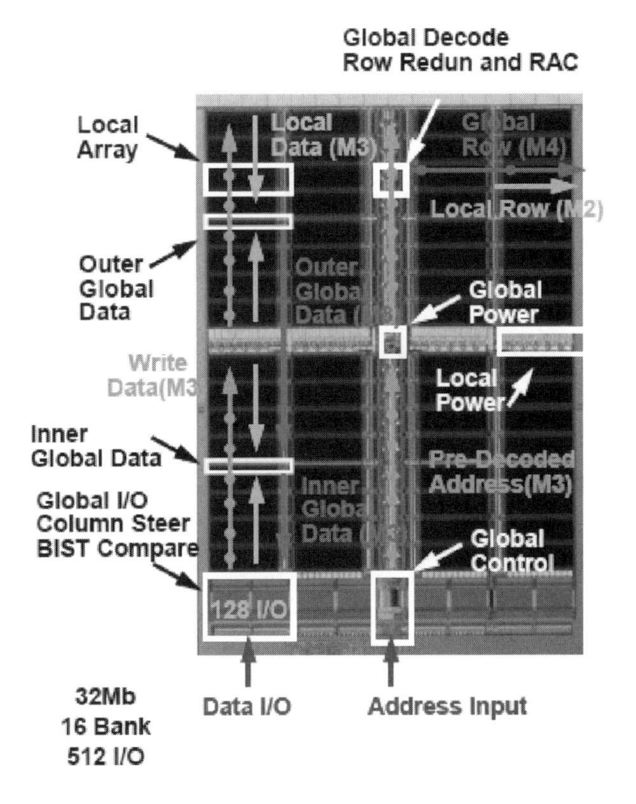

Fig. 5.15 Routing showing the details of address delivery from the macro pins to the local array kernel

Figure 5.15 shows the details of address delivery from the macro pins to the local array kernel. Addresses are received at the edge of the macro, predecoded and driven up the center of the macro. A single global row decoder is selected and drives 1 of 128 M4 master wordlines to the local array kernels shown horizontally in red. The local kernel level shifts the master wordline and performs the final 1 of 4 decode. One of 512 local wordlines, shown horizontally in light blue, are driven outward and periodically stitched with M2. Row redundancy and allocation logic used during BISR are also included in the global row decoder. Write data bits are received across the edge of the macro, and rebuffered on M3 to the local array kernels. Read data bits are gathered hierarchically, first on local data lines shown vertically in light blue, and then transferred to global data lines shown vertically in red. The read data is latched after passing through redundant data-bit steering, and driven of the macro on the next clock. Independent read and write paths are required for gapless 500 MHz multi-bank performance. Also included in the I/O kernel is the column redundancy allocation used during built-in self test and repair. Within a local kernel bitlines are twisted to cancel neighbor noise. Four interleaved sense amps share a local buffer to receiver write data and transmit read data as shown in Fig. 5.16.

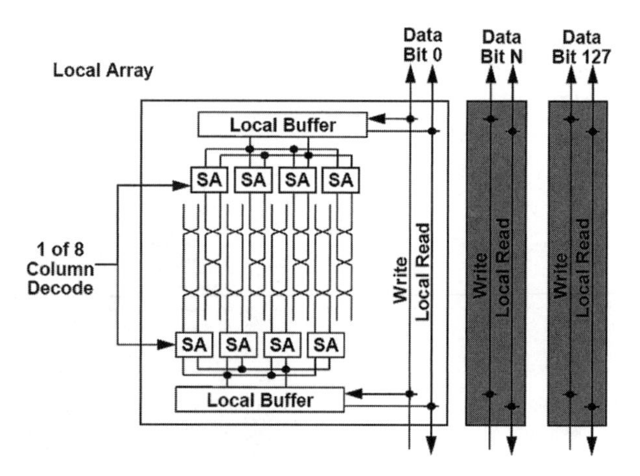

Fig. 5.16 Interleaved sense amplifier

5.5.2 Compilation

To begin compilation, it is perhaps inappropriate to speak in terms of growing an embedded DRAM. We begin with a maximum-sized DRAM and shrink it to the desired size. Across this broad range of potential sizes we find qualitative as well as quantitative changes. The DRAM topology already described is valid for the maximum-configured size, in both bank-depth and data-width. Reducing the bank-depth or data-width causes two different types of changes: simple deletion of circuit blocks and outright alteration of topology. Topological changes in the datawidth dimension are simple and easily contained, but those required in the bank-depth dimension are more complex, and require three different versions of the array support area.

The essential element of achieving two-dimensional programmability can be realized, without losing density or performance, through careful floor planning and by assigning specific functions to the intersection points. All of the major blocks are conventionally designed and laid out, as with previous generations of full-custom DRAM. Normally intersections where tiles of major blocks meet are handled by hand on an ad hoc basis, which is inconsistent with automatic growth/shrinkage. With careful thought and understanding of the design, it is possible to assign functions to these intersections that allow automation of macro construction. In particular, those functions that are required by a given topology need to be placed in blocks consistent with that topology. For the most part, this allows us to eliminate unused circuitry on smaller macros.

When there are eight or fewer banks, two global data buffer structures are no longer needed. Furthermore, the signals driven from the supports up into the array no longer need to be buffered at the midpoint. There is a data-out clock tree which is three-stages deep with more than eight banks, only two stages deep with three to eight banks, and simply driven for one or two banks. A set of CAS read timing signals are also buffered and given tree distribution for more than eight banks, just tree distributed for three to eight banks, and simply driven for one or two banks. Finally, DC test modes are buffered at each global data buffer when there are more than three banks. Such changes in topology are accommodated by having three separate schematics for the array section of the macro. One schematic covers 1 or 2 banks, one for 3 through 8 banks, and one for 9 through 16 banks. These schematics have an identical interface, and either smaller array can be substituted for the largest array. To produce any intermediate density, the next larger array is substituted, an appropriate number of arrays are pruned out, and empty space is compacted away.

Similar considerations are found in the data-width dimension. First, each tile may have 146 or 128 data bits. The master schematics are drawn with the larger tiles, and the smaller may be substituted with unused bits handled in a cleanup stage. There is also a clock tree issue in the data-width dimension, similar to that found in the bank-depth. Many (dozen+) signals are driven directly where one or two tiles are on one side of the spine or the other, but have a buffer tree for three or four tiles. This is accommodated by correctly selecting the fill cells that fit between the data-path tiles. Most of these fill cells contain miscellaneous circuitry, but between data-path tiles for words 2 and 3, and 5 and 6, extra buffers are added for those critical signals. When these alternate fill cells are used, the signals driven from the control circuitry are routed to a different set of wiring channels to get to the buffers. The outputs of the buffers then use the original wiring channels to deliver treed signals to the loads.

5.5.3 Direct Write

Before we can reduce cycle time we must first study the components of a DRAM's cycle. Figure 5.17 shows the three components of DRAM cycle: (1) precharge: wordline (WL) deactivation to WL activation; (2) signal development: WL activation to sense amplifier enable; (3) write-back: sense amplifier enable to WL deactivation. Precharge time, which includes wordline level shifting and bitline (BL) restore, was reduced with segmentation, improving wordline slew and enabling tighter wordline to BL precharge separation. Improved BL precharge was previously addressed by the ground sense scheme shown at ISSCC 2002 [23]. Further reductions in random cycle time must come from signal development and write-back, both a function of the DRAM cell's fundamental time constant (RC). Typically DRAM designs attempt to read 90% and write-back 90% of the cell's charge capacity, each operation requiring approximately 2.3× RC (90%). As the cycle time is pushed below the sum of the precharge plus 90% read and writeback (4.6× RC), the cell can no longer transfer 90% of the charge. In this situation, significantly more time is required to write opposite data than to read/refresh the previous state.

To address this write cycle limit, Taito [24] and Pilo [25], both introduce techniques for utilizing the full write-back interval by eliminating the sense amp enable to bit-switch delay; however, both are still limited by writing the opposite state to the cell. The direct write scheme extends the duration for writing new cell data to include both signal development and write-back time intervals, making refresh the new cycle limiter. To accomplish the direct write, data is placed onto the BL before

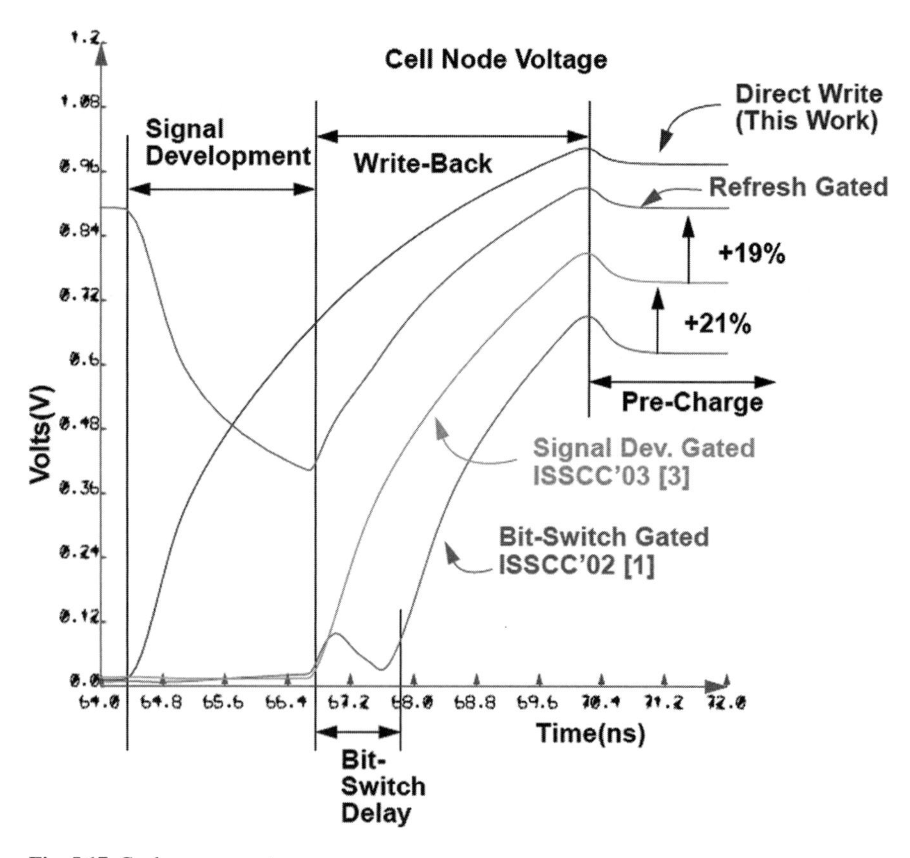

Fig. 5.17 Cycle components

the wordline is activated, relying on BL twisting [22] to cancel noise coupled onto adjacent BLs.

For sense amps with a common set node, writing the BLs prior to amplification, will feedback through the X-couple PFETs and adversely pre-amplify neighboring sense amps as shown in Fig. 5.18. To prevent pre-amplification, the set node for each amplifier must be isolated, requiring a dedicated PFET set device shown in Fig. 5.19. To allow write data to flow from the write buffer (Fig. 5.20) directly to the selected BL, the bit-switch must also be opened prior to wordline activation. Because the bit-switch gates are shared across the local kernel, pre-selection could adversely increase the BL load for data bits which are masked during the write (DRAM reads are destructive, a masked bit must be sensed and written back). Extra load on the BL during a masked write will degrade the transfer ratio and BL signal. To minimize this loading effect, a PFET bit-switch was utilized (see "PFET Bit Switch" in Fig. 5.19, direct write sense amp). With BLs precharged to GND, the load from the read circuitry is not seen by the BL during signal development, even though the bit-switch gate is selected (V_{gs}=0, V_{ds}=0). In fact the device does not turn on until

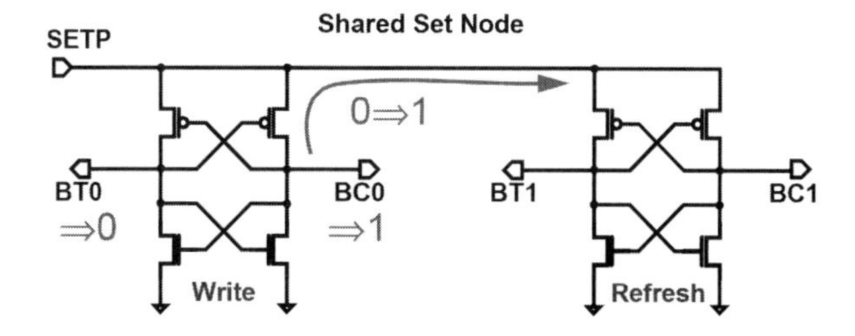

Fig. 5.18 Premature amplification

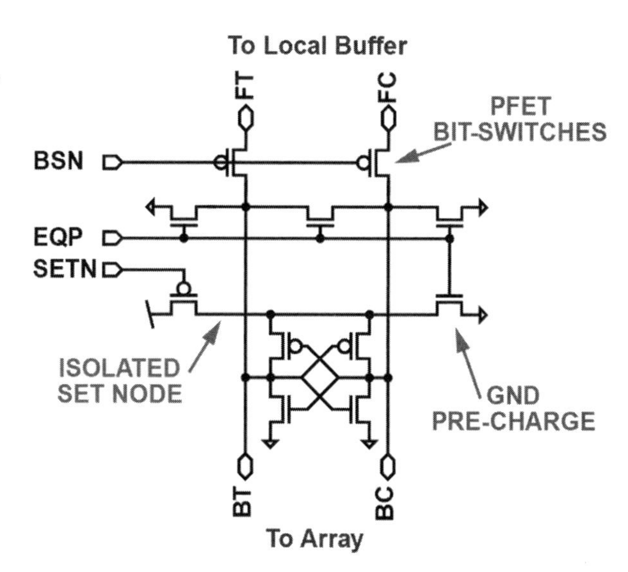

Fig. 5.19 "PFET Bit Switch" in direct write sense amplifier

after the sense amp sets and the BL rise a PFET Vth above ground as shown in Fig. 5.21. As an added benefit, pre-selecting the bit-switch improves read performance by allowing the circuit to pass a full high level from the sense amp to the read circuit as soon as the sense amp sets. Utilizing a PFET bit-switch enables layout efficiencies, allowing the bit-switch to be placed adjacent to the cross-coupled amplifier PFETs, which in turn are adjacent to the dedicated PFET set device. This minimizes overlay effects by bounding BL nodes with poly silicon, reducing sense amp mismatch. The area impact of the direct write was estimated to be less than 1.5%

5.5.4 Pipelining

To support 500 Mhz pipeline operation shown in Fig. 5.22, a precharged, predecoded address bus was chosen, minimizing skew, switching noise, and power. Addresses are predecoded in 2 bit groups during clock buffering, balancing

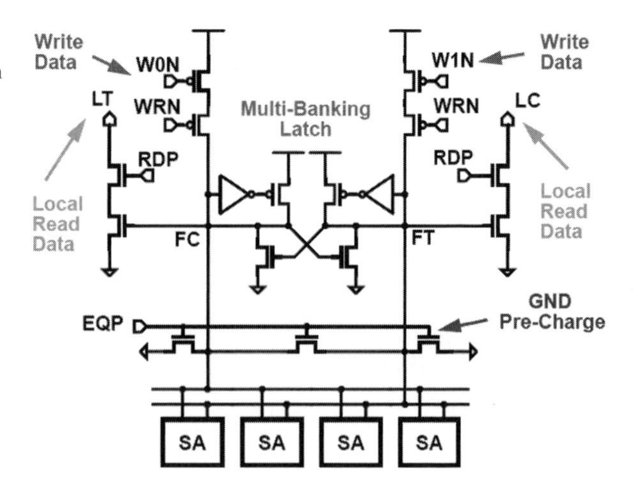

Fig. 5.20 Local buffer to allow write data to flow from the write buffer

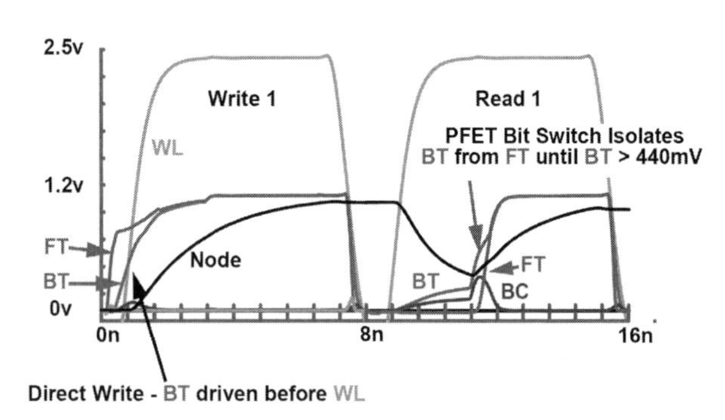

Fig. 5.21 Direct write simulation

setup/hold times. The predecode outputs are precharged to enable forward path optimization and remote self timing. Local refresh Address counters in each bank improves availability by allowing user control of bank address during refresh cycles. Local bank select pipe counters can be configured to support a 1, 2, or 4 stage pipeline for 8, 4, and 2 ns clocks, respectively, all maintaining the 8 ns random bank cycle time. To meet the 2 ns data-path cycle, a local bank address pipe stage was inserted to capture the read bank address, 1 cycle early. The read clock is controlled with a DLL to ensure consistent delivery of data across voltage, temperature, and process. The hierarchical data path was broken into local and global segments, the local segment utilizing precharged, self resetting circuitry. To stabilize the power supply and reduce possibility of malfunction due to switching noise, extensive deep trench decoupling capacitors were throughout the design, with each array kernel providing 0.845 nF of decoupling. Deep trench cap are also offered as an ASIC library element and offer at least 4× more capacitance per unit area than planar caps.

To support column redundancy with a wide I/O interface (×256), it is more efficient to implement redundancy at the data-bit level, as shown in ISSCC 2002 [22],

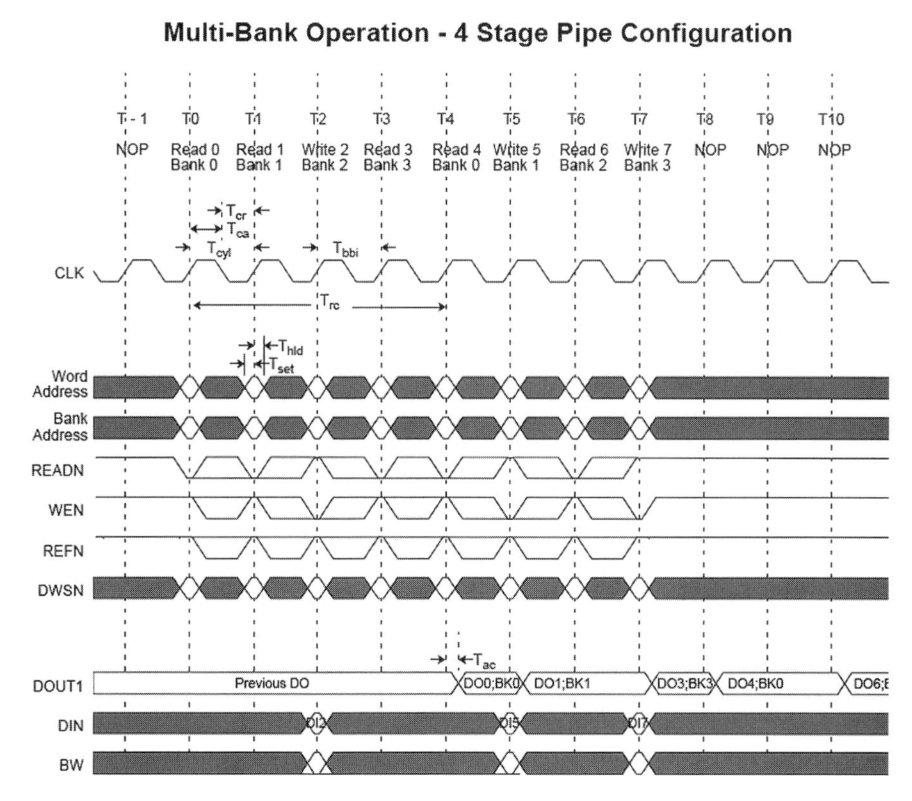

Fig. 5.22 Pipeline timing

in contrast to conventional bitline replacement which occurs at the column decoder. Previous shifting designs used the row activation time to setup the column shifting, however, the multi-banking column redundancy solution must switch at the faster multi-banking rate and therefore must set up in a fraction of the time. For speed, the repair solutions are stored in local latches (one per data bit per bank) and are rapidly selected by the incoming bank address in time to set shifting switches prior to driving write data to the subarray (see Fig. 5.23). A bank's column solution is stored in the local latches as string of 0 s and 1s, with a 1 indicating the failing location.

In order to support direct write and multi-banking, the data steering required additional pipelining. The complexity is introduced by direct write, requiring the data to arrive at the local buffer prior to wordline rise. A bank's column redundancy solution must be applied to a first set of write steering switches shown in Fig. 5.23, steering, delivering write data to the array kernels prior to WL activation. The solution is then piped according to the number of configured stages, to a second set of read steering switches, time aligning the solution with the bank's read data. The write steering solution must be valid in cycle 0, independent of pipe depth.

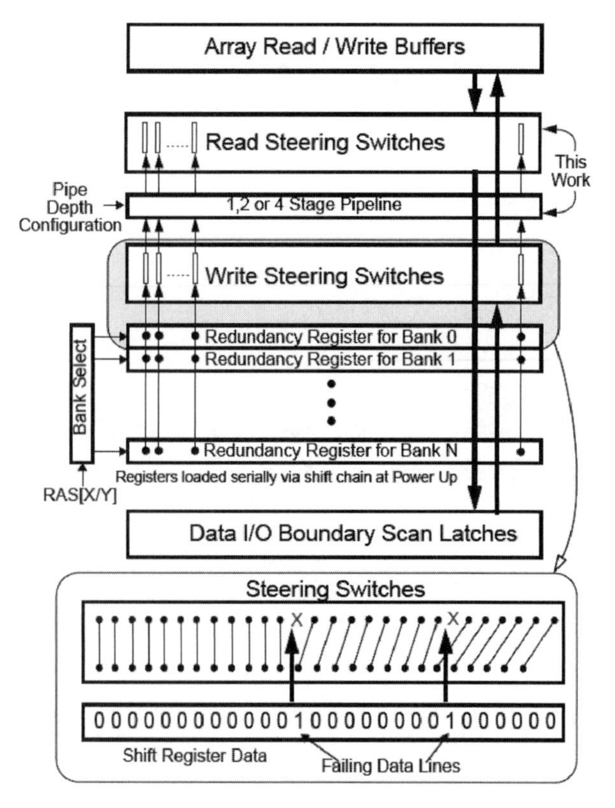

Fig. 5.23 Steering

The arrival of read data from the array, however, is a function of pipe depth. Read latency to the steering is three cycles for the four-stage pipe mode, one cycle for the two-stage pipe mode, and zero cycles for the one-stage pipe mode. To manage the cycle delay for two- and four-stage pipe modes, a variable stage pipeline for the read solution was added, with the depth of the pipe controlled by the pipe mode.

5.5.5 BIST Enhancements

The repair solution is stored locally in a redundancy register. The redundancy register is loaded at power-up via a serial shift chain from the remote eFuse compression/decompression circuits. The compression/decompression reduces the number of fuses required to store the format needed by the steering. Remote eFuse compression was described at ITC 2002 [26].

Electronic fusing replaces laser fusing used in prior ASIC technologies. The electronic fusing system consists of multiple embedded macros that all work together to create a full BISR system and methodology. The key macros involved in the BISR test flow are the fuse control macro (FCM), redundancy fuse macros, and

BIST/RAL circuitry in the RAMs. The manufacturing test flow for memory repair is arbitrated via the FCM which enables the RAM BISTs. First the FCM is clocked with initialization instructions that set it up for that specific chip. This allows any chip configuration to be used with a single FCM design. The FCM initialization allows it to determine the number of RAMs and eFuses available on the device for controlling clock counts and shift register lengths. After initialization, the RAM BIST engines are run and failing addresses are stored in the macro's local redundancy registers. When BIST is complete, the FCM clocks the redundancy registers to unload and compress the repair solutions. When compression has completed, a high-voltage supply is asserted and fuses are programmed in a serial fashion.

There are three uses for the redundancy eFuse macros: primary, secondary, and tertiary. They are personalized by how they are connected to the fuse controller and provide for up to three repair passes. The majority of the fuses and repair actions are used during the primary repair pass. Due to this, the majority of the eFuse macros are assigned as primary repair macros with far fewer assigned to secondary and tertiary repairs. Primary repair is done at wafer and captures all static (stuck-at) fails and any dynamic fails that show up at the wafer test conditions. Secondary repair is available at wafer or module after primary repair and decompression is complete.

Secondary repair is accomplished by re-running BIST with the primary solution loaded into the redundancy registers. New fails are added to the preexisting primary repair data, creating a new repair solution. These new fails are often due to testing at a different temperature or after module build and/or stress. At the completion of secondary BIST testing, the FCM begins secondary compression. The primary solution is decompressed from the fuse macros and compared to the updated repair data in the redundancy registers. The delta (bitwise XOR) is compressed into the secondary fuse macros and then fused. Upon decompression, the primary and secondary fuse solutions are recombined (bitwise XOR) to reproduce the final repair solution. Tertiary fusing is performed in a similar fashion storing a delta from the primary and secondary repairs.

Operation of the fuse controller is set up via serially loaded opcodes (instructions for the state machine similar to the software-controlled BIST). Each opcode designates a specific operation that will be performed by the fuse controller, from accessing the redundancy fuse macros, arbiJSSC trating BIST tests, or commands to the fuse control macro itself. The programmability of the fuse controller allows for software adjustment of the BISR flow and the addition or removal of diagnostic operations as needed.

5.6 On Processor Embedded DRAM Cache

5.6.1 Current Level of Integration

Logic-based embedded DRAM has matured into a wide range of applications (Fig. 5.24), most notably the on-chip cache for IBM's BlueGene/L supercomputer

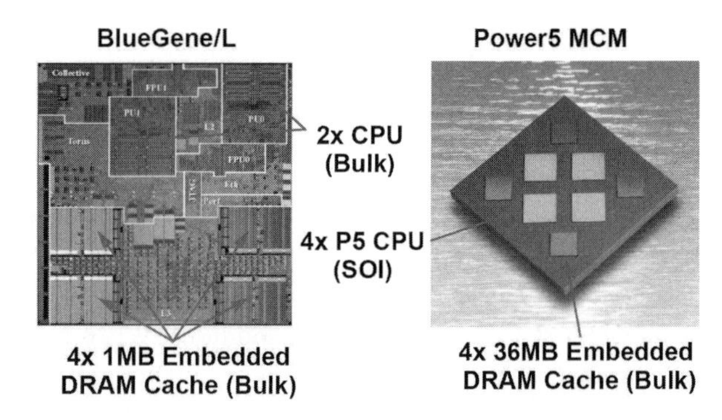

Fig. 5.24 Current eDRAM applications

[27], currently ranked #1 on the top 500 supercomputer list [28]. Integration of embedded DRAM for caching enabled BlueGene/L to achieve world-class performance, while utilizing only a modestly performing bulk processor technology. IBM's dual-core Power5 microprocessor [29], in contrast, leverages SOI technology to achieve multi-GHz CPU frequencies, and exploits bulk technology embedded DRAM to provide a massive single-chip 36 MByte external L3 cache. Imagine the performance attainable if embedded DRAM were available on-chip with the high-performance SOI microprocessor. In order to achieve this next level of integration, two requirements must be met: first, embedded DRAM technology must be integrated on the same high-performance SOI technology platform currently utilized by microprocessors; second, DRAM performance must be enhanced to service the aggressive latency requirements of multi-GHz processors.

5.6.2 SOI Embedded DRAM Technology Features

To satisfy the first requirement, DRAM has been embedded on a 65 nm, partially depleted SOI, high-performance technology platform (Fig. 5.25) [30]. The technology offers both 1.12 nm thin oxide and 2.35 nm thick oxide devices. The DRAM cell is fabricated on SOI and utilizes the technology's standard 2.35 nm thick oxide floating body NFET. A silicide strap forms a low-resistance connection to a 20 fF deep trench capacitor. High-trench capacitance is achieved by digging through the SOI and buried oxide (BOX), 4.5 μm into the substrate. Conveniently, the buried oxide provides isolation between the SOI and substrate, eliminating the oxide collar required by bulk technologies to protect against vertical parasitic leakage. The details of this mechanism are described in the following section.

5.6.3 Collar Process Elimination

Figure 5.26 shows the bulk and SOI embedded DRAM process cross sections. Bulk trench technologies must deal with parasitic leakage created by the vertical device formed between the node and N-Band, gated by the trench sidewall. To mitigate this

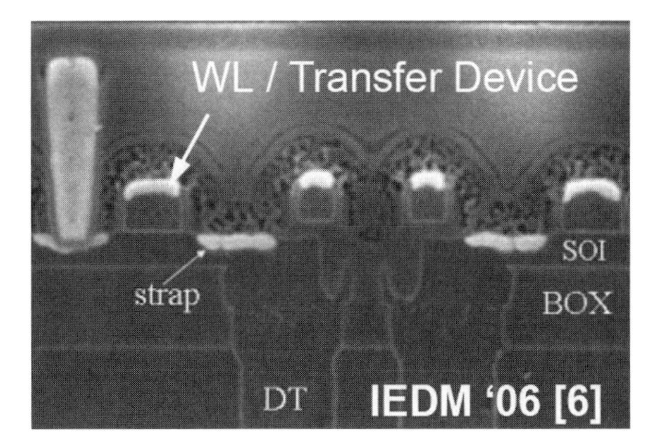

Fig. 5.25 SOI DRAM SEM

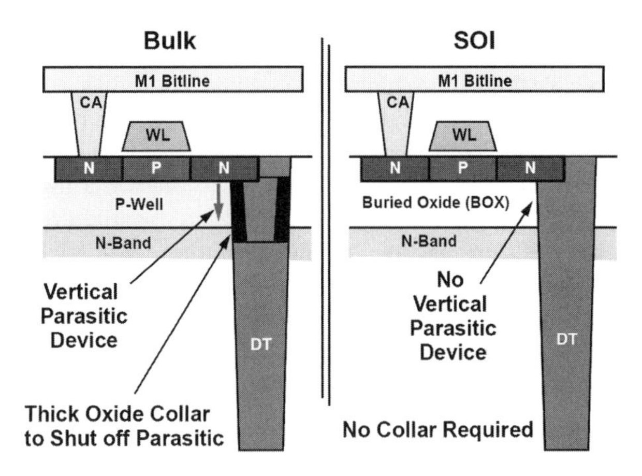

Fig. 5.26 SOI versus bulk DRAM

leakage, bulk technologies must form a thick oxide collar at the top of the trench. In SOI, the bulk P-Well is replaced by the buried oxide, isolating the node from the N-Band, and precluding the need for a collar. Eliminating the collar not only reduces process complexity but also increases the cross-sectional area of the trench, further reducing parasitic resistance to the storage capacitor. These factors combine to drop the overall eDRAM process cost adder from 15% for bulk to 7% for SOI. In addition, the absence of wells in SOI realizes lower source-drain capacitances and enables layout area reduction by eliminating well tie-downs and allowing smaller NFET to PFET spacing.

5.6.4 Floating Body Effects

Integrating a DRAM cell into SOI technology does, however, introduce a concern: without a well, the body of the eDRAM cell's SOI device is floating. The potential of the body is determined by a number of mechanisms. Primarily, these mechanisms are coupling, junction leakage, impact ionization, thermal generation, and

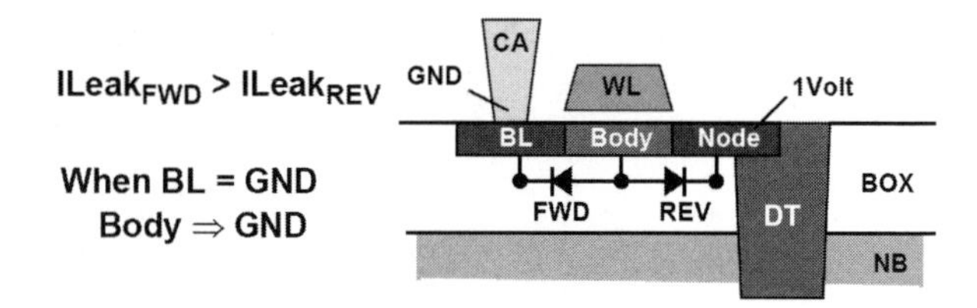

Fig. 5.27 SOI body leakage

recombination. Some of these can be exploited, others must be carefully managed. During DRAM cell write-back, the array's wordline and bitline are both driven high, coupling the body of the transfer device high, acting to lower its threshold, and enabling a higher voltage to be written into the cell. When write-back is complete, the wordline is lowered, trapping charge on the storage capacitor. At this time, leakage mechanisms take over, specifically junction leakage at the forward and reverse bias diodes formed between the body and source–drain regions (Fig. 5.27). If the device's body floats high, its V_t will be reduced, degrading the cell's retention characteristics due to subthreshold leakage. The high body potential also increases vulnerability to bipolar current when the source is quickly discharged.

For these reasons, it is desirable to prevent a high body situation on the cell access device. Recognizing that a forward-biased junction has higher leakage than a reverse-biased junction, holding the bitline low will cause the body to float closer to ground, thus reducing subthreshold leakage. To realize this scenario, bitline ground precharge is implemented to maintain a low body potential. Additionally, periods of bitline high are limited by the short page depth associated with this embedded DRAM's wide I/O design.

5.6.5 Array Floating Body

Figure 5.28 shows two graphs of net body charge, with coupling effects removed. The top graph shows the weak reverse bias leakage slowly charging the floating body high on the eDRAM cell's NFET. This would occur for a design like that of commodity DRAM, where the bitline is kept high for a long period of time, to keep a "page" of data open. When the bitline is lowered, high subthreshold leakage and bipolar current then rapidly drain charge stored in the DRAM cell. The stronger forward bias leakage can also be seen discharging the body. The bottom graph in Fig. 5.28, SOI body charge, shows frequent bitline grounding, as used in this chapter's design; this maintains the cell device's body charge desirably near ground.

5.6.6 SOI Macro Architecture

Architecturally the macro is divided into upper and lower halves, each half consisting of four 292 Kbit subarrays (Fig. 5.29). In the central region, addresses are

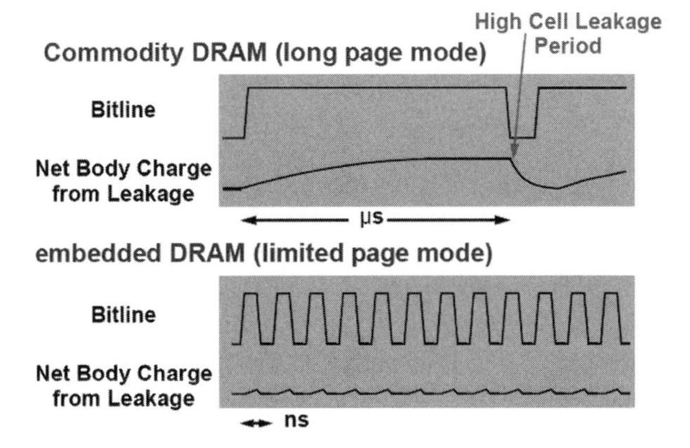

Fig. 5.28 SOI body charge

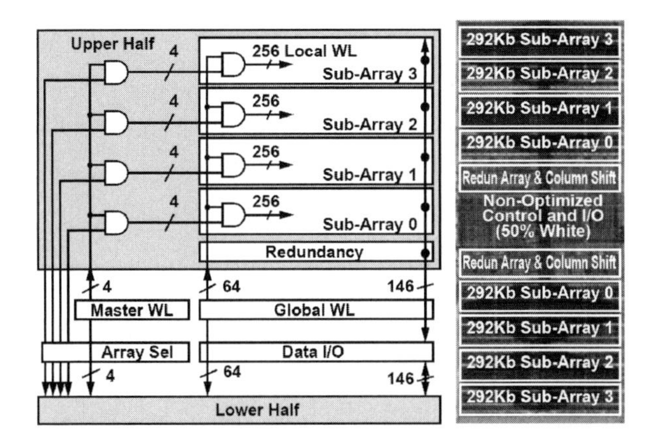

Fig. 5.29 SOI macro architecture

decoded into array selects, master wordlines, and global wordlines, which are delivered over the subarrays on Metal 4. Locally, master wordlines are enabled by a single array select and combined with the global wordlines to select 1 of 256 local wordlines. One-hot late-select column-select signals support 16-way cache associativity. Eight are driven to the upper array, eight to the lower array, selecting 1 of 16 columns. The central region also includes control logic and the 146 bit data I/O. It should be noted that this region reused an existing SRAM I/O design and is not area optimized.

5.6.7 Pyramid Row

Figure 5.30 shows the unconventional orthogonal location of the wordline driver as introduced by Weiss et al., at ISSCC 2002 [31]. Metal 4 is used in a pyramid-like pattern to jumper the orthogonal wordline driver to the east/west Metal 3 wordline strap. The pyramid pattern, easily distinguishable in the micrograph on the right side

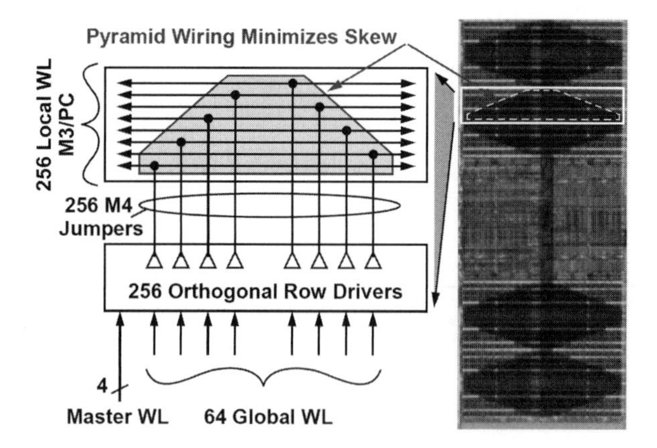

Fig. 5.30 SOI orthogonal row driver

of Fig. 5.30, minimizes wordline skew by using the shortest Metal 4 jumper on the longest Metal 3 runs. The orthogonal wordline configuration saves area by decoupling the wordline driver layout from the array wordline pitch and avoids replicating wordline predecode in each subarray.

5.6.8 Short Bitline

To satisfy the second requirement for on-chip embedded DRAM caches, macro performance must be improved to meet the latency requirements of a multi-GHz microprocessor. Where area overhead is not a concern, reducing bitline length is an accepted method for improving DRAM performance. Figure 5.31 compares two different eDRAM array designs: one with 128 cells per bitline, and a second with 32 cells per bitline. This figure highlights three performance gains: first, the lower bitline capacitance of the 32 cell bitline results in a reduced read time constant and

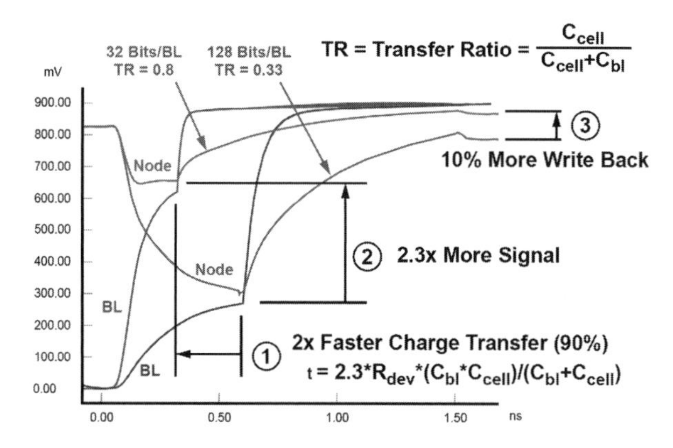

Fig. 5.31 Charge transfer

faster signal development; second, the high transfer ratio of the short bitline design creates a large bitline voltage swings, providing more signal; and third, reduced cell charge transfer during a read minimizes charging time to replenish the cell during write-back.

5.6.9 Overhead

Short bitline performance is attractive, but must be weighed against increased area overhead. Each time a bitline is cut in half, sense circuits and data buffers must be doubled. The high-performance direct write embedded DRAM presented at ISSCC 2005 [7] requires 11 transistors for each sense amp, plus reference cells and bitline twisting, totaling a 14% area overhead. A 128 bits per bitline version would suffer 27% overhead; extrapolating to 32 cells per bitline would result in an unacceptable 80% overhead. To make matters worse, implementing the sense amp design of Ref. [23] in SOI would require further overhead for body-tied SOI devices, required to prevent history-induced sense amp mismatch. To extract overall benefit from a high cell to bitline transfer ratio, a low overhead sensing scheme is required.

5.6.10 Micro Sense Amplifier (μSA)

Figure 5.32 proposes a three-transistor scheme to minimize sense amplifier overhead. The hierarchical scheme relies on the high transfer ratio during a read to create a large voltage swing on a local bitline; large enough, in fact, to be sampled with a single-ended amplifier. The amplifier has been minimized by decomposing its three basic operations – read, write "0" and write "1" – each being implemented with a single transistor. Each micro sense amplifier services a short 32 cell local bitline, labeled LBL in Fig. 5.32. The single-ended nature of this sensing scheme allows a

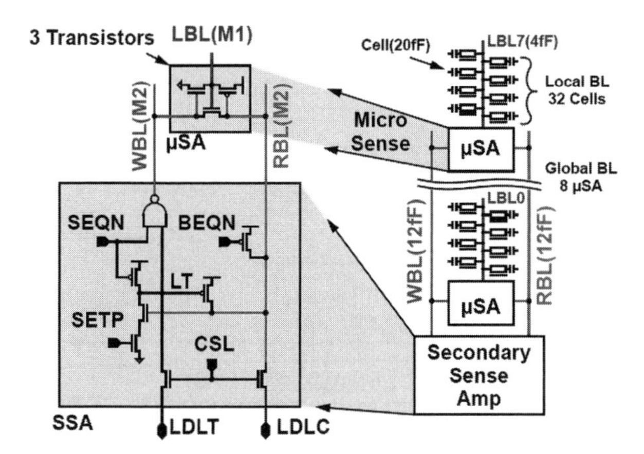

Fig. 5.32 Micro sense amplfier

relaxed Metal 1 pitch, reducing the line-to-line capacitance by a factor of three. The short bitline, relaxed pitch, and low SOI junction capacitance combine for a local bitline capacitance of only 4 fF. When coupled with a 20 fF cell, an 83% transfer ratio is realized, ideally producing over 800 mV of signal – more than enough to turn on a 250 mV V_t transistor.

Each three-transistor micro sense amplifier is configured with an NFET for reading, a PFET for writing a full one, and an NFET for writing a full zero. The micro sense amplifier transfers data to/from a secondary sense amplifier (SSA) via two global read/write bitlines, labeled RBL and WBL. The Metal 2 global bitlines, routed in parallel to the Metal 1 local bitlines, control the read/write operations to all micro sense amplifiers in parallel. Each secondary sense amplifier services eight micro sense amplifiers in a hierarchical fashion.

5.6.11 Tertiary Sense Amplifier

The micro sense amplifier adds an extra level of hierarchy and necessitates a third-level (tertiary) sense amplifier. The bidirectional tertiary sense amplifier shown in Fig. 5.33 is responsible for transferring data between the Metal 4 global datalines and the selected secondary sense amplifier. One of eight secondary sense amplifiers is selected by the column address, which controls the one-hot column select lines (CSL). In a set-associative cache, the column address could be used as a late way select.

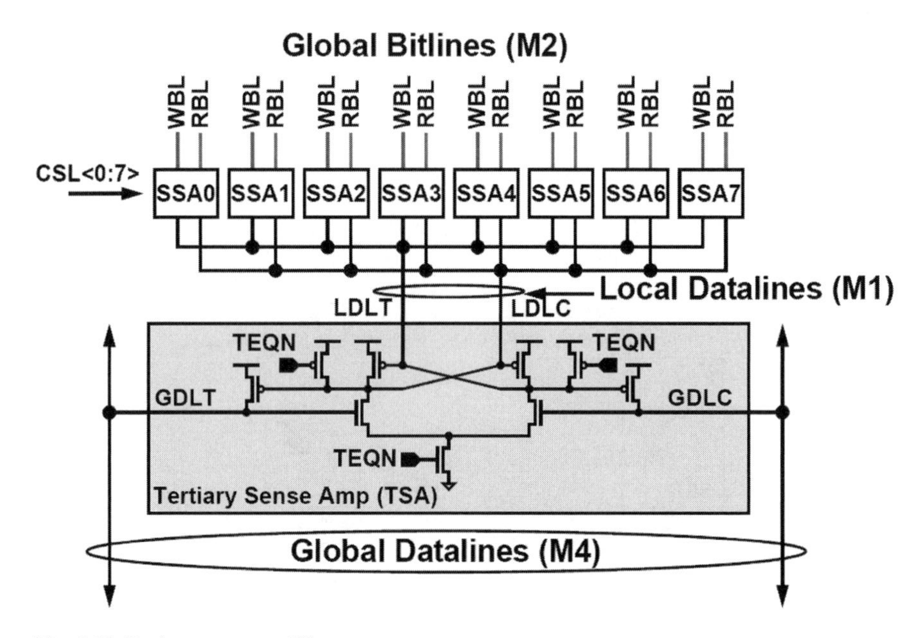

Fig. 5.33 Tertiary sense amplifier

During a write, either the true or complement global dataline is remotely driven high. In turn, the tertiary sense amplifier pulls either the true or complement local dataline low, which is then passed to the selected secondary sense amplifier. During a read, the global data lines are left in the low precharge state until the selected secondary sense amplifier pulls either of the local datalines low. The data passes through the tertiary sense amplifier and pulls the corresponding global dataline high. It should be noted that a read or write to the subarray is determined by the timing of the global datalines and does not require dedicated read/write control signals.

5.6.12 Operation Truth Table

Table 1 shows the logical operation of the micro sense amplifier. All cycles start in the precharge condition with read and write global bitlines held high, precharging the local bitline to ground. Subarray selection triggers the sense amplifier equalize signal SEQN to go high (Fig. 5.32), forcing the write bitline low. This enables the micro sense amplifier read transistor, and tri-states the local bitline, indicated by a "Z" in the truth table. A write "1" is achieved by driving the read bitline low – as indicated by the arrows in Table 1 – while a write "0" is achieved by driving the write bitline high, indicated by the arrows. Logically, the local bitline remains low in the read "0" state and only transitions high in a read "1" state, as indicated by the arrow.

5.6.13 Write Waveforms

Figure 5.34 shows a simulation of write operation using the micro sense amplifier. To write a "1," the selected secondary sense amplifier pulls the read bitline low, forcing the local bitline high, writing a "1" into the selected cell. To write a "0," the selected secondary sense amplifier drives both the read bitline and write bitline

Table 1 Operational Truth Table

Operation	WBL	RBL	LBL
Pre-Charge	1	1	0
Write '1'	0	0	1
Write '0'	1	1	0
			SETP
Read '1'	0	1Z	1Z
Refresh '1'	0	0	1
Read '0'	0	1Z	0Z
Refresh '0'	1	1	0

Z = Hi Impedance SETP

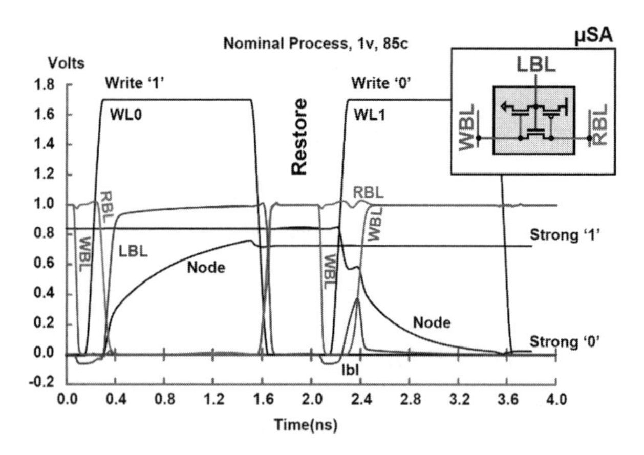

Fig. 5.34 Micro sense write simulation

high, in turn driving the local bitline low and writing a "0" into the selected cell. It should be noted that due to the micro sense amplifier's unique configuration, writing opposite data onto the local bitline is accomplished without contention, eliminating any technology P to N ratio dependence.

5.6.14 Read Waveforms

The process of reading using the micro sense amplifier is also relatively straightforward, but contains one unexpected component. As shown in Fig. 5.35, the micro sense amplifier is first taken out of precharge by lowering the write bitline. The read bitline, however, is held in the precharge state until an interlock circuit detects the write bitline is low. This extended precharge absorbs any line-to-line coupling from

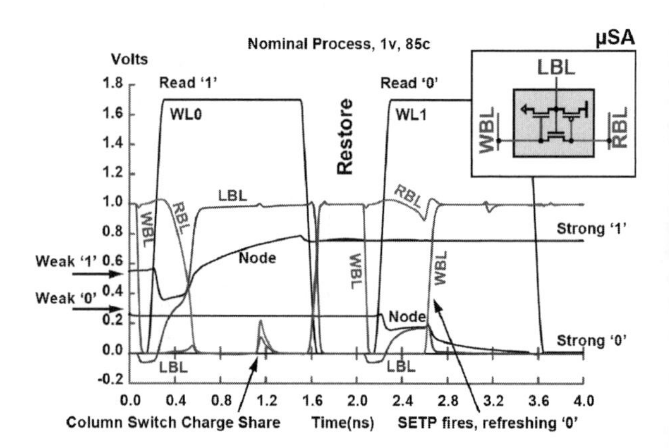

Fig. 5.35 Micro sense read simulation

the write bitline to the read bitline, preventing any premature amplification. When a wordline is activated, read data is transferred from the cell to the local bitline. For a stored "1," the local bitline rises at least 1 threshold above the read transistor, weakly pulling the read bitline low. Here is the novel component of this operation: when the read bitline falls below the threshold of the micro sense amplifier PMOS, feedback drives the local bitline to a full high level. This amplifies the local bitline, refreshes the cell, and strongly drives the read bitline low, which can be seen as a slope change on both the local bitline and the read bitline. It should be noted that the refresh of a "1" is self-timed, and requires no external timing control.

For a stored "0," the local bitline and cell are very close in potential, therefore the local bitline remains low, leaving the read transistor off and the read bitline high. The read bitline remains high until the external timing signal SETP triggers the secondary sense amplifier to evaluate the read bitline. With the read bitline still in the high state, the secondary sense amplifier drives the write bitline high, forcing the local bitline low, refreshing the "0."

5.6.15 Cycle Limits

Historically, the cycle time of NFET DRAM arrays are determined by writing a high level into the cell. Figure 5.36 shows the evolution of embedded DRAM architectural advances aimed at reducing cycle time. These advances include the sense-synchronized write [24, 25] and direct write [23] described at ISSCC 2003 and 2004, respectively. Similar to prior work, this chapter's tertiary, secondary, and micro sense amplifiers are configured to be bidirectional and support direct write of data prior to wordline activation, without disrupting reads of adjacent sense amplifiers and data bits. By significantly improving the refresh cycle time, the micro sense architecture reduces cycle time down to the cell's fundamental ability to write opposite data.

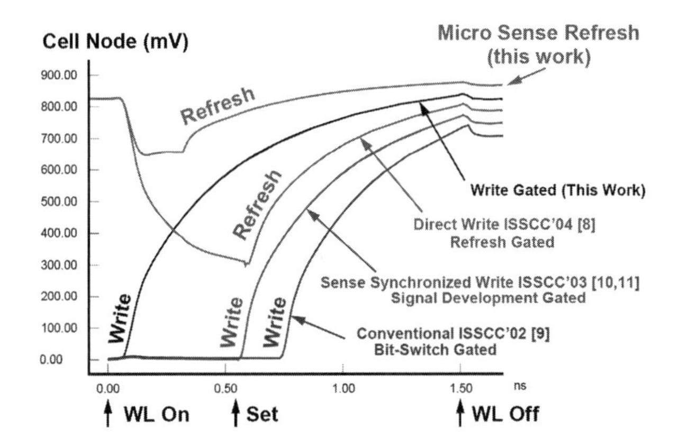

Fig. 5.36 Architectural advances

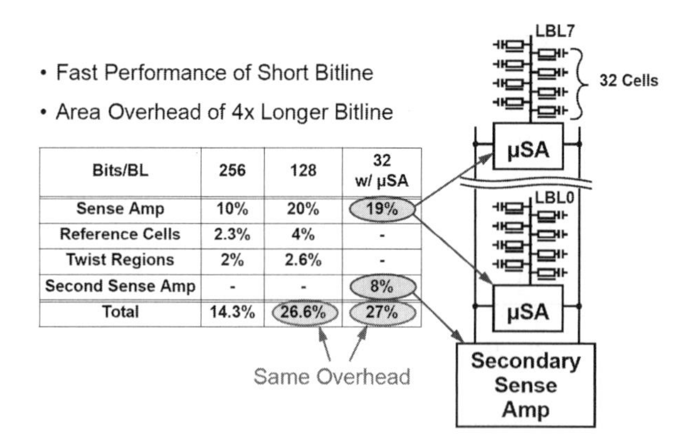

Bits/BL	256	128	32 w/ μSA
Sense Amp	10%	20%	19%
Reference Cells	2.3%	4%	-
Twist Regions	2%	2.6%	-
Second Sense Amp	-	-	8%
Total	14.3%	26.6%	27%

Fig. 5.37 Area overhead

5.6.16 Micro Sense Amplifier Overhead

As mentioned previously, the performance of short bitline designs must be weighed against their increased area overhead. To quantify this trade-off, Fig. 5.37 compares the area overheads of three different eDRAM array designs: (1) a direct write array with 256 cells/bitline [23]; (2) the same design style, but with 128 cells/bitline; and (3) the micro sense amplifier-–based array described in this chapter, with 32

Technology	65nm PD-SOI (Tox=1.12nm/2.35nm)
Cell Size	0.23μmx0.55μm(0.1265μm²)
Retention	40us @ 105c, 98.7% Avail w/ CR [12]
Macro Size (Calculated)	568μmx1170μm 0.665mm² (0.317x SRAM)
Organization	1K Row x 16 Col x 146 (2Mb)
Sub-Array	Hierarchical w/ Micro Sense 32 Rows/Local BL 8 Local BL/Global BL 1200 Bits/WL
Redundancy	2 - 32 WL Row Arrays Static Column Shifting
Supply	1.0V / 1.7V (WL High)
Performance	2ns Random Cycle 1.5ns Random Access
Power(Nom)	AC only - 76mW (0.84x SRAM) Stdby + Refresh - 42mW (0.18x SRAM)

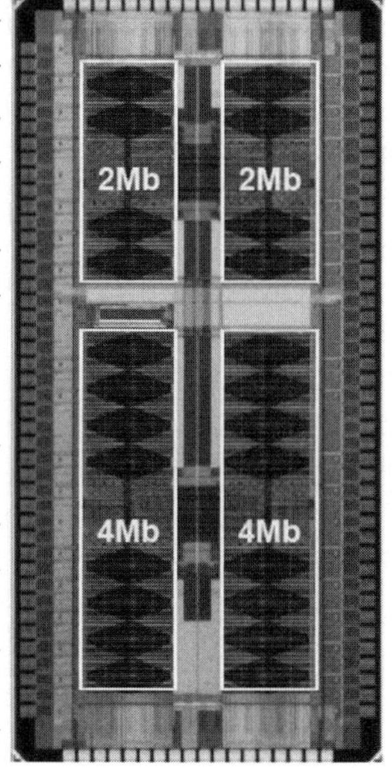

Fig. 5.38 SOI macro features

cells/bitline. As seen in the total area overhead row, the three-transistor micro sense amplifier design achieves its breakthrough performance with the same 27% area overhead associated with a bitline four times its length.

5.6.17 SOI Macro Features

Figure 5.38 shows a photo micrograph of the 13.5 Mbit test chip, containing two 2 Mbit macros and two 4 Mbit embedded DRAM macros. Figure 5.38 also shows a features table, including the array's 40 s retention specification. Using concurrent refresh (described by Kirihata et al. ISSCC 2004 [13]), array availability is 98%. As listed in the features table, typical keep-alive power (static power plus the power cost of periodic refresh) is 45 mW – one-fifth the power of SRAM in the same technology.

5.7 Embedded DRAM Summary

We have seen a continuous advance of embedded DRAM performance as outline in this chapter. Combining the newly found performance with advantages of DRAM in area and power will make it the workhorse for applications requiring large amounts of contiguous memory such as microprocessors, gaming, and networking applications. Semiconductor manufacturers that invest in DRAM production will also benefit from yield learning attributed to highly diagnosable arrays that are large enough to measure and learn to 6 sigma quality levels.

References

1. J. Dreibelbis, J. Barth, H. Kalter, and R. Kho, "Processor Based Built-In Self Test for Embedded DRAM," IEEE J. Solid-State Circuits 33, No. 11, November 1998, 1731–1740
2. T. Yabe, S. Miyano, K. Sato, M. Wada, R. Haga, O. Wada, M. Enkaku, T. Hojyo, K. Mimoto, M. Tazawa, T. Ohkubo, and K. Numata, "A Configurable DRAM Macro Design for 2112 Derivative Organizations to be Synthesized Using a Memory Generator," IEEE J. Solid-State Circuits 33, No. 11, November 1998, 1752–1757.
3. NeoMagic Corporation, 3250 Jay St., Santa Clara, CA 95054. The NeoMagic memory and logic graphics processors were introduced in 1993; see http://www.neomagic.com/about/ history.asp.
4. G. Giacalone, R. Busch, F. Creed, A. Davidovich, S. Divakaruni, C. Drake, C. Ematrudo, J. Fifield, M. Hodges, W. Howell, P. Jenkins, M. Kozyrczak, C. Miller, T. Obremski, C. Reed, G. Rohrbaugh, M. Vincent, T. von Reyn, and J. Zimmerman, "A 1 MB, 100 MHz Integrated L2 Cache Memory with 128b Interface and ECC Protection," Proceedings of the IEEE International Solid-State Circuits Conference (ISSCC), 1996, pp. 370–371.
5. A guide to IBM embedded DRAM offerings can be found at http://www.ibm.com/chips/ techlib
6. S. Crowder, R. Hannon, H. Ho, D. Sinitsky, S. Wu, K. Winstel, B. Khan, S. R. Stiffler, and S. S. Iyer, "Integration of Trench DRAM into a High Performance 0.18 um Logic Technology

with Copper BEOL," International Electron Devices Meeting, Digest of Technical Papers, 1998, pp. 1017–1020.
7. T. Obremski, "Advanced Non-Concurrent BIST Architecture for Deep Sub-Micron Embedded DRAM Macros," Ph.D. Dissertation, University of Vermont, Burlington, May 2001.
8. N. Watanabe, F. Morishita, Y. Taito, A. Yamazaki, T. Tanizaki, K. Dosaka, Y. Morooka, F. Igaue, K. Furue, Y. Nagura, T. Komoike, T. Morihara, A. Hachisuka, K. Arimoto, and H. Ozaki, "An Embedded DRAM Hybrid Macro with Auto Signal Management and Enhanced on Chip Tester," IEEE International Solid-State Circuits Conference, Digest of Technical Papers, 2001, pp. 388–389, 469.
9. R. Matick, et al., "Logic-based eDRAM: Origins and rationale for use", IBM Jour. Res & Dev. Vol. 49, No 1, January 2005, pp. 145–165.
10. E. Cohen, et al., "A 64B CPU Pair: Dual and Single-Processor Chips", 2005 ISSCC Dig. Tech. Papers, 2005, pp. 106–107.
11. S. Naffziger, et al., "The implementation of a 2-core Multi-Threaded Itanium-Family Processor", 2005 ISSCC Dig. Tech. Papers, 2005, pp. 182–183.
12. R. H. Dennard, "Field Effect Transistor Memory," U.S. Patent 3,387,286, June 4, 1968.
13. T. Kirihata, P. Parries, D. Hanson, H. Kim, J. Golz, G. Fredeman, R. Rajeevakumar, J. Griesmer, N. Robson, A. Cestero, M. Wordeman, and S. Iyer, "An 800 MHz Embedded DRAM with a Concurrent Refresh Mode," Proceedings of the IEEE International Solid-State Circuits Conference (ISSCC), 2004, Digest of Technical Papers, 1, 2004, pp. 206–523.
14. M. Kumar, M. D. Steigerwalt, B. L. Walsh, T. L. Doney, D. Wildrick, K. A. Bard, D. M. Dobuzinsky, P. A. McFarland, C. E. Schiller, B. Messenger, S. E. Rathmill, A. R. Gasasira, P. C. Parries, S. S. Iyer, S. E. Chaloux, and H. L. Ho, "A Simple and High-Performance 130 nm SOI EDRAM Technology Using Floating-Body Pass-Gate Transistor in Trench-Capacitor Cell for System-on-a-Chip (SoC) Applications," Proceedings of the IEEE International Electron Devices Meeting (IEDM), 2003, Technical Digest, 2003, pp. 17.4.1–17.4.4.
15. E. B. Eichelberger and T. W. Williams, "A Logic Design Structure for LSI Testability," J. Design Automat. Fault- Tolerant Comput. 2, May 1978, 165–178.
16. J. Dreibelbis, J. Barth, H. Kalter, and R. Kho, "Built-In Self Test for Embedded DRAM," Proceedings of the IEEE North Atlantic Test Workshop, West Greenwich, RI, 1997, pp. 19–27.
17. R. McConnell, U. Moller, and D. Richter, "How We Test Siemens' Embedded DRAM Cores," Proceedings of the International Test Conference, 1998, pp. 1120–1125.
18. R. Aitken, "On-Chip Versus Off-Chip Test: An Artificial Dichotomy," Proceedings of the International Test Conference, 1998, p. 1146.
19. J. Dreibelbis, J. Barth, Jr., R. Kho, and T. Kalter, "An ASIC Library Granular DRAM Macro with Built-In Self Test," IEEE International Solid-State Circuits Conference, Digest of Technical Papers, 1998, pp. 74–75.
20. H. A. Bonges III, R. D. Adams, A. J. Allen, R. Flaker, K. S. Gray, E. L. Hedberg, W. T. Holman, G. M. Lattimore, D. A. Lavalette, K. Y. T. Nguyen, and A. L. Roberts, "A 576 K 3.5 ns Access BiCMOS ECL Static Ram with Array Built-in Self Test," IEEE J. Solid-State Circuits 27, No. 4, April 1992, 649–656.
21. P. Jakobsen, J. Dreibelbis, G. Pomichter, D. Anand, J. Barth, M. Nelms, J. Leach, and G. Belansek, "Embedded DRAM Built In Self Test and Methodology for Test Insertion," Proceedings of the International Test Conference, 2001, pp. 975–984.
22. J. Barth, et al., "A 300 MHz multi-banked DRAM Macro featuring GND Sense, bit-line twisting and direct reference cell write," IEEE International Solid-State Circuits Conference, vol. XLV, February 2002, pp. 156–157.
23. J. Barth, et al., "A 500 MHz Multi-Banked Compilable DRAM Macro with Direct Write and Programmable Pipeline", 2004 ISSCC Dig. Tech. Papers, 2004, pp. 204–205.
24. Y. Taito, et al., "A High Density Memory for SoC with a 143 MHz SRAM Interface Using Sense- Synchronized- Read/Write", IEEE International Solid-State Circuits Conference, vol. XLVI, February 2003, pp. 306–307.

25. H. Pilo, et al., "A 5.6 ns Random Cycle 144 Mb DRAM with 1.4 Gb/s/pin and DDR3-SRAM Interface", IEEE International Solid-State Circuits Conference, vol. XLVI, February 2003, pp. 308-309.
26. M. Ouellette, et al., "On-chip repair and ATE-independent fusing methodology", IEEE International Test Conference Proceedings, October 2002, pp. 178–186.
27. S. Iyer, et al., "Embedded DRAM: Technology Platform for Blue Gene/L chip", IBM Jour. Res & Dev. Vol. 49 NO. 2/3 MARCH/MAY 2005, pp. 333–350.
28. Top 500 Supercomputer Sites, "TOP500 List – November 2007," http://www.top500.org/list/2007/11/100.
29. J. Clabes, et al., "Design and Implementation of the Power5 Micro Processor", 2004 ISSCC Dig. Tech. Papers, 2004, pp. 56–57.
30. G. Wang, et al., "A 0.127 μm^2 High Performance 65 nm SOI Based embedded DRAM for on-Processor Applications", 2006 IEDM, 2006.
31. D. Weiss, et al., "The on-chip 3-MB Subarray-based third-level cache on an Itanium microprocessor", 2002 ISSCC Dig. Tech. Papers, 2002, pp. 112–113.

Chapter 6
Embedded Flash Memory

Hideto Hidaka

Abstract Chapter 6 first introduces the expanding variety of applications and requirements for embedded nonvolatile memory especially in microcontroller applications, then describes how and why embedded flash memory has expanded the functions and applications supported by process, device, and circuit technology evolutions. Embedded-specific flash memory technologies focused on the floating-gate and charge-trapping devices with split-gate and 2Tr cell concepts are overviewed in Section 6.2. Descriptions on basic embedded flash design concepts and examples of actual embedded flash designs along with challenges and future targets for embedded flash memory are provided in Section 6.3.

6.1 Application and Technology Trend in Embedded Nonvolatile Memories

6.1.1 Introduction

Nonvolatile memory, which keeps stored data for more than 10 years while the power supply turns off, is applied to many aspects of human life and industry. Nonvolatility is preferred especially in remote local systems in eliminating the problems in power supply and battery backup in order to achieve system simplicity and maintainability. It also helps mitigate environmental concerns against extensive use of batteries. Recently multiple frequent updates of stored parameters, program codes, and system reconfigurability functions have been extensively utilized with embedded rewritable nonvolatile memory. These application trends favor embedded nonvolatile memory in system-on-chip (SOC) and microcontroller unit (MCU) products.

Currently the major application of embedded nonvolatile memory is the flash memory for code storage in microcontroller (MCU) products. This application,

H. Hidaka (✉)
Renesas Technology Corporation, Itami, Japan
e-mail: hidaka.hideto@renesas.com

K. Zhang (ed.), *Embedded Memories for Nano-Scale VLSIs*, Series on Integrated Circuits and Systems, DOI 10.1007/978-0-387-88497-4_6,

called flash-MCU, has achieved a large market volume for embedded memory applications, only second to embedded SRAM which is indispensable for general CMOS logic. Recently the flash-MCU market has been expanding into many increasing segments in automotive and smart-IC card applications. Moreover in some cases, additional data storage with multiple-time rewrite capability is needed. This code-data configuration of flash-MCU has helped increase the market volume. Most embedded application fields require nonvolatility for 10–20 years and rewrite capability ranging up to 10^6 (one million) times according to the practical uses in code and data storage. In the past applications requiring more than million times of rewrite capability have resorted to SRAM and DRAM by battery backup features, with drawbacks in maintainability and environmental concerns for the battery disposal. This situation is changing by the multiple rewrite capability offered by either sophisticated flash memory technology or newly emerging nonvolatile technology beyond flash memory.

Currently most widely used embedded nonvolatile memories are ROM technologies as classified in Fig. 6.1. Electrically programmable ROM (EPROM), electrically erasable and programmable ROM (EEPROM), and flash memory utilizing either a floating-gate or a charge-trapping technology are the most practical devices at present. They are the proven technologies used in products with reliability requirement. Most of nonvolatile memories now productized are ROM and alterable-ROM. The distinction between alterable-ROM and RAM lies in their performance. RAM requires unlimited number of read and write operations, but actually 10–20 years lifetime with rewrite capability up to 10^{16} suffices. Typically flash memories are limited to less than 10^6 times of rewrite endurance and much slower write (2–4 orders in time) than existing RAM because of the physical limitation in SiOx/SiNx system.

In recent years emerging nonvolatile memory has been widely explored to be realized by physical mechanisms like ferroelectric polarization (FeRAM, see Chapter 8), magnetization orientation (MRAM, see Chapter 7), and so on. Some of these will offer much improved performance over ROM in write speed and rewrite endurance. The International Technology Roadmap for Semiconductors (ITRS) in 2007 mapped eight types of physical memory, as illustrated in Fig. 6.2. Since 2001, five nonvolatile memories, NAND, FeRAM, MONOS, MRAM, and

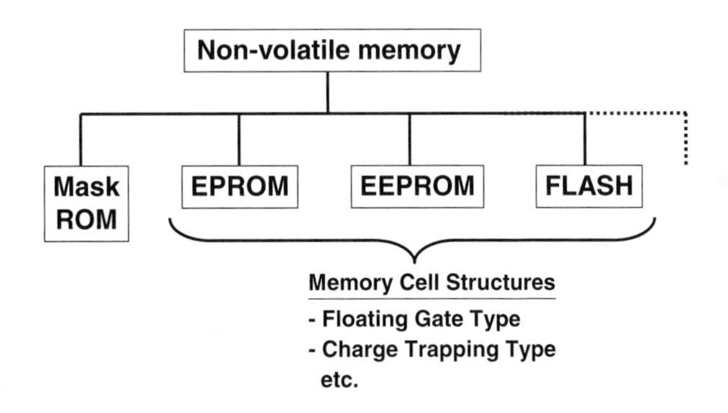

Fig. 6.1 Conventional nonvolatile memory devices

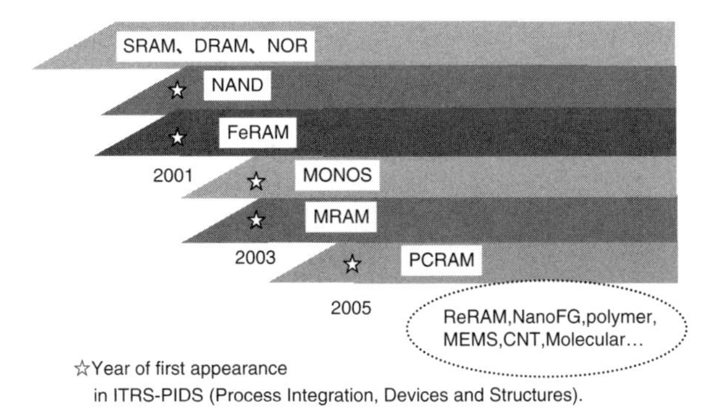

Fig. 6.2 Memory roadmap as illustrated according to ITRS, 2007 (International Technology Roadmap for Semiconductors) [28]

PCRAM (Phase Change Memory) have newly appeared in the roadmap, and it introduces even newer devices in the emerging research device section including resistance RAM (ReRAM), polymer memory, microelectro-mechanical system (MEMS) memory, molecular memory, carbon nano-tube (CNT), and FIN-FET memory. A key question remains in whether new physical memory will come into reality for embedded uses, and how it goes into technological convergence hopefully.

Figure 6.3 lists candidates for near-term embedded nonvolatile memory technologies, as compared with currently available floating-gate NOR flash memory. One of the important features in the emerging nonvolatile memory is suitability for embedded applications including cost and CMOS compatibility. Also much improved rewrite capability with lower power poses an attractive challenge for emerging memories in nonvolatile RAM. The definition of ROM and RAM is becoming ambiguous because recent development of technologies offers properties

F: min. feature size

	NOR-Flash	FeRAM	MRAM	Phase Change
material	SiOx/Poly-Si	PZT,SBT	CoFeB etc.	GST
cell size(F2)	10	40	30	20
read	NDRO	DRO	NDRO	NDRO
read speed	20ns	50ns	<10ns	20ns
write speed	1μs	50ns	<10ns	100ns
N. of writes	1E 5	1E 8∼12	>1E 16 (∞)	1E 6∼12
power in write (pW•sec/bit)	∼1E 5	30	40 (field SW) / 20 (Spin)	20–1000
advantage	conventional material	high-speed	high-speed	simple structure
challenge	N. of writes	scalability	field disturbance	thermal disturbance

Fig. 6.3 An example of nonvolatile memory comparison

in-between, such as 10^{6-12} of P/E (program and erase), high-speed read, and low-energy rewrite per bit. A remarkable feature realized by emerging nonvolatile memory is orders-of-magnitude lower energy per bit in write than flash memory, as shown in this figure. The low-voltage switching performance and high-speed write operation contribute to this remarkable advantage. This property in low-power rewrite is essential for realizing nonvolatile RAM (NV-RAM) applications where NV-RAM will encounter with frequent write operations. Actually this item may be another way of defining NV-RAM requirement, in contrast with ROM. In summary, we see a clear direction in realizing non-volatile RAM, requiring a high-speed write and a large number of rewrite endurance. However, the emerging memory devices still need tremendous works for applications to real products.

6.1.2 Market and Application Overviews for Embedded Nonvolatile Memories

The real market for embedded nonvolatile memory is dominated by embedded flash memory in MCU (micro controller unit) applications resulting in the largest market for embedded memory except embedded SRAM. Embedded flash memories show a wide variety of applications due to the needs for nonvolatile storage for code and data. A memory-centric view of innovation and its future projection observed in the MCU reminds us of a very wide and deep meaning of embedded alterable non-volatile memory functions. The point is why and how the flash-MCU has penetrated the market, owing to the technology and circuit developments to create application values.

Figure 6.4 briefly describes the embedded flash memory applications from the perspectives of MCU applications. MCUs are now widely applied to a number of embedded control applications, as shown in this market breakdown. In addition to the steady growth in the mainstream product lineup, recently, market has been

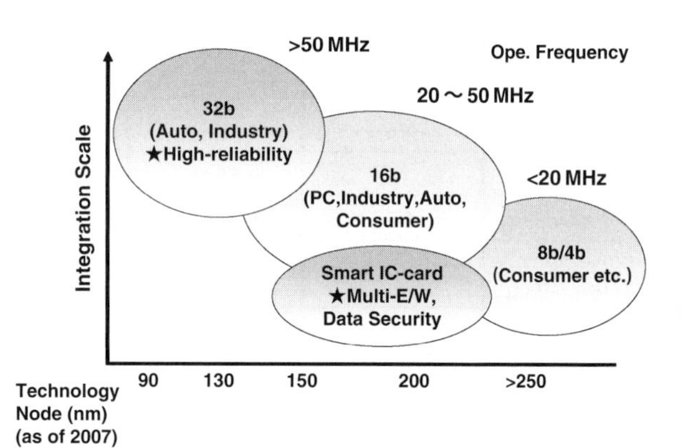

Fig. 6.4 MCU market by applications [21] (© 2008 IEICE)

expanded significantly driven by automotive and smart-card applications. These two segments are also the forefronts of embedded flash memory technologies because their usage conditions are very stringent, requiring high-reliability under high-temperature, large numbers of program/erase, and very high degree of data security in the embedded high-density memory applications. In almost all the embedded control applications flash MCUs are in large production.

Figure 6.5 describes the evolution of MCU products with regard to the use of the embedded memory. Since 1980s, the embedded memory in MCU has evolved significantly every decade from mask-ROM, one-time programmable ROM (OTP), and finally flash memory. In the end of 1980s, OTP was incorporated on the technical basis of EPROM, thus reducing the cost of MCU chip by the use of plastic package, instead of ceramic package with UV window for programming. In the middle of 1990s, there emerged strong demands for applying the flash memory in MCU to save total cost and to reduce turnaround time of MCU system development. After the introduction of flash memory in MCU, embedded flash memory has improved its features like single power supply operation and reliability.

Beginning with replacing mask-ROM for prototyping applications, we have seen a great leap in market penetration of embedded flash memory, thanks to the overall cost reduction through design, production, and inventory control by programmability in flash-MCU. Moreover the "embedded-ness" for flash memory is required for high performance and data security. Here the overall cost advantage by the economy of scale exceeds the disadvantage of higher wafer process cost for incorporating flash memory, which is a key to make flash-MCU the most successful embedded memory business ever except embedded SRAM.

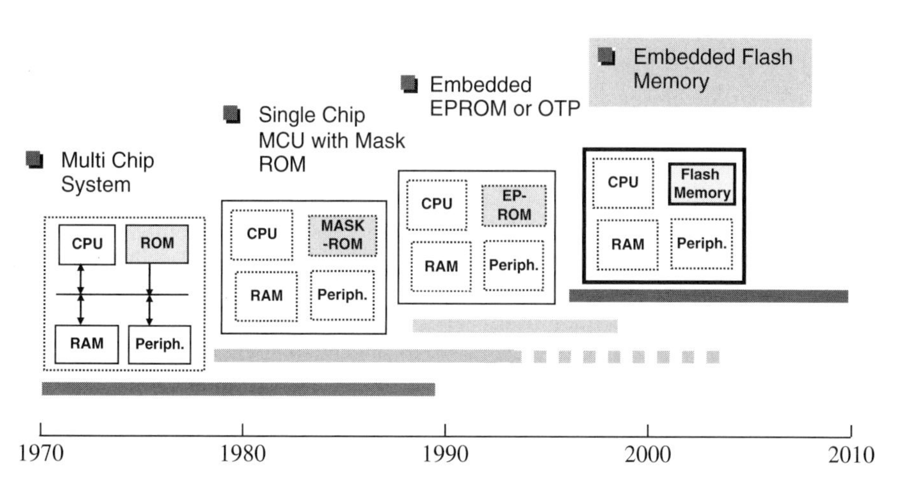

Fig. 6.5 Evolution of MCU by on-chip memory [19] (© 2008 IEEE)

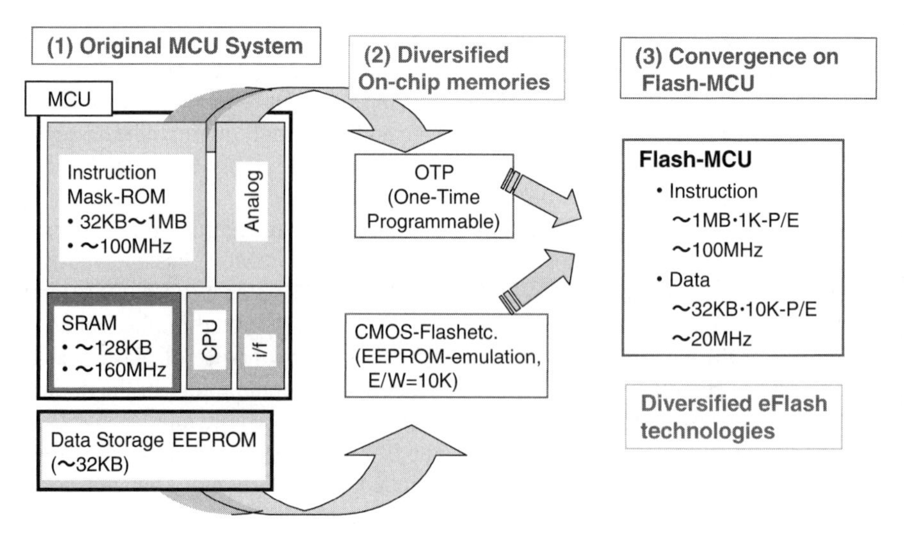

Fig. 6.6 Convergence on flash-MCU [18, 22] (© 2008 IEEE)

This history indicates a converging and diverging scenario in retrospect. The on-chip flash memory merged code and data ROM, providing a standard small system configuration to trigger a market expansion, as shown in Fig. 6.6. At first this was limited to prototyping uses for debugging systems because of higher chip cost. Then after developing applications for multiple-time programmability in real-time control and achieving overall cost advantage as in automotive application, all the MCU markets now focus on flash-MCU solutions. One important contributor to this convergence is that flash-MCU succeeded in incorporating EEPROM emulation for data ROM in addition to code ROM, and another is embedded flash memory has attained high speed competitive to mask-ROM configuration. However, embedded flash memory technology to support flash-MCU is much diversified today, which shows another dimension of divergence to consider next.

In reality high-density flash memory technologies introduced by large-capacity discrete data memory do not meet all the requirements by embedded applications as they are. This is the main reason why we have much diversified embedded-specific flash memory technologies. Differentiators to consider in embedded flash memory are area factors in small memory capacity, operating voltage, speed, reliability, and EEPROM emulation strategy.

A convergence in embedded flash technology for further cost disruption by the economy of scale is favorable for the industry, but actually we are split into several solutions. Although MONOS is expected to see a promising future because of some advantages in reliability, the key for success is how to implement effective learning process in the course of technology development. Therefore, focus on learning is a key to the convergence of embedded flash technology.

The benefits of introducing embedded flash memory in MCU products for users in system design are

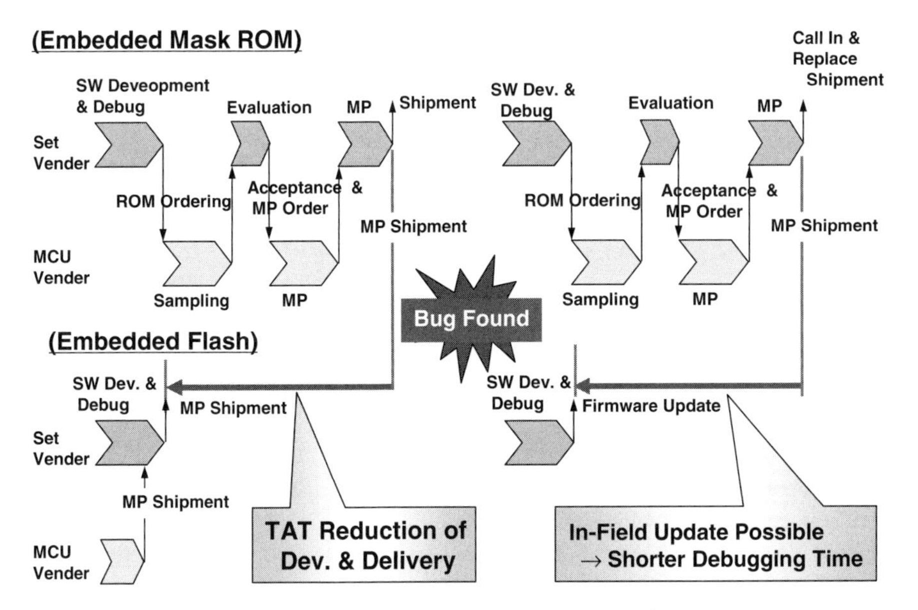

Fig. 6.7 System development cycle is shortened by flash-MCU [19] (© 2008 IEEE)

(1) Short system development turnaround time (Fig. 6.7)

Embedded mask-ROM scheme takes longer time from the start of software development to shipping of a system because a program modification and debugging require iterations of mask-ROM revisions to be shared by the system vendors and MCU vendors. Flash-MCU products are shipped with unprogrammed (blank) data to customers during the development of software. Flash memory is programmed electrically after system assembly, and the developments of total system and program can proceed in parallel, resulting in shortened turnaround time of system development. And the program recovery time after debugging is shorter because of the recovery in field. The total system development cost and time are reduced.

(2) Production control

Product set variation is often offered by the MCU firmware. If the mask-ROM MCU is used, each product chassis must be manufactured with each different mask-ROM data, which causes complex inventory management. The flash-MCU eliminates this complication by programming the MCU on the common chassis before the set assembly.

(3) Software development paradigm shift

By introducing single-chip MCU in the system development, hardware control is replaced by software control, and the product lineup is unified. Here a new problem has arisen in the turnaround time for software development as a bottleneck because the software gets more complex according to system requirements. The embedded flash-MCU alleviates this problem by introducing learning mechanism by parameter updates in the system.

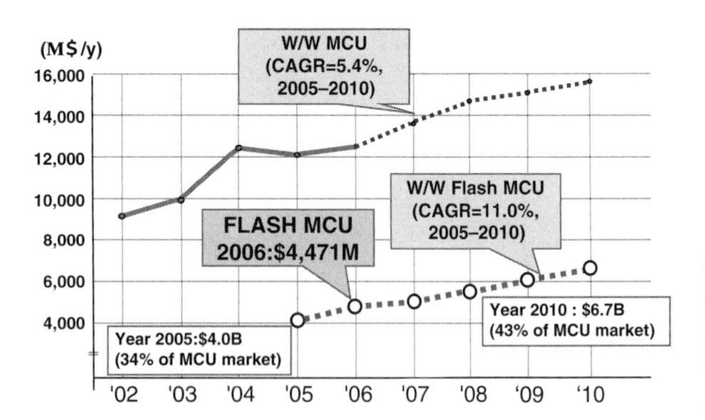

Fig. 6.8 Market growth of MCU and flash-MCU [22] (© 2008 IEEE)

Microcontroller with embedded flash memory storage has seen an up-surge in real-time control application markets. The programmable code storage provided by on-chip flash memory contributes to reductions of production cost and to expansion of real-time adaptive control applications, realizing a value innovation with remarkable cost/value advantage. Since the advent of flash-MCU (microcontroller with embedded flash memory) in 1990s, expansion of real-time control applications for MCU has been accelerated. The compound average growth rate (CAGR) of the whole MCU is about 5%. The flash-MCU accounts for more than 30% in this whole MCU market with growth rate of 11% (Fig. 6.8). In the year 2010, flash-MCU is predicted to gain more than 50% of the whole MCU market. This indicates that embedded flash memory is rapidly being accepted and has become indispensable in the MCU products. Actually almost all the MCU market segments now adopt embedded flash solutions.

Market components for flash-MCU in Fig. 6.9 show the major application of flash-MCU is automotive, such as power-train, car information, chassis, and car audio, and it accounts for more than 50% of flash-MCU applications in 8, 16, and 32 bit products. Precise control of automotive engines is now motivated by environmental concerns and fuel efficiency that are regulated by government. This is the main reason electrical adaptive control scheme has been widely adopted in cars, which accounts for the expansion of MCU applications in this segment of the market.

6.1.3 Automotive Application Examples

Automotive application examples of flash-MCU depicted in Fig. 6.10 indicate that today's electrically equipped car extensively uses MCUs for electrical control, most of which are flash-MCU for locally storing the control parameters and measured data [19]. The trend in today's automotive control is to save the fuel cost, to get more security system, and to be more connective to the outer information. These

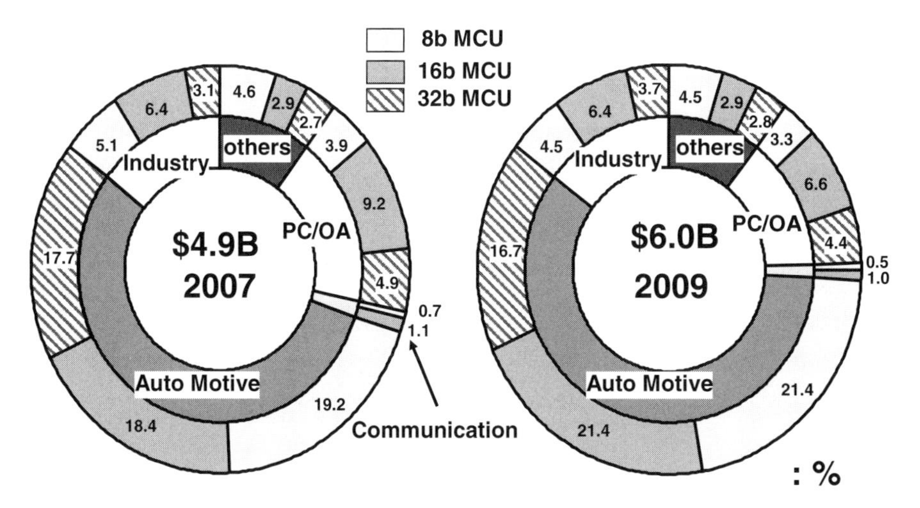

Fig. 6.9 Applications of flash-MCU [19] (© 2008 IEEE)

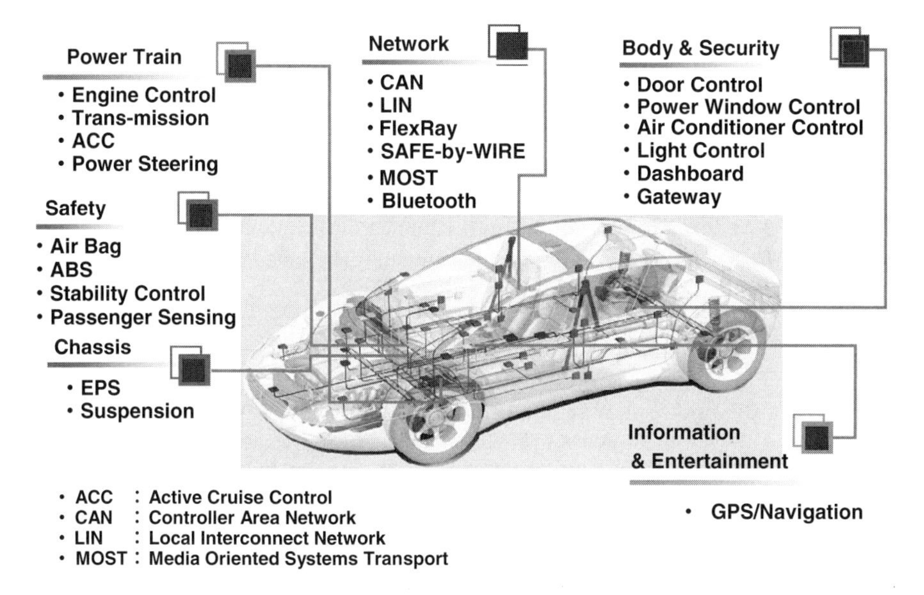

Fig. 6.10 Flash-MCU applications in a car [19] (© 2008 IEEE)

new functions add to uses of flash MCUs from 8b to 32b products, resulting in 30–100 units of flash MCUs being used in a single car.

Figure 6.11 shows an example of engine control scheme by flash-MCU. Several sensors for crank angle, air flow, and knocking phenomenon are connected to 32-bit flash-MCU through application-specific linear ICs. According to measured data from these sensors flash-MCU controls fuel injection, the timing of ignition plug,

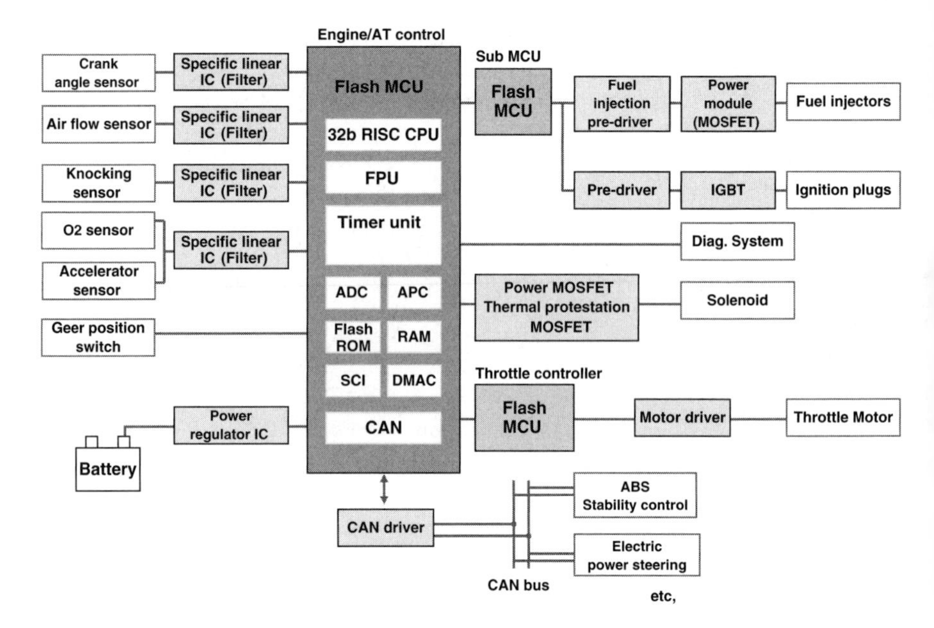

Fig. 6.11 Flash-MCU applications in a car engine control [19] (© 2008 IEEE)

throttle motor, etc., to be in the most suitable condition. These local distributed real-time control in automotives are beginning to dominate MCU applications.

The CPU performance trend in Fig. 6.12 and ROM capacity trend in Fig. 6.13 indicate ever-expanding technology requirements. The performance requirements

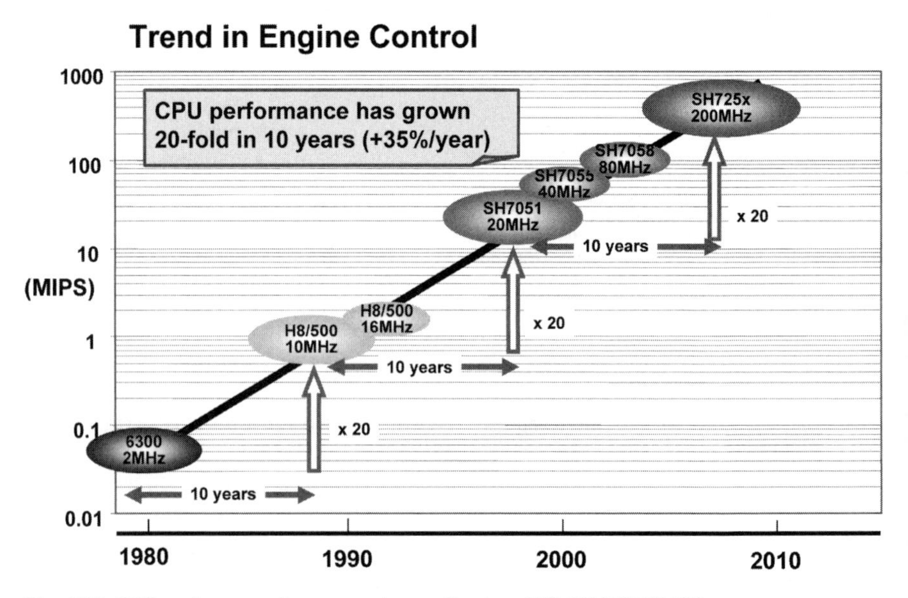

Fig. 6.12 CPU performance for automotive applications [19] (© 2008 IEEE)

in automotive applications are growing by 20-fold in 10 years, 35% per year. This growth of computing power in automotive applications is supported by device scaling and design for reliability. On the other hand, the on-chip ROM capacity grows 23% per year to support the growth of on-chip code storage for program statements in automotive applications. Flash-MCU for automotive should incorporate scaled CMOS process with embedded flash memory to achieve high-density memory as well as high-performance CMOS logic.

The examples of embedded flash memory requirements for automotive applications in Fig. 6.14 reveal actual challenges that embedded flash technology faces. From performance to match CPU to reliability under a very wide temperature range and low cost, embedded flash memory for MCU is challenging in many aspects of

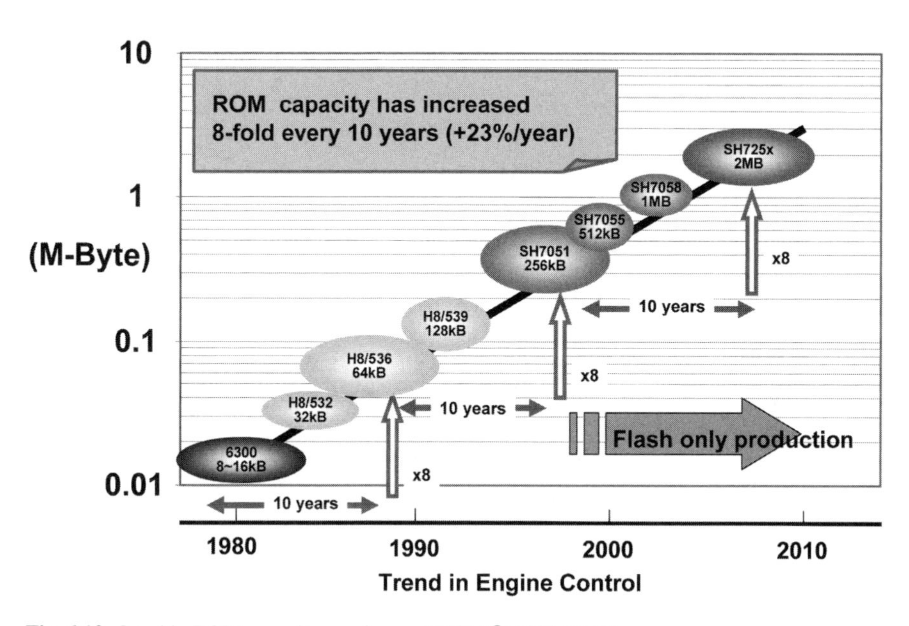

Fig. 6.13 On-chip ROM capacity requirement [19] (© 2008 IEEE)

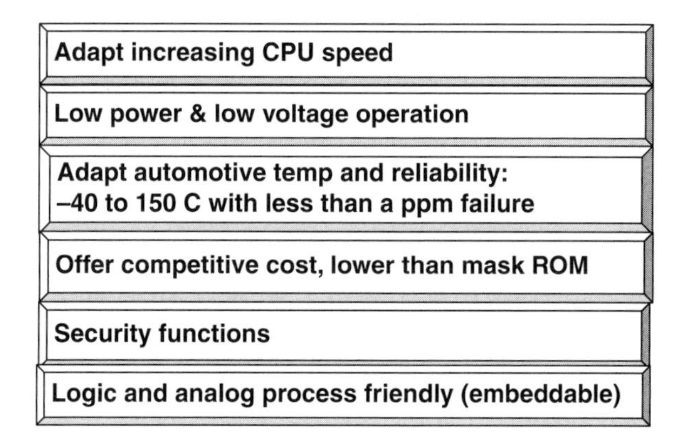

Fig. 6.14 Requirements for embedded flash memory in automotive uses

today's semiconductor memory. Also data security function is becoming a prevailing factor in the requirements for MCU in general.

6.1.4 Requirements for Embedded Flash Memory Applications and Trends

Figure 6.15 summarizes the requirements for embedded flash memory in MCU. From high performance required in automotive and industry to high-temperature operation for automotive and to small macro size for smaller memory capacity in cost-sensitive applications like AV, PC/OA, and consumer appliances, the requirements vary significantly according to each application. As for the program and erase endurance up to 100 K cycles is strongly required for multiple updates of control parameters.

With the growth of available on-chip flash memory capacity, the on-chip system solutions have been validated in many small systems. From prototyping uses to real-time parameter updates, security data memory, and overall cost reduction for code storage in MCU, the flash-MCU market has been expanding steadily. It also finds new market drivers in automobile, PC, consumers, and smart-IC cards. Flash-MCU has won one of the most successful business in embedded memory applications, second only to embedded SRAM.

We observe three important points in the trend of embedded flash memory technologies for MCU applications driven by above application requirements

		Automobile			Industry	PC/OA	Consumer
		Power Train	Body	AV			
MCU	Performance (frequency)	~300 MHz	150~ 200 MHz	100 MHz	~300 MHz	25~ 50 MHz	20~ 100 MHz
	Power	0.5 mA /MHz	0.5 mA /MHz	0.25 mA /MHz	1 mA /MHz	0.5 mA /MHz	0.25 mA /MHz
	Temp.(Ta)	−40 ~ 125 C			max 85 C	−20 ~ 85 C	
FLASH	Density	8 MB	2 MB	2 MB	1 MB	2 MB	1 MB
	P/E cycle	Program Area : 1K cyc./Data Area : 100K cyc.(EEPROM)					
	Small Cell	✓					
	Small Macro			✓		✓	✓
	Fast Access	✓			✓		

Fig. 6.15 Requirements of flash-MCU technology for different applications. Important factors in each application are marked [19]

Fig. 6.16 Trend of embedded flash memory in MCU [22]

(1) Converging into Flash-MCU solution

all-in-one solution

low-cost validation for in-system uses

(2) Multiple eFlash technologies to meet diversified needs;

high-temp, high-speed, multiple-P/E, byte-wide access,

area merit @small capacity

(3) Embedded flash incorporates multiple functions

high-speed modes, data security functions etc.

(Fig. 6.16). As the penetration of on-chip flash proceeds for in-system uses beyond prototyping uses, MCU market is now converging into flash-MCU solutions. However, diversified embedded flash memory technologies are now competing in the market to meet a wide variety of needs, together with multiple on-chip functions based on new circuit technologies.

Embedded-specific nonvolatile memory technologies significantly deviate from dominant discrete memory technologies, NOR and NAND flash memories, because of the required embedded-specific requirements like host-logic CMOS compatibility, performance, cost with smaller capacity of on-chip memory, and reliability. The major technology requirement for embedded nonvolatile memory is minimum additional cost, including fewer process steps, voltage reduction for area penalty, to the base CMOS logic, rather than mere cell size scalability. The extra mask steps should be as small as possible and the area penalty reduction in small capacity of memory (10–100 KB of memory capacity) frequently encountered in embedded applications is focused on smaller periphery circuitry. High-voltage operations in conventional flash memory for discrete memory products do not meet this demand. Therefore these important factors for embedded applications often do not allow the direct import of technology from discrete high-density flash memory like NOR and NAND. Also quite often embedded uses require faster access time to match the on-chip processing speeds, which may further deviate from discrete flash memory in technology and design. The embedded memory may be defined as “memory with minimum extra cost adder to existing CMOS.”

One important factor for reducing the production cost of embedded nonvolatile memory products is that we can unify a number of product line-ups because multiple numbers of mask-ROM data variations and product configurations can be supported by merely altering the data in the programmable nonvolatile memory, achieving the overall production cost reduction.

The uniqueness of nonvolatile memories in the embedded memory spectrum is that it stores, in nonvolatile manner, information like code and data such as program code, system boot code, system configuration register, firmware, system reset vector, frequently updated parameters and coefficients, the state before power-down, etc. To secure these functions, nonvolatile memories have additional features like awareness against any errors by power up and down operations and data security. In reality the flash memory necessitates somewhat different designs of memory IP for code and data mainly in program/erase endurance and access time. Because of above requirements and additional features for high speed, low power, high reliability, and low EMI, etc., flash memory designs show a wide variety of circuit technologies.

Two important functions specific to embedded nonvolatile memory, code(program) storage and data storage, pose diversified requirements in access time and program/erase endurance. Also various application fields pose stringent conditions for use in wide range of temperature and program/erase endurance. Actually the flash memory market has been dominated by NOR and NAND flash memory technologies optimized for discrete memory products with smaller cell size and sacrificing the performance and functionalities. Although the mere import from these technologies offer shortened learning curve and time to market, these are not optimized for embedded uses in cost, performance, and value creation opportunities.

Actually the base technology node available for embedded nonvolatile memory somewhat lags behind the most advanced CMOS logic process because the embedded technology has seen time-consuming optimization of the technology, according to the actual product applications. This situation will be required to improve if we look for solutions in power limitations in SOCs by nonvolatile memories. This may be the true challenge for embedded technology.

The future challenges and opportunities for research and development of embedded nonvolatile memory technologies are summarized in Section 7.3, after discussions on current flash memory and MRAM (in the near-term future) devices in Sections 6.2, 6.3, 7.1, and 7.2.

6.2 Embedded Flash Memory Technology

This section gives an overview of the evolution of embedded-specific flash memory technology, aside from large-scale discrete flash memory. Focus is on the floating-gate devices from 1Tr to split-gate and 2Tr cell structures and how charge-trapping structures such as SONOS technology improves the properties of floating-gate technology to support embedded uses, and foresees the future trends in scalability and functionality for embedded flash memory.

Currently most widely used embedded nonvolatile memories are ROM technologies as classified in Fig. 6.17. EPROM, EEPROM, and flash memory utilizing either a floating-gate or a charge-trapping technology are the most practical devices at present. "Flash memory" is one kind of EEPROM originally named after the flash erasure operation in a sector, a large block of bit cells (of the order of K bits), at one time to enhance the performance in bulk rewrite due to the structural restrictions in high-density cell arrays. On the contrary, EEPROM often refers to a property of byte-wide program/erase. In this section "embedded flash memory" is referred to broadly in the sense of alterable embedded nonvolatile memory constructed by Si/SiOx/SiN/poly-silicon systems regardless of how the erasure function is operated, for convenience of notation.

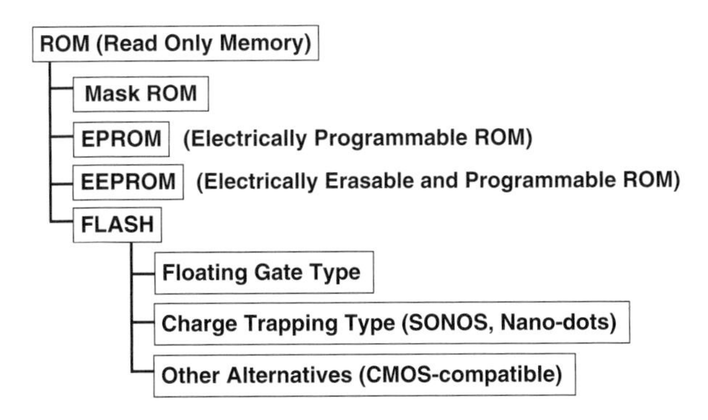

Fig. 6.17 Embedded flash memory and related devices

All the flash memory structures store signal charge by electrons or holes in a potential well. The charge injection (program) and erasure to this potential well are essential operations in these types of alterable memory. The signal charge can be electrons or holes, and how these two kinds of charge are controlled is important in understanding the operations of flash memories. Most of the flash memory technologies feature block-oriented program and erase, which comes from the basic physical limitations in the cell operations. It has faster bulk data update, but actually this scheme may pose a restriction in byte-wise data manipulations favored in embedded applications. EEPROM device generally complements this flash property in byte-wise rewrite in a small amount of memory capacity, in spite of a drawback in area and operation speed.

Each flash memory technology is often measured by the program/erase (P/E) endurance and byte-wise data manipulation capability in P/E operations. OTP, multiple-time programmable (MTP, 100–1 K), flash (1–10 K), and EEPROM (–100 K) are usual terms denoting P/E endurance performance. EEPROM usually emphasizes on the small byte-wise units of P/E in byte, unlike flash memory programmed and erased in blocks of K bits. Because each flash technology feature comes from the physical structures (memory cell operation principle and array organization), we have difficulties in supporting required multiple functions on a chip in embedded applications by one flash technology. For example, an embedded flash technology supporting a large capacity with intermediate P/E endurance also have to support one million P/E usually supported by EEPROM on the same chip by a precise P/E algorithm and/or different memory array designs, emulating the EEPROM functions. The requirement for the code and data storage on the same chip has been one of the challenges for one-chip solutions by embedded flash memory technologies.

Flash memory technologies are also classified by physical structures for signal charge storage. Typical physical structures include floating poly-silicon gate, charge-trapping layer by $SiNx/SiO_2$ system, or nanocrystalline discrete charge traps, etc. These are illustrated in Figs. 6.18a,b and are representative of the physical operation mechanisms and actual device implementations. Devices in Fig. 6.18a represent important development in embedded floating-gate devices, whereas Fig. 6.18b depicts major charge-trapping devices now under development.

(1) A 1Tr stacked-gate cell is extensively utilized in embedded uses as well as discrete memory products, for its high density and matured technology (Fig. 6.18a-1) [36].
(2) An early EPROM trial on a split-gate structure (1.5Tr cell) demonstrated a very fast/low-voltage programming operation owing to the so-called source-side injection (SSI) mechanism [29]. This has established an embedded-specific device structure idea in flash memory technology (Fig. 6.18a-2).
(3) In addition to the split-gate structure, poly-to-poly tunneling operation realizes fast erasing and small sector size [31]. This structure, SST Super-Flash cell^{TM}, has become widely used technology of embedded flash memory (Fig. 6.18a-3).
(4) A 2Tr cell adding a transfer gate to a 1Tr cell and utilizing FN/FN (Fowler-Nordheim tunneling) mechanisms for P/E offers a good compromise between simple structure and lower-power operation (Fig. 6.18a-4) [12, 38].
(5) Charge-trapping SONOS structures offer other possibilities in flash memory technologies. 1Tr, 1.5Tr, and 2Tr cells have been tried on (Fig. 6.18b-5,6,7) [9, 14, 37].
(6) Nano-dot gate flash memory (Fig. 6.18b-8) improves the scalability of floating-gate device by discrete charge storage scheme [46, 55]. This is one of the candidates for future embedded flash memory.

These are discussed in Sections 6.2.1 and 6.2.2. The major device trend is from the expansion and variations of existing floating gate to SONOS and nano-dot charge-trapping structures for structural simplicity and scalability. Figure 6.19 lists some of important literatures for embedded flash memory technologies.

	(1)	(2)	(3)	(4)
	1Tr NOR cell	**1.5Tr cell (split-gate)**	**1.5Tr cell (SuperFlash^{TM})**	**2Tr cell**
Program	**CHE**	**SSI**	**SSI**	**FN**
Erase	**FN (poly-sub)**	**–**	**FN (poly-poly)**	**FN (poly-sub)**
Device Structure	Control Gate, Floating Gate (N+-poly), N+, N+, S, D, P-substrate	Select Gate, Floating Gate, N+, N+, S, D, P-substrate	Floating Gate, Word-line, N+, N+, S, D, P-substrate	Control Gate, Access Gate, Floating Gate, N+, N+, N+, S, D, P-substrate
Reference	**[36](1998) etc.**	**[29] (1982)**	**[31] (1994)**	**[38](1997),[52](1999)**

Fig. 6.18a Embedded floating-gate flash devices

	(5)	(6)	(7)	(8)
	1Tr SONOS (NROM™)	1.5Tr SONOS (split-gate)	2Tr SONOS (PMOS)	1.5Tr Nano-dot
Program	CHE	SSI	CHE	SSI
Erase	HH	HH	FN	FN
Device Structure	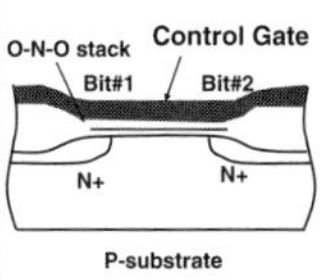 O-N-O stack, Control Gate, Bit#1, Bit#2, N+, N+, P-substrate	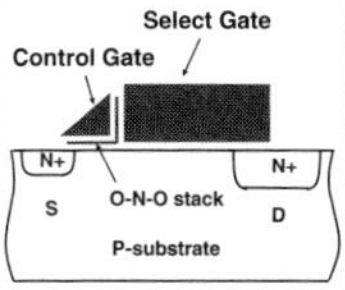 Select Gate, Control Gate, N+, N+, S, O-N-O stack, D, P-substrate	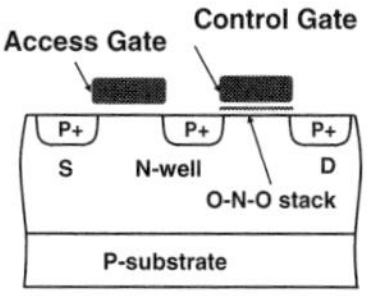 Access Gate, Control Gate, P+, P+, P+, S, N-well, D, O-N-O stack, P-substrate	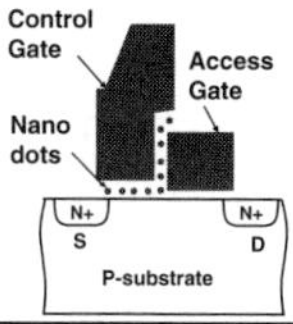 Control Gate, Access Gate, Nano dots, N+, N+, S, D, P-substrate
Reference	[14] (1999)	[9] (1997)	[37] (2006)	[55] (2007)

SSI : Source Side Injection
CHE : Channel Hot Electron Injection
HH : Hot Hole Injection
FN : Fowler-Nordheim tunneling

Fig. 6.18b Embedded charge-trapping flash devices

Structure	N. of poly/ SiN	P/E	Year	Ref.	Comment
split gate	2p	SSI/-	1982	[29]	EPROM w/1st SSI
split gate	3p	SSI/-	1986	[54]	EPROM w/SSI 1st named
split gate	2p	SSI/FN	1991	[34]	Embedded MCU products
split gate	2p	SSI/FN	1993	[25]	HIMOS
split gate	2p	SSI/PIP	1994	[31]	SST SuperFlash™
split gate	3p	SSI/PIP	2006	[10]	Erase gate
split-gate/SONOS	2p	FN/FN	1991	[47]	Split-gate/SONOS
split gate/SONOS	2p+SiN	SSI/BTBT-HH	1997	[9]	side-wall gate/SONOS
2Tr	2p	FN/FN	1999	[52]	read @0.9V
2Tr	2p	FN/FN	2001	[12]	read @1.2V
3Tr	2p	FN/FN	2000	[27]	3Tr-NAND
2Tr /SONOS	1p+SiN	CHE/FN	2006	[37]	NeoFlash
split gate/Si-NC	2p+NC	SSI/FN	2007	[55]	Split-gate Nanocrystal

Fig. 6.19 Embedded flash memory in literatures

6.2.1 Floating-Gate Flash Technology

The stacked-gate structure forming a floating gate to store signal charge has been explored and utilized in EPROM, EEPROM, and flash memory technologies, and now is the dominant technology of choice for embedded uses as well as discrete

NOR and NAND flash memory technologies. In this section embedded-specific floating-gate flash technologies are summarized.

6.2.1.1 1Tr-NOR Cell and Operations

Figure 6.20 describes conventional floating-gate 1Tr cell schematically and in SEM photographs. The floating-gate structure which stores charge in the floating polysilicon gate to alter the threshold voltage of n-channel memory transistor is in extensive uses in stand-alone discrete high-density flash memory devices such as NOR and NAND flash memories.

The 1-Tr NOR cell structure has been widely used also for embedded code storage with up to 10^4 times of program/erase typically. The device structure has stacked gates, control gate over floating gate forming an N-channel transistor cell. The bottom oxide is tunneling oxide and the top oxide acts as a charge-blocking layer for data retention.

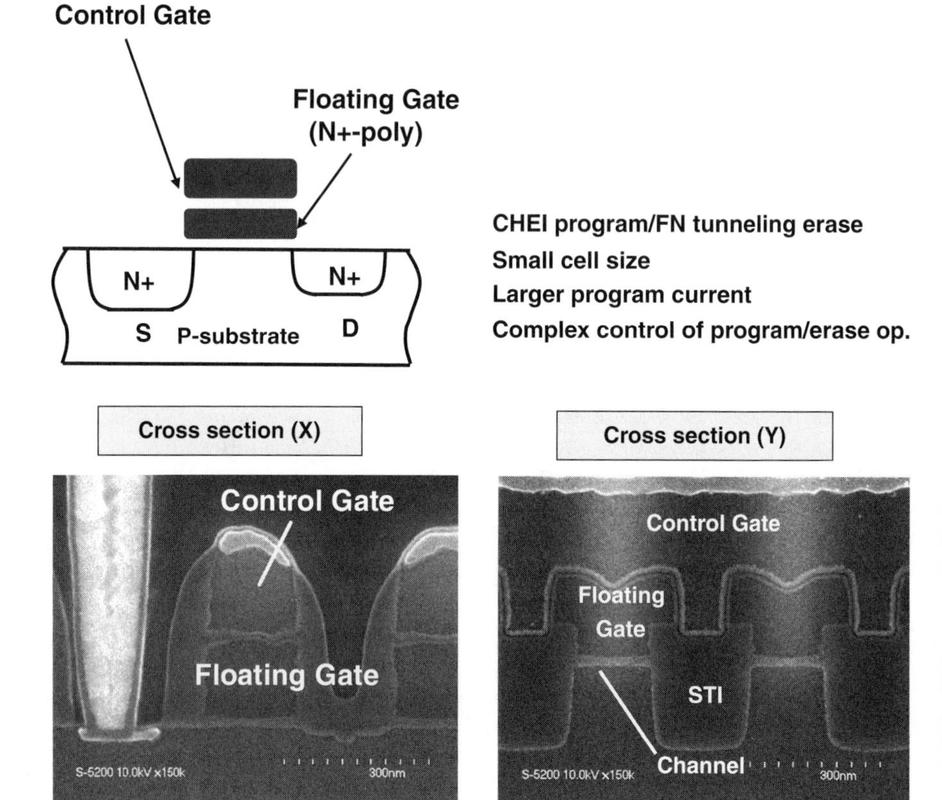

Fig. 6.20 Floating-gate 1Tr-NOR cell structure

The program/erase operations typically utilize channel hot electron (CHE) injection from the drain of the cell transistor and Fowler-Nordheim (FN) electron tunneling mechanism through the bottom tunneling oxide to the drain or channel (substrate), as shown in Fig. 6.21a. In the program operation, the source-to-drain current generates the channel hot electron, part of which is injected into the floating gate through the bottom oxide and negatively charges the floating gate, heightening the threshold voltage of the memory transistor. On the other hand, the erase operation is done at once in a sector, typically in K bits, for saving the erase time, in a "flash erasure" manner. All the memory cells in a selected block are erased, regardless of the previous stored data. The FN electron tunneling from the floating gate to drain or channel/substrate decreases the threshold voltage of the memory cell transistor toward the depression state. This 1Tr cell is applied very widely to embedded uses, because it has a small cell size of less than $10F^2$ (F: minimum feature size) and enjoys a mature state of technology by accumulated learning in discrete memory and MCU industries.

Typical features of 1Tr-NOR flash device listed in Fig. 6.21b shows limitations in this technology. By a cell read current in the range of 1–10 μA, up to 25 MHz random-access read operation is widely achieved, while it ranges up to 80 MHz based on the speed-enhanced cell and architecture optimization. The program/erase endurance is typically limited to 10 K cycles. The limiting factors for the data retention and P/E endurance in floating-gate cell are stress-induced leakage current (SILC) caused by damages introduced by hole conduction and charge trap gener-

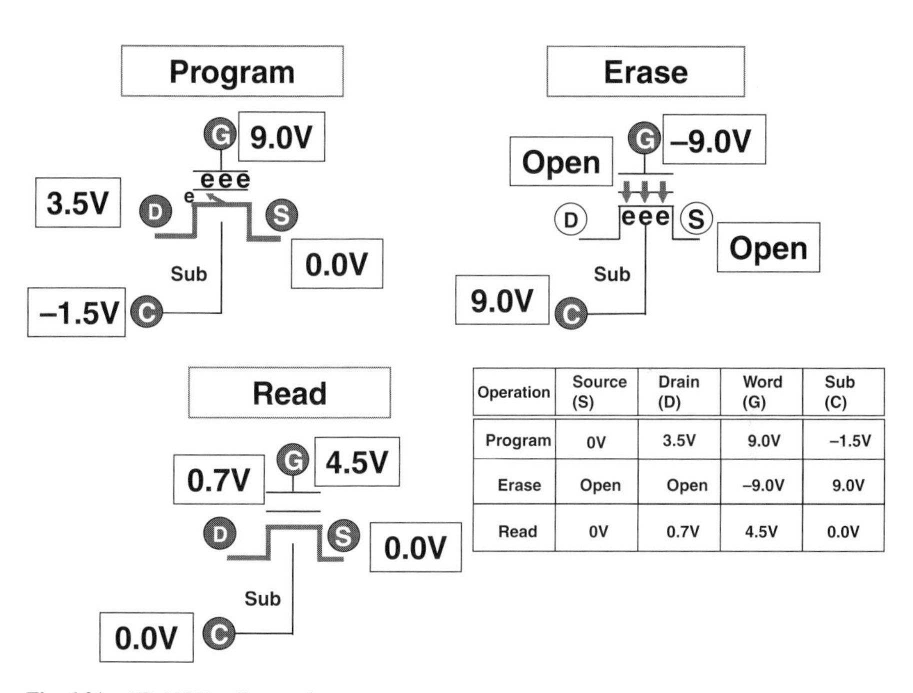

Operation	Source (S)	Drain (D)	Word (G)	Sub (C)
Program	0V	3.5V	9.0V	−1.5V
Erase	Open	Open	−9.0V	9.0V
Read	0V	0.7V	4.5V	0.0V

Fig. 6.21a 1Tr-NOR cell operation

Program Time = 100us/cell, Program Current = 100uA/cell (CHEI)
Erase Time = 500ms/cell, Erase Current < 0.1uA/cell (FN)
P/E cucles = 10K(max), Iread = 1–10uA/cell

Fig. 6.21b Typical features of 1Tr-NOR cell

ation in the tunneling oxide (SiO_2). This limits the scaling of the tunneling oxide thickness around 90–100 A, limiting the voltage scalability and read performance.

The drawbacks in 1Tr-NOR cell, especially for embedded applications, are as follows:

(1) The over-erasure, which causes decreasing threshold voltage of the cell *V*th(cell) over to the depression state, should be prohibited because the leakage current through unselected memory cells with reduced *V*th(cell) on the same bit-line with selected read cell degrades the read current due to increased cell leakage. This over-erasure problem inherent in the 1Tr-NOR cell array necessitates a precise control of the erased *V*th(cell) distribution by verify-after-erase operation and fine tuning of *V*th(cell) in the erase operation to prevent the unselected cells in read in the tail of the *V*th distribution from coming into weakly ON state. Typically the *V*th(cell) is 6 V/2 V for program/erase states at the centers of their distributions.
(2) Because of the inevitable requirements for enhancement mode *V*th(cell) of erased cells the control gate for word-line should be boosted up to around 5 V above externally applied power supply in the read operation. The boosted word-line introduces area-consuming, inefficient charge-pumping circuitry, making the 1Tr cell unsuitable for small-capacity, low-power embedded applications.
(3) The program current is rather large, for example, more than 100 μA/cell, because of inefficient channel hot electron injection mechanism. The efficiency is in the order of 10^{-6}, which is drastically enhanced by introducing the source-side injection mechanism described in the following parts in this section.

In reality various modes of disturb in the memory cell array inherently limit the device structure and operations more or less in every type of flash memory cell array. Three notable disturb modes in 1Tr-NOR flash memory array are

(1) Gate disturb in program: at the unselected cells in the selected word-lines (CG)
(2) Drain disturb in program: at the unselected cells in the selected bit-lines
(3) Drain disturb in read: at the unselected cells in the selected bit-lines.

Finely tuned device structures and operating conditions including disturb inhibit voltages are essential in the proper operation of the memory cell array. In the actual design of flash memory array care should be taken of the various disturb modes.

The 1Tr-NOR floating-gate flash technology suffers from scalability issues in channel length because of source-drain leakage, and tunnel oxide thickness because of program/erase endurance and data retention, leading to small read currents not meeting high-speed on-chip access time requirements. Additionally a large program current for CHEI sets an obstacle in achieving low-power program operations. In the overall technology construction, high-voltage generation and transfer, over-erasure problem, and disturb elimination mainly restrict the actual device structure, operating conditions, and chip-area efficiency in 1Tr cell technology. Because of these factors 1Tr-NOR cell inherently does not meet the embedded requirements for very high speed read with large signal margin, low-power P/E and read operation, compact peripheral circuitry and simple program/erase algorithms for smaller chip area penalty. However, because of the learning effect in major discrete flash memory manufacturers and of the high-density nature of the memory cell 1Tr-NOR has been most commonly used in embedded flash applications.

One way of alleviating the scaling limitation of 1Tr cell is poly-to-poly tunneling erase scheme to allow for thicker oxide structure by concentrated electric field to eliminate the hole conduction [39]. A two-channel cell in Fig. 6.22 [30] also mitigates the influence of damage introduced in program operation to the read current by separating the read current path from the program current path suffering from damage, extending the P/E limitation to 100 K–1 M cycles.

For low-voltage, low-power operations 1.5Tr split-gate and 2Tr structures have been explored as described in the following sections. P-channel device provides a much enhanced program current with higher efficiency, which is another way of improving the device performance. An extensive discussion on p-MOS cell is found in Ref. [5].

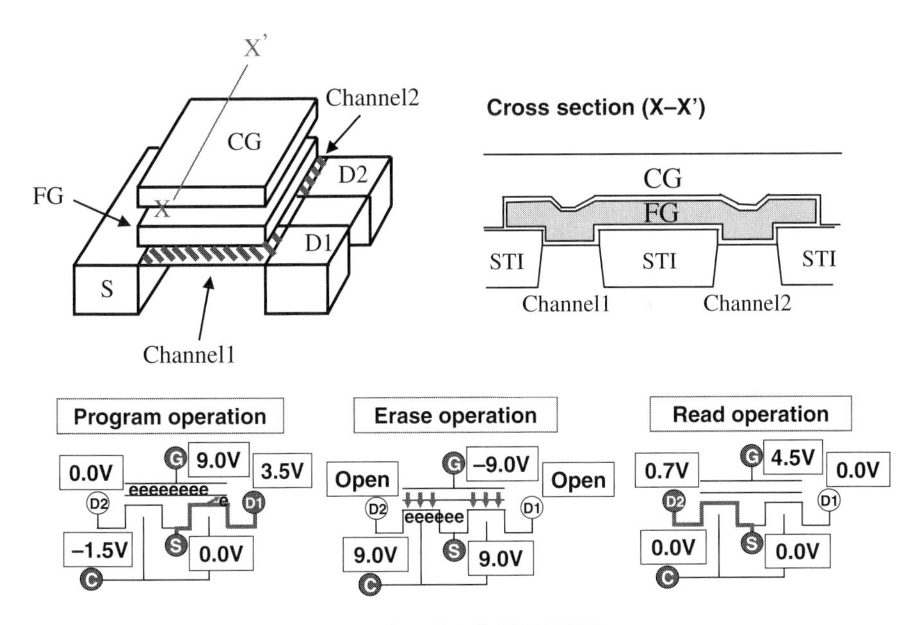

Fig. 6.22 A two-channel cell and operation [30] (© 2008 IEEE)

6.2.1.2 Split-Gate Cell (1.5Tr-Cell)

One important variant in the floating-gate technology for aggressive scaling and performance enhancement in embedded applications is the "split-gate" structure shown in Fig. 6.23. A split-gate structure is a divided gate spaced each other by a gap of 10–100 nm with independent voltage bias control for efficient carrier generation at the channel and injection into the storage node. The idea of a split-gate is to use source-side injection mechanism for program operation dated back to early 1980s. In 1982, the floating-gate type EPROM with the split-gate structure was proposed to demonstrate lower programming voltage and significantly lower channel current than the channel hot electron program scheme, shown in Fig. 6.23(1) [29]. In 1986, another device structure was demonstrated for EPROM with a side-wall gate for controlling cell current. The main gate is composed of a stack of a floating gate and a control gate as shown in Fig. 6.23(2) [54]. Both the floating gate in memory part and the floating side-wall gate are coupled to the control gate to provide the programming current efficiently, which also demonstrated an efficient programming scheme, named "source-side injection". In 1993, another device structure was introduced for a flash EEPROM, as shown in Fig. 6.23(3). In addition to the split-gate structure a program gate (PG) helps enhance the program efficiency. Figure 6.23(4) describes another split-gate structure, the SST Super-Flash cell, which utilizes a poly-poly FN tunneling mechanism for erase operation.

Figure 6.24 describes the source-side injection mechanism in the split-gate cell compared with the channel hot electron injection in stacked-gate 1Tr-NOR cell.

	(1)	(2)	(3)	(4)
	1.5Tr cell (split-gate)	**2Tr cell**	**1.5Tr cell**	**1.5Tr cell (SuperFlash™)**
Program	**SSI**	**FN**	**FN**	**SSI**
Erase	–	**FN (poly-sub)**	**FN (poly-sub)**	**FN (poly-poly)**
Device Structure	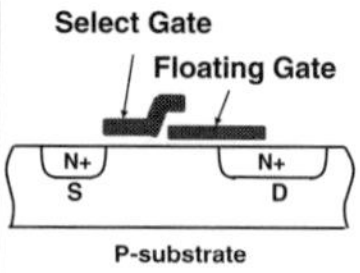 Select Gate, Floating Gate, N+, N+, S, D, P-substrate	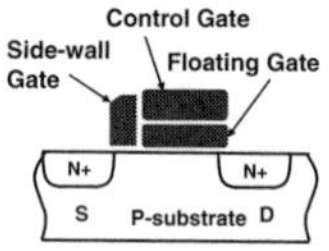 Control Gate, Side-wall Gate, Floating Gate, N+, N+, S, P-substrate, D	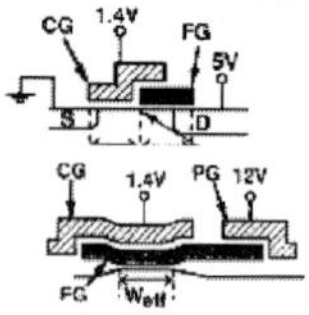 CG, 1.4V, FG, 5V, S, D, CG, 1.4V, PG, 12V, FG, W_{eff}	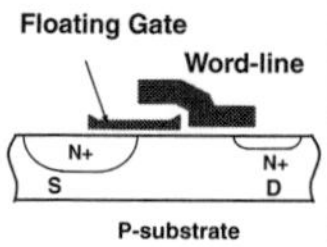 Floating Gate, Word-line, N+, N+, S, D, P-substrate
Reference	[29] (1982)	[54] (1986)	[25] (1993)	[31] (1994)

SSI : Source Side Injection
CHE : Channel Hot Electron Injection
HH : Hot Hole Injection
FN : Fowler-Nordheim tunneling

Fig. 6.23 Development of split-gate devices utilizing source-side injection

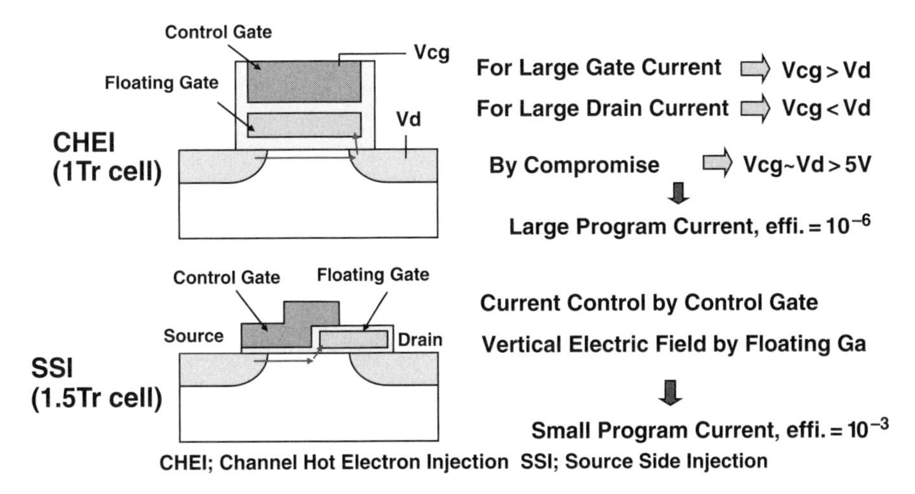

Fig. 6.24 Source-side injection mechanism

The stacked-gate structure utilizes the hot electron generation in silicon and the injection of electron into the floating gate. For enhancing hot electron generation, gate voltage must be lower than drain voltage. However, for increasing gate current to charge the floating gate, gate voltage must be larger than drain voltage. Since two voltage conditions conflict each other, a compromised condition of larger gate and drain voltage is used with lower efficiency for program in the order of 10^{-6}, causing a large program current.

A split-gate structure realizes a very efficient programming of the order of 10^{-3}. By adjusting the control-gate voltage, it is possible to minimize channel current for programming. In addition, with a strong vertical electric field by the floating gate, channel electrons are injected into the floating gate efficiently. In the split-gate structure the freedom of gate bias conditions maximizes the lateral and vertical field peaks overlapping at the same position at the gap. This realizes a favorable efficient hot electron generation and injection into the floating gate. The gap position is at the source side of the memory transistor, hence the name "source-side injection" in contrast with drain side injection in the CHEI in 1Tr stacked gate cell.

Figure 6.25 shows an example operation of a split-gate device listed in Fig. 6.23(4). During the program operation word-line is biased to the voltage around threshold voltage, while high voltage is applied to the source. Since the floating gate is coupled to this overlapped source region, the floating gate is also biased to high voltage, depending on the coupling coefficient Cs-fg, the capacitance between source and floating gate. Electrons from the drain region is accelerated in the channel, and injected into the floating gate, according to the source-side injection mechanism. In the erase operation word-line is biased up to around 14 V, and electrons stored in the floating gate are ejected to the word-line by a tunneling mechanism. The sharp edge of floating-gate poly-Si enhances the electron tunneling by intensified electric field around the edge. This structure allows the use of relatively thick oxide and is able to alleviate the stress-induced leakage current (SILC) in FN tunneling to drain/channel [39]. This is important to enhance the program/erase (P/E) endurance. The P/E limiting factor in 1Tr-NOR cell is damages in SiO_2 by hole conduction in P/E operations inducing SILC (Stress-Induced Leakage Current), which is mitigated by using only electron conductions in P/E operations in this structure. Note that in Fig. 6.25 the source and drain notations are directed to the read operation, which is reversed from Fig. 6.24.

In the read operation word-line and the drain are biased to Vcc and the current flowing through the memory cell provides read signal that is sensed by the sensing circuitry. Read current does not flow when the floating gate is programmed and it flows when the floating gate is erased in sensing the stored cell data.

The actual program and erase characteristics of a split-gate cell in Fig. 6.26 [25] show exponential dependence on the program and erase time, respectively, providing fast program/erase operations, by virtue of the freedom of setting bias conditions. In the lower left of the figure, threshold voltage changes as a function of program time with program gate voltage dependence is demonstrated. Drain is biased to 5 V, CG to 1.4 V, and PG to 10–15 V. By this PG bias, floating-gate potential gets higher by capacitive coupling to open the channel under FG. Threshold voltage increases exponentially with respect to program time, and the threshold voltage shift of 4 V after 2–10 μs is observed depending on the program gate voltage Vpg. The erase operation by the poly-to-poly tunneling of electrons from FG to CG also shows exponential dependence on the erase time, as shown in the lower right figure.

As a simulated electrical potential along the channel assuming a dual-gate structure with different oxide thickness in Fig. 6.27 [29] shows, smaller Vg1 is better for carrier injection, resulting in program current reduction in the split-gate cell.

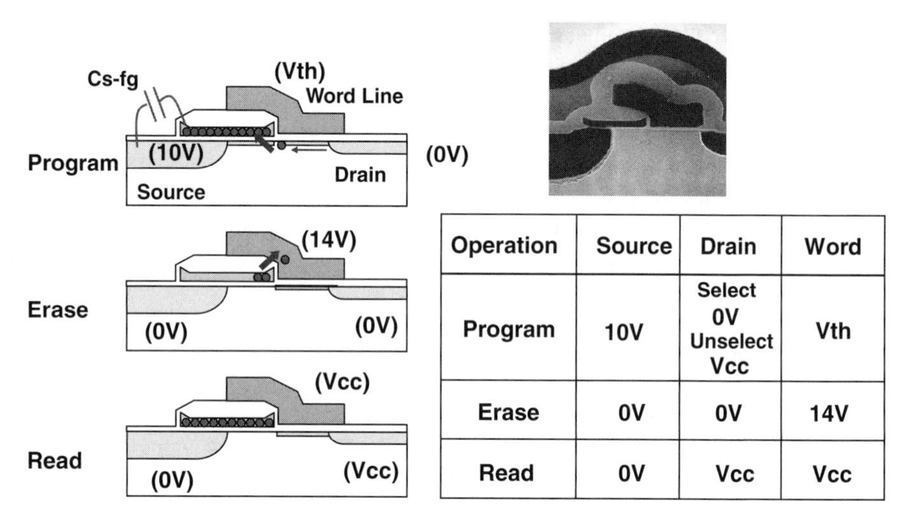

Operation	Source	Drain	Word
Program	10V	Select 0V Unselect Vcc	Vth
Erase	0V	0V	14V
Read	0V	Vcc	Vcc

Fig. 6.25 A split-gate cell structure and operations [31] (© 2008 IEEE)

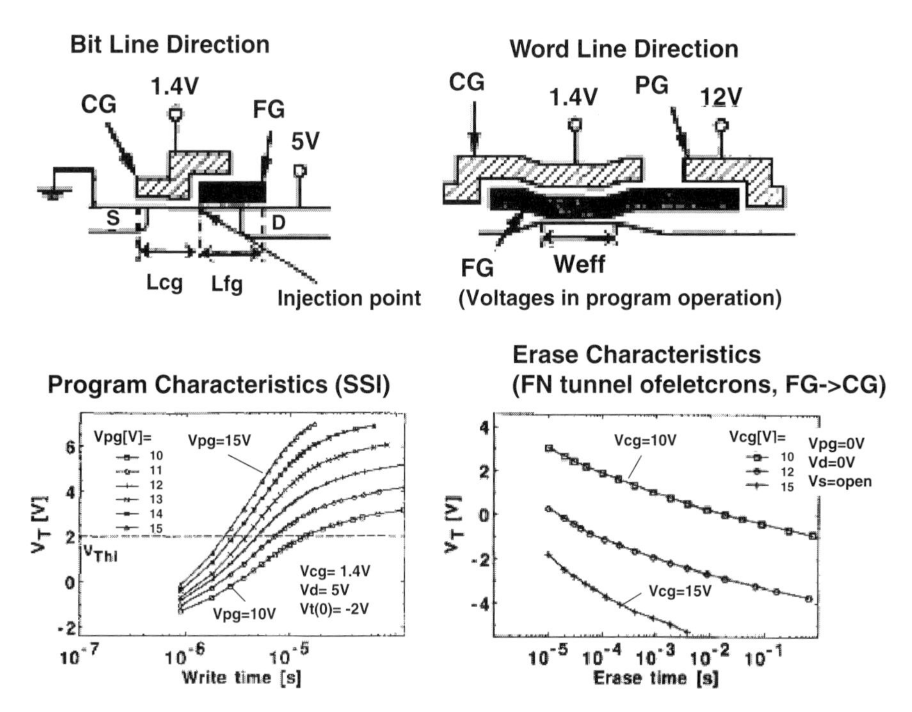

Fig. 6.26 A split-gate cell structure and program/erase performance [25] (© 2008 IEEE)

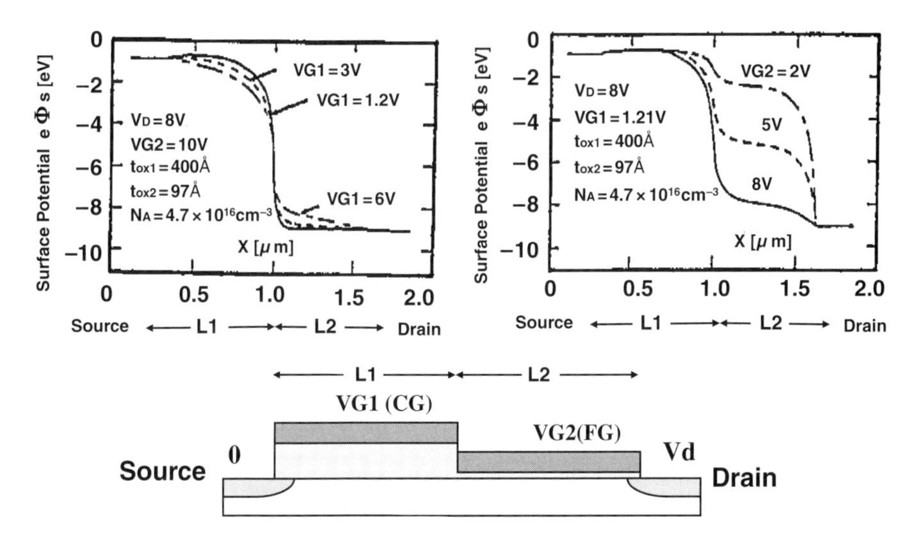

Fig. 6.27 Enhanced electric fields in split-gate structure [29] (© 2008 IEEE)

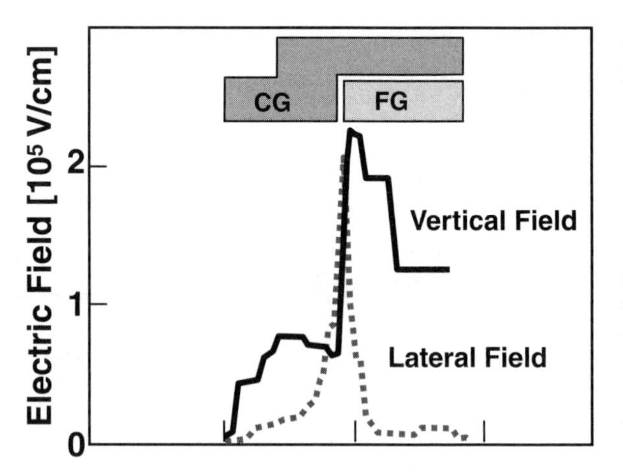

Fig. 6.28 Coincident electron generation and injection points [24] (© 2008 IEEE)

A simulated electric field along the channel of the split-gate structure in Fig. 6.28 [24] shows that lateral electric field enhances carrier energy, and vertical field helps inject electrons into a floating gate. In the case of the split-gate structure, both electric fields have the maximum field strength almost at the same point near the gap between the floating gate and the control gate, revealing why the split-gate structure demonstrates very efficient carrier injection for programming with limited channel current.

A direct comparison of gate current by source-side injection and channel hot electron injection of MOSFET in Fig. 6.29 [8] indicates that SSI achieves larger gate current than CHEI, although control-gate voltage is much smaller. More than three orders of magnitude larger gate current is observed, which demonstrates an efficient programming by SSI.

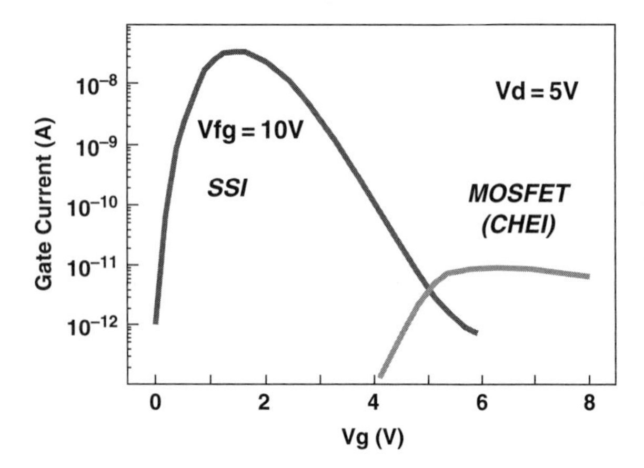

Fig. 6.29 Efficient electron injection by source-side injection [8] (© 2008 IEEE)

The disturb problems specific to various split-gate devices that restrict the structural parameters and operating voltages as in the case of 1-Tr NOR cell are extensively addressed in references [10, 25, 26]. Careful structural design and considerations in operation scheme are essential also in the split-gate cell array.

In summary the split-gate structure achieves a very efficient programming operation at high speed and low voltage, enhancing the current efficiency in program from 10^{-6} by CHEI to 10^{-3}. It also eliminates the over-erasure problem realizing high-speed read with nonboosted word-lines and simple program/erase procedures.

6.2.1.3 2Tr Cell for Low-Voltage, Low-Power Operations

Another variation from the basic 1Tr floating-gate device for small-capacity embedded applications is 2Tr-cell, which inserts a transfer gate in the memory cell of 1Tr cell in order to eliminate the over-erasure problem to achieve high-speed and low-voltage operations.

In Fig. 6.30 a 2Tr structure utilizing the cell transistor stack for the transfer gate of the memory cell is shown [13]. This type of memory cell usually pursues low-voltage read operation by nonboosted word-line voltage, in some cases exploiting the thin-oxide logic transistor for the transfer gate of the cell. Typically 0.9–1.2 V read operations are achieved [13, 38, 52]. A low-power program/erase operation is achieved by using the FN tunneling mechanism for program/erase [33]. The program voltage is lower than 1Tr cell since high threshold voltage is not necessary in program state because of negative erase threshold voltage. 2Tr cell is a reasonable practical compromise that takes advantage of matured 1Tr cell structure with an additional equivalent transfer gate transistor or with a transistor available in the baseline CMOS technology for performance enhancement. The 2Tr cell has been a favored choice in embedded applications for many years.

In summary the floating-gate technology has been utilized extensively because of the maturity by the discrete memory development, but because of the drawbacks in voltage scalability and CMOS compatibility required by embedded uses some of the industry segments have preferred split-gate structures. Intrinsic limitations in the floating-gate structures have emerged as the limiting factors for ever higher reliability requirements and lower cost. The split-gate cell benefits from the

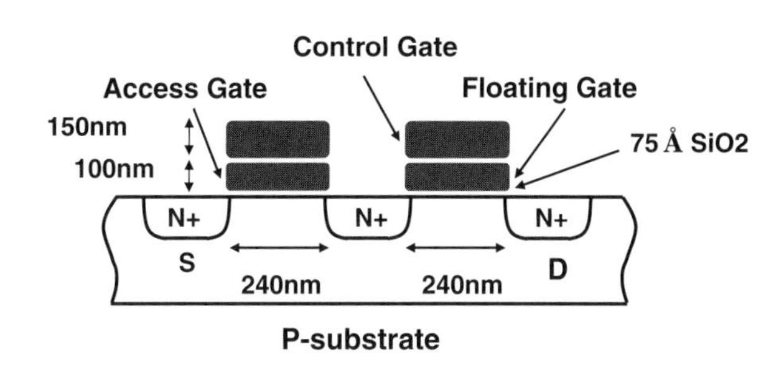

Fig. 6.30 A 2Tr cell structure [13, 38, 52]

high-speed/low-voltage source-side injection especially for embedded uses. A nice collection of extensive reviews on floating-gate flash memory technology is found in Ref. [7].

6.2.2 Charge-Trapping Flash Technologies, SONOS and Nano-dot

One of the disadvantages of the floating-gate technology is if it contains a point defect between the floating gate and the surrounding conductors such as substrate and poly-silicon control gate, the whole charge stored in the floating gate is lost through this point defect. The defects related to SILC (Stress Induced Leakage Current) is the major limiting factor in the scaling of the tunneling oxide thickness in the floating-gate technology, which limits the operating voltage scaling and the read access speed. This problem is mitigated by introducing discrete localized charge-trapping structures such as SONOS and Si nano-dots. SONOS (Silicon-Oxide-Nitride-Silicon) or MONOS (Metal-Oxide-Nitride-Oxide-Silicon) structure offers a discrete charge-trapping mechanism where one point of defect does not influence the whole system of charge storage.

A typical structural dimension in MONOS device is shown in Fig. 6.31. The bottom oxide is tunneling oxide and the top oxide is a blocking oxide to suppress charge injection from the electrode for good charge-retention characteristics. For operation voltage reduction and read access performance in an embedded LSIs, SONOS structure is considered to be favorable because of this thin-film structure.

Figure 6.32 illustrates the defect-resistant nature of charge-trapping structures as compared with a floating-gate structure. Since the injected electrons distribute uniformly over the conducting floating gate, a conductive defect in the oxide film around floating gate causes loss of all the stored charge. In contrast in SONOS structure, since charges are stored by spatially separate traps in the nitride film and at the interface of nitride and oxide, only charges near defects are lost and all of the other charges are preserved, providing a defect-resistant nature.

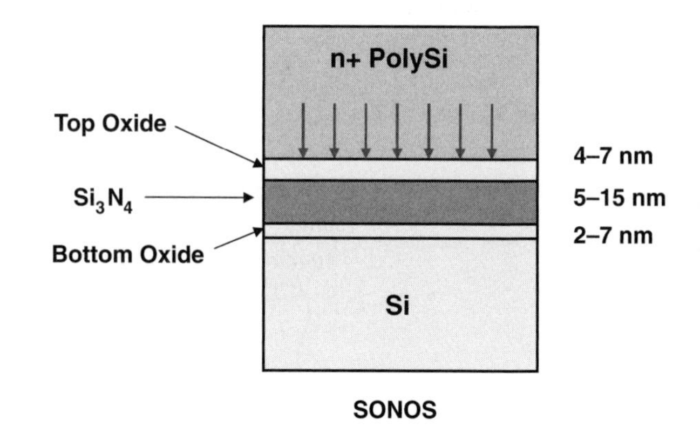

Fig. 6.31 A SONOS structure

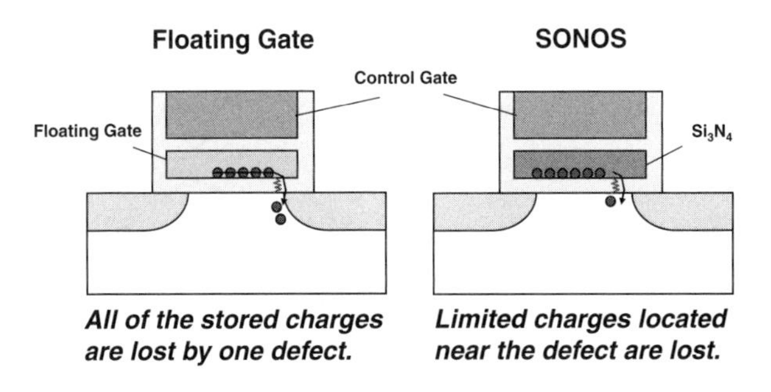

Fig. 6.32 Floating-gate and SONOS structures

Although SONOS technology has more than 30-year history of research, its advantage has only recently demonstrated by discrete memory product in NROM™ with 2 bits/cell storage scheme [14]. By virtue of the structural simplicity in vertical dimensions, high-density NAND is projected to adopt SONOS technology [50].

6.2.2.1 SONOS Basics

Figure 6.33 shows the basic SONOS operation conditions. In program operation, control gate is biased up to Vp while keeping the source voltage of the selected cell at ground. Electrons are injected from the substrate into the nitride film through the thin tunnel oxide and captured by SiNx and SiOx/SiNx interface traps. For erase operation, CG voltage is set to 0 V, and Vp is applied to the substrate. During erase operation, both source and drain are floating. Electrons in the nitride traps are ejected to the substrate or holes are injected into the nitride. For read operation, drain and control gate are biased at Vcc to read the read signal. Since small current FN tunneling is used both for the program and erase operation, large current does not flow, resulting in low-power program/erase operations.

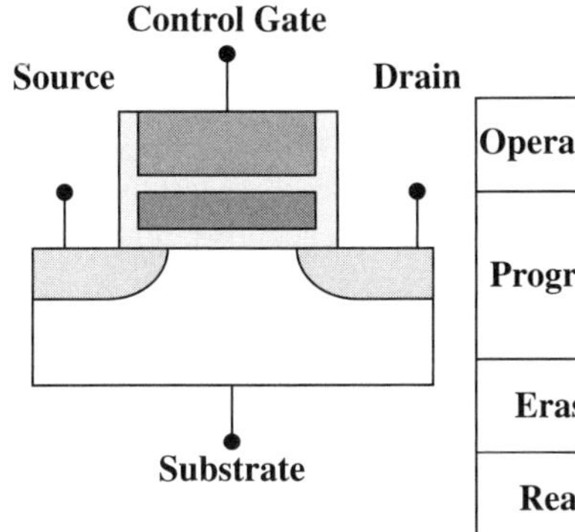

Operation	Source	Drain	CG	Sub
Program	Select 0V Unselect Vp	Open	Vp	0V
Erase	Open	Open	0V	Vp
Read	0V	Vcc	Vcc	0V

Fig. 6.33 Basic SONOS operations

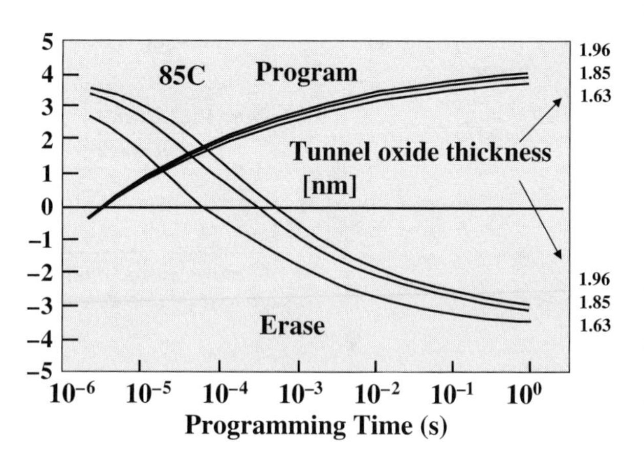

Fig. 6.34 Program and erase characteristics of an SNOS cell [45] (© 2008 IEICE)

The SONOS performance directly inherits from former semiconductor-nitride-oxide semiconductor (SNOS) device structure well researched for a long time since 1980s. Figure 6.34 exemplifies program (write) and erase characteristics of an SNOS memory cell [45]. Initial cell has a natural threshold voltage of Vthi. By electron injection into the nitride film by FN tunneling, threshold voltage increases, and threshold voltage shift shows exponential dependence on programming time. Erase is performed also by using FN tunneling of holes into the nitride film through the thin tunneling oxide. Threshold voltage decreases exponentially with erase time. Figure 6.35 shows an example of the data-retention characteristics of an SNOS memory cell at 125°C [56]. Memory cells have a tunnel oxide with the thickness of 1.6 and 2.1 nm. A typical SNOS device shows charge decay logarithmically, depending on the holding time. The SONOS memory device characteristics resemble these program/erase and charge-retention characteristics of the SNOS device. Typically the controlled charge decay in well-designed SONOS cell achieves the data retention for 20 years at 125°C.

6.2.2.2 Recent SONOS Trends

Figure 6.36 lists recently reported flash memory cell structures based on charge-trapping SONOS devices. Conventional SONOS devices store charges uniformly above its channel by using FN tunneling mechanism. However, recently a number of localized charge storage devices have been proposed to exploit the SONOS advantages.

A split-gate device published in 1997, shown in Fig. 6.36(1), consists of select gate and memory gate formed by a side-wall spacer of the select gate transistor. The memory gate channel length is shortened by the side-wall process and the source-side injection is employed for programming. For erase operation, hot hole generated by band-to-band tunneling (BTBT) is used. In 1999, a SONOS device enabling two-bit storage per cell named nitrided ROM (NROM) (Fig. 6.36(2)) was reported. In

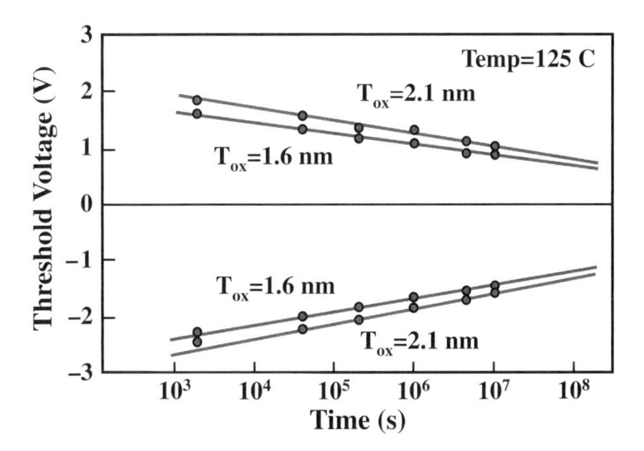

Fig. 6.35 Data-retention characteristics of a MNOS cell [56] (© 2008 IEEE)

	(1)	(2)	(3)	(4)
	1.5Tr SONOS (split-gate)	1Tr SONOS (NROM™)	Twin-MONOS (split-gate)	2Tr SONOS (PMOS)
Program	SSI	CHE	SSI	CHE
Erase	HH	HH	HH	FN
Device Structure	Select Gate, Control Gate, N+, N+, S, O-N-O stack, D, P-substrate	O-N-O stack, Control Gate, Bit#1, Bit#2, N+, N+, P-substrate	Select Gate, Control Gate, Control Gate, N+, N+, S, O-N-O stack, D, P-substrate	Access Gate, Control Gate, P+, P+, P+, S, N-well, D, O-N-O stack, P-substrate
Reference	[9] (1997)	[14] (1999)	[20] (2000)	[37] (2006)

Fig. 6.36 Development of charge-trapping SONOS cells

2000, a cell using two side-wall spacers for two bits of data storage (Fig. 6.36(3)) were reported. These devices are expected to be new players in embedded flash technologies for MCU and SoC by their easy adaptability in standard CMOS logic process. SONOS device also merits from the 2Tr cell configuration in Fig. 6.36(4) by a very simple structure achieving cost-competitiveness. These devices have a distinctive feature that memory structure is simple, resulting in the process friendly technology with fine CMOS, and are expected to be suitable memory technology for embedded applications.

6.2.2.3 SONOS Cell with Localized Charge Trapping

Aside from the two bits/cell operation, NROM™ proposes a practical scheme of operating SONOS device, which has found a commercially successful discrete

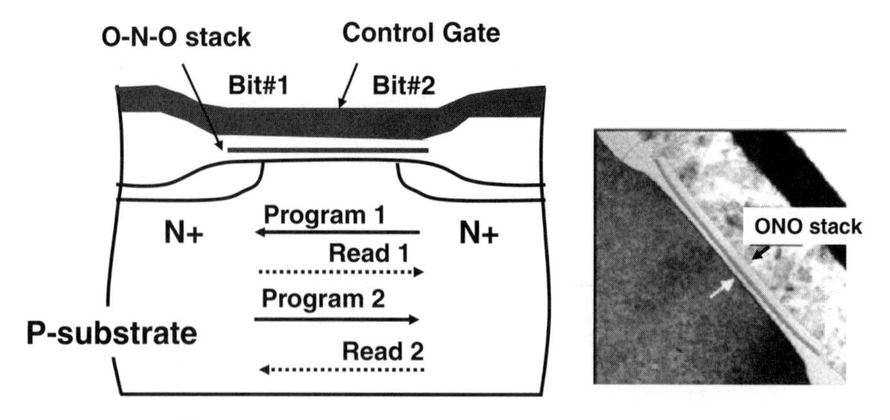

Fig. 6.37a NROM^{TM} (nitrided-ROM) structure [41] (© 2008 IEEE)

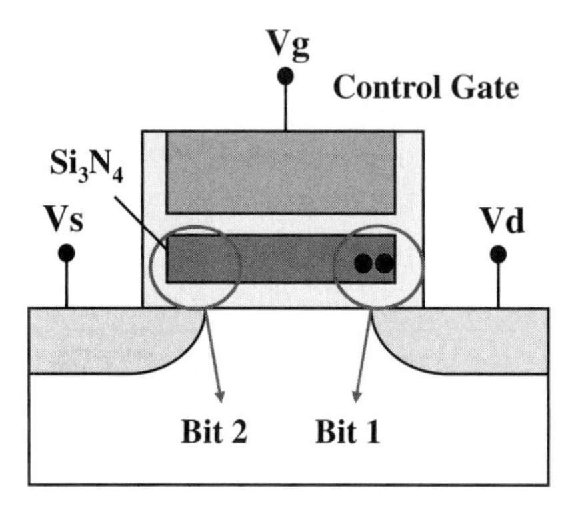

		Program	Erase	Read
Bit 1	Vg	9.0V	–5V	3.0V
	Vd	4.5V	5V	0V
	Vs	0V	0V	1.5V
Bit 2	Vg	9.0V	–5V	3.0V
	Vd	0V	0V	1.5V
	Vs	4.5V	5V	0V

Fig. 6.37b Operations of NROM^{TM}[15]

memory product. Based on this scheme, embedded SONOS device will be discussed. Figure 6.37a depicts the NROM^{TM} structure alongside the word-line direction [41]. ONO stack structure consists of bottom oxide, nitride, and top oxide with thickness of 7, 5, and 9 nm, respectively, and the buried diffusion layer beneath the thin LOCOS oxide act as a bit-line.

As shown in Fig. 6.37b program operation is performed by channel hot electron injection, and erase operation done by hot hole injection. For programming Bit-1, high voltage (9 V) is applied to the gate and the drain is biased to 4.5 V for inducing drain avalanche. Erase is performed by applying negative voltage (–5 V) to the gate and relatively high voltage (5 V) to the drain. Generated holes by BTBT (band-to band tunneling) are injected into the nitride film by negative electric field induced by the gate electrode, causing electron and hole recombination in the SiNx film. Read operation is performed by applying Vdd to the source, while the drain is grounded. If electrons are injected into the nitride film near the drain, threshold voltage is measured to be increased under this voltage condition. By changing the role of the source and the drain, it is possible to inject electrons to Bit-2 without affecting the Bit-1. For programming, 4.5 V is applied to the source, while keeping the drain ground. Erase is performed by biasing the gate at –5 V, while raising the source voltage to 5 V. Read operation is also performed by changing source and drain. Through this symmetrical operation, the memory cell is able to program and erase each of 2 bits in a cell independently. This is the principal mechanism of two-bit storage of this NROM™ device, utilizing the properties of localized charge trapping.

In the program operation described in Fig. 6.38a channel hot electron generation is applied to SONOS as in the floating-gate NOR structure. High gate voltage and drain voltage are needed for hot electron generation, resulting in a large program current compared with source-side injection mode. Figure 6.38b shows the erase operation using hot hole injection. A negative bias is applied to the gate and large positive voltage is applied to the drain, and it causes band-to-band tunneling in the silicon near the drain. Thus, holes are generated and injected into the nitride film, and recombination between trapped electrons and holes occur in the nitride film. Elimination of electrons lowers threshold voltage of the memory cell, resulting in the erase state. Figure 6.38c describes a dynamic behavior in program and erase operations with bit#1 and bit#2 operated independently of each other.

SONOS storage device has had an inherent drawback in data-retention performance because of the shallow potential barrier and existence of interface traps at the SiOx/SiNx interface [8]. Efforts in material and structure research are being made to improve the data-retention performance. Moreover, localized SONOS storage scheme suffers from the stored charge distribution problem because stored charges are isolated in the charge traps. The injected electrons distribute over the SiO/SiN

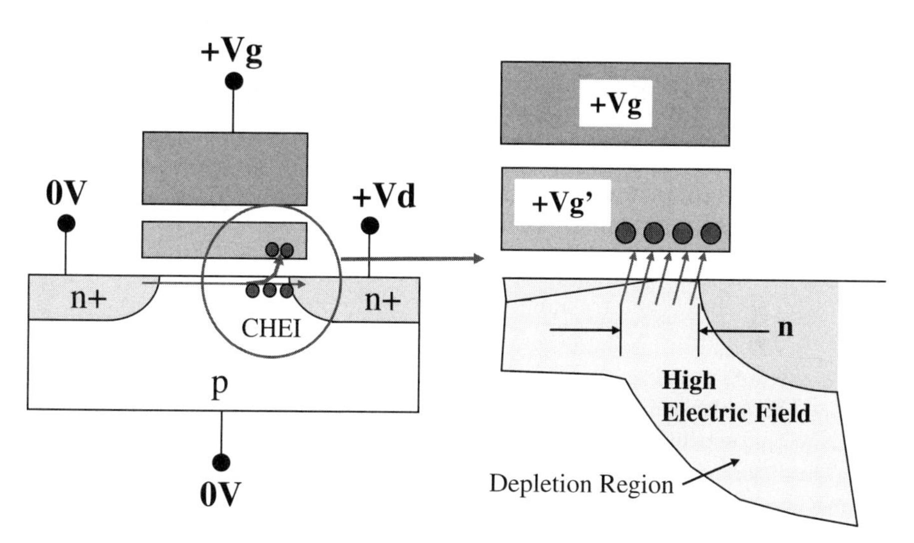

Fig. 6.38a Program mechanism of NROM™ (nitrided-ROM); CHEI (channel hot electron injection)

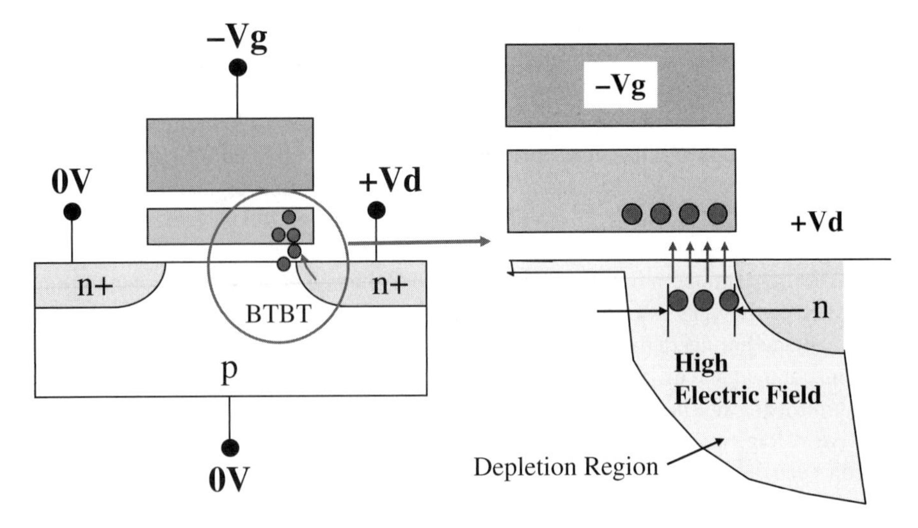

Fig. 6.38b Erase mechanism of NROM™ (nitrided-ROM);, BTBT (band to band tunneling).

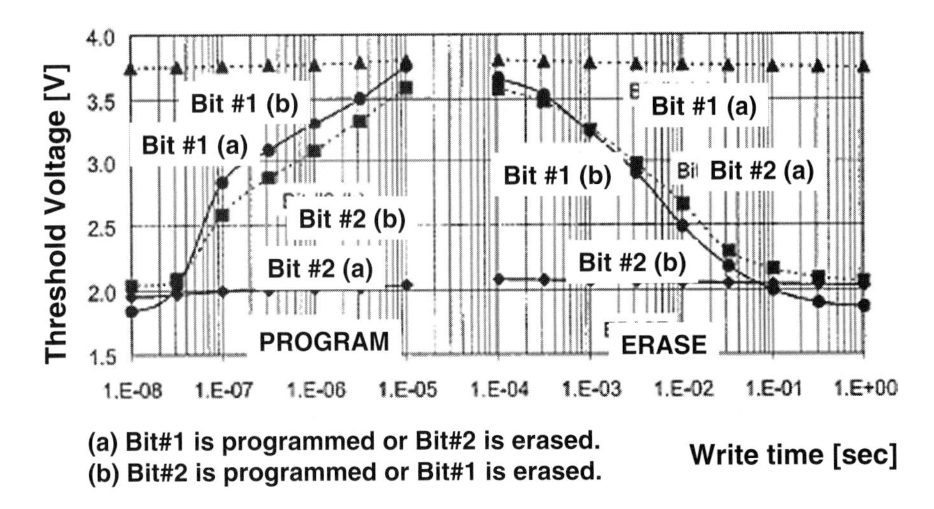

Fig. 6.38c Program/erase speed of NROM [15] (© 2008 IEEE)

interface as well as in traps in SiNx film. Because of the nonconducting property of SiNx and SiOx the spatial distribution of injected electrons by program operation is ideally to be compensated by holes in erase operation with the same spatial distribution. The difference in charge distributions for electrons over the whole SiN element both in horizontal and vertical directions has loomed up as an important factor in limiting the P/E endurance [16, 17].

Figure 6.39 shows an example of charge distribution simulation according to device operation mode, P/E bias conditions, and structural parameters in NROM™

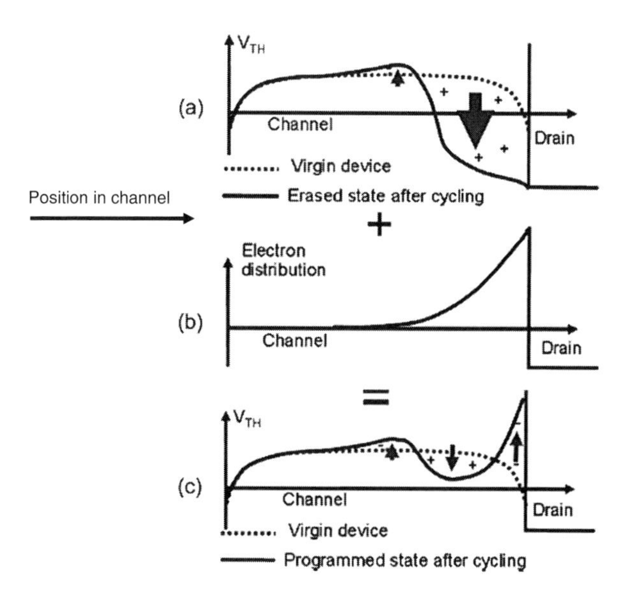

Fig. 6.39 A charge distribution in NROM™-like SONOS structure [17] (© 2008 IEEE)

type of SONOS cell, at the charge storage region near drain [17]. The CHE-injected electrons tend to be distributed narrowly near the drain, which BTBT/HH erase does not compensate spatially. In the data-retention period after program, lateral charge redistribution in the SiNx film degrades the data retention [17].

Optimized program and erase conditions to match the programmed electron distribution with hole distribution for erase shows dramatically improved performance in P/E cycles, especially under the high-temperature conditions, as shown in Fig. 6.40 [16]. Investigations into the charge behavior in program/erase operations offer much improved performance in SONOS devices.

6.2.2.4 Split-Gate SONOS Cell

The split-gate structures have been also tried on SONOS structures [47]. Figure 6.41 shows an implementation of SONOS in a split-gate structure [9], which is another example of the device using localized charge storage. In this split-gate type memory cell using a side-wall gate as a memory gate, the ONO dielectric film is formed both on silicon substrate and side-wall of the select gate. Source-side injection is employed for programming and BTBT hot hole injection operation is implemented for erase operation.

The operating bias conditions of this cell is listed in Fig. 6.41. In program operation 9 V is applied to the memory gate, while select gate is biased at 1 V, a little higher than threshold voltage of the select gate. Thus, it is possible to control programming current as small as possible. Electrons are injected near the corner of the triangle side-wall gate by the source-side injection mechanism. For erase operation,

	Vs [V]	Vd [V]	Vg [V]	Time
A Prog.	0	5	9.5	1us
A Erase	0	5	−7	100us
B Prog.	−2	1.5	10	20ms
B Erase	4	8	−2	10us
C Prog.	−0.5	4.5	9	1us
C Erase	3	7	3	50us

Conditions of program/erase above are denoted as ; Ap, Ae, Bp, Be, Cp, Ce in the graph showing Vth shift.

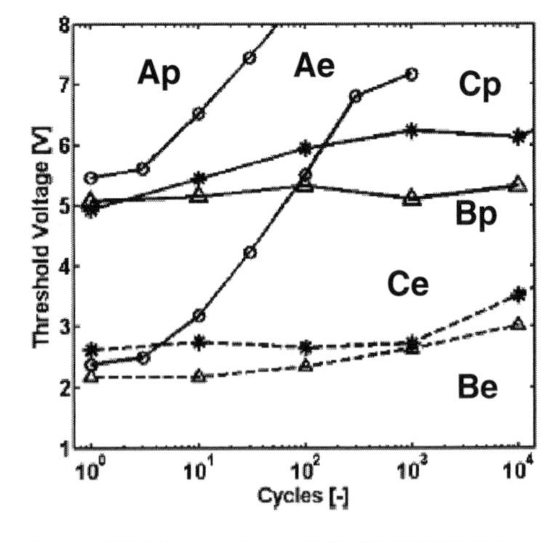

Fig. 6.40 A P/E endurance performance according to P/E bias conditions [16] (© 2008 IEEE)

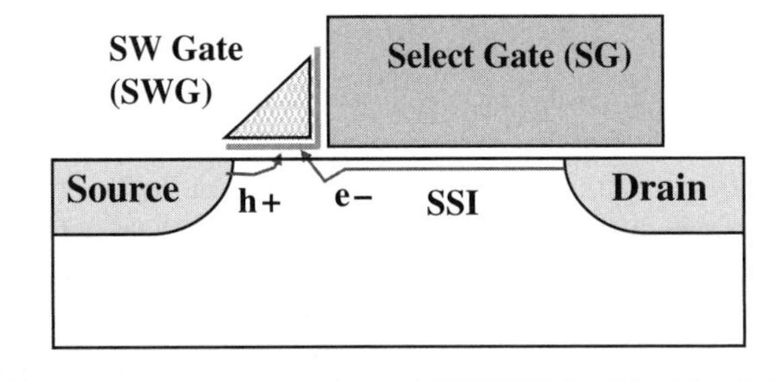

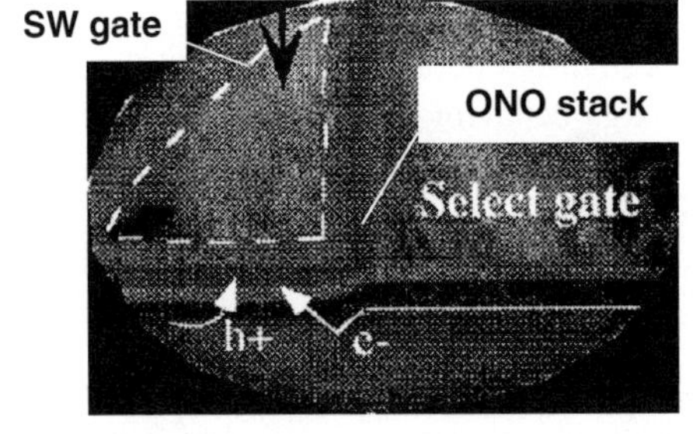

	Vs	Vswg	Vsg	Vd
Program	5	9	1	0
Erase	7	–9	0	0
Read	0	1.8	1.8	1.8

Fig. 6.41 A split-gate SONOS cell structure and operating conditions [9] (© 2008 IEEE)

deep negative voltage is applied to the gate and high positive voltage is applied to the source for hot hole generation and injection.

In Fig. 6.42 are shown typical program and erase characteristics. One of the striking points here is a much faster programming than FN tunneling program in Fig. 6.34. It is possible to program a cell within micro-second time. It shows very fast threshold voltage rise at the early stage of programming, and threshold voltage becomes saturated. Erase is also faster compared with FN tunneling cell. Increasing

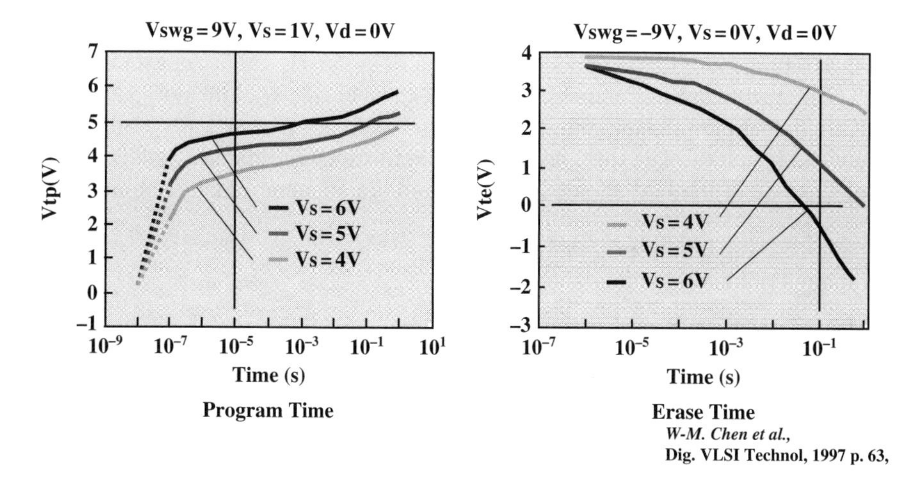

Fig. 6.42 Program and erase performance of a split-gate SONOS cell [9] (© 2008 IEEE)

source bias raises erase speed due to the enhanced BTBT hot hole generation and injection at increased source bias.

Because of the requirements for high-speed low-voltage operations in embedded applications, channel hot-electron (CHE) or source-side injection (SSI) for program and band-to-band hot-hole (BTBT/HH) injection from the substrate for erase are preferred choices of physical mechanisms for embedded SONOS devices.

In the basic structure of the split-gate SONOS cell, we have two candidates as shown in Fig. 6.43. By using double poly-Si layers for transfer gate TG (select gate) and memory gate MG (control gate) above SONOS, both MG-first and TG-first structures are available in split-gate SONOS devices [55]. Although this reference considers structures in silicon nano-dot device this can be extended to conventional SONOS devices. The embedded applications often prefer TG-first structure because this offers a CMOS-logic-compatible high-performance transfer gate transistor structure. Because the TG-first structure is not possible by the structure of floating-gate devices, it can be an advantage for SONOS in embedded applications.

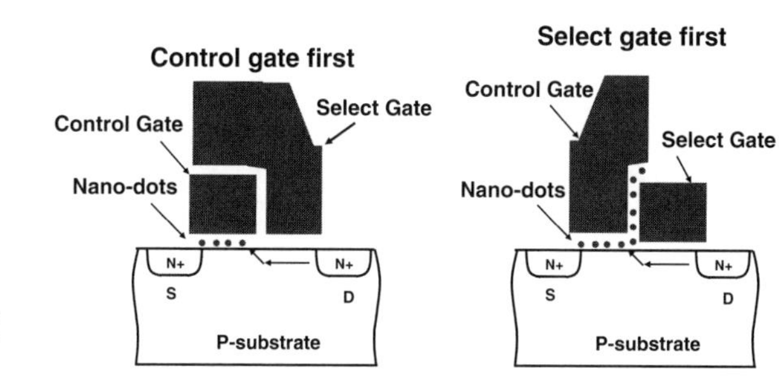

Fig. 6.43 MG-first and TG-first structures for split-gate memory cells [55] (© 2008 IEEE)

One interesting structure specific to SONOS is side-wall memory gate structure in the TG-first configuration previously shown in Fig. 6.41. This is an extension of the TG-first structure in Fig. 6.43. By a short-channel side-wall memory gate with CMOS-compatible transfer gate, this structure offers a small cell size with high-speed read operation. The split-gate with side-wall structure validates SONOS technology as one of the best choice for high-speed scalable embedded flash memory technology.

6.2.2.5 2Tr SONOS Cell

The 2Tr cell concept has been applied also to SONOS device (Fig. 6.44) [37]. A very simple structure with single-poly, CMOS-compatible structure and CHEI/FN of electrons for program and erase realizes an over-erasure free cell at low cost. Hot electron injection efficiency in p-channel devices can be higher than in n-channel devices and channel tunneling erase in p-channel device does not produce hot hole injection because of the larger hole barrier height, reinforcing the reliability [37]. P-channel cells often offer significant advantages over n-channel cells [6].

The cell size is 0.76 μm^2 by a 0.18 μm CMOS technology, which operates at an access time less than 40 ns at Vcc=1.8 V in read operation. This technology offers another fit for low-capacity simple embedded flash memory solution.

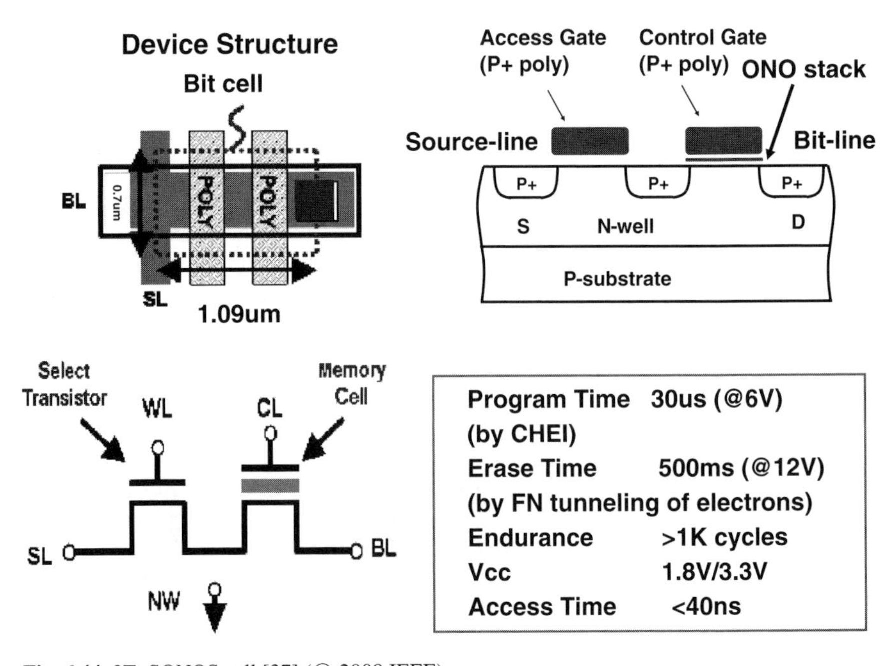

Fig. 6.44 2Tr SONOS cell [37] (© 2008 IEEE)

6.2.2.6 Nano-dot Memory Device

Recently Si nano-dot nonvolatile memory has attracted much interest, as another candidate for charge-trapping flash memory [1, 46]. This technology utilizes a nano-dot crystalline Si structure for the charge storage structure instead of nitride film. As an example of the nanocrystal (Fig. 6.45a) [1] shows, typically Si nanocrystals of 5–10 nm diameter act as a relatively dense and deep-level charge-trapping structure with localized charge storage. As in the case of SONOS this structure is tolerant against the point defect thus improving the scalability of bottom oxide thickness, lowering the operating voltage and enhancing the read access time. Additionally a Coulomb blockade mechanism is expected to confine the trapped charge in nano-dots, improving the scalability of the cell in vertical dimensions. Typically nanocrystal is fabricated by LP-CVD (low-pressure chemical vapor deposition) to deposit small amounts of poly-Si on top of the tunneling oxide to act as nucleation sites. The deposition process is stopped before continuous poly-Si layer is formed, producing hemispherical Si islands.

An advanced split-gate implementation of nano-dot memory in Fig. 6.45b [55] employs an SSI of electrons for program and FN tunneling of electrons to CG for erase. Programming time to 3 V threshold voltage window takes 5 μs at CG voltage of 10 V and erase time is within 1 ms at CG voltage of 14 V. The achieved P/E endurance is 10 K cycles.

Nano-dot technology is expected to improve over the conventional floating-gate devices in tunneling-oxide scalability, data retention, and P/E, but these advantages are not clearly achieved in real device implementations today. However, the accelerated data retention and P/E endurance show a promising progress in technology development [44]. Nano-dot technology is yet to prove the advantages of its own and to compete with conventional mature flash memory technologies for commercialization.

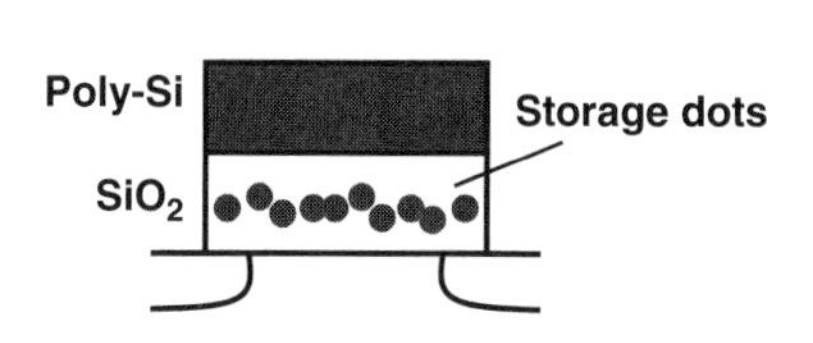

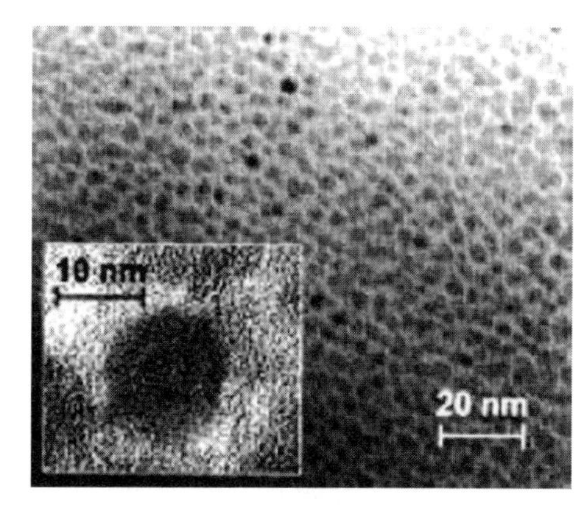

Fig. 6.45a A nanocrystal structure and a memory cell [1] (© 2008 IEEE)

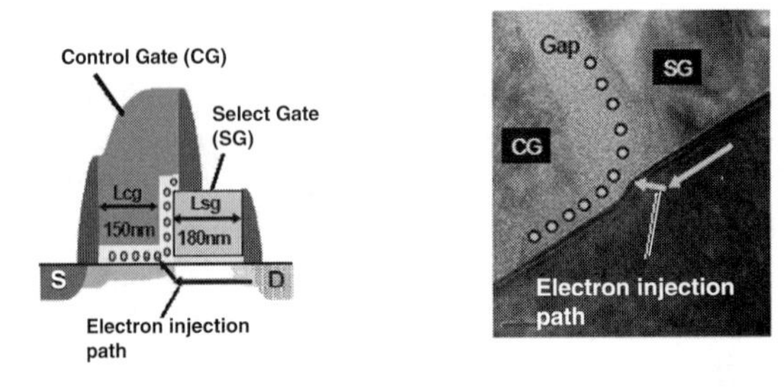

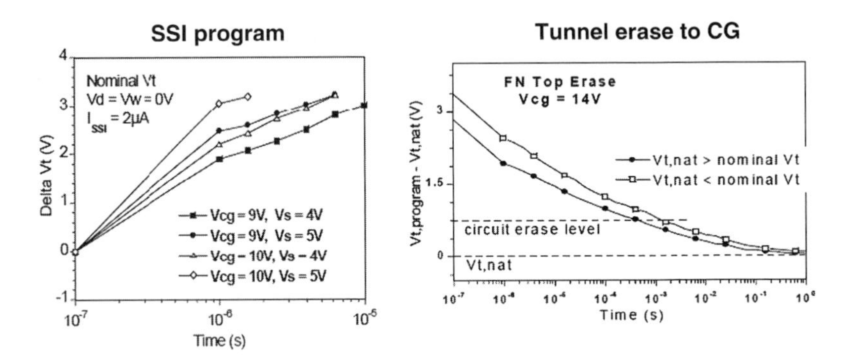

Fig. 6.45b Structure and performance of a split-gate nano-dot memory [55] (© 2008 IEEE)

6.2.2.7 Advantages and Challenges in Charge-Trapping Flash Devices

SONOS technology now enjoys growing interest for embedded uses as well as discrete high-density NROM and NAND flash memory because of the structural simplicity for lower cost and enhanced reliability by defect-resistant localized charge traps. The benefit of defect-resistant nature is attractive for high-reliability requirement in the embedded applications; however, it can only be exploited if inherent data retention and charge distribution/redistribution problems are solved by research in program/erase and data-retention mechanisms. SONOS is yet to prove basic performance in data retention and program/erase endurance, and to prove potential scalability for high-density embedded memory.

In comparison with the floating-gate technology, SONOS has the possibility of realizing simple and reliable CMOS-compatible structure, as listed in Fig. 6.46. The simple thin-film nature of SONOS finds a promising TG-first structure with sidewall memory gate for a high-density split-gate scheme, as shown in Fig. 6. 41. This is not attainable by a thick-film floating-gate structure. Nano-dot technology inherits all the advantages of SONOS device in principle, but is yet to prove the predicted advantages and further benefits over SONOS.

Fig. 6.46 Advantages of floating-gate and SONOS technologies

◆ Floating-gate flash
– good data retention, mature technology
– established high-density cell

◆ Charge-trapping SONOS flash
– simple in process integration
lower cost
better CMOS-compatibility in structure and performance
aplicable to TG-first and/or side-wall split-gate structure
– robust by localized charge mechanism
reliability, Tox/Vop scalability and performance
– charge distribution in the storage node usable for localized charge cell

6.3 Embedded Flash Memory Design

This section discusses embedded flash memory designs along with the benefits of embedded flash memory in many aspects of applications. After an introduction to embedded flash memory benefits followed by design considerations and a basic design example, typical embedded flash memory designs are reviewed and the future trends in the technology and design of embedded flash memory is also provided.

6.3.1 Benefits of Embedded Flash Memory and Design Considerations

The benefits of embedded nonvolatile memory come from both embedded nature and nonvolatility as follows:

(1) Because the off-chip memory access path and drivers are eliminated, high-speed, low-power operations are achieved, and data security is easy to implement with less accessible internal data bus. Also high-density physical packaging, enhanced reliability, reduced EMI, and lower system cost are provided by the embedded memory solutions.
(2) The flexibility of design gives optimal designs in memory capacity, interface configuration, memory functions, operating voltage, etc., for each application. This is favored by chip designers, but tends to increase the number of macro variety, which poses a tough challenge in optimizing the number of variations. "One for all" way of thinking tends to lose the benefits of flexible embedded memory environment.
(3) The nonvolatility attribute gives further opportunity of low-power design. From eliminating the battery-backup SRAM to low stand-by power by powering off the system, attentions are more focused on zero-power stand-by nonvolatile

memory because today's advanced scaled embedded SRAM often sees a large stand-by current due to the scaled MOSFET in the memory cell.

On the other hand, the disadvantages and concerns in embedded nonvolatile memory are

(1) Embedded memories are nonstandard products, in contrast with the stand-alone discrete memory device.
(2) Single-source may degrade stable product supply operations.
(3) Very low or very high memory density is not cost-effective against separate die solutions by logic and memory.
(4) Memory density is 1–2 generations of technology behind the discrete products because of the requirements for CMOS compatibility and low process cost.
(5) Increased test cost, to be solved by parallelism in time and BIST (Built-In Self Test) in tester cost.

In view of these disadvantages embedded flash memories are validated in the market only if they support standard products such as MCUs.

With the advantages above, embedded nonvolatile memory is capable of meeting various demands in a wide range of applications. Figure 6.47 lists a number of embedded flash memory applications as classified by functions, applied products, and by uses in the stage of development and production.

Embedded nonvolatile memory provides functions of code and data storage, backup storage, system boot, and trimming information storage for memory and analog parts of the chip, etc. By applications we find code and data storage dominant in MCU (micro controller unit), updatable coefficient parameter storage in

- By functions
  - Code storage; system boot, user program, firmware, look-up table
  - Data storage; EEPROM emulation, shadow storage, frequently updated parameters and coefficients, state before power down
  - Trimming information storage etc.
- By applications
  - MCU (micro controller unit); for code and data storage
  - DSP (digital signal processor); for coefficient storage
  - Smart-IC cards; data storage
  - RF-ID; data storage
  - Reconfiguration register; FPGA(FieldProgrammable Gate Array) etc.
- By uses
  - proto-typing; verify system concepts
  - system development; program debug and updates
  - early productions; program updates
  - volume productions; production and inventory control

Fig. 6.47 Embedded flash memory applications

DSP (digital signal processor), data storage in smart-IC card and RF-ID tag, and configuration storage in FPGA (field programmable gate array) and reconfigurable logic, etc. By uses in the stages from prototyping to volume production, embedded flash memory acts as easy-to-change ROM storage for easily verifying system concepts, fixing program bugs, and supporting program updates as well as production and inventory control by unified product lineup.

Figure 6.48 exemplifies key considerations in embedded flash memory designs. The area cost of the peripheral circuitry for flash memory is significant at small capacity of memory, and the cost for CMOS compatibility at large capacity of memory is enormous. The robustness against various design and usage conditions is essential for inherent multi-purpose objects of embedded memory. Because embedded flash memory in MCU and MPU is inherently required a high-speed to cover mask-ROM performance to meet CPU speed, performance-oriented design is required in many aspects of applications, which is quite different from discrete memory products. Taking advantage of the above-mentioned benefits to meet the expanding application demands, embedded flash memory design has multiple points to note.

Inherent high-speed and low-power read property and data security achieved by embedded nature can be further augmented by various design implementations. For high-speed read operation, keys are larger signal, nonboosting WL, and sensing architectures. Also data reliability as well as high-temperature and low-leakage product strategy are important factors in embedded applications. In some cases the EMI generated by the on-chip oscillator for charge-pumping circuits in high- and negative-voltage generators for flash memory operations is to be reduced for RF signal processing environments on the same chip.

Figure 6.49 illustrates various aspects in the design of embedded flash memory to meet varied demands in the embedded environments. This is to show various circuit techniques to support wide range of applications. High-speed and low-voltage operations are leading factors in embedded flash memory trend. For example, 100 MHz random-read at 1.5 V operation is state-of-the-art in the current products. Data reli-

(1) Cost effectiveness
  area cost at small capacity, CMOS cost at large capacity
(2) Re-usable in various applications and design environments
(3) Robustness against various use environments
(4) High performance
  – Larger signal in read; memory cell/structure
  – Non-boosted WL in read; memory cell design
  – High-speed access mode (by page/burst/cache)
  – Hierarchical architecture
(5) Low power
  – Charge-pumping circuit power at stand-by
  – Use of non-volatile storage for control in power On/Off
(6) Low EMI
  – Charge-pumping circuit slew

Fig. 6.48 Embedded flash memory design considerations

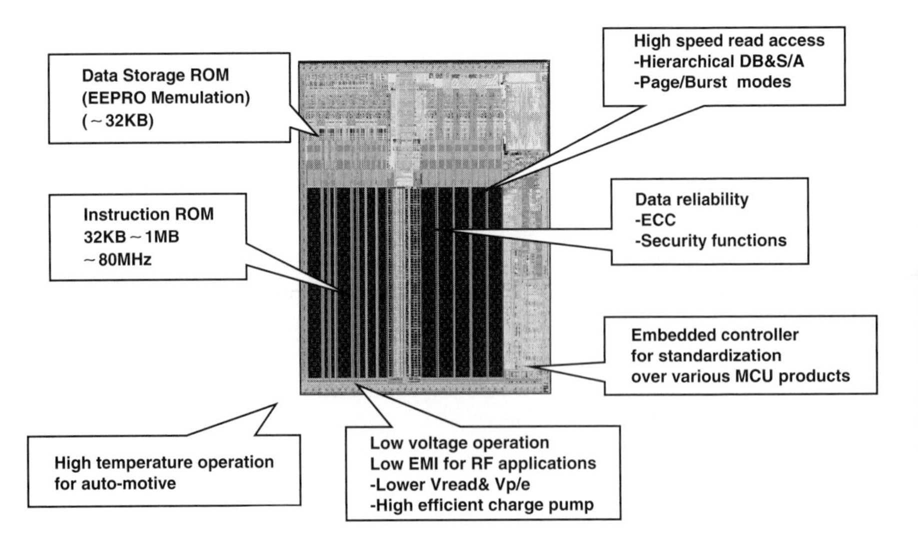

Fig. 6.49 Design aspects of embedded flash memory [22] (© 2008 IEEE)

ability as well as high-temperature and low-leakage product strategy are important factors in embedded applications. Also a low-EMI charge-pumping scheme is required by on-chip RF circuitry. Embedded flash related circuitry is now being developed in every aspect of ever wider application requirements. The requirements are getting higher and broader because of expanding uses of embedded flash memory. Ever-challenging situations will be met by scaling the device and circuit development.

6.3.2 Basic Flash Memory Design with Floating-Gate 1Tr-NOR Cell

Figure 6.50 shows a brief sketch of flash memory macro organization. In addition to the flash memory array, multi-parallel data latch for wide internal bandwidth, program/erase voltage generator and sequence controller, together with internal timer and control circuitry, are supplied for composing a self-contained flash memory macro.

A cross-section view of stacked-gate 1-Tr NOR memory cell and a hierarchical bit-line architecture are shown in Fig. 6.51. The drain of the memory cell is connected to the sub-bit-line (SBL) by M1(first-level metal) and the source of the memory cell is extended by a N+ diffusion line to be connected to the ground potential. The control gate is extended to be a word-line formed by CoSi-salicided poly-Si. The sub bit-lines are selectively connected to the main bit-lines (MBL) by M3 through the select gates (SG). In order to achieve a fast read access, the SBL consists of metal wiring, and CoSi-salicided word-line is shunted by M2 metal line.

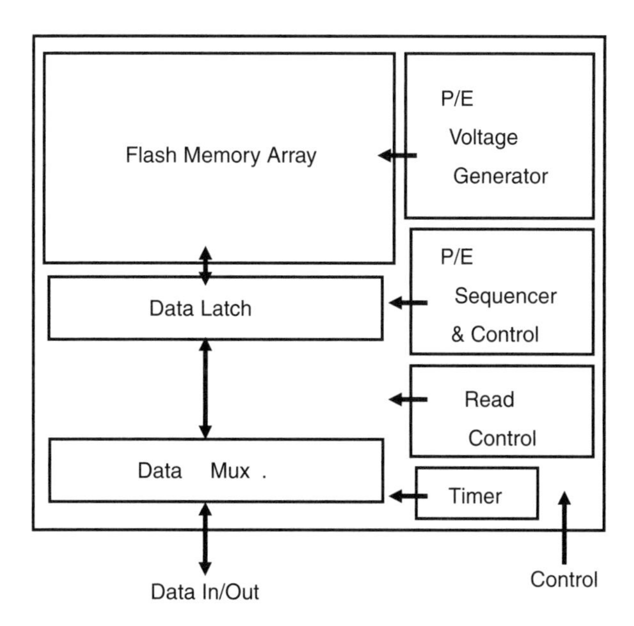

Fig. 6.50 A flash memory macro organization

Because a fast read access is often required for providing code steps for computing state machines in the CPU, the read access path design is important. Simple one-chip MCU solutions require one-cycle read for on-chip flash memory. An example of the read access path in Fig. 6.51 shows a hierarchical design controlled synchronously by address latch enable (ALE) signal. Addresses from CPU are latched by the ALE signal which initiates the read operation of the memory cell. The memory cell current is transferred to the sense amplifier and latched after amplified. The ECC function acts on the output data for data reliability.

The memory array organization and operation of flash memory are strongly affected by the basic operation principle and its limitations of the memory cell. By observing this structure we can see

(1) A memory cell is controlled by a word-line/control-gate line, bit-line, source-line, and substrate well, thus this is a four-port device electrically.
(2) Because of the high-voltage operation requirements, high-voltage generation and biasing mechanism strongly affect the array structure and circuits.
(3) The flash erase operation by FN tunneling to the substrate well necessitates the bulk erase operation in a sector, usually in a large group in K bits with a common well, for enhancing the erase throughput.
(4) Program operation is done along a selected word-line. The disturb problem often limits how small part of a word is programmed and to leave the rest of the bits. Thus, the memory cell performance in disturb immunity strongly affects the operation of the memory cell array in program and erase.

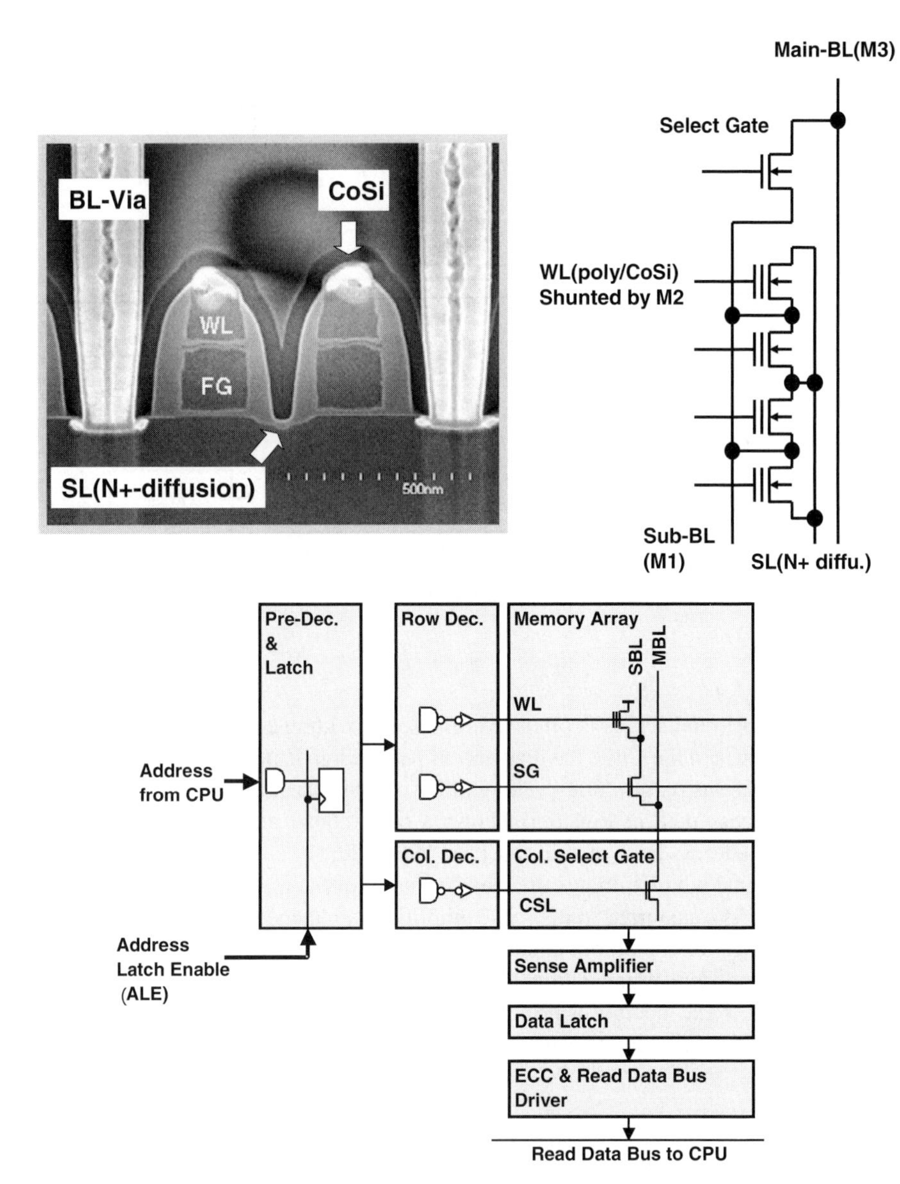

Fig. 6.51 A cross-section view of 1Tr-NOR cell and array organization [19] (© 2008 IEEE)

(5) Data read operation is by selecting the word-line at a boosted voltage level and applying the read bias to the drain side by the bit-line and detecting the current flowing through the source.

As shown in Fig. 6.52 the flash memory structure is divided into multiple blocks for code storage, as well as a few small-capacity EEPROM emulator blocks for data

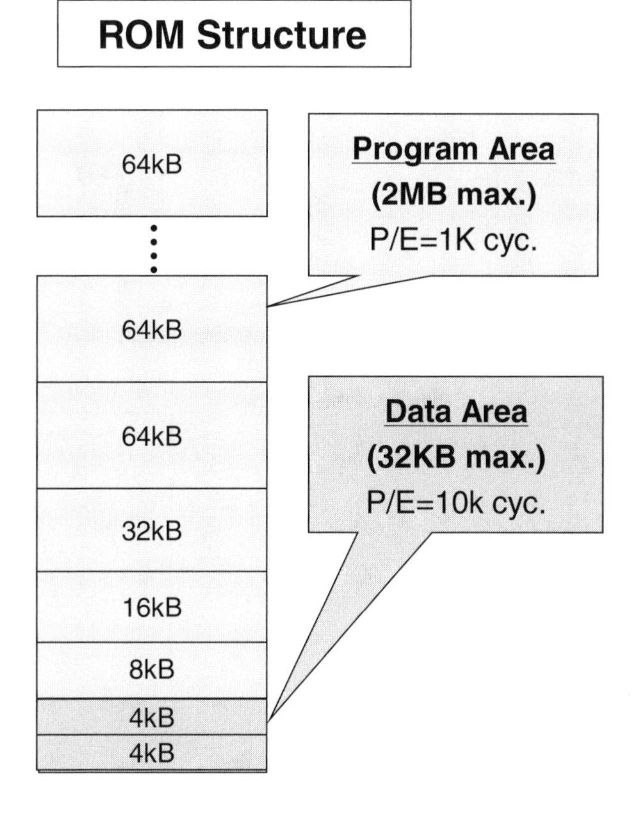

Fig. 6.52 Basic logical organization of a flash memory macro [22] (© 2008 IEEE)

storage. These heterogeneous block divisions are essential in matching the physical structure of embedded flash memory module with logical operation for user convenience in program/erase operations, etc.

A dual-voltage word-line scheme with dual path design in Fig. 6.53 shows a speed enhancement [23]. Having two-way word driver to separate low-voltage read access path from high-voltage program/erase access path, it achieves faster access using low-voltage thin gate oxide transistors in the read access path.

The on-chip ECC logic often degrades the read access time. As shown in Fig. 6.54, this problem is coped with by inserting a cycle for data check and correction in the access path only when an error occurs. One wait cycle is inserted at the output with corrected data by the ECC logic.

Figure 6.55 shows a breakdown of typical on-chip access time. The random access initiated by the row address latch signal takes over 20 ns. The column access time started by reading the column address latch signal is much shorter, which can be exploited by page-mode operations.

Figure 6.56 tabulates an example of specification for an embedded flash memory macro. Using a floating-gate 1-Tr. NOR cell technology embedded in a 150 nm CMOS process with four layers of metal, 80 MHz cycle time is achieved along

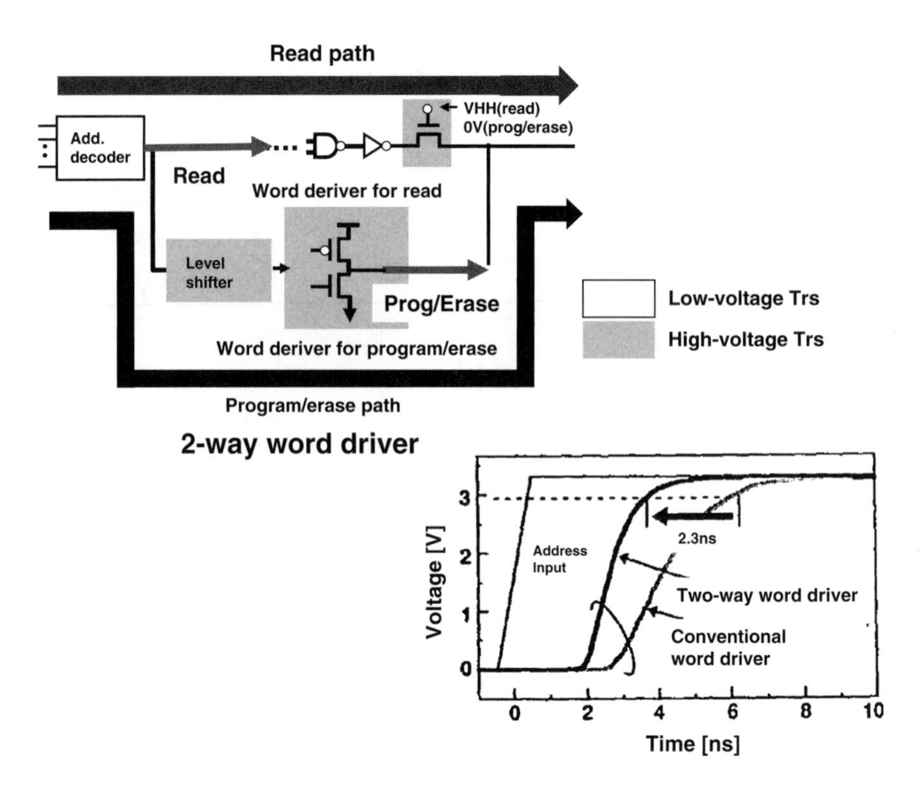

Fig. 6.53 A fast read access path by two-way word driver [23] (© 2008 IEEE)

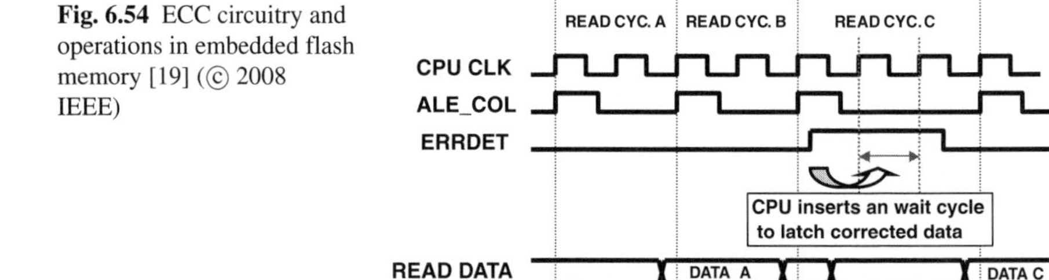

Fig. 6.54 ECC circuitry and operations in embedded flash memory [19] (© 2008 IEEE)

with embedded ECC. For the user convenience, flash-ROM structure is divided into multiple blocks for code storage, as well as a few small-capacity EEPROM emulator blocks.

In general, faster access time by larger signal sensing and lower Vth state prefers split-gate or 2Tr cell structures. Nonboosting word-line has an advantage in enhancing the access speed. In the boosted word-line scheme, the thick-oxide MOS devices

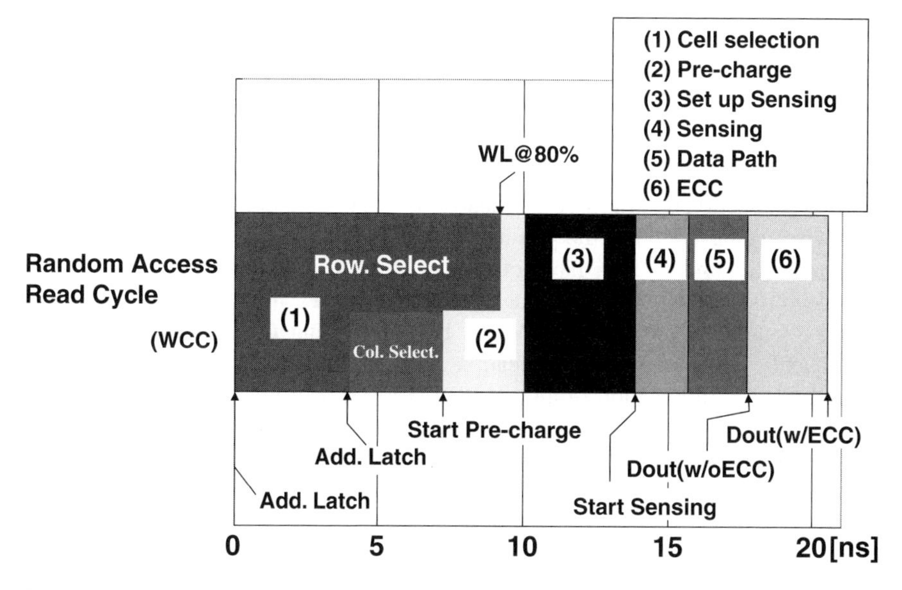

Fig. 6.55 Access time components in embedded flash memory (© 2008 IEEE)

Fig. 6.56 Example features of embedded flash macro

Process	0.15um/FG-NOR, 4LM/CoSi
High Performance	80MHz Random-access
	64bit-wide read data bus
High Reliability (for Automotive)	Support 150°C-Tj operation
	ECC(SEC for 64bit)
Usability	2MB(max.) Instruction Code Area (P/E 1k cyc.)
	Data Area for EEPROM emulation (P/E 10k cyc.)

are necessary for controlling high-voltage supply and transfer and they limit the operation speed in the critical access path.

6.3.3 Embedded Flash Memory Design Examples

The major advantages of embedded flash memory are performance, low power, and data security on the chip. Real designs of embedded flash memories range from high-speed, large-capacity program code storage to a small amount of multiple updatable data memory. More and more embedded-specific applications are being developed to meet the demands for relatively small amount of alterable-ROM capacity and for low-power nature of nonvolatile memory. In this section, representative designs to exploit favorable attributes of embedded flash memory are reviewed.

6.3.3.1 High-Performance Flash-MCU Design

Embedded flash memory macro has been most applied to general-purpose micro-controller (MCU) products. Since the early implementation of flash memory on RISC MCU with up to 512 K-byte of split-gate flash memory in 1990s [34, 35, 36], flash memory has been incorporated into MCU products very extensively, as is reviewed in Section 6.1. In MCU applications, the access and cycle time of the flash memory often have to meet the performance of CPU.

Automotive applications have offered a great deal of market expansion for embedded flash memory. Frequent parameter updates in the real-time control have exploited flash memory in MCU, thus expanding the application markets significantly. The automotive application requires proper operations under hostile environments in temperature range (–40 to 150°C), extremely low (zero) failure rates, high-performance, interface like car area network (CAN) specific to automotive uses. Figure 6.14 lists requirements for flash-MCU in automotive applications.

Figure 6.57 shows a recent implementation of high-speed MCU for automotive applications. A 80 MHz random-access floating-gate NOR flash memory macro with 1 M-byte of capacity for code storage is implemented on a 160 MHz RISC MCU employing a 150 nm CMOS technology. Analog circuits like AD-converter and communication ports such as CAN (car area network interface) are implemented for automotive uses.

A typical organization and internal bus structure of MCU in Fig. 6.58 shows a modern MCU design framework and how the MCU is organized on a com-

- M32R–FPU 160MHz (0.15um)
- FLASH 1MB/80MHz (FG–NOR)
- RAM 64kB
- Timer, ADC
- DMACx10ch
- Communication Interface
- Tjmax = 150C

DMAC: Direct Memory Access Controller
DRI: Direct RAM Interface
SIO: Serial Interface
RTD: Real Time Debugger
SDI: Scalable Debug Interface
NBD: Non Break Debug

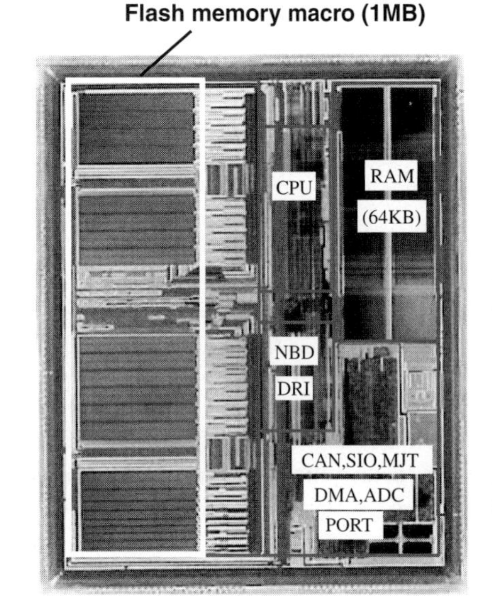

Fig. 6.57 A high-performance MCU for automotive applications [22] (© 2008 IEEE)

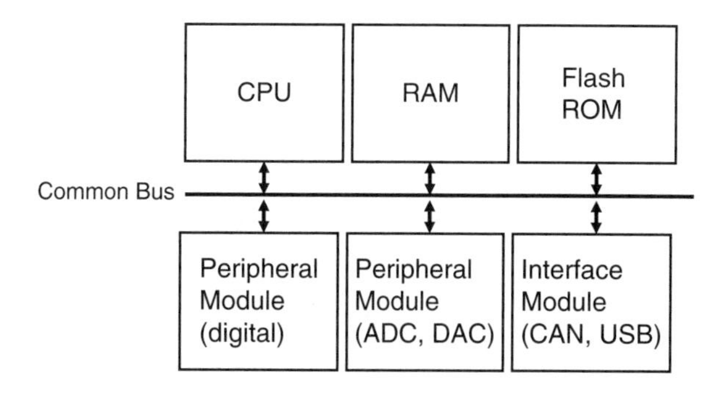

Fig. 6.58 An internal structure of MCU

mon bus platform architecture with modular design in each macro. The modular design methodology is important because multiple families of products are developed based on the same technology of flash memory. Besides CPU, memory (flash-ROM and SRAM) and clocking circuitry like PLL, modern MCU incorporates many IP macros for analog signal processing like ADC, DAC, and interface protocols such as USB and CAN (car area network). These IP macros reside on a unified bus platform. ROM and RAM interfaces are organized to form directly accessible instruction and data ports. For the flash memory supplying the ROM storage some data protection circuit schemes are provided for prohibiting erroneous data write to the memory cell. Also data security function by encryption is getting more important in such application as smart-IC cards.

Figure 6.59 shows another flash-MCU design intended for automotive control [11]. A flash memory module with 43 MHz random access with 2 M-byte capacity is achieved by an optimized floating-gate NOR flash memory design. A careful design of the memory array employing a 130 nm technology provides a stable operation of the MCU at 170 MHz at –40 to 150°C. The sensing circuitry in Fig. 6.59 shows a symmetrical referencing approach with the mid-point of sensing determined by device ratio of the current mirror, providing a device mismatch resistant performance.

A 90 nm MCU has been announced with high-speed embedded flash memory by using a SONOS cell [51]. The performance is as high as 200 MHz with 10 ns access operation for the flash memory macro with 2.5 M-byte capacity for automotive power-train applications. This indicates evergrowing demands for high-performance embedded flash memory for core control of modern cars.

Figure 6.60 shows a low-voltage 2Tr cell array [12]. FN tunneling program and erase operations are implemented for low-power program/erase. Together with a low-power read operation at 1.2 V with nonboosted word-line scheme, this is a low-power approach by a direct extension of stacked floating-gate flash cell technology. A cell size of 0.78 μm^2, about 1.5× of the corresponding 1Tr cell, is employed. A read operation of 0.98 mW/MHz with an access time of 70 ns at 1.2 V is achieved.

- 170MHz MCU for Automotive (130nm)
- FLASH 2.125MB/23.5ns (FG – NOR)
- 2GB/s read throughput by burst mode
- Tjmax=150C

Main Applications:
- Engine Management
- Transmission Control

- Sensor
  Lambda, Knock, Air Mass, Pressures, Temperatures, Voltages, Throttle Position, ...
- Real-Time Processing
- Actuator
  Ignition, Throttle, Injectors, ...

1 mm

Die photo

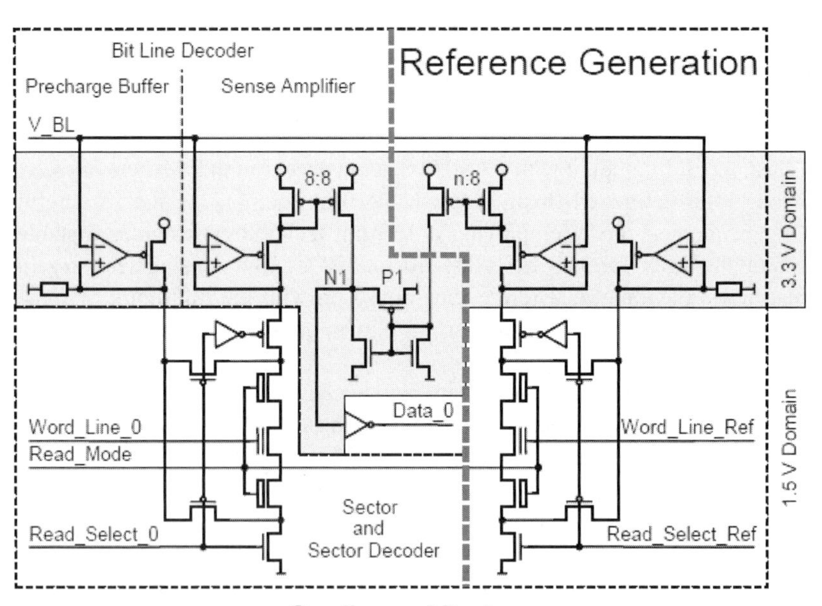

Sending architecture

Fig. 6.59 A flash memory macro in MCU for automotive applications [11]. Low-power oriented design of embedded flash memory with 2Tr cell (© 2008 IEEE)

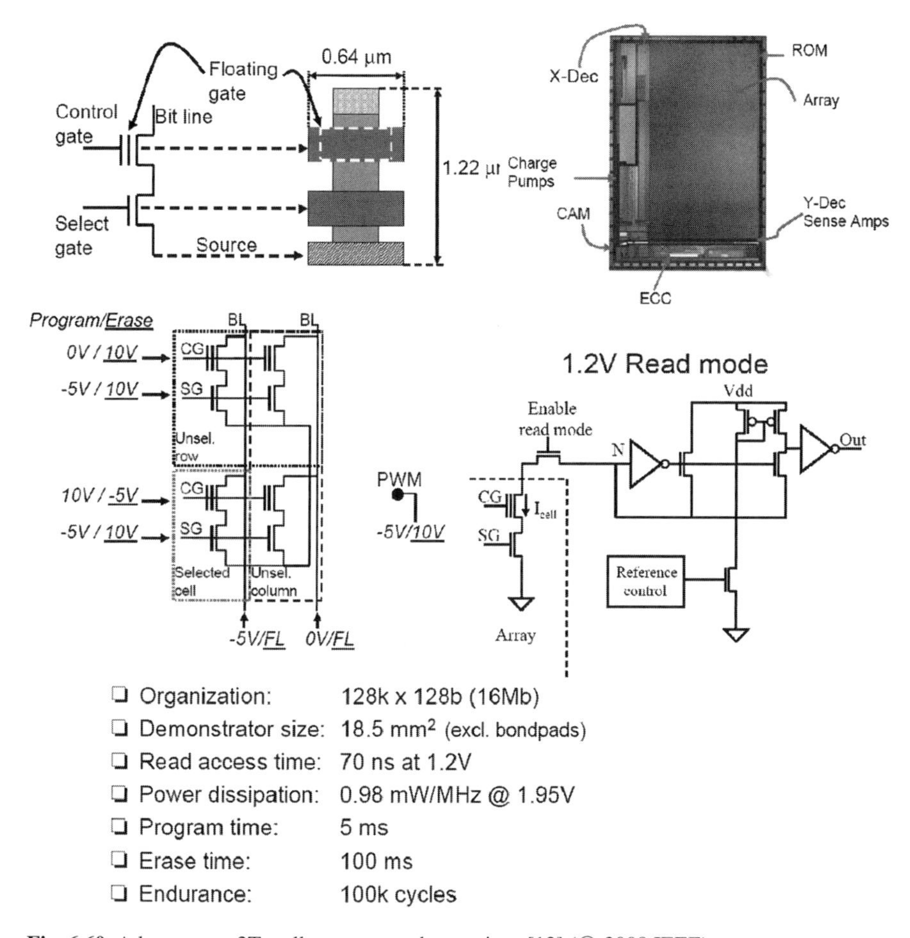

Fig. 6.60 A low-power 2Tr cell structure and operations [12] (© 2008 IEEE)

6.3.3.2 Embedded Flash Memory Design for Reconfigurable Logic

Embedded flash memory has also found uses in reconfigurable logic applications. Figure 6.61a and b describes a reconfigurable processing unit utilizing on-chip flash memory storage for embedded FPGA (field programmable gate array) to provide processor instruction extensions, bus-mapped coprocessors, and flexible I/O configuration [2, 3]. A 2 Mbit × 4 flash memory module provides 128 bit I/O, 3 content-specific ports with 40 ns access time. An example benchmark for face recognition shows a 8.5× speed-up and 6.7× energy gain over an equivalent RISC+DSP architecture [2]. In another application flash memory storage controls the connections among multiprocessor system to optimize the cross-bar interconnection configuration by multiple processors [4]. These examples highlight important aspects of embedded flash applications to help reorganize processing systems to enhance the hardware efficiency according to specific application.

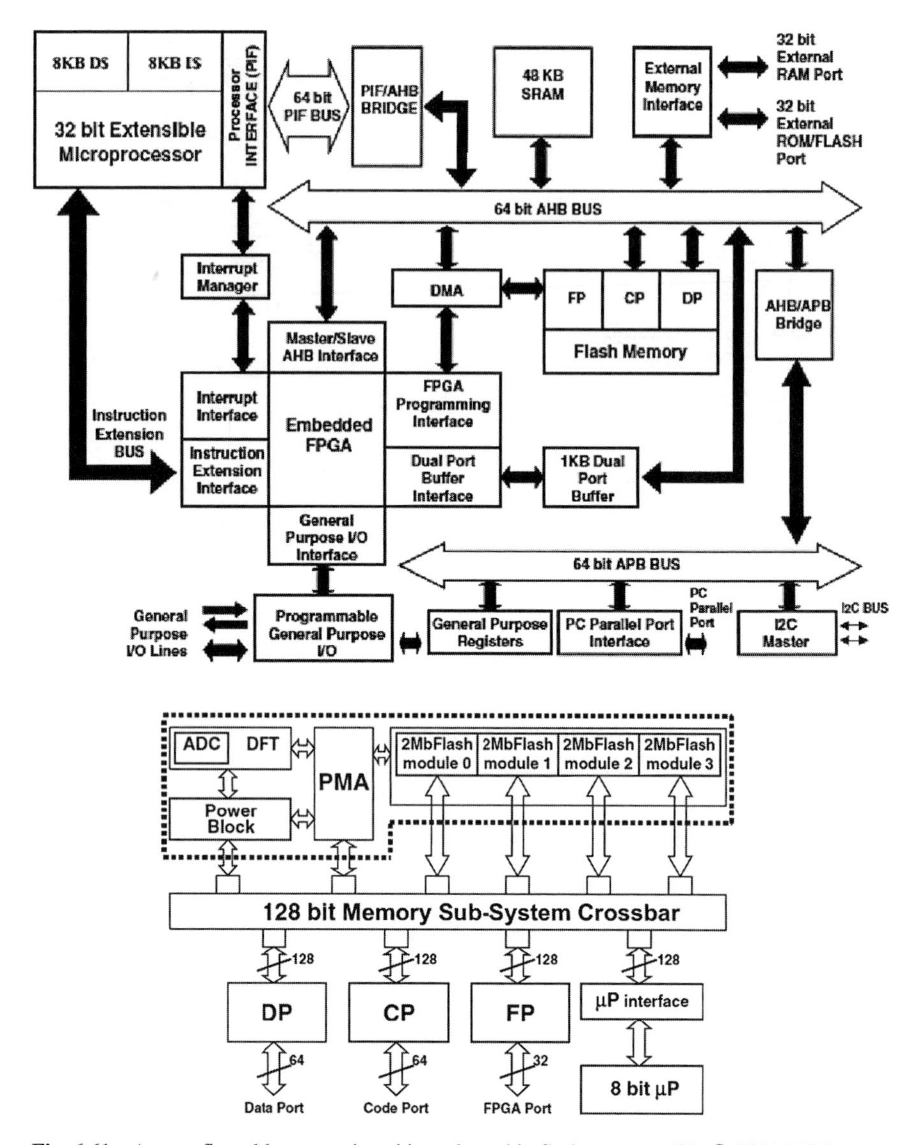

Fig. 6.61a A reconfigurable processing chip and on-chip flash structure [2] (© 2008 IEEE)

6.3.3.3 Fully CMOS-Compatible Nonvolatile Storage Design by CMOS Flash

Completely CMOS-compatible (CMOS-inclusive) nonvolatile memories have attracted attentions for low-density storage, although they deviate from the high-density flash memory technology. They are intended for small-capacity nonvolatile data storage for applications in advanced SOC to small low-cost controller chips. The main applications are redundancy control for embedded SRAM and DRAM, analog circuit trimming, chip-ID, security key storage, etc. Especially advanced

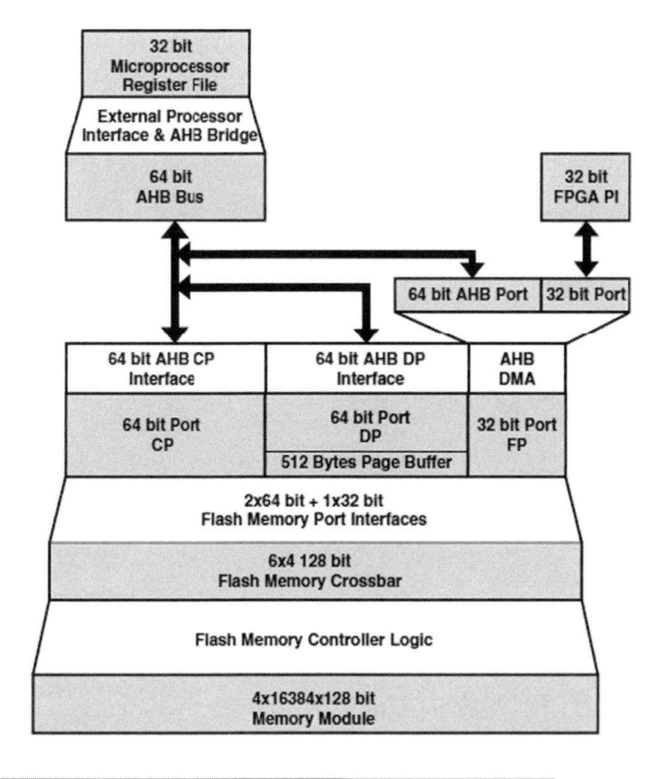

TECHNOLOGY AND DEVICE CHARACTERISTICS	
Process	0.18μm 2P 6M CMOS Tunneling oxide: 10nm Flash cell size: 0.35μm^2
Flash Memory (4x)	256Kb x 9 Sectors Word: 128b Program Throughput: 1MB/s Typ. Read Rate: 400MB/s
SRAM memory	I$: 8kB (64b wide) D$: 8kB (64b wide) Buffers: 4x256B (8b wide)
Chip size	8.4x8.4mm^2
e-FPGA size	8.2mm^2
Customizable I/O	24 general-purpose inputs 24 general-purpose outputs (tristate) 8 general-purpose bidirs
Power supply	2.7-3.6V (I/O), 1.6-2.0V (core)

Fig. 6.61b A reconfigurable processing system chip, memory hierarchy, and overall features [2] (© 2008 IEEE)

CMOS technology for SOCs necessitates redundancy program for on-chip SRAMs and trimming functions for analog circuits, which makes simple CMOS-inclusive OTP ROMs indispensable for proper yield in productions.

A "CMOS flash memory" is composed of only the thick-oxide transistors for I/O circuitry in CMOS (Fig. 6.62) [43]. This is a direct implementation of flash memory by available CMOS devices. Serially connected two PMOS capacitors, MC1 and M3, form a floating-gate charge storage node. A large coupling control capacitor MC1 is for CHEI program operation, while a small tunneling capacitor M3 is for FN tunneling erase. Together with an NMOS transistor M2 for program and read operation, this basic 1Tr-2Cap structure acts as a CMOS flash memory cell.

The challenges in composing this memory cell is to cope with the high-voltage by existing CMOS devices. Implemented in a 0.25 μm CMOS technology this memory cell achieves 1 K times of P/E, P/E operations in the orders of 100 ms, more than 10 years of data retention, and 2.5 V read operation, with a rather large cell size of 50 μm^2. The scaling of gate oxide below 100 A significantly degrades the

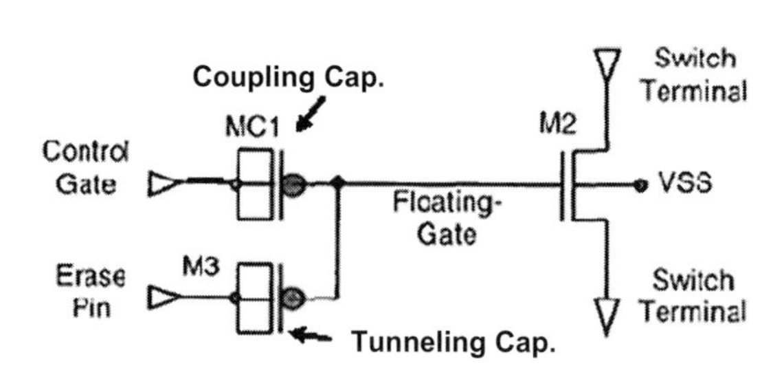

- CHEI program
- Erase by FN tunneling of electrons.
- MC1 for control/coupling
- MC3 for tunneling for deep erase
- P/E in 100ms order

TABLE 1: Cell Voltages

	New-cell			
	Source	Drain	Control Gate	Erase Gate
Erase	0	0	0	10
Program	0	5.5	6.5	0
Read	0	1.5	2.25-2.75	0

TABLE 2: Device Summary

Technology	0.25 micron, 2.5 volt, logic, CMOS
Normal Device Gate Oxide	50 A
Cell & High Voltage Gate Oxide	100 A
Cell Size	50 sq. um
Cell Current (erased)	> 1E-5 A
Cell Current (programmed)	< 1E-12 A
Endurance	1000 cycles
Data Retention	> 10 years

Fig. 6.62 CMOS-compatible flash memory implementation [43] (© 2008 IEEE)

data-retention performance, limiting the use of this device in the scaled SOC technology with thinner I/O devices.

Various circuit implementations and array architectures have been developed for higher reliability, testability, and performance for the CMOS flash memory. Figure 6.63a [48, 49] is an example intended for security applications. The security applications benefit from the embedded memories because the data crossing the chip I/O is vulnerable to external attacks. This array design adopts complementary twin cell organization based on a 1Tr-2C cell to improve the operational margin.

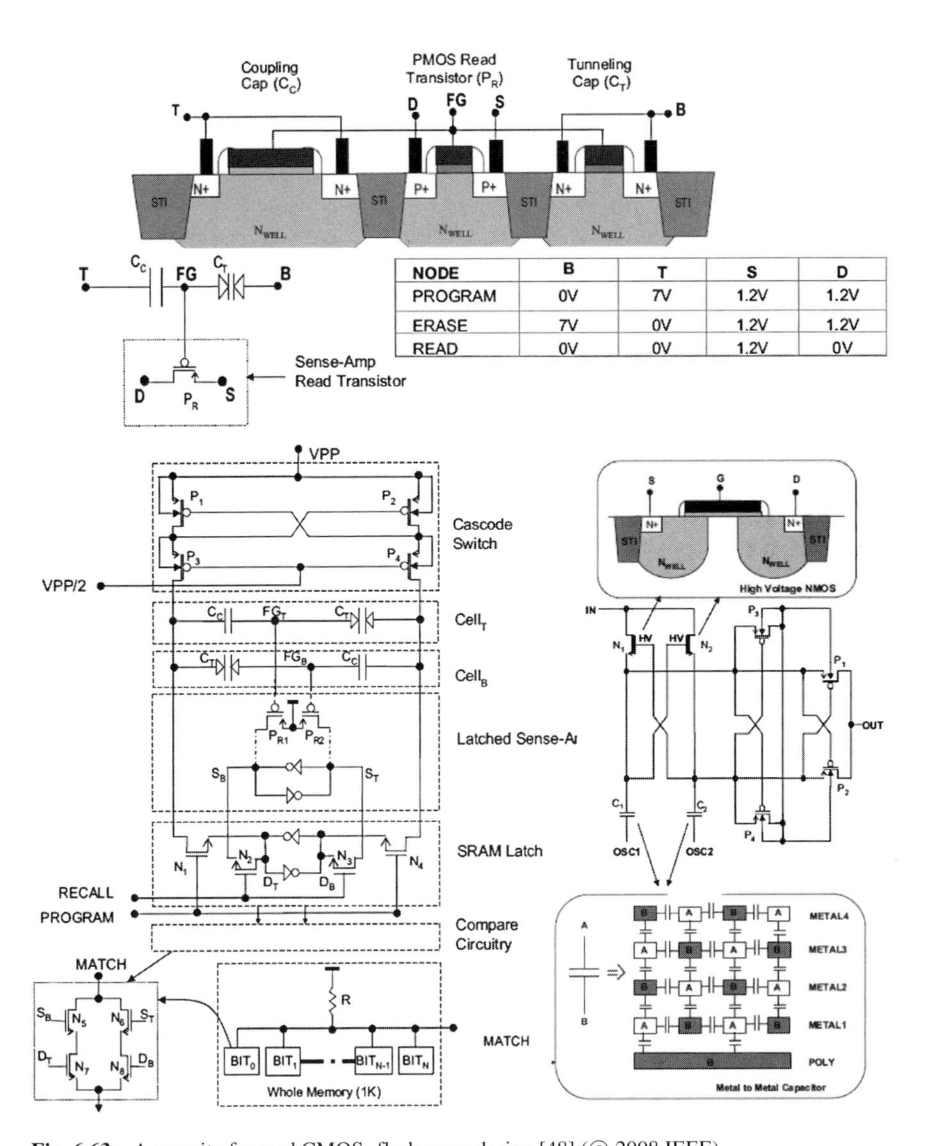

NODE	B	T	S	D
PROGRAM	0V	7V	1.2V	1.2V
ERASE	7V	0V	1.2V	1.2V
READ	0V	0V	1.2V	0V

Fig. 6.63a A security focused CMOS- flash array design [48] (© 2008 IEEE)

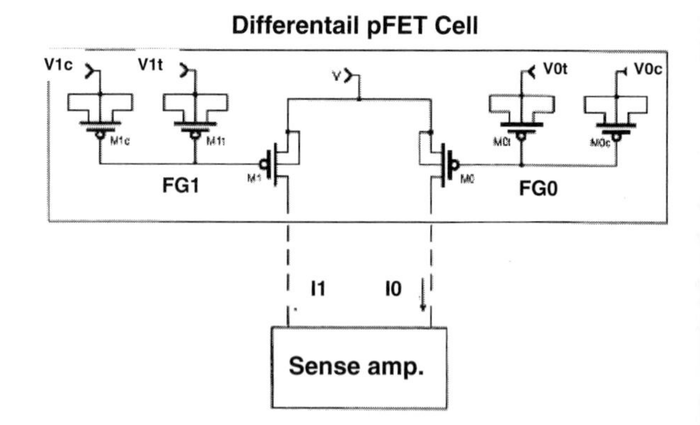

Fig. 6.63b A security-intended CMOS-flash array design [40] (© 2008 IEEE)

The memory cell employs 7 nm thick oxide available for 3.3 V chip I/O circuitry in CMOS and is programmed/erased by FN/FN tunneling of electrons at an internally generated 7 V supply. With additional features for security identification by matching the flash data with the SRAM ID data in a bit-by-bit manner, this IP macro provides an easy-to-use implementation of flash memory for security applications in standard 0.13 μm CMOS technology.

Figure 6.63b shows another implementation of CMOS flash intended for RF-ID tag applications with a 256 bit capacity of flash memory [40]. Here 1Tr-2C cell is paired for differential operation to improve the read margin. The read operation is carried out at 0.9 V for low power. The memory cell employs 7 nm thick oxide and is programmed/erased by FN/FN tunneling of electrons at an internally generated 8 V supply. A standard 0.25 μm CMOS process provides a promising application of CMOS flash to meet the required lowest chip cost at $0.1/chip for RF-ID tag uses.

6.3.3.4 Fully CMOS-Compatible Nonvolatile Storage Design by OTP (Fuse)

Another family of CMOS-inclusive nonvolatile memory is one-time programmable (OTP) type of devices. An anti-fuse structure by the breakdown (rupture) of the gate dielectric of MOS transistor offers one-time electrically programmable ROM, which also realizes CMOS-compatible programmable devices. Figure 6.64 [42] is a design example incorporating this OTP memory in 65 nm CMOS process for SOC. More than three decades of resistance change in the unprogrammed and programmed cell is obtained. Stable data write operation by oxide rupture at 6.5 V supply is implemented with additional features by separate ports for read/write for improved performance, write disturb elimination, and a stable program function to compensate for the leakage in the scaled transistors in advanced CMOS technology. An 8 Kbit macro-utilizing a 65 nm pure CMOS logic technology is demonstrated, with the cell size of 15.3 μm^2 and macro-size of 0.244 mm^2 for 8 Kbit of memory.

Other CMOS-inclusive OTP ideas include laser programmable fuse, electrical fuse by blowing metal line by excessive current, and an EM fuse utilizing

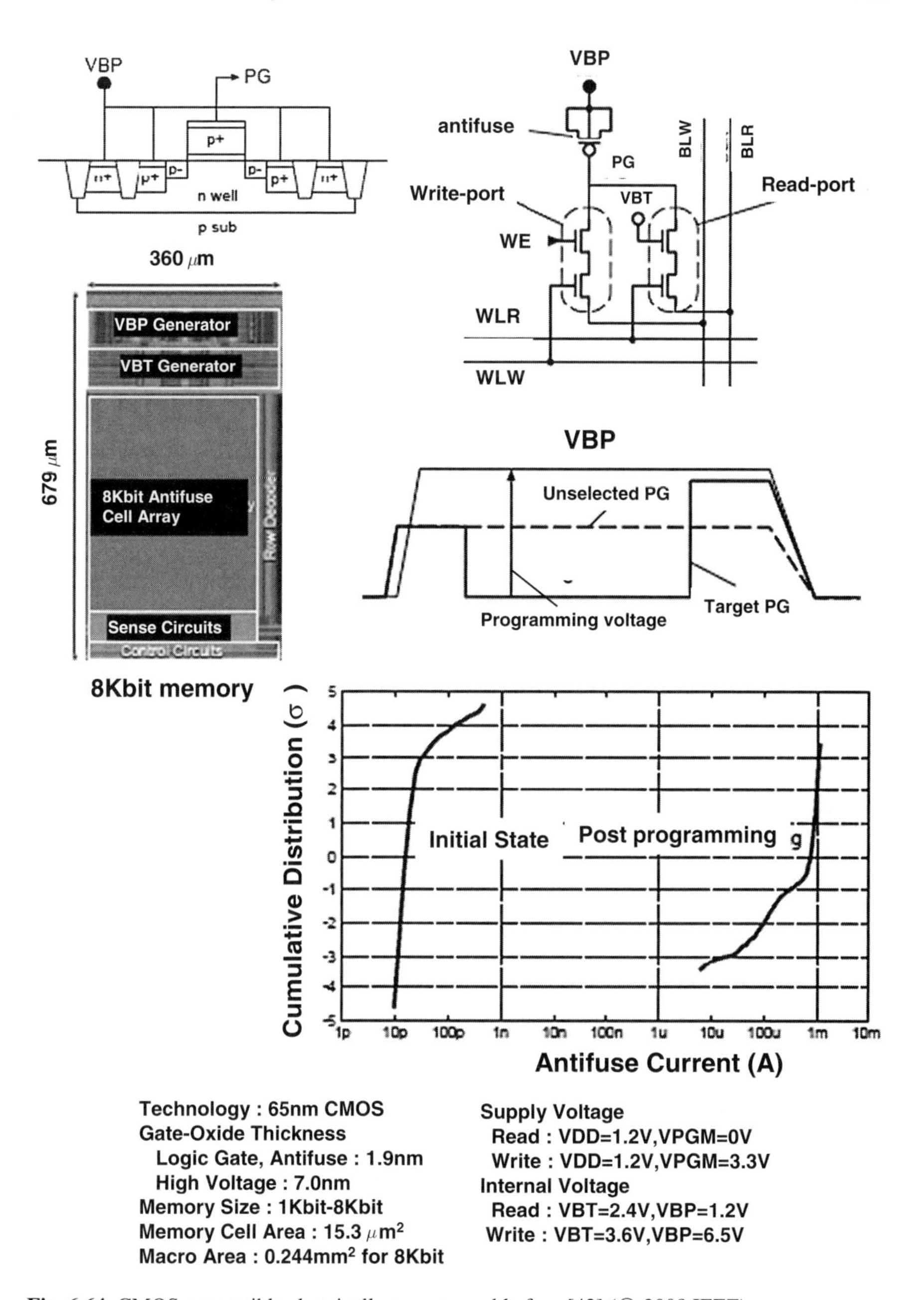

Fig. 6.64 CMOS-compatible electrically programmable fuse [42] (© 2008 IEEE)

electro-migration of CoSi or NiSi silicide available in CMOS [32, 53]. Although OTP memory is not an alterable memory deviating from flash memory, its importance has increased in CMOS-SOC implementations for uses in redundancy program and circuit trimming.

6.3.4 Future Trend in Embedded Flash Memory Technology and Design

A quick look at Figs. 6.18, 6.19, 6.23, and 6.36 listing embedded-specific flash memory technologies reminds us of quite a long history with numerous literatures of proposed technologies, most of which have had little acceptance in the real market so far. Diversified flash memory technologies for embedded uses provide a wide spectrum of possible application-specific technology choices. However, a large amount of production volume is essential in realizing cost effectiveness. The deviation from standard flash technology for discrete high-density flash memory products may pose a big challenge in embedded-only technologies. A convergence of technology for embedded applications is desirable, with diversified design variations to cover a wide range of applications.

After the convergence into flash-MCU product concept, currently diversified embedded flash memory technologies have a true challenge of convergence for cost innovation, standardization, and effective learning schemes. The future flash technology will keep its advantages in accumulated learning or one of emerging memory technologies will well surpass by the performance/cost establishing a dominant design in a short period of time. Speculations on how the flash memory technology meets the varied requirements of embedded uses and on possible technological solutions provide us possible future trends.

The requirements for embedded aspects of flash memory are summarized as:

(1) CMOS process compatibility in terms of structure and performance. Small number of additional mask steps in the flash process.
(2) Small area penalty by flash-specific peripheral circuitry.
(3) Multiple functions on the chip, for example, high-speed large capacity storage for code and medium-speed small-capacity EEPROM storage for data.
(4) Low-power and high-speed read operations.
(5) Reliability under a wide range of operating temperature.

Out of various memory cell devices and structures, the problem-solving path in most of the embedded flash applications tells us that possible future trends are

(1) Split-gate structure for high-speed program as well as low-power and low-voltage program/read operations is suitable for embedded applications. Floating-gate type split-gate structure is going to be most widely used for embedded applications.

(2) Charge-trapping SONOS and nano-dot technologies realize simple, low-cost structures and are defect-resistant, providing an advantage in high reliability requirements by embedded uses.
(3) TG-first structure for CMOS logic compatibility is important. The combination of charge-trapping device and split-gate structure is expected to be a suitable technology for embedded applications in the near future.

Based on these considerations, a split-gate discrete charge-trapping cell with TG-first structure is desirable, for example, in scalability and performance. Drastic gate engineering by high-*k*/metal-gate technology will result in accelerated program/erase operations with reduced operating voltage. These new structures are getting more and more important.

The low power nature of nonvolatile on-chip memory requires careful power control, especially in the stand-by state. A bold assumption in embedded flash applications is that all the on-chip systems operate on nonvolatile state machines which store all the circuit states and data before power-off and then resume immediately after power-on. All the parts of operations are executed on the nonvolatile memory in operation. In addition to slow program and erase operation, flash memory needs a long time to recover after power-on for setting up the high-voltage supply by charge-pumping operations. It may hinder an instant-on (and instant-off) system performance demanded in the future. This is a challenge that may require emerging nonvolatile RAM technology.

On the design front, improvement in design methodology, design verification system, testability, manufacturability, on-chip data protection and security functions, multiple storage configurations to accommodate various types of data such as code, data, boot program, hardware trimming/repair information, etc., is the focus of design efforts in future embedded flash memories.

References

1. Blauwe J D (2002) Nanocrystal Nonvolatile Memory Devices. IEEE Trans. Nanotechnol., 1(1):72–77
2. Borgatti M, Calì L, Sandre G D, Forêt B, Iezzi D, Lertora F, Muzzi G, Pasotti M, Poles M, Rolandi P L (2003) A 1GOPS Reconfigurable Signal Processing IC with Embedded FPGA and 3-Port 1.2 GB/s Flash Memory Subsystem. Dig. Tech. Papers ISSCC:2.7
3. Borgatti M, Cali L, De Sandre G, Foret B, Lertora I F, Muzzi G, Pasotti M, Poles M, and Rolandi P L (2003) A reconfigurable signal processing IC with embedded FPGA and multi-port Flash memory. Proc. Des. Autom. Conf.:691–695
4. Borgatti M, Auricchio C, Pelliconi R, Canegallo R, Gazzina C, Tosoni A, Rolandi P L (2003) A multi-context 6.4 Gb/s/channel on-chip communication network using 0.18 μm^2 Flash-EEPROM switches and elastic interconnects. Dig. Tech. Papers ISSCC:466–467
5. Brewer J E and Gill M (ed.) (2008) Nonvolatile Memory Technologies with Emphasis on Flash. IEEE Press:337–371
6. Ibid.:337
7. Ibid.: Chapters 3–8, and 10–11
8. Brown B D and Brewer J E (ed.) (1998) Nonvolatile Semiconductor Memory Technol. IEEE Press: 22

9. Chen W-M, Swift C, Roberts D, Forbes K, Higman J, Maiti B, Paulson W, and Chang K-T (1997) A Novel Flash Memory Device with Split Gate Source Side Injection and ONO Charge Storage Stack (SPIN). Symp. VLSI Technol. Dig. Tech. Papers:63–64
10. Cho C Y-S, Chen M-J, Chen C-F, Tuntasood P, Fan D-T, and Liu T-Y (2006) A Novel Self-Aligned Highly Reliable Sidewall Split-Gate Flash Memory. IEEE Trans. Elec. Devices, 53(3):465–472
11. Deml C, Jankowski M, and Thalmaier C (2007) A 0.13 μm 2.125 MB 23.5 ns Embedded Flash with 2 GB/s Read Throughput for Automotive Microcontrollers. Dig. Tech. Papers ISSCC:26.4
12. Ditewig T, Cuppens R, Chen K-L, Frowijn V, Jetten F, Kalkman W, Malabry M, Slenter A, Storms M, Tandan N, Teuben S and Grácio J (2001) An Embedded 1.2 V-Read Flash Memory Module in a 0.18 μm Logic Process. Dig. Tech. Papers ISSCC:34–35
13. Duuren M, Schaijk R, Slotboom M, Tello P, Goarin P, Akil N, Neuilly F, Rittersma Z, and Huerta A (2006) Performance and Reliability of 2-Transistor FN/FN Flash Arrays with Hafnium Based High-K Inter-Poly Dielectrics for Embedded NVM. Non-Volatile Semiconductor Memory Workshop:48–49
14. Eitan B, Pavan P, Bloom I, Aloni E, Frommer A, and Finzi D (1999) Can NROM, a 2-bit, Trapping Storage BVN Cell, Give a Real Challenge to Floating Gate Cells? Proc. Int. Conf. Solid State Devices and Materials Proc.:522–524
15. Eitan B, Pavan P, Bloom I, Aloni E, Frommer A, and Finzi D (2000) NROM: A Novel Localized Trapping, 2-Bit Nonvolatile Memory Cell. IEEE Ele. Device Lett. 21(11): 543–545
16. Furnémont A, Rosmeulen M, Zanden K, Houdt J V, Meyer K D, and Maes H (2007) New Operating Mode Based on Electron/Hole Profile Matching in Nitride-Based Nonvolatile Memories. IEEE Elec. Device Lett., 28(4):276–278
17. Furnémont A, Rosmeulen M, Zanden K, Houdt J V, Meyer K D, and Maes H (2007) Root Cause of Charge Loss in a Nitride-Based Localized Trapping Memory Cell. IEEE Trans. Elec. Devices, 54(6):1351–1359
18. Hatanaka M and Hidaka H (2007) Value Creation in SOC/MCU Applications by Embedded Non-Volatile Memory Evolutions. Proc. Tech. Papers, Asian Solid-State Circuits Conf.:38–41
19. Hatanaka M and Hidaka H (2008) Embedded Flash Memory for MCU/SOC. ISSCC2008 Embedded Memory Forum, Feb. 2
20. Hayashi Y, Ogura S, Saito Y, and Ogura T (2000) Twin MONOS Cell with Dual Control Gate. Symp VLSI Tech. Dig. Tech. Papers:122–123
21. Hidaka H (2006) Embedded Memory Challenges for Innovations. Proc. System LSI Workshop, IEICE Japan, Nov. 27
22. Hidaka H (2007) Embedded NV memory design. ISSCC2007 NV-Memory Forum, Feb. 11
23. Hiraki M, Tanaka T, Shinagawa Y, Fujito M, Kawai Y, Mishina D, Ohshima T, Abe S, Kubota H, Yamaki T, Tamura S, Shiba K, Kuroda K, Ohsuga H, Masujima K, Matsubara K (1999) A 3.3 V 90 MHz Flash Memory Module Embedded in a 32b RISC Microcontroller. Dig. Tech. Papers ISSCC:116–117
24. Houdt J V, Heremans P, Deferm L, Groeseneken G, and Maes H E (1992) Analysis of the Enhanced Hot-Electron Injection in Split-Gate Transistors Useful for EEPROM Applications. IEEE Trans. Elec. Devices, 39(5):1150–1156
25. Houdt J V, Haspeslagh L, Wellekens D, Deferm L, Groeseneken G and Maes H E (1993) HIMOS – A high efficiency flash E2PROM cell for embedded memory applications. IEEE Trans. Elec. Devices 40(12):2255–2263
26. Houdt J V, Wellekens D, Haspeslagh L (2003) The HIMOS Flash Technology: The Alternative Solution for Low-Cost Embedded Memory. Proc. IEEE, 91(4):627–635
27. Ikehashi T, Noda J, Imamiya K, Ichikawa M, Iwata A and Futatsuyama T (2000) A 60 ns Access 32 k Byte 3-Transistor Flash for Low Power Embedded Applications. Symp. VLSI Technol. Dig. Tech. Papers:162–165
28. International Technology Roadmap for Semiconductors (2007) http://www.itrs.net

29. Kamiya M, Kojima Y, Kato Y, Tanaka K and Hayashi Y (1982) EPROM Cell With High Gate Injection Efficiency. Tech. Digest Int. Elec. Device Meet.:741–744
30. Kawai S, Hosogane A, Kuge S, Abe T, Hashimoto K, Oishi T, Tsuji N, Sakakibara K, and Noguchi K (2008) An 8 kB EEPROM-Emulation DataFLASH Module for Automotive MCU. Dig. Tech. Papers ISSCC:508–509
31. Kianian S, Levi A, Lee D, and Hu Y-W (1994) A Novel 3 Volts-Only, Small Sector Erase, High Density Flash E2PROM. Symp. VLSl Technology Dig. Tech. Papers:71–72
32. Kothandaraman C, Iyer S K, and Iyer S S (2002) Electrically Programmable Fuse (eFUSE) Using Electromigration in Silicides. IEEE Elec. Device Letts., 23(9):523–525
33. Kuo C, Yeargain J R, Downey III W J, Ilgenstein K A, Jorvig J R, Smith S L, Bormann A R (1982) An 80 ns 32 K EEPROM Using the FETMOS Cell. IEEE J. Solid-State Circuits SC-17(5):821–827
34. Kuo C, Toms T, Weidner N M, Choe H, Shum D, Chang K-M and Smith P (1991) A Micro-controller with l00 K Bytes Embedded Flash EEPROM. VLSI-TSA:138–140
35. Kuo C, Weidner M, Toms T, Choe H, Chang K-M, Hanvood A, Jelemensky J, and Smith P (1992) A 512-kb flash EEPROM Embedded in a 32-b Microcontroller. IEEE J. Solid-State Circuits:574–582
36. Kuo C, Chrudimsky D, Jew T, Gallun C, Choy J, Wang B, and Pessoney S (1998) A 32-Bit RISC Microcontroller with 448 K Bytes of Embedded Flash Memory. Int. NonVolatile Memory Technol. Conf.:28–33
37. Lee H M, Woo S T, Chen H M, Shen R, Wang C D, Hsia L C and Hsu C C-H (2006) NeoFlash^R – True Logic Single Poly Flash Memory Technology. Tech. Digest of Non-Volatile Semiconductor Memory Workshop:15–16
38. Liu W, Chang K T, Cavins C, Luderman B, Swift C, Chang K M, Morton B, Espinor G, and Ledford S (1997) A 2-Transistor Source-Select(2TS) flash EEPROM for 1.8 V-Only Applications. Non-Volatile Semiconductor Memory Worshop:4.1.1–4.1.3
39. Liu X; Markov V, Kotov A, Dang T N, Levi A, Yue I, Wang A and Qian R (2006) Endurance Characteristics of SuperFlash Memory. 8th Int. Conf. Solid-State and Integr. Circuit Technol.:763–765
40. Ma Y, Pesavento A, Nguyen H, Li H, Paulsen R (2006) Reliability and Qualification of a Floating Gate Memory Manufactured in a Generic Logic Process for RFID Applications. Non-Volatile Semiconductor Memory Workshop:44–45
41. Maayan E, Dvir R, Shor J, Polansky Y, Sofer Y, Bloom I, Avni D, Eitan B, Cohen Z, Meyassed M, Alpern Y, Palm H, Kamienski E S, Haibach P, Caspary D, Riedel S, and Knöfler R (2002) A 512 Mb NROM Flash Data Storage Memory with 8 MB/s Data Rate. Dig. Tech. Papers, ISSCC:100–101
42. Matsufuji K, Namekawa T, Nakano H, Ito H, Wada O, and Otsuka N (2007) A 65 nm Pure CMOS One-time Programmable Memory Using a Two-Port Antifuse Cell Implemented in a Matrix Structure. IEEE Asian Solid-State Circuits Conf.:212–215
43. McPartland R J, and Singh R (2000) 1.25 volt, low cost, embedded flash memory for low density applications. Dig. Tech. Papers, Symp. VLSI Circuits, 2000:158–161
44. Min H-C, Yater J, Kang S-T, Gasquet H, and Chindalore G (2007) Reliability Study of Split Gate Silicon Nanocrystal Flash EEPROM. Non-Volatile Semiconductor Memory Workshop:75–76
45. Minami S and Kamigaki Y (1991) Tunnel Oxide Thickness Optimization for High-Performance MNOS Nonvolatile Memory Devices. IEICE Trans. E74(4):875–884
46. Muralidhar R, Steimle R F, Sadd M, Rao R, Swift C T, Prinz E J, Yater J, Grieve L, Harber K, Hradsky B, Straub S, Acred B, Paulson W, Chen W, Parker L, Anderson S G H, Rossow M, Merchant T, Paransky M, Huynh T, Hadad D, Chang K-M, and White B E Jr (2003) A 6 V Embedded 90 nm Silicon Nanocrystal Nonvolatile Memory. Tech. Dig. IEDM:26.2.1–26.2.4
47. Nozaki T, Tanaka T, Kijiya Y, Kinoshita E, Tsuchiya T and Hayashi Y (1991) A 1-Mb EEP-ROM with MONOS Memory Cell for Semiconductor Disk Application. IEEE J. Solid-State Circuits, 26(4):497–501

48. Raszka J, Advani M, Tiwari V, Varisco L, Der Hacobian N, Mittal A, Han M, Shirdel A, and Shubat A (2004) Embedded Flash Memory for Security Applications in a 0.13 μm CMOS Logic Process. Dig. Tech. Papers ISSCC:2.4
49. Rosenberg J (2005) Embedded flash on a CMOS logic process enables secure hardware encryption for deep submicron designs. Non-Volatile Memory Technology Symposium:19–21
50. Shin Y, Choi J, Kang C, Lee C, Park K-T, Lee J-S, Sel J, Kim V, Choi B, Sim J, Kim D, Cho H-J, and Kim K (2005) A novel NAND-type MONOS memory using 63 nm process technology for multi-gigabit flash EEPROMs. IEDM Tech. Dig.:327–330
51. Stenzl W and Hupper J (2008) Zuverlässig und schnell: Monos-Flash. EE-Times Europe, March 17. http://eetimes.eu/germany/206904002
52. Takahashi K, Doi H, Tamura N, Mimuro K, Hashizume T, Moriyama Y, and Okuda Y (1999) A 0.9 V Operation 2-Transistor Flash Memory for Embedded Logic LSIs. Symp. VLSI Technol.:21–22
53. Uhlmann G, Aipperspach T, Kirihata T, Kothandaraman C, Li Y Z, Paone C, Reed B, Robson N, Safran J, Schmitt D, Iyer S (2008) A Commercial Field-Programmable Dense eFUSE Array Memory with 99.999% Sense Yield for 45 nm SOI CMOS. Dig. Tech. Papers ISSCC:406–407
54. Wu A T, Chan T Y, Ko P K and Hu C (1986) A Novel High-Speed, 5-Volt Programming EPROM Structure with Source-Side Injection. Tech. Dig. Int. Elec. Device Meet.:584–587
55. Yater J A, Kang S T, Steimle R, Hong C M, Winstead B, Herrick M, Chindalore G (2007) Optimization of 90 nm Split Gate Nanocrystal Non-Volatile Memory. Non-Volatile Semiconductor Memory Workshop:77–78
56. Yatsuda Y, Nabetani S, Uchida K, Minami S, Terasawa M, Hagiwara T, Katto H, Yasui T (1985) Hi-MNOS II Technology for a 64-kbit Byte-Erasable 5-V-Only EEPROM. IEEE Trans. ED, 32(2):224–231

Chapter 7
Embedded Magnetic RAM

Hideto Hidaka

Abstract In Chapter 7, magnetic RAM (MRAM) technology is introduced as a key technology candidate for creating new applications such as nonvolatile RAM. After an introduction of the history and basic principles of MRAM, we look into MRAM technology and basic design as well as on various memory cell architectures in Section 7.1. Then overviews on representative MRAM design examples, possible applications, and future challenges of MRAM are provided in Section 7.2. Finally nonvolatile memory frontiers and challenges are discussed in Section 7.3 as a conclusion of Chapters 6 and 7. The focus of this chapter is to highlight and summarize important concepts of the new technology.

7.1 Embedded Magnetic RAM Technology and Design

This section provides an overview of brief history and principles of magnetic random access memory (MRAM), MRAM technology, and basic designs. In this section the physics of magnetism and magnetic materials are only briefly described, avoiding an incomplete discussion on the vast background physics and technology.

7.1.1 MRAM Basics

7.1.1.1 History of MRAM

Since the early stages of computer systems, magnetism has contributed to the development of storage media in the forms of magnetic core memory, magnetic tape, magnetic bubble memory, and hard disk drive (HDD). Data storage by magnetization orientation in magnetic media is a basic physical mechanism of magnetic memories. The scalability and cost aspects have made HDD almost only one magnetic

---

H. Hidaka (✉)
MCU Technology Division, Renesas Technology Corporation, Tokyo, Japan
e-mail: hidaka.hideto@renesas.com

K. Zhang (ed.), *Embedded Memories for Nano-Scale VLSIs*, Series on Integrated Circuits and Systems, DOI 10.1007/978-0-387-88497-4_7,

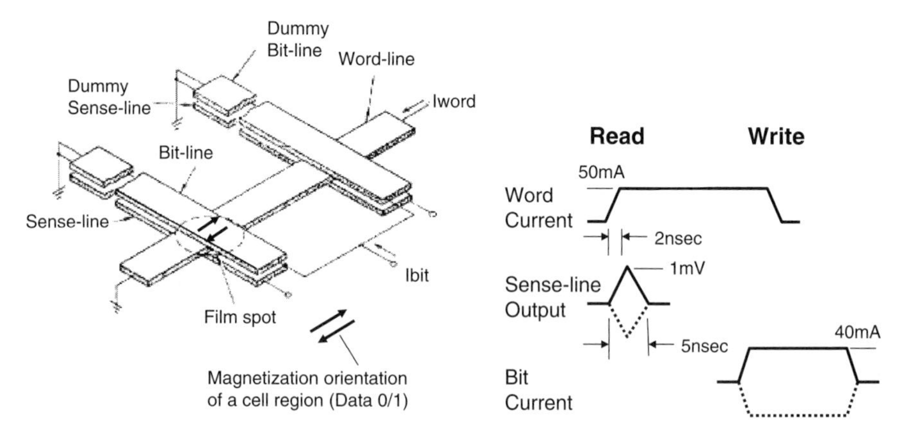

Fig. 7.1 Proposed plated wire magnetic memory in 1966 [16] (© 2008 IEEE)

storage medium still alive in today's computer and consumer systems. The MRAM concept originating in the trial on integrating magnetic memory on the semiconductor substrate dates back to 1960s. The thin-film magnetic memory, "plated wire memory scheme," exemplified in Fig. 7.1 [16] was to integrate a thin-film magnetic memory on a common substrate for monolithic integration with the peripheral circuitry by silicon chips, which was actually a multi-chip module integration project with thin-film magnetic layers on the substrate.

In this memory device data write is conducted by a summed magnetic fields generated by the selected word-line and bit-line for the selected ferromagnetic memory cell region at the cross-point. The data read is done by sensing the small signal potential around +/–1 mV by the electromotive force owing to the electromagnetic induction by a fixed word-line current (field), depending on the stored data as represented by the magnetization orientation in the selected cell. Since this is sometimes a destructive read out (DRO) operation, because a word-line current may flip the magnetization of the cell region, a data rewrite operation after the sensing operation is necessary to restore the data. This memory has a capacity of 64 words $\times$24 bits with 250 ns cycle time. Integrated selection and current provider circuits are used, face-down bonded to aluminum wiring on a common substrate with the magnetic film and matrix wiring. Memory dimensions, including circuitry, are 3" $\times$ 6" $\times$ 0.200". The memory storage element is a very thin magnetic film, with word and bit switching current requirements at 50 and 40 mA, respectively. Power consumption is 4 W. [16]

The thin-film magnetic memory concept in 1960s found a drawback in scalability because the read signal from the memory cell by the electromagnetic induction when applying a magnetic field is rather small, of the order of millivolts, and the signal gets smaller when the device is scaled down. Hence, how to read the magnetization orientation in a cell with a wide margin has been a main challenge to overcome. Therefore the MRAM concept has explored the scalable data read principles.

The magneto-resistive effects, anisotropic resistivity depending on the magnetization, with scaled device dimensions have offered solutions for improving this situation. The use of giant magneto-resistive (GMR) effects found a promising solution to this scalability problem, which realized the first commercial MRAM [11] described in Figs. 7.2 and 7.3. In the GMR device the resistance of a current flowing in the plane of a ferromagnetic metal is modulated by several percentages by the giant magneto-resistive effect, according to the parallel or anti-parallel magnetization of the stacked two ferromagnetic layers separated by a nonmagnetic metal spacer layer (Fig. 7.2a). The GMR effect comes from the in-plane resistance modulation by spin-dependent scattering in spin-polarized electrons at the interface of two ferromagnetic layers. In parallel magnetizations the spin-polarized conduction electrons see small scattering at the interface, providing a low resistance for in-plane current, while in anti-parallel magnetizations the scattering raises the in-plane resistance.

A pseudo-spin-valve (PSV) structure was devised for a GMR memory cell with read/write capability (Fig. 7.2a and b). A PSV is composed of a storage ferromagnetic layer with larger switching field and a sensing ferromagnetic layer with smaller switching field. Both layers are separated by a nonmagnetic metallic spacer layer. According to the magnetization orientations of the two ferromagnetic layers, this system has four states of magnetization as shown in Fig. 7.2b. The data write is done by applying a sufficiently large current, positive or negative, to switch the magnetization of the storage layer. Actually a selective write operation is carried out by the vector sum of the Hword and Hsense (plus optional Hbias). Only the memory cell at the cross-point of selected word and bit is applied a sufficient switching field while keeping unselected cells unswitched (Figs. 7.2c and 7.3). The data read is by applying a relatively small sensing field to flip the magnetization of the sensing layer from positive to negative (Fig. 7.3), thus generating a rising or falling resistance change according to the stored data, by the steep property around the origin point in Fig. 7.2b.

Because the ferromagnetic metal has quite a low resistance, a GMR-MRAM cell cannot be connected to a MOS selection transistor with much higher resistance for a sufficient signal margin. GMR-MRAM array has a string of GMR memory cell elements each with 10–100 ohm of resistance, which are serially connected in a column providing a read current path (Fig. 7.3). Applying a fixed magnetic field on the selected memory cell modulates the serially connected total resistance in a column to be sensed for the data of the selected cell. The chip performance is 50 ns for read/write for a 1 Mbit device fabricated in a 150 nm technology [11].

The GMR-MRAM concept much improved over the former plated wire magnetic memory concept in the read signal margin and scalability, owing to the GMR effect. But due to still small GMR ratio (several percentages) and the low resistivity of GMR element, GMR-MRAM was encountered by the difficulty in integrating on the high-density baseline CMOS. An intrinsic CMOS compatibility problem has become important. For the resistance-modulation-based memory cell like MRAM to be integrated with selection MOS-Tr in a one-to-one manner to secure the read signal margin, the resistance range of the cell resistor is required to match that of

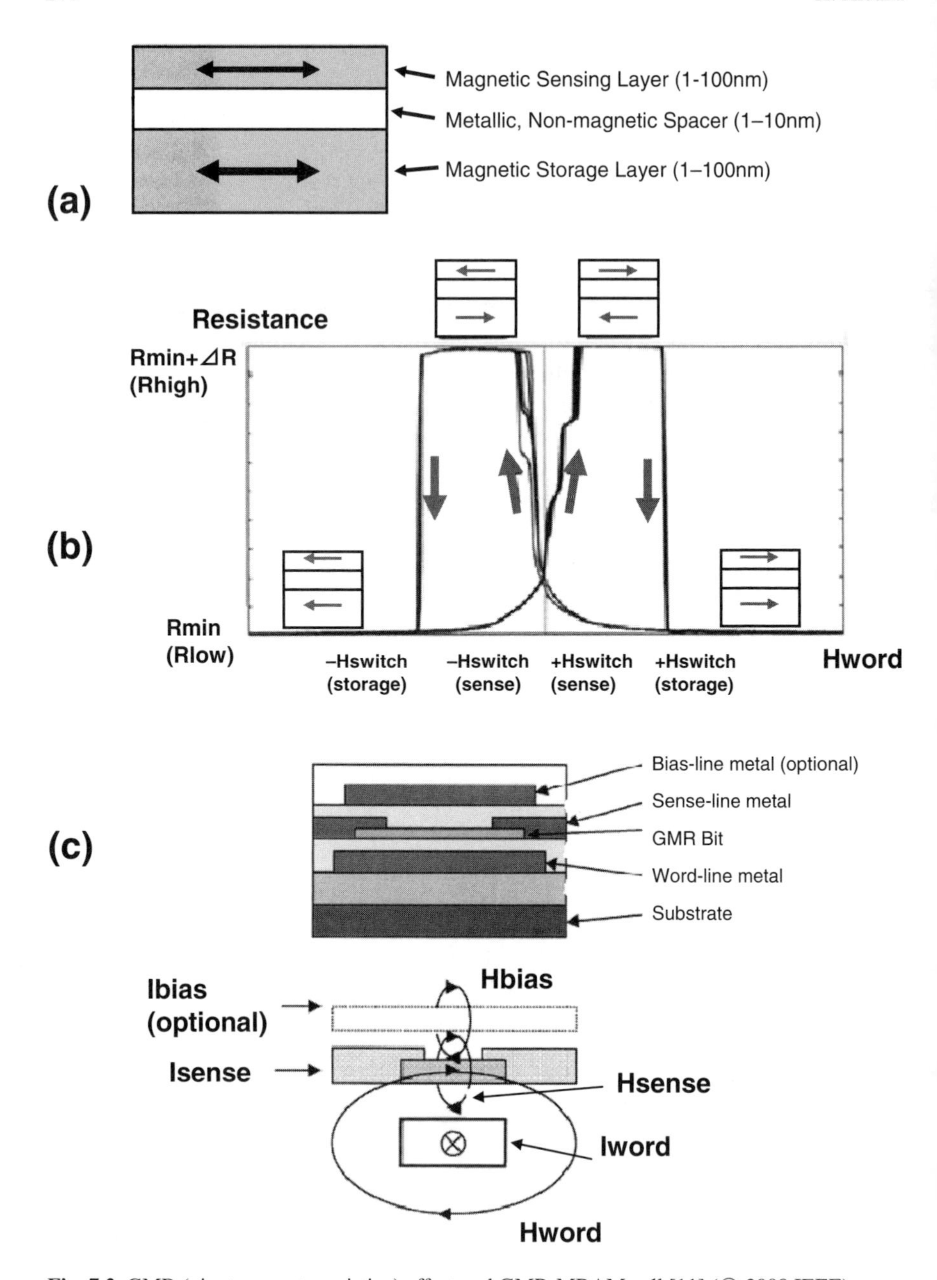

Fig. 7.2 GMR (giant magneto-resistive) effect and GMR-MRAM cell [11] (© 2008 IEEE)

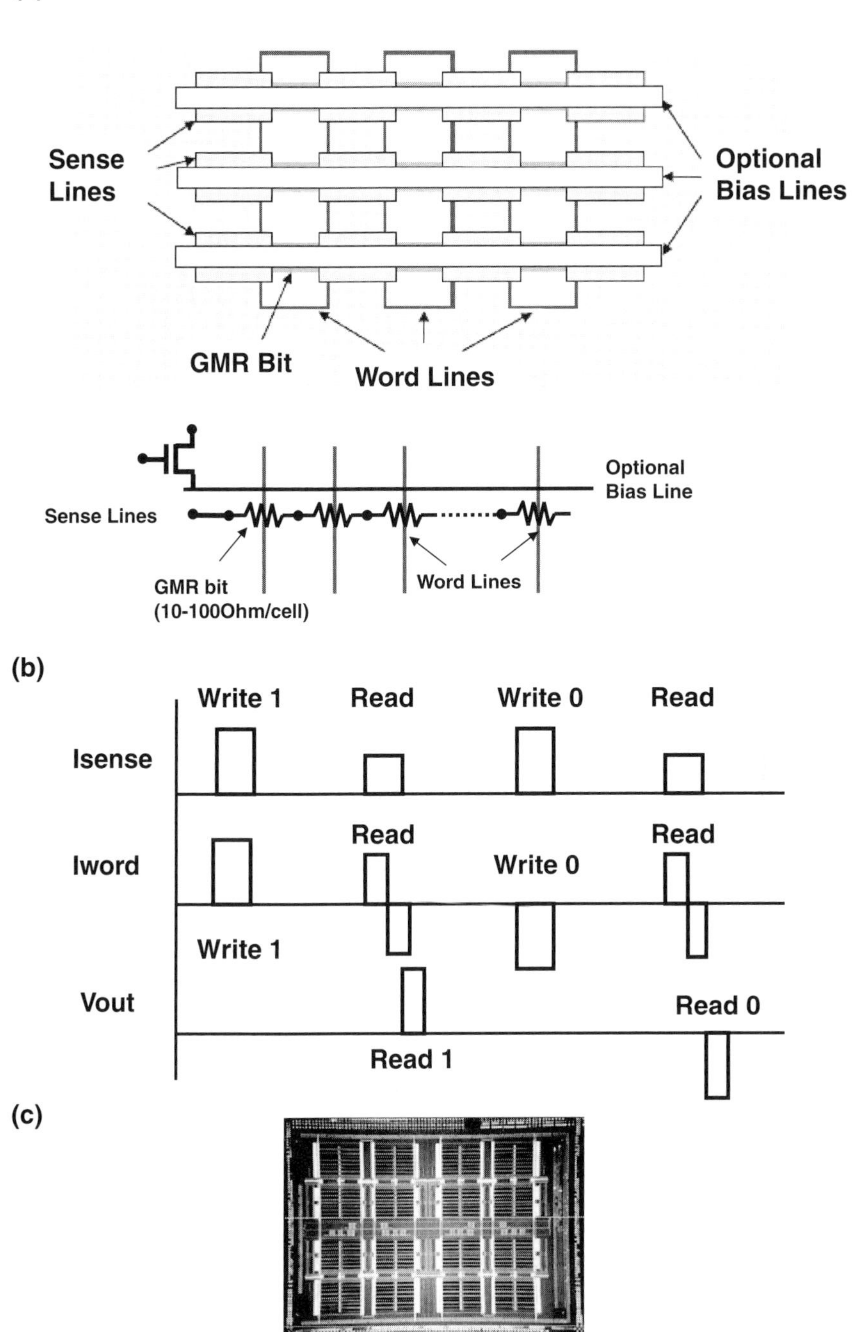

Fig. 7.3 GMR-MRAM cell array and operations [11] (© 2008 IEEE)

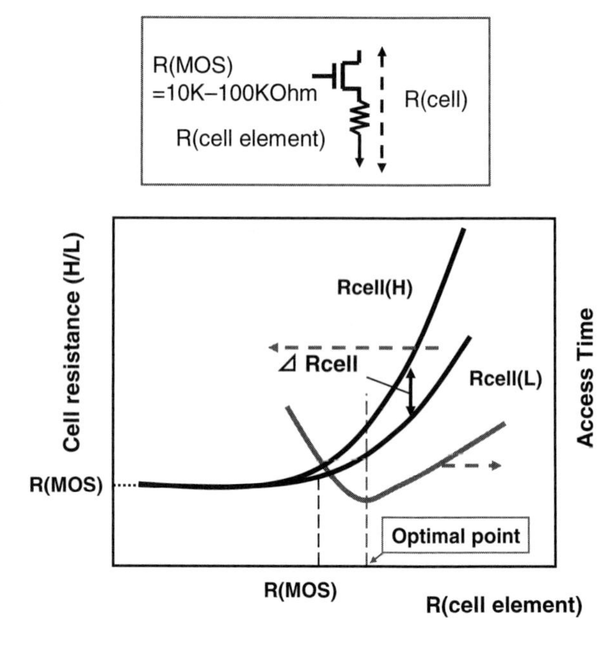

Fig. 7.4 Suitable cell element resistance in a CMOS-resistive memory. In a CMOS-resistive memory with 1Tr-1R cell, read access time is faster for lower cell resistance and larger cell resistance amplitude. The select-MOS resistance degrades the cell resistance amplitude significantly if the cell element resistance is low compared with the select-MOS resistance

the cell selection MOS transistor in view of the resistance change to be sensed in an optimized access time. The higher the cell resistance is, the access time becomes slower because of low current drivability, and the lower the cell resistance is, the resistance of the cell selection MOS-Tr degrades the total cell resistance ratio, making the sensing operation slower, as shown in Fig. 7.4. Therefore matching the resistance of the cell with the selection MOS Tr resistance is a key to implement resistance modulation based memory cell. GMR single-bit cell typically with a few 10-ohm resistance does not meet this requirement.

In 1990s, tunneling magneto-resistive (TMR) effect found a fit to the CMOS integration with better than 20% TMR ratio and more than *k* ohm order of cell element resistance comparable with the channel resistance of the memory cell selection MOS transistor. After exploratory works and commercialization in 1990s of the GMR and TMR sensor heads for HDD (hard-disk drive) products, TMR-MRAM was established in concept. Early trials on TMR-MRAM showed successful integrations on existing baseline CMOS with selection transistor [3, 20]. After the year 2000, efforts have been put in solving MRAM-specific problems not seen in HDD head, like device variability such as switching current distributions. TMR-MRAM has borne new technologies after inheriting much from GMR-MRAM technologies. Improved etching techniques, TMR stack structure, the toggle cell conception, and cladding structures have been major achievements we have seen so far. Future projections of MRAM include spin-torque switching for scalability and value creation in the market by the NV-RAM concept. Figure 7.5 describes a brief history of MRAM especially in what problems have been solved.

Year of Events	*Achievements*
1960s	
- Plated-wire thin-film magnetic memory	**monolithic integration**
1990s	
- GMR MRAM (MR=several %)	**read signal scalability, commercialized**
2000–2006	
- TMR-MRAM (MR=20-200%) Commercialization (2006)	**high MR & 1Tr-1R cell, single-domain magnetization, toggle-cell, Cladding**
Future prospects	
- SpinTorque Transfer	**Iwrite scalability**
- Perpendicular TMR	**Iwrite scalability & thermal disturbance immunity**
- NV-RAM applicatlons	**CMOS compatibility in flash/SRAM technology**

Fig. 7.5 A brief history of MRAM, milestones and timelines

7.1.1.2 Principle of TMR-MRAM

The TMR (tunneling magneto-resistive) effect is provided by a magnetic tunneling junction (MTJ) composed of two ferromagnetic layers separated by a tunneling barrier. The TMR effect comes from the difference of state density in spin-polarized electrons in these two layers of magnetized ferromagnet separated by a tunneling barrier. When both ferromagnetic layers have parallel magnetizations the state density has more overlap than in the anti-parallel case, providing higher electric conduction (Fig. 7.6).

Thus, the MTJ resistance has two distinct values, Rmax and Rmin, according to two states of magnetizations, in parallel and in anti-parallel. The TMR ratio defined by TMR=(Rmax–Rmin)/Rmin is typically 20–200% depending on the ferromagnetic materials, tunneling barrier material and thickness, and bias voltage along the MTJ. The MTJ device has seen a tremendous improvement of TMR ratio and its bias dependence in the last 30 years since the first discovery of TMR effect in 1975 [10], which continues to provide an ever-promising outlook for MRAM technologies.

Considering the split energy state density in spin-polarized ferromagnetic materials separated by a tunneling barrier, more overlap of states at Fermi-level Ef leading to more conduction of current (lower electric resistance) is observed in parallel magnetization than in anti-parallel conditions. Properties in magnetic materials and tunneling barrier for realizing a high TMR ratio, higher spin polarization, small spin scattering in the tunneling barrier, etc. have been explored extensively. NiFe/AlOx/NiFe and CoFeB/AlOx/CoFeB systems find 20–50% TMR ratio at room temperature and are commonly used for HDD head and MRAM. Recently, use of MgO tunneling barrier has shown much higher TMR ratio, as large as $>200\%$.

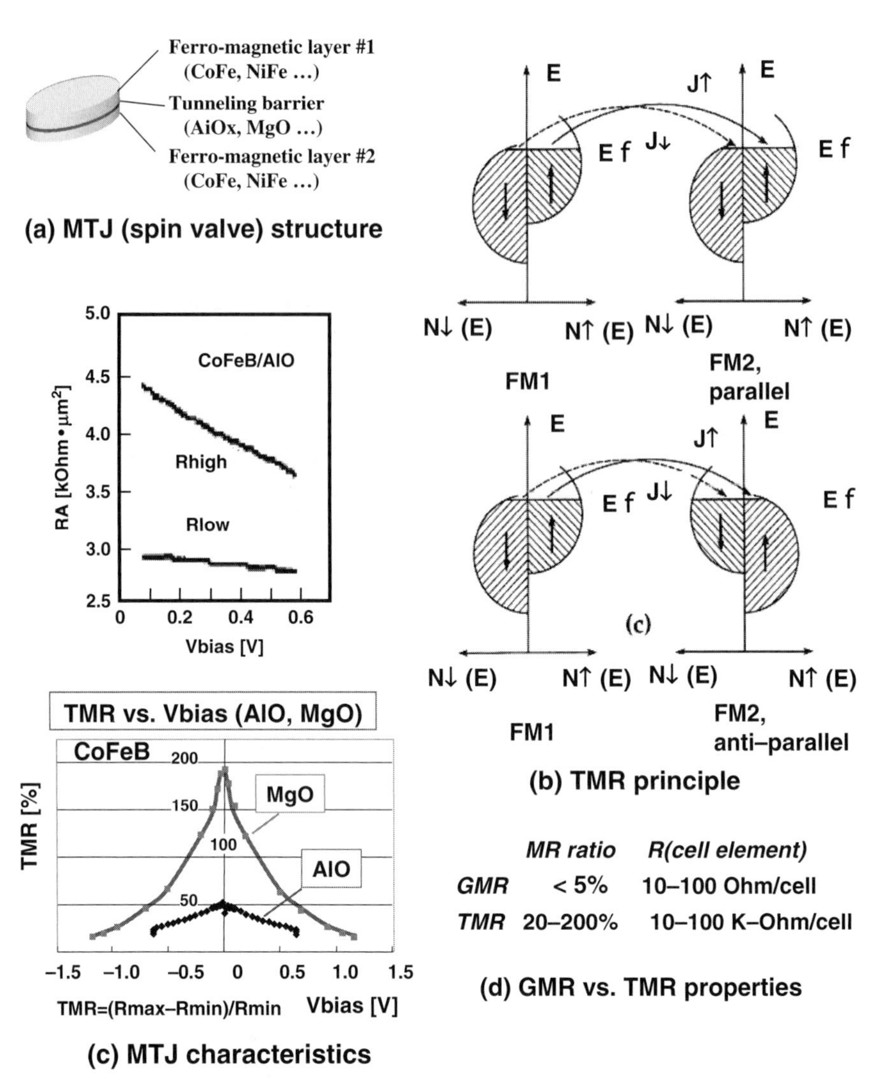

Fig. 7.6 Principle of TMR effect and MTJ characteristics [24] (© 2008 IEEE)

Material explorations have been extended to half-metal materials showing more than 1000% in TMR ratio.

The steep dependence of TMR on the bias voltage as shown in Fig. 7.6 limits the actual operating conditions of MTJ device in MRAM applications. The MTJ resistivity (RA) is varied according to the thickness of the tunneling barrier layer to adjust the MTJ resistance to match the minimum size MOS selection transistor resistance ranging in 10 k–100 k ohm/cell.

In TMR-MRAM, storage data of the cell is read through the TMR effect at the MTJ junction, while data write is done by magnetization reversal operations by either field switching or spin-torque switching principles. Because of the collectively cooperative behaviors of spin, fine-patterned ferromagnetic element provides a bi-stable memory in principle because scaled ferromagnetic element with oval shape realizes a two state stability with nearly uniform magnetizations (Fig. 7.7). Moreover the scaled cell dimensions below 1 μm^2 naturally contributes to single-domain magnetization of the magnetic elements providing a bi-stable behavior of the cell. In principle, TMR-MRAM is a nonvolatile bi-stable memory composed of ferromagnetic element switchable either by applied magnetic field or spin-torque

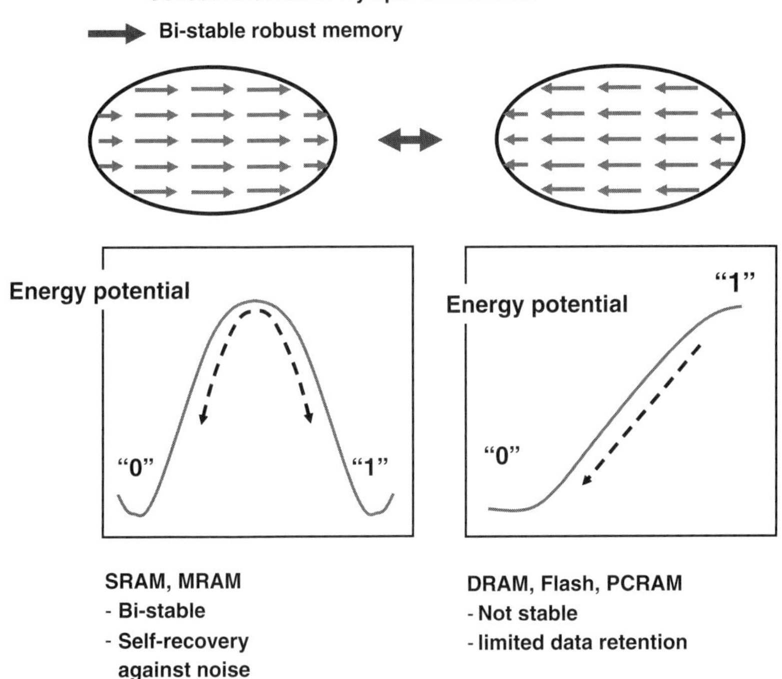

Fig. 7.7 Principle of TMR-MRAM

transfer induced by the current through magnetic materials. The MRAM property of nonvolatile bi-stability provides a memory principle eliminating the problems of data retention in flash memory, thus realizing a nonvolatile RAM.

To write a data by altering magnetization orientation of the free layer, either field switching or spin-torque transfer principle is exploited. The variablility control of the fine-patterned TMR elements in the switching field or current has been a focus of practical MRAM development, because this determines the write operation margin and immunity against thermal/field disturbance. This is an MRAM-specific problem because a GMR or TMR magnetic sensor used in HDD system is a one-device product not suffering from on-chip variations in multiple numbers of fine-patterned magnetic elements. The data write operation is described in Section 7.1.2.

7.1.2 MRAM Integration and Basic Design

7.1.2.1 MRAM Cell Structure and Integration

The actual MTJ structure in Fig. 7.8 finds a somewhat complicated multi-layer structure. Besides the essential three-layer MTJ composed of two ferromagnetic layers, one for free magnetization and the other with fixed (pinned) magnetization, and a tunneling layer, an underlying structure to form a pinned layer with closed-flux structure without magnetic field interference and barrier metal layers for thermal process budget, is necessary for the normal operation and process margin of the MRAM cell.

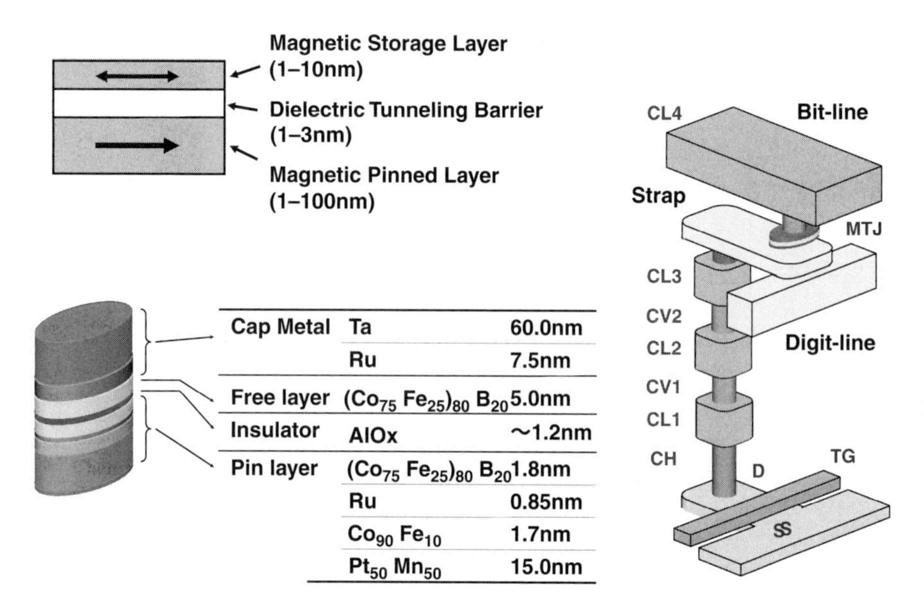

Fig. 7.8 Example structure of MTJ and MRAM cell

One important consideration in the field-switching MRAM cell is how to write a data selectively to a selected memory cell by summed x and y magnetic fields applied by the bit-line and digit-line, and not to the unselected and half-selected cells. The selectivity in the data write operation in field-switching MRAM cell is expressed in the asteroid curve, which plots the magnetization hysteresis curve in two dimensions (Fig. 7.9). Because of the rotational magnetization behavior, the hysteresis in x-direction narrows when applied y field increases. While the selected memory cell is given a data-reversing field by x and y fields getting outside the asteroid curve, the unselected cell must stay inside the asteroid, not to be reversed. In particular, half-selected cells on the x- or y-axis should not pass the reversing boundary of asteroid curve.

These selective write conditions limit the applied fields in the hatched region in the asteroid curve. Since the memory cells have certain variations in the switching field due to the fabrication fluctuation, this operating region gets narrower for multiple bits of memory. This restriction poses the field-switching MRAM cell a limitation on the device variations. Since the conception of MRAM, every effort has been put to improve the magnetic device variations to provide MRAM with sufficient margin for fabrication. This has been the major focus in MRAM through 2000s because magnetic disk or magnetic head sensor does not have patterned multiple elements. Improvement of fine-patterning techniques, ferromagnet shape, and the invention of the toggle-mode operation [14] are major achievements so far.

Basic TMR-MRAM structure integrated in the baseline CMOS shows a CMOS-compatible solution quite different from flash memories. A cross-section view of MRAM cell shown in Fig. 7.10 [23] reveals a simple add-on structure in existing CMOS. An MTJ element is inserted between different levels of metal (Cu)

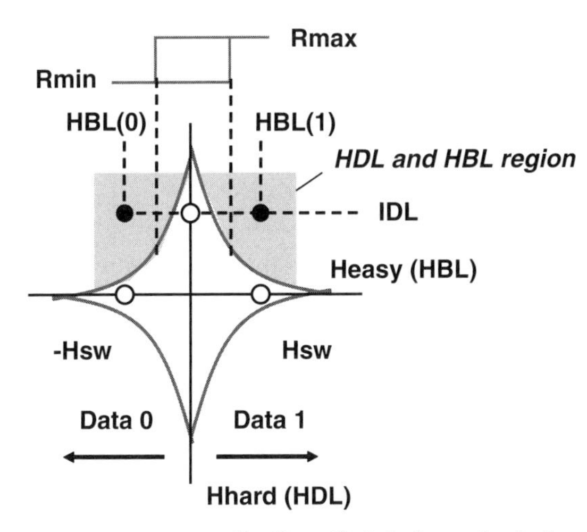

Fig. 7.9 An asteroid curve plotting 2D hysteresis of patterned ferromagnet

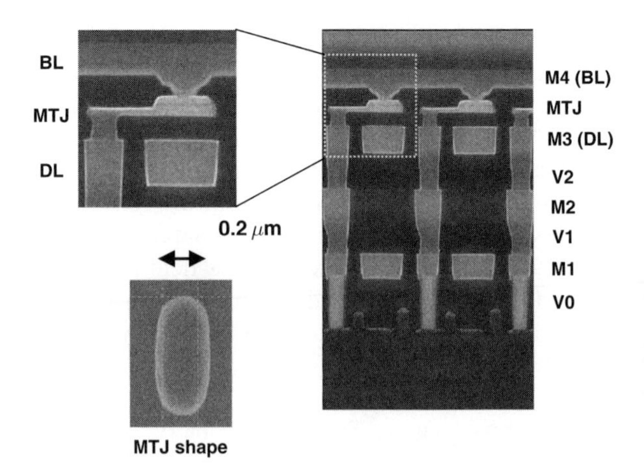

Fig. 7.10 TMR-MRAM cell structure [23, 24] (© 2008 IEEE)

interconnects. In this case the MTJ is formed between third and fourth metal layers M3 and M4, together with a local interconnect layer. The current conducting lines digit-line (DL) and bit-line (BL) are formed of Cu interconnects to conduct the switching current in x and y directions. Although MRAM cell can be easily embedded in the CMOS structure, the cell size scaling is limited by the metal interconnect rules because field-switching TMR-MRAM needs metal interconnects in x and y directions per cell for write current conduction.

7.1.2.2 Basic TMR-MRAM Design Considerations

MRAM operating margins are closely related to properties of magnetic materials/ devices including magnetic field disturbance by internally generated fields by on-chip currents and externally applied fields, magnetic device variability in the fabrication process, thermal disturbance, write margin under device variability and thermal disturbance margin, and read signal margin as briefly listed in Fig. 7.11.

Figure 7.12 shows how MRAM circuit features fit embedded memory applications. The operating voltage is low enough to match the CMOS logic environment for high performance. Area advantage comes up because it does not need charge-pumping circuitry for highly positive and negative-voltage generations unlike flash memory. This is important for small-capacity memory frequently encountered in

(1) Magnetic field disturbance	
by DL/BL (half-select field)	**- toggle cell, cladding, spin-torque**
by external environment	**- field shield structure in package**
(2) Switching field variance	**- etching techniques, toggle cell**
(3) Thermal disturbance	**- larger Ms, SAF free layer, ECC ...**
(4) Iwrite setting	**- consider thermal disturbance**
(5) Read signal margin	**- larger TMR, higher Vbias in read**

Fig. 7.11 Important factors in MRAM operating margins and solutions

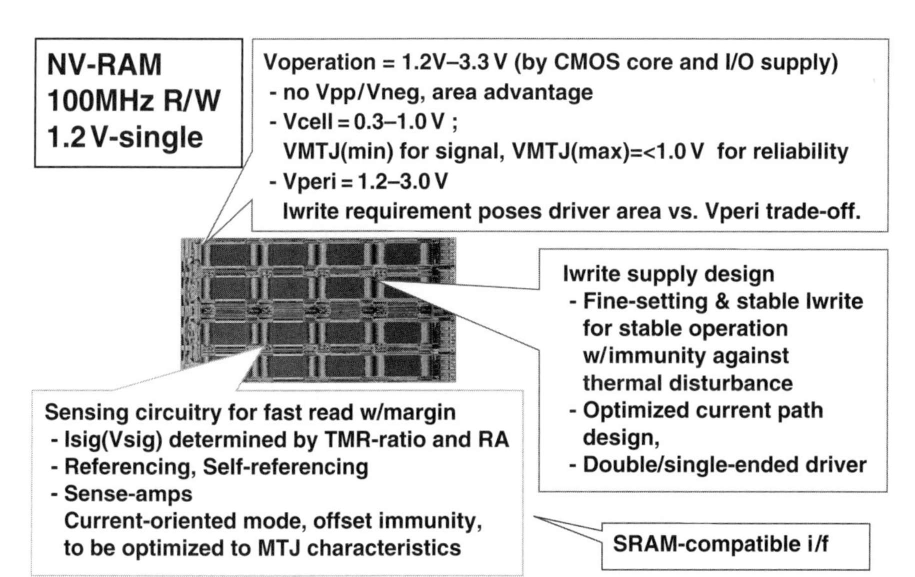

Fig. 7.12 MRAM circuit observation [7] (© 2008 IEEE)

embedded applications. Because MRAM has no sector divisions for erase operations unlike flash memory and it also has symmetrical access time for read and write, fully SRAM-compatible specifications and interface are achieved.

In the basic architecture and circuits of TMR-MRAM, important points in design to meet the requirements for embedded uses are

(1) Sensing architecture and circuits
As MRAM cell has a relatively small resistance change ratio for data 0 and 1 (20–100%), a signal sensing scheme with sufficient margin and high-speed sensing operation is essential. A current-oriented sensing scheme with a device tuning mechanism finds a fast sensing solution. Also a dummy cell scheme is a design of choice for enhancing the signal margin. Array noise factors like asymmetrical data and power lines configurations are also to be eliminated for stable sensing operations. In setting the memory cell operation voltage it should be considered that the bias dependence of the TMR ratio and the reliability of the tunneling barrier limit the maximum bias voltage at the MTJ junction. Practically around 0.5 V at the MTJ is a design choice.
(2) How to generate and convey relatively large write currents
Since the field-switching MRAM employs a rather large switching current in the order of 5–10 mA at peak per line, a stable generation and conveyance of the switching current is a very important factor in stable write operation. The design of large current drivers for write operations imposes area versus voltage trade-offs in the periphery circuitry. As the supply voltage gets higher, the driver transistor size gets smaller with disadvantage by relaxed design rule.

Use of higher supply voltage, for example, 3.3 V for I/O circuitry in some cases provides the reduced area for current drivers. Fine setting and stable supply of write current are important together with optimized current path design with minimum resistance.

(3) Magnetic disturbance on the chip
On-chip power lines and signal lines sometimes generate significant magnetic fields to disturb the memory cell operation. A careful design in view of on-chip magnetic fields generated by currents in the interconnect lines is essential in stable memory cell operations. A full-chip magnetic field simulation scheme is favored in MRAM chip design.

(4) Compact array
Since the high-sensitivity sensing architecture and large current drivers tend to lower the area efficiency of MRAM macro, area-efficient array organization and circuitry are often required. Array division and selection of operating voltage for each circuit part are quite important in the optimized design to meet embedded-specific requirements for a low-voltage operation and small-area penalty especially in lower capacity of memory.

7.1.2.3 Basic TMR-MRAM Design

A design of a 1 Mbit TMR-MRAM macro [23] to meet the above-mentioned requirements is described. Figure 7.13 is a brief introduction of the module designed for embedded uses employing a 130 nm CMOS technology. By utilizing hierarchical architectures, 100 MHz performance for both read/write is achieved at 1.2 V single supply. Overall, SRAM-compatible specifications and interface are successfully achieved by this MRAM design. The area breakdown in Fig. 7.13 shows large local current drivers account for as much as 33%. One way to alleviate this area penalty is to employ higher power supply like 3.3 V supplied for I/O circuitry in write current drivers.

Figure 7.14 describes an MRAM array organization. A dummy row, folded bit-line architecture, distributed current driver, and sense amplifiers are well accommodated in this design of MRAM memory mat. The dummy cell configuration is preferred because of the balanced symmetrical sensing architecture. This dummy cell scheme combines data 0 and 1 dummy cells in parallel to generate the reference current by averaging the reference current by data 0 and 1 dummy cells. In this design, the generated reference current is shared by 2 bits of memory cells for sensing each data by the sense amplifiers SA0 and SA1.the data bus shunt SH and bus swap switch SW0 and SW1 function to achieve this averaging and referencing scheme.

A sensing circuitry for MRAM described in Fig. 7.15 reflects on a relatively small resistance change in current MTJ. A solution for high-speed sensing is to implement a current-oriented sensing scheme. Because canceling device variation is a key to this type of sensing circuitry, device variation consideration and tuning mechanism are essential in the design. As the MTJ performance is improved

1 Mbit, 130 nm-CMOS/4 LM, 1Tr-1MTJ cell
CoFeB/AlOx (TMR = 48%@Vbias = 0 V)
Cell size = 0.81um2, Module size = 3.88 mm2/1 Mb
1.2 V-single, 100 MHz (R/W), Idd(max)=50 mA@100 MHz (Write)
Divided-WL/BL, Distributed current driver for low-voltage

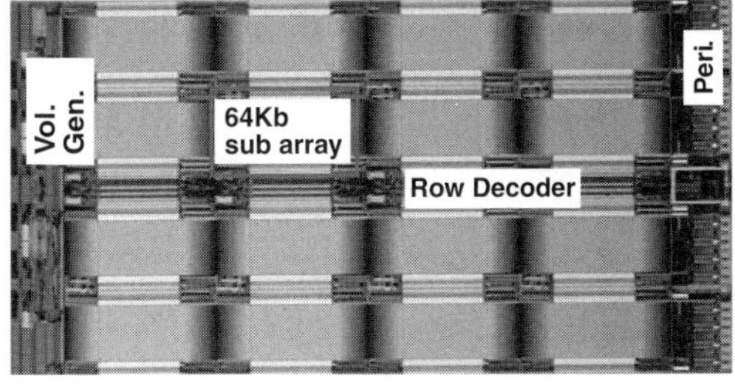

Area breakdown

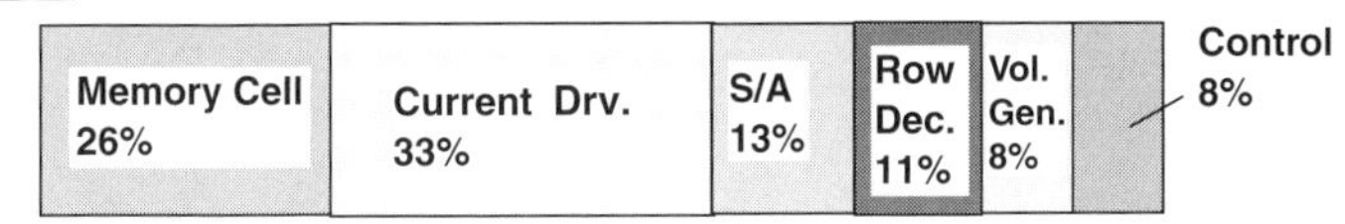

Fig. 7.13 An embedded 1 Mbit MRAM macro and area breakdown [7, 23] (© 2008 IEEE)

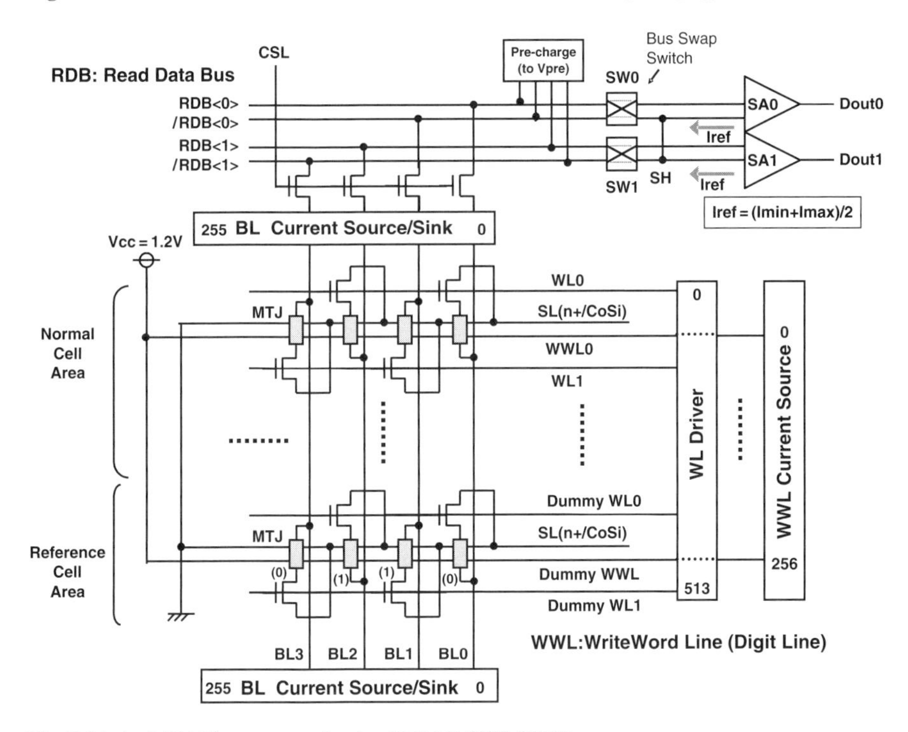

Fig. 7.14 An MRAM array organization [23] (© 2008 IEEE)

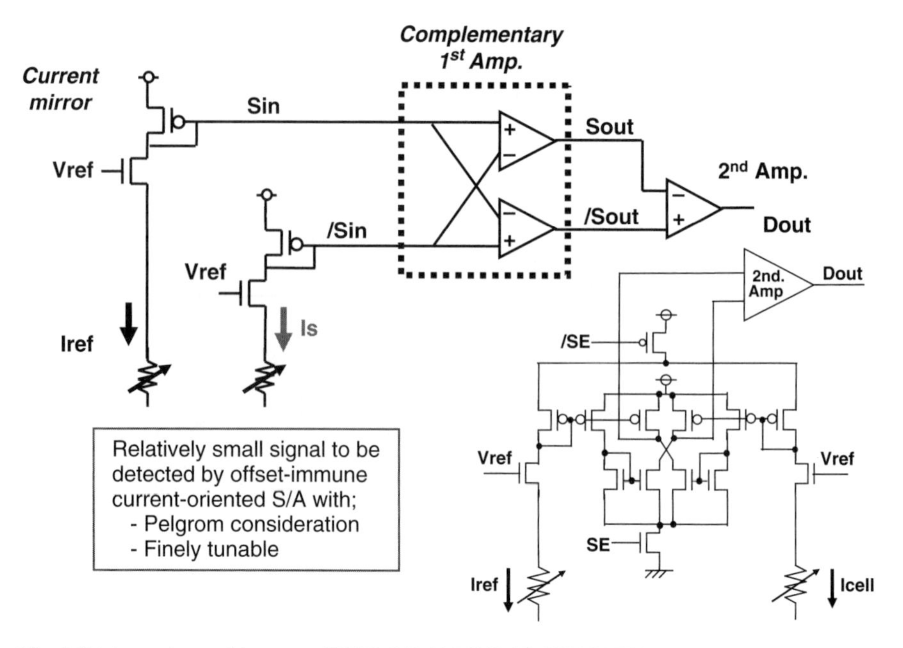

Fig. 7.15 A sensing architecture of TMR-MRAM [23] (© 2008 IEEE)

by adopting MgO tunneling barrier in the future, the sensing architecture may be changed to simpler DRAM and SRAM-like scheme for enhanced area efficiency.

Figure 7.16 shows a distributed local write current driver circuit for bit-line which realizes a very short current path from power supply to ground and eliminates serially connected common transistor or long common data current line, providing a current line with small stray capacitance and resistance. This contributes to low-voltage, high-speed write operation. Here the write current is controlled by the gate voltage level of the current control transistors, which is subject to a fine tuning mechanism. Actually a 100 MHz write cycle at 1.2 V operation is achieved. Likewise the digit-line drivers are distributed each by digit-line. In the digit-line the single-ended driver is adopted because the digit-line current can be in single direction (Fig. 7.14), eliminating the both-ended driver scheme further reducing the impedance for smaller driver size and much reduced area penalty.

The above-mentioned device performance, array architecture and circuit techniques contribute to realize a very high performance and low-voltage TMR-MRAM macro. The access time shmoo plots in Fig. 7.17 show sufficient margins for 100 MHz operation in read/write of a 1 Mbit MRAM macro. A read access time of 8 ns at worst is achieved owing to the low-noise and high-speed array architecture and sensing scheme. Write operation is much faster than read operation because the switching of magnetization orientation is inherently fast in the range of less than 1 ns theoretically. Out of 3 ns of write access time, less than 1 ns

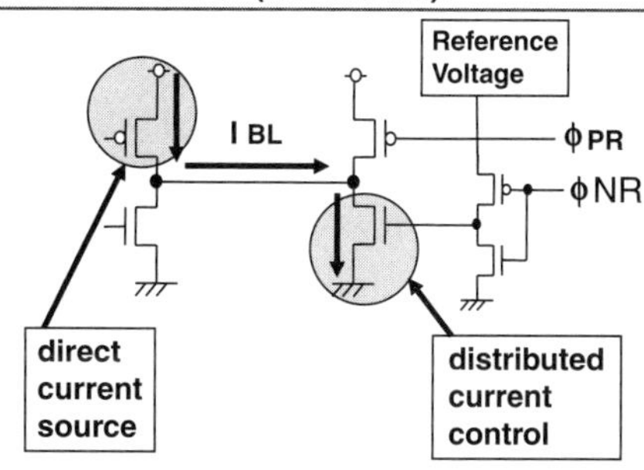

Fig. 7.16 A distributed current driver scheme for bit-line and digit-line [23] (© 2008 IEEE)

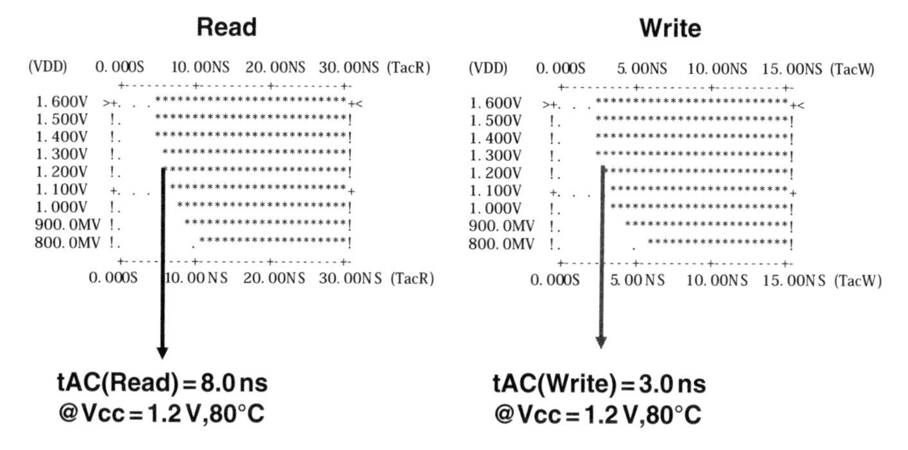

Fig. 7.17 Access time shmoo plots of TMR-MRAM

is dedicated to intrinsic cell switching and the remaining 2 ns is for periphery circuit operation.

Figure 7.18 describes the most remarkable attribute of MRAM in nonvolatile memories, limitless write. In this evaluation of the endurance characteristics on 1 Mbit die, we have observed no failures up to 10^{14} times of write. This test is suspended because of the test time limitation. MRAM's predicted endurance performance seems to be valid. Limitless number of write endurance is expected to be achieved because the spin flip does not see any fatigue or accumulated change of physical states and the collective bi-stable behavior helps resist noise due to local anomaly in the cell magnet to some degrees.

Figure 7.19 shows an estimate of the chip area breakdown in a typical flash memory compared with an MRAM. Because of area-consuming inefficient charge-pumping circuits for high-voltage and negative-voltage generations and transfer,

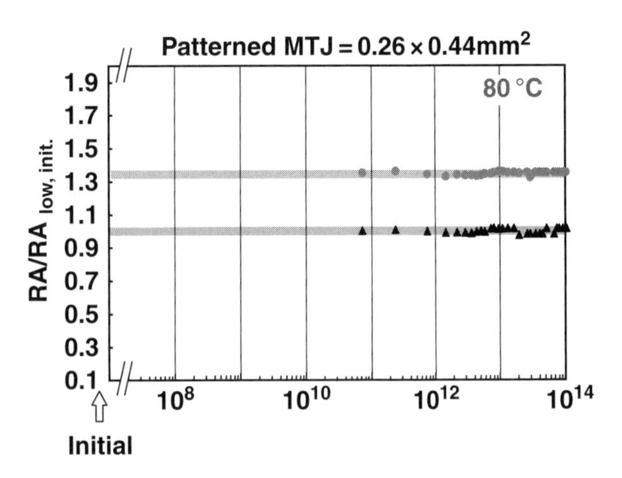

Fig. 7.18 Endurance in write cycling [6, 7] (© 2008 IEEE)

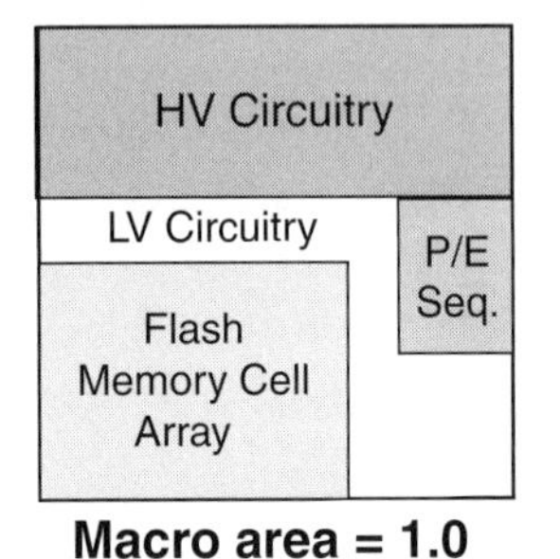

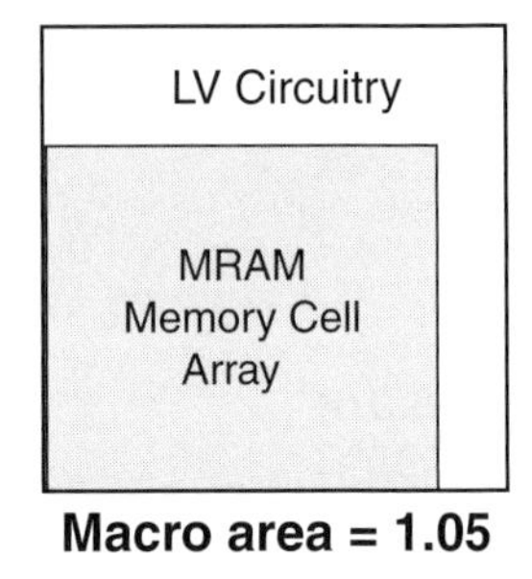

Fig. 7.19 Area breakdown, flash, and MRAM modules for 1 M-byte at 90 nm

flash memories suffer from a large area penalty. In contrast, an MRAM needs no high voltage in its operation and operates at 1.2 V single power supply in this design. This provides MRAM an advantage in the total macro size. A comparison shows us that even an MRAM cell size is quite larger, the macro size is about the same as a flash memory in a 90 nm technology.

For the read performance improvement, MgO (magnesium oxide) for tunneling barrier with much higher MR ratio than AlOx provides a promising solution. Although MgO is required in future spin-torque transfer MRAM for high-efficiency spin-torque transfer, conventional field-switching MRAM also merits from much improved MTJ performance by MgO for improving read access time. Up to 200 MHz read performance is expected by the simulation in a 130 nm

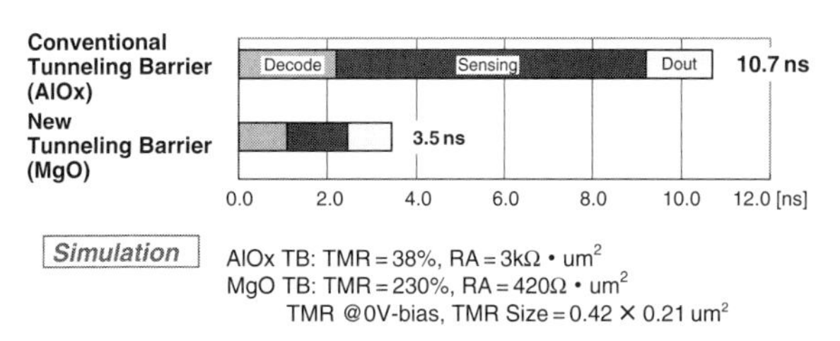

Fig. 7.20 An access time breakdown of 1 Mbit TMR-MRAM [7] (© 2008 IEEE)

CMOS technology, owing to the improved MTJ performance by MgO, as shown in Fig. 7.20.

Figure 7.21 describes the progress of MRAM technology development in cell size trend. Based on the TMR technology applied in the HDD industry, TMR-MRAM was conceived with relatively large MR ratio and appropriate memory cell resistance to be utilized in large-capacity memory devices. TMR-MRAM has come up to be productized at 0.18 μm technology node, after solving MRAM-specific problems not existing in HDD head technology. Beyond 90 nm generation, we will be more focused on the scalability issues by using magnesium oxide (MgO) for tunneling barrier and spin-torque switching principle.

7.1.2.4 MRAM Cell Architectures

Several experimental MRAM cell architectures shown in Fig. 7.22 represent current efforts employing existing memory circuit knowledge and newly developed concepts. Typical MRAM cell structures range from field-switching (a)–(e) to

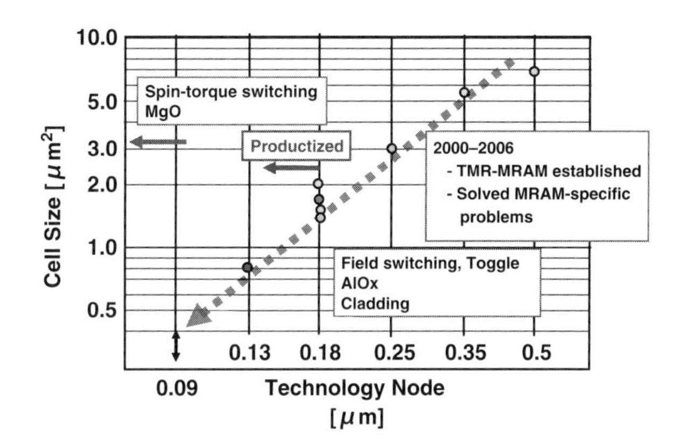

Fig. 7.21 Cell size trend of TMR-MRAM [7] (© 2008 IEEE)

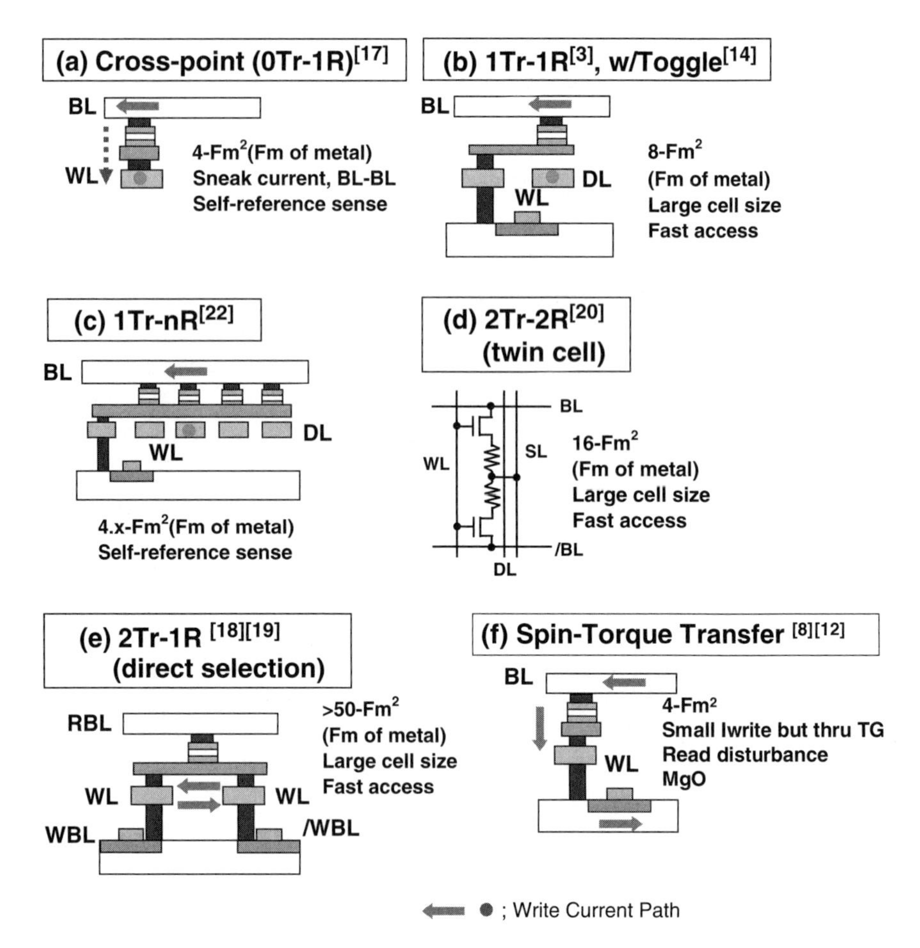

Fig. 7.22 MRAM cell architectures

spin-torque switching (f). (a) 0Tr-1TMR, (b) 1Tr-1TMR, and (c) 1Tr-nTMR cell structures show variations in composing a cell by an MTJ and corresponding selection transistors.

By virtue of MRAM's unlimited number of write endurance, a self-referencing scheme by writing a known data to the selected cell and sensing the cell resistance change before and after the write for judging the stored data is quite viable. This is utilized in earlier trials on the cross-point array in Fig. 7.22(a), but this architecture suffers from "sneak current" between bit-lines in read and write operations. Currently 1Tr-1MTJ cell Fig. 7.22(b) is the mainstream scheme, with or without toggle-mode concept. This is now seeing commercial products in the market. A 1Tr-nR architecture in Fig. 7.22(c) connects plural MTJs in parallel to be selected by one transistor at once. The read operation is conducted in a self-referencing manner. This architecture does not have a sneak current problem, unlike the cross-point configuration. Although the effective TMR ratio is reduced owing to the parallel connection of MTJs, this architecture may see a future by the expected MgO development with very large TMR ratio. A twin-cell structure in Fig. 7.22(d) composed of two 1Tr-1MTJ cells is a popular configuration in various types of memory for enhancing the operating margin. Also a 2Tr-1MTJ cell for the selective application of the write field to the selected cell show in Fig. 7.22(e) is to proposed for improving the write margin.

7.1.2.5 Spin-Torque Switching MRAM

The spin-torque transfer (STT) phenomenon was found to be effective in switching the stored data in MRAM cell [8, 12, 21]. In the MTJ structure, the spin torque is transferred when electrons with polarized spin from the free magnetization layer comes into a fixed layer through the tunneling barrier (Fig. 7.23). In an MTJ with a pinned layer (FM1), an MgO tunnel barrier layer (spacer) and a free layer (FM2), under a bias voltage Vb (>0), when a spin-polarized electron flows from F2 to F1, the spin direction rotates according to the directions of magnetic moment $\boldsymbol{M}1$ and $\boldsymbol{M}2$. The rotation of spin of the electrons in FM1 and FM2 layer generates a spin torque, $d\boldsymbol{M}1/dt$ and $d\boldsymbol{M}2/dt$, to the magnetic moment $\boldsymbol{M}1$ and $\boldsymbol{M}2$, by an exchange coupling of spins. When the given torque is large enough by the supply current, magnetization of FM2, $\boldsymbol{M}2$, is reversed. Then, the magnetization of FM1 and FM2 transform from parallel to anti-parallel alignment (from low-resistance state to high-resistance state). When the bias voltage is reversed, a spin-polarized electron flows from FM1 to FM2, which generates a parallelizing spin torque to FM2, then flips the magnetization in FM2 if a sufficient current is supplied [8, 21].

Because this process is low in efficiency, a current level of 10^{6-7} A/cm^2 is required for complete switching of the cell magnetization. The spin-torque transfer phenomenon has been conceived as a noise factor in hard-disk drive head sensor with a very thin tunneling oxide for a large sensing current, but in the positive use for cell switching in MRAM it provides promising prospects as follows:

(1) Switching current is drastically reduced, to 1/10 of the field-switching cell, and it is scalable by physical dimensions.
(2) Half-select disturbance problem is eliminated.

In the spin-torque MRAM, the data read operation is conducted in the same manner as the field-switching MRAM by using the TMR effect of the MTJ elements. By eliminating the digit-line the cell size is reduced to about half of the field-switching MRAM cell.

Beyond 65 nm, STT principle will dominate in MRAM because of the scaling difficulty in the switching current density at the current driving lines in the field-switching cells. On the other hand, the current problems and challenges in STT-MRAM include [7]

(1) Switching current reduction is essential because the minimum size cell selection transistor must provide the switching current. This restricts the cell size reduc-

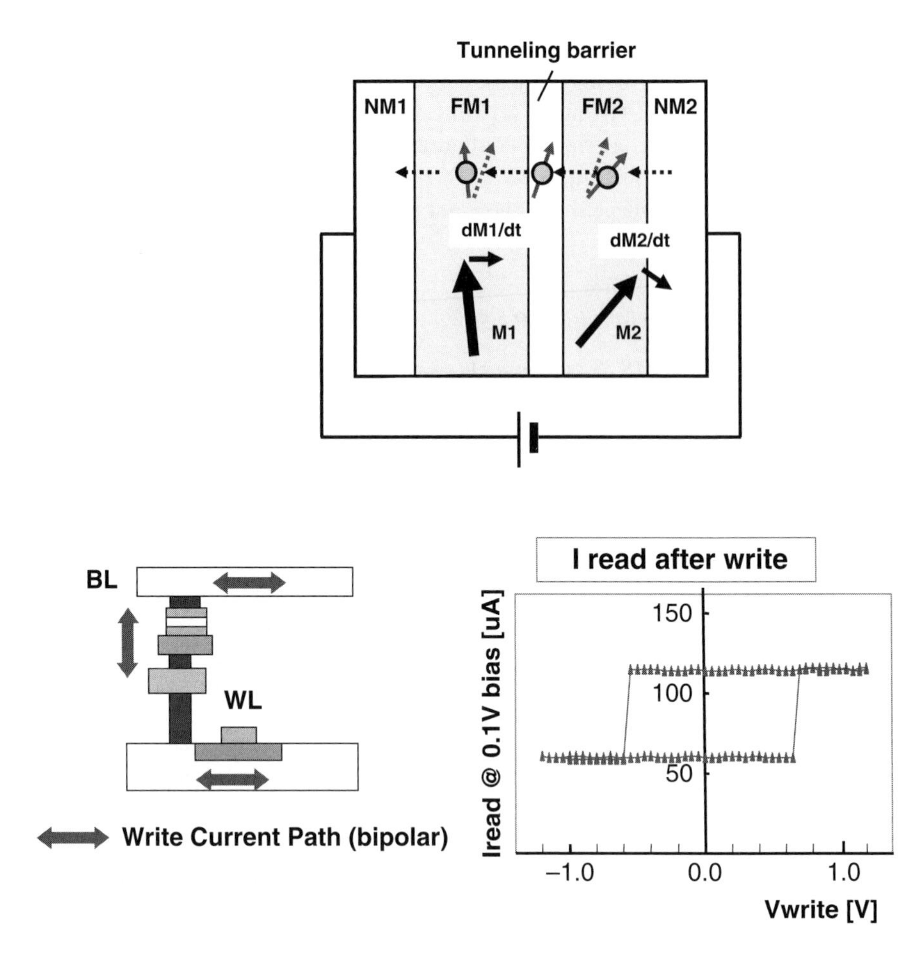

Fig. 7.23 An STT (spin torque switching) cell and operation [7, 8, 21] (© 2008 IEEE)

tion and/or fast switching. Today the switching current of 200 μA/cell has been achieved [12], but this cannot be provided by the minimum size cell selection transistor. Research in material and structure for reducing the switching current is now underway.

(2) As the free-layer dimension is coming down to the range of less than 0.1 μm, thermal disturbance error problem is looming up, especially during read disturbance. Some structural improvement is now explored.

(3) The required high-temperature process as high as 400°C for MgO formation may not be compatible with underlying CMOS. Lower temperature process is another challenge.

7.2 MRAM Design Examples and Applications

In this section, various aspects of MRAM designs are reviewed in their design concepts and implementation features (Section 7.2.1). Then potential MRAM applications are discussed (Section 7.2.2).

7.2.1 MRAM Design Examples

7.2.1.1 A 4 Mbit MRAM with Toggle Mode of Operation

Figure 7.24 shows a commercial TMR-MRAM with 4 Mbit of memory capacity, which incorporates a toggle mode of cell operation, a dummy column sensing architecture, and a current mirror sensing scheme. The toggle mode has shown a write operating margin superior to the conventional cell. By incorporating a closed-flux synthetic anti-ferromagnet (SAF) structure for the free layer placed 45-degree inclined to the current lines, the rotational applications of field switches the cell between binary states very stably. A wider window of write current selection is obtained by this structure as shown by the asteroid curve in Fig. 7.24 without half-select problems [1, 14, 15].

This cell operation needs only a unipolar switching current both at word-line and bit-line, reducing the complexity of the current driver circuitry resulting in a smaller chip size. By employing a 0.18 μm baseline CMOS technology, a 4 Mbit 25 ns cycle MRAM with cell size of 1.55 μm^2, die size of 4.5 × 6.3 mm^2, and 1.8 V/3.3 V power supply is achieved.

The dummy column architecture merits from a "pseudo-symmetrical" sensing operation and high-density memory cell array. Dummy cell columns are located not far from each memory cell column by a divided array manner. Sensing architectures combine two dummy cells with data 0 and 1 for generating a mid-point reference level corresponding to a selected cell, as shown in Fig. 7.25(a).

The current driver schemes in Figs. 7.25(b1), (b2), and (b3) show representative bit-line driver schemes. In Fig. 7.25(b1) [14] a shared current generator providing a write current, which is distributed and supplied to a selected bit-line through a transfer switch, is shown. This is a hierarchical scheme meriting from the easy control of the current adjustment. Figure 7.25(B2) [15] shows a distributed current driver scheme locating a current driver Tr (m0) at each column of bit-line. The supply current is adjusted by the Vwriteref given to the gate of the current driver. This scheme is area efficient, which is also employed in the distributed driver in Fig. 7.16 [23], with the improved ground noise immunity by gate isolation in the current driving period [15]. Figure 7.25(B3) is another trial in the current driver scheme to reduce the effective metal resistance of the current lines by using a FORK configuration [9].

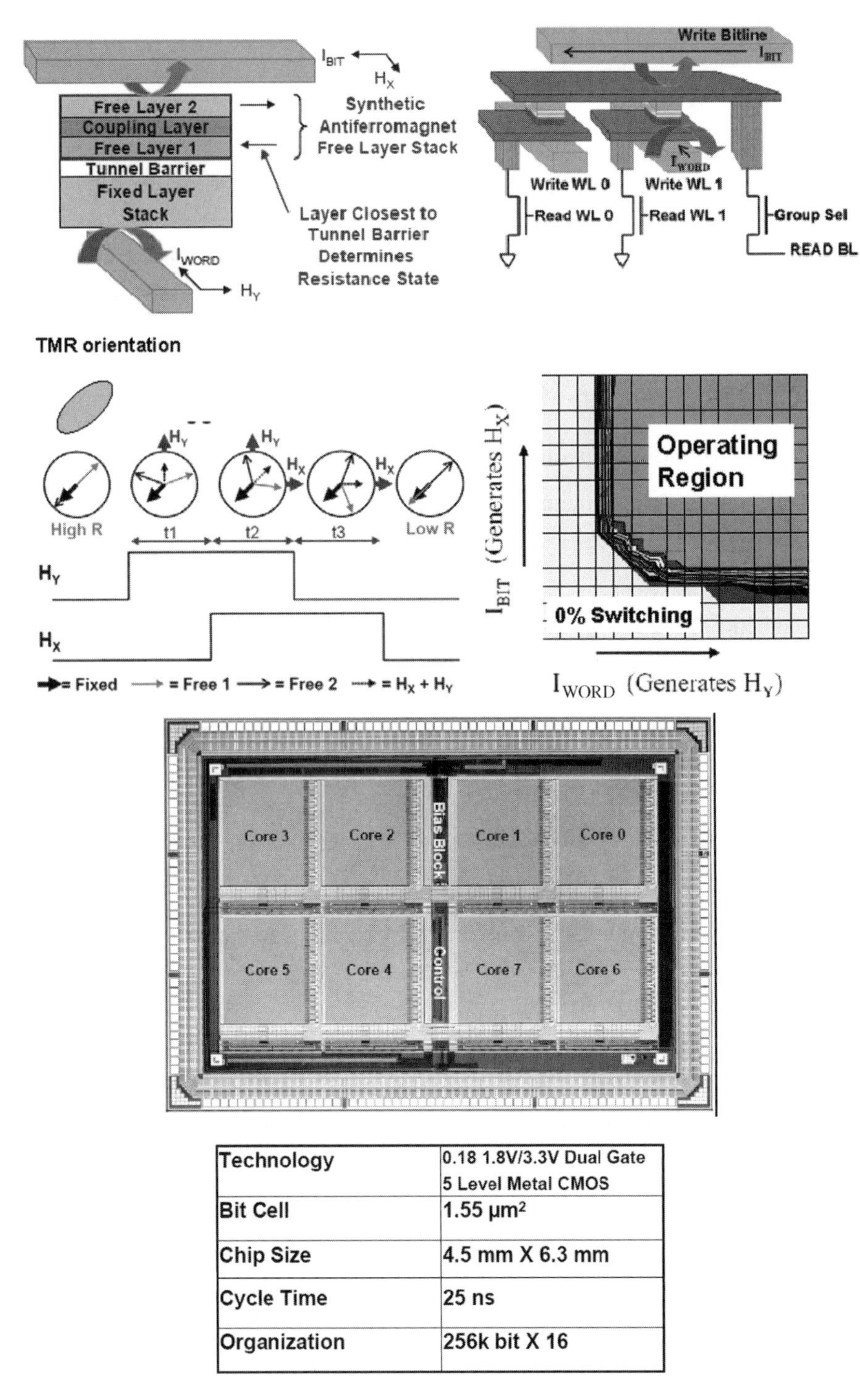

Technology	0.18 1.8V/3.3V Dual Gate 5 Level Metal CMOS
Bit Cell	1.55 μm^2
Chip Size	4.5 mm X 6.3 mm
Cycle Time	25 ns
Organization	256k bit X 16

Fig. 7.24 A toggle-mode TMR-MRAM [14] (© 2008 IEEE)

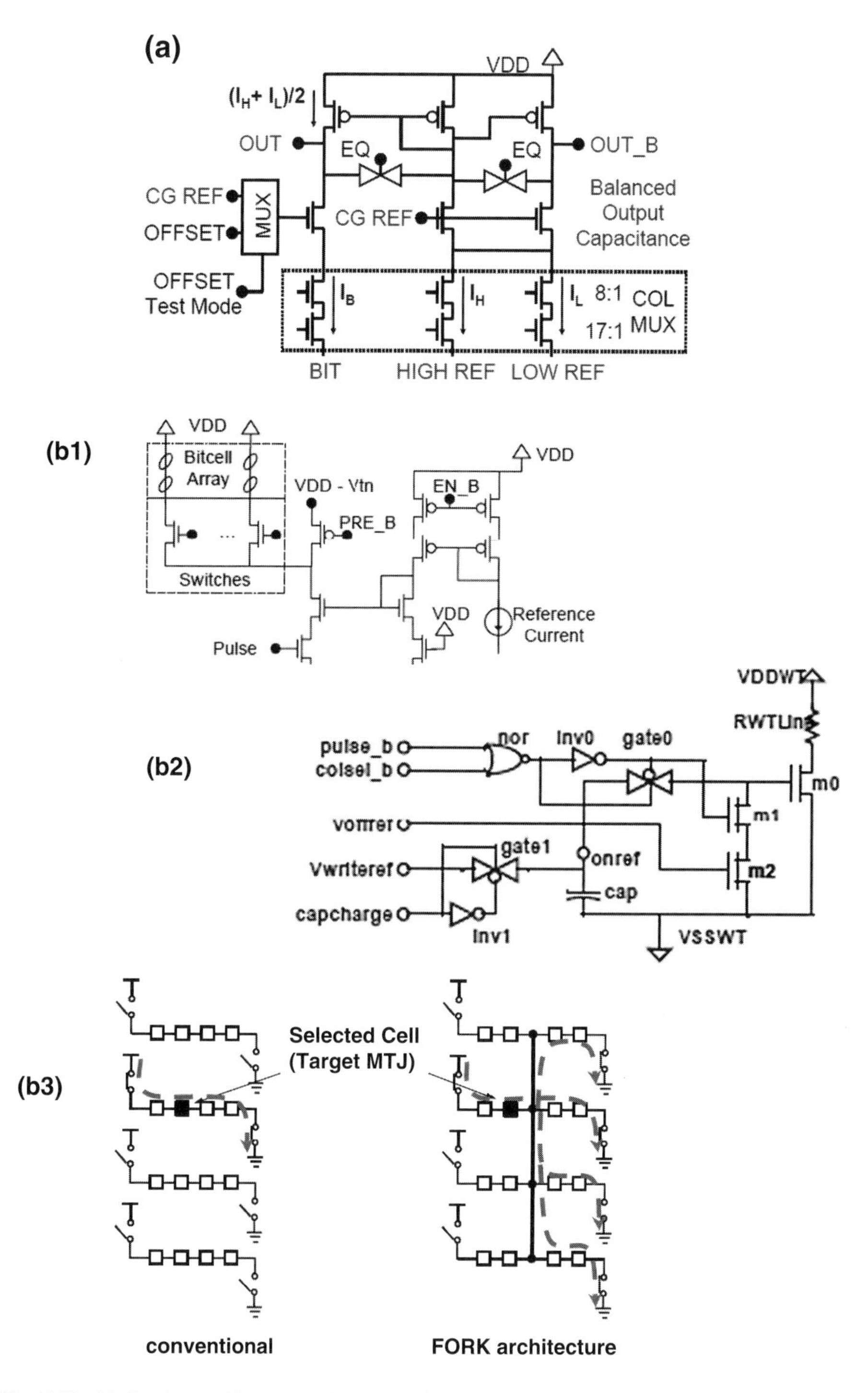

Fig. 7.25 **(a)** Sensing architecture [15]. **(b1)–(b3)** Current driver schemes [1, 9, 14] (© 2008 IEEE)

7.2.1.2 A 16 Mbit MRAM with Dummy Column Architecture

Figure 7.26(a) and (b) show another design of TMR-MRAM device employing a 1Tr-1MTJ cell. In the bit-line write current supply, a self-boosting mechanism is implemented at the bit-line selection gate to get sufficiently conductive. The sensing

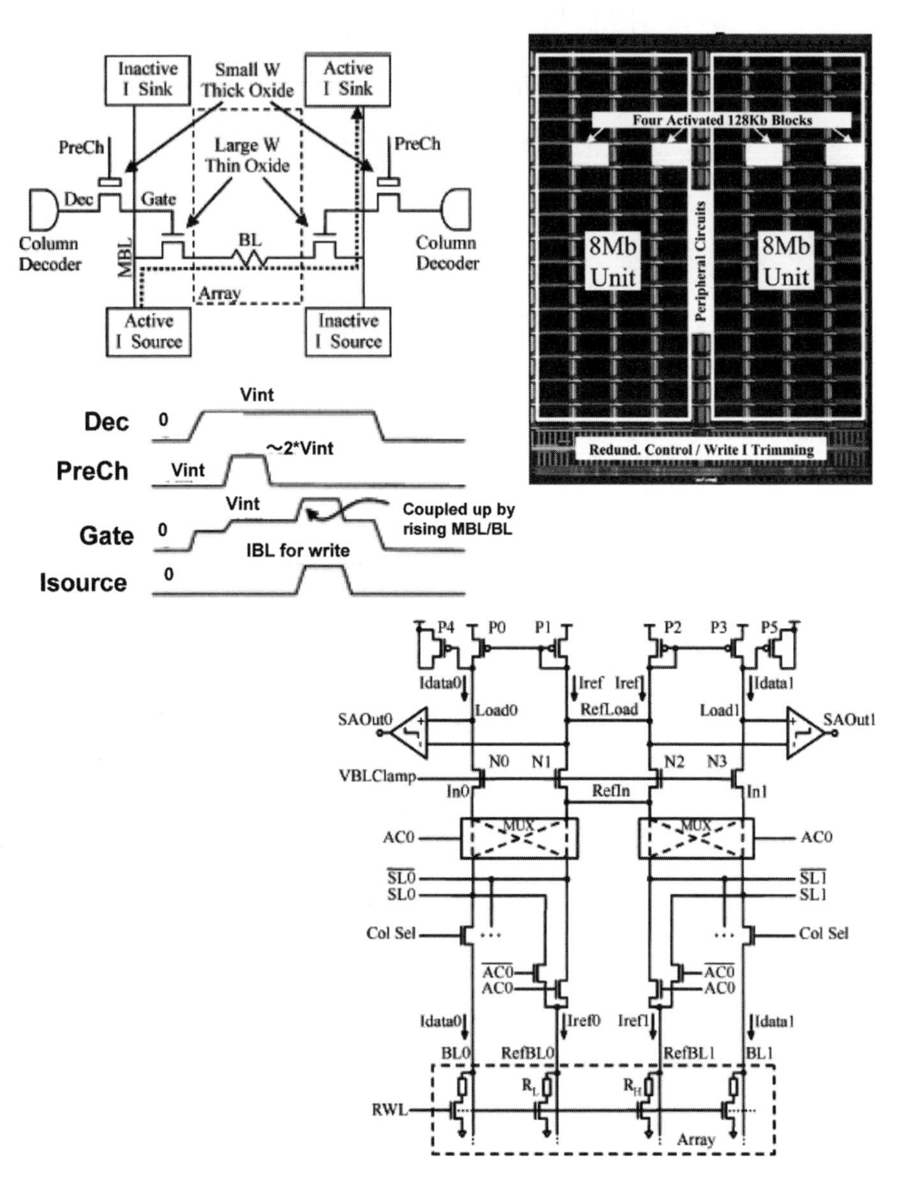

Fig. 7.26 **(a)** A current driving scheme, a sensing scheme, and die photo of a 16 Mbit MRAM design [2, 5] (© 2008 IEEE)

Fig. 7.26 **(b)** Features of a 16 Mbit MRAM design [2, 5] (© 2008 IEEE)

Base technology	1-poly, 3-Cu, 0.18μm CMOS
MRAM process adder	3 levels (VA, MA, MTJ)
Cell size	1.16μm (WLpitch) * 1.22μm (BLpitch) = 1.42μm²
Chip size	7.9mm x 10mm = 79mm²
Supply voltage	2.3-3.3V external, 1.8V internal
Write cycle time Average write current	30ns 80mA
Read cycle time Average read current	30ns 25mA
Standby current	32μA (at 40°C)
Deep power down current	5μA (at 40°C)
User interface	x16 asynch. SRAM-like
Package	48pin BGA (asynch. SRAM-like)
Test modes and metal mask options	External SA reference, Unregulated 1.8V external supply, 8Mb twin cell, Toggle write

architecture adopts a mixed/shared dummy cell scheme using reference cells with data 0 and 1. By a 0.18 μm CMOS technology with 1.42 μm^2 cell size and three Cu layers, this 16 Mbit device realizes a 30 ns read/write cycle time and a 79 mm^2 chip size [2, 5].

7.2.1.3 A 1Tr-4MTJ MRAM with Self-Reference Sensing [22]

In pursuing a high-density MRAM cell array, a self-reference sensing scheme has been first applied to a cross-point cell array [17] with sneak current paths between bit-lines, suffering from a read margin problem and write current supply problem in the memory arrays.

A cell architecture to share a cell selection transistor by multiple cells alongside a bit-line, 1Tr-*n*R cell, has turned out to provide a reasonable implementation of high-density MRAM array. Figure 7.27 shows a 1Tr-4R cell implementation employing a self-referencing sensing scheme. First read and second read after fixed data write operations provide two read status to be compared and judged for stored data detection. A 1.5 V operation by a 130 nm CMOS technology achieves a 66 MHz operation, as compared with 100 MHz operation assuming the same conditions by a 1Tr-1R cell architecture.

7.2.1.4 A New Field-Switching Scheme with A 2T-1MTJ Cell [18, 19]

Figure 7.28 shows another field-switching cell structure and a very high-speed MRAM architecture. A very straightforward way of selectively applying magnetic field to only a selected memory cell is provided by a 2Tr-1MTJ cell scheme, where

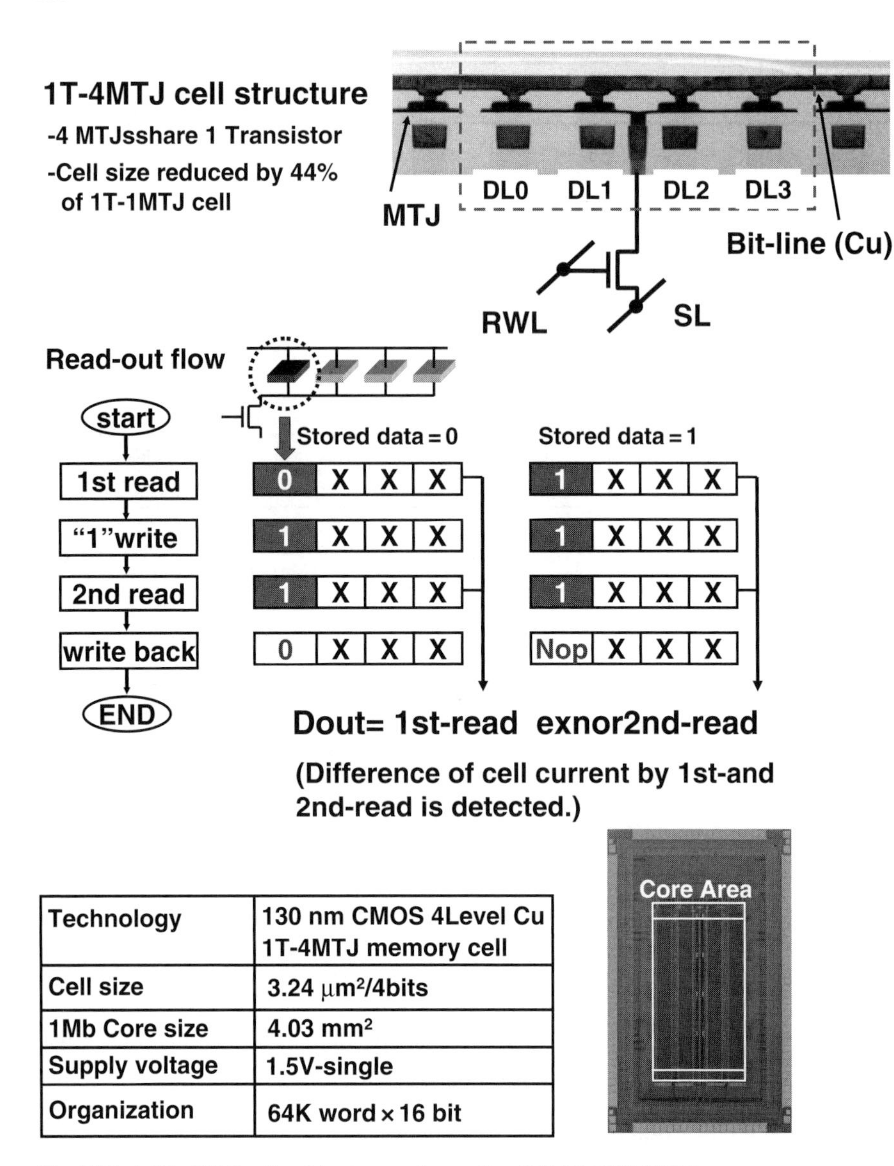

Technology	130 nm CMOS 4Level Cu 1T-4MTJ memory cell
Cell size	3.24 μm^2/4bits
1Mb Core size	4.03 mm^2
Supply voltage	1.5V-single
Organization	64K word × 16 bit

Fig. 7.27 A 1Tr-4MTJ cell architecture and operation [22] (© 2008 IEEE)

two cell transistors transfer the write current applied to a pair of selected bit-lines. Therefore only the selected cell is applied a write field, without any half-select cells. Less than 1 mA of switching current is achieved by this structure. The cell size is affected by the switching current because the current determines the minimum size of the cell transistors. Around 50–100 F^2 cell size and a very high speed read architecture employing dedicated read bit-lines achieve a 250 MHz read/write operation

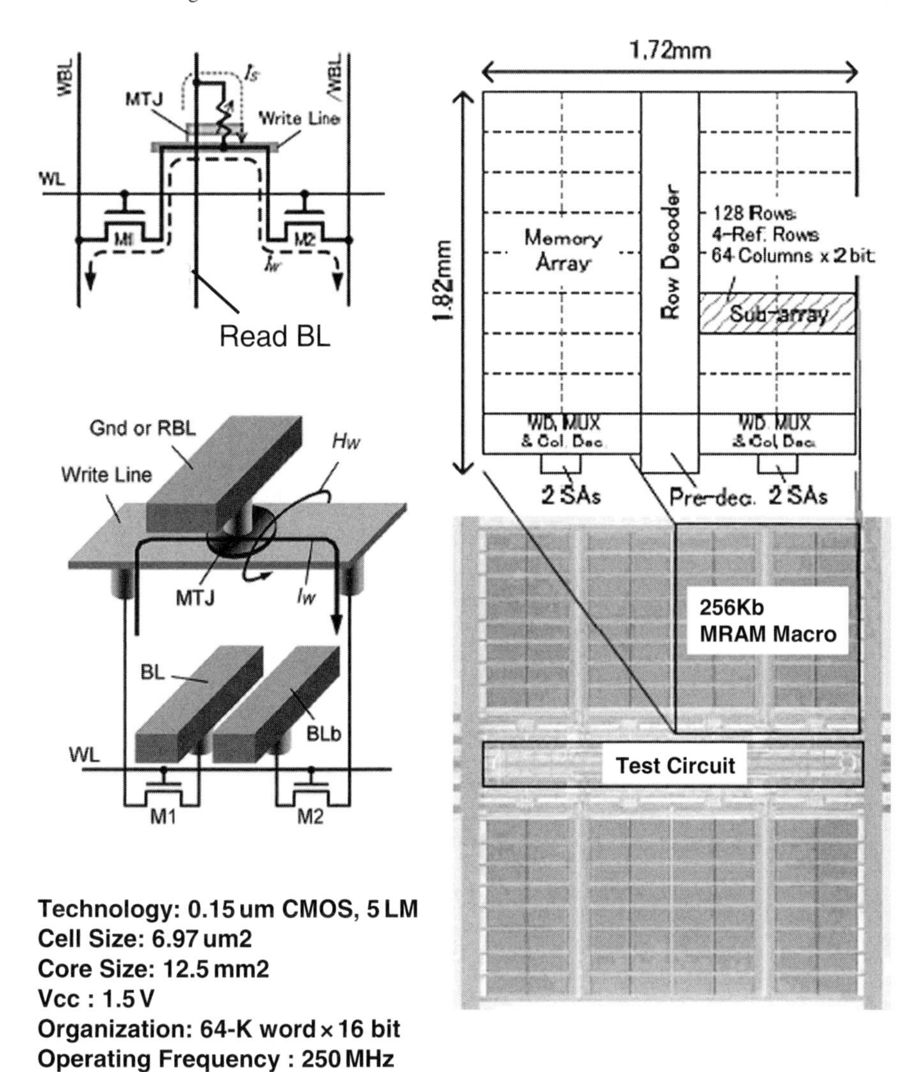

Fig.7.28 A field-switching scheme with a 2T-1MTJ cell [18, 19] (© 2008 IEEE)

in a 1 Mbit device, well validating a technology replacing SRAM in performance and cost [18, 19].

7.2.2 MRAM Applications and Future Challenges

A distinctive property offered by MRAM is non-volatile random access memory (NV-RAM) function. The MRAM attributes are high-speed universal memory, with random access cycle over 200 MHz for read and write, and the CMOS-compatible structure to overcome limitations in current embedded SRAM and embedded flash memory. Because today's MCU is inherently required to integrate SRAM and flash memory on the chip, MRAM is considered to fit MCU applications because MRAM unifies SRAM and flash, both of which often impose conflicting restrictions in one-chip integration. This is a major advantage of embedded MRAM. Therefore MCU is considered to be a key application of embedded MRAM technology.

Figure 7.29 is a proposed embedded universal NV-RAM realized by MRAM. MRAM inherently permits overwrite to switch the cell data regardless of the previous stored data and write operation is in bit-by-bit manner, thus realizing an SRAM-compatible interface. This clearly shows the distinction of MRAM from flash memory. MRAM will replace existing ROM, flash, and SRAM offering a physically homogeneous and logically unified organization intended for cost reduction, performance enhancement, and design simplification.

Figure 7.30 lists capabilities offered by MRAM. By enabling factors coming from high-speed NV-RAM, applications like low-power buffers and cache, time-

- Memory consumes more than half of SOC area
- ROM, Flash, and SRAM merged by one technology for area / cost reduction
- Simplified system architecture enhances performance
- Break-through eSRAM& eFlash limitations

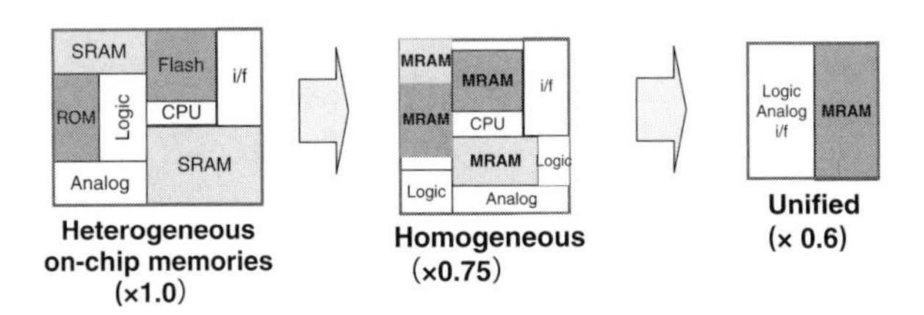

Fig. 7.29 Universal embedded memory

(1) High-speed NVbuffer/cache
← *High-speed, unlimited write*
(2) Power control (on-chip/in system)
← *Non-volatile memory, fast write*
(3) Instant on/off
← *Fast upload write, fast power-on (no charge-pumps.)*
(4) Maximum programmability
by flexible instruction set, universal architecture
← *Universal memory*

Fig. 7.30 Capabilities offered by NV-RAM

Fig. 7.31 Possible embedded MRAM applications

(1) Drive recorder for automotive
by fast & unlimited write.
(2) HDD controller
for high-speed and reliable data manipulation, fast boot.
(3) PCs; for instant-awake capability.
(4) Robotics and motor control.
(5) Military and avionics.
(6) Universal memory
for cutting the stand-by power and break-through technological limitations in embedded memory.

domain power control, instant on/off features, and new architectural creations are expected to come. Based on these attributes by MRAM, limitless application spectrum is appearing, as exemplified in Fig. 7.31. MRAM will find fits especially in power-aware real-time control applications, fast updatable nonvolatile storage, and universal memory in SOC.

MRAM development is encountered by many technology and circuit challenges for integrating magnetism into existing silicon CMOS platforms. Although both technical fields are well established, efforts are required for establishing MRAM technolgy, in immunity against magnetic field disturbance, thermal process, and variability of fine-patterned magnetic elements. From memory cell technology to circuits and architectures to meet NV-RAM demands, every effort is paid in every aspect of incorporating magnetism on the silicon chip.

7.3 Embedded Nonvolatile Memory Frontiers and Challenges

In this section current status and future directions of embedded nonvolatile memory in light of how on-chip programmability function is implemented to create values in MCU/SOC are discussed [6, 7]. A number of innovation capabilities are attributed to values by programmability and to technological evolution of nonvolatile memory.

One remarkable trend in the semiconductor memory roadmap as summarized in the ITRS (Fig. 6.2, Section 6.1) is that the emerging memories deviate from discrete memory technologies. This comes from embedded-specific demands for low-voltage operation and CMOS logic compatibility. Another clear direction is the demand for nonvolatile RAM, requiring high-speed write and a large number of rewrite endurance.

One important aspect in the evolution of LSI in the design and cost structure has been the on-chip programmability. Actual evolution of on-chip programmability has come to be a main factor in performance and cost optimization in LSI implementation. The on-chip programmability function has laid the basis for design and cost structures in LSI as described by Fig. 7.32. Beginning with the ROM-based logic operated by stored instructions in the CPU, alterable/reconfigurable logic organizations have emerged at the second stage, where the main players are flash-MCU and reconfigurable logic products. The flash-MCU innovation has had its root in low-

	Innovation	Main Enabler	Product	Effect
1	Memory-based Logic	ROM Program Register-based comput.	MPU, MCU	Program-driven Logic
2	Alterable Logic	SRAM/Flash	Flash-MCU, FPGA	Re-configurable Production, Inventory, Delivery Efficiency
3	Universal Memory	NV-RAM	Unified-Memory/ MCU, SOC	Technology Break-thru, Power-Reduction, Design Cost, Re-usable?

History of Programmability
- generality vs. performance optimization (←integration level, power constraint)
- design cost-down, SW vs. HW co-design etc.

Fig. 7.32 VLSI evolutions by programmability on the chip

ering the overall production cost by eliminating the complicated production control for embedded mask-ROM products.

The third stage realized by NV-RAM will see a much broader possibility of innovations. Enabling factors by high-speed NV-RAM are expected to support applications in time-domain power control, instant on/off features and new architectural creations for design revolutions by the unified memory concept.

Figure 7.33 shows the impact of NV-RAM in the design space. The dedicated hardware and general-purpose processors are at the opposite ends in this programmability versus performance relation. NV-RAM is expected to bridge the discrepancy between these two designs in performance/cost space to realize the best of the two worlds by lowering total cost and by enhancing the programmability and performance. This design space is for both dedicated hardware and proces-

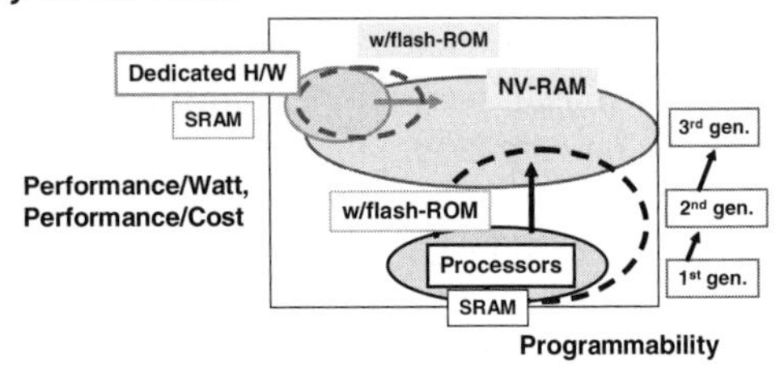

Fig. 7.33 Impact of non-volatile RAM

Universal Memory (ROM/RAM) by NV-RAM

- Simplified/low-cost/low-power system design
- Maximum degree of design freedom

Solve emb-Flash/SRAM problems in;

- CMOS compatibility beyond 90nm node
- High-reliability and low Istby@HT (by N.V.)

Fig. 7.34 Breakthrough required for emerging nonvolatile memory technologies

sors to be exploited in the system design optimization. The key for success is development of NV-RAM-specific cost reduction factors in the applied system to realize the economy of scale.

One target for emerging embedded nonvolatile memory is to bring the technology trend into convergence by two aspects of development (Fig. 7.34). First, NV-RAM to realize universal memory unifying ROM/RAM will cause a tremendous shift in the cost structure and system design to offer value-creation opportunities. Second, the upcoming embedded nonvolatile memory is expected to get through limitations in the scaled down embedded flash/SRAM technologies in CMOS compatibility and leakage power control. If these two factors are reasonably realized by one technology, it will provide the next focal point of convergence in embedded memory innovations.

A tentative example of memory-centric view on the MCU product generations (Fig. 7.35) reflects the evolution of programmability. After the second stage of MCU

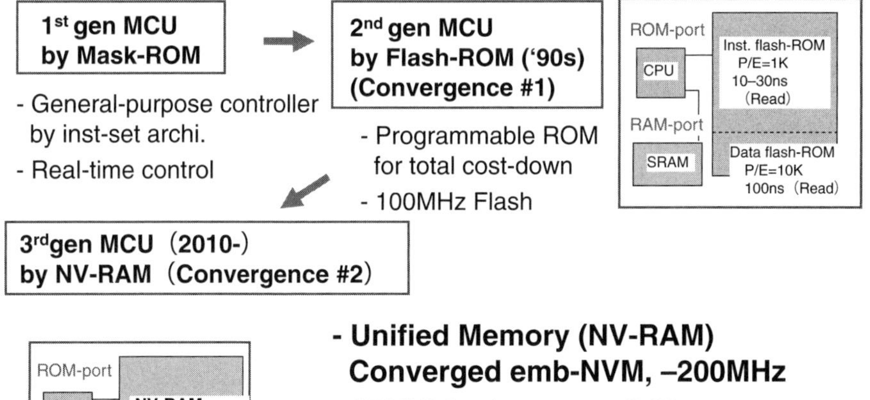

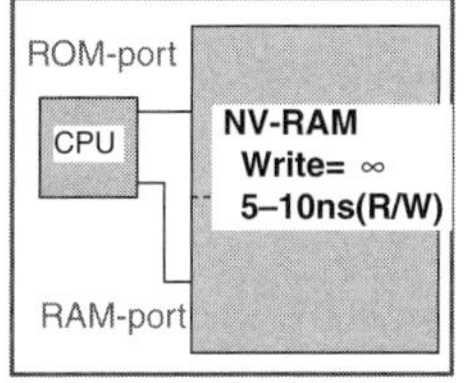

Fig. 7.35 MCU innovations by nonvolatile memory technologies [7] (© 2008 IEEE)

by flash memory, NV-RAM is expected to establish the third stage of MCU products. By utilizing NV-RAM features, the third-generation MCU will fit a wide variety of distributed real-time control and signal processing applications, especially in power control and high-speed secure data updates in the system. One candidate for NV-RAM is MRAM. According to the current development status of MRAM, we foresee the third-generation MCU to be productized with MRAM at 130 to 90 nm technology nodes and in pervasive use in 2020 when NV-RAM value drives the next convergence for innovation. When the value created by NV-RAM will be validated in real applications, we will see a convergence of embedded memory technology based on NV-RAM value and technological necessity for embedded integration. This may bring about the next convergence after the flash-MCU innovation.

The embedded flash innovation in MCU has been important in both value creation and cost reduction (Fig. 7.36) in the value innovation framework [13]. Beginning with replacing mask-ROM for prototyping applications, we have seen a great leap in market penetration of embedded flash memory, thanks to the user merits and overall cost reduction. Moreover, the necessity for embeddedness comes from required high performance and data security. Here the expected overall cost advantage by the economy of scale exceeds the disadvantage of higher wafer process cost for incorporating flash memory. This has been a key to make flash-MCU the most successful embedded memory business ever except SRAM, bringing about a value innovation pattern expanding the market.

The challenges for future embedded nonvolatile memory are considered to be threefold, as shown in Fig. 7.37. First a new value creation is to be initiated by NV-RAM concept. Second, we see a possibility of overcoming the technological limitations in existing embedded memory by a truly embeddable memory technology like MRAM. If these two factors are achieved by one technology, this is surely a focus of convergence necessary for the next innovation. The third point is R&D activity to support such an embedded memory innovation should address how to establish a learning mechanism in a situation quite deviated from the stand-alone memory developments. Additionally, in some cases interdisciplinary efforts are to be employed in bridging semiconductor and magnetism, for example. A key is how to connect and collaborate in the flat world today. These threefold efforts should be

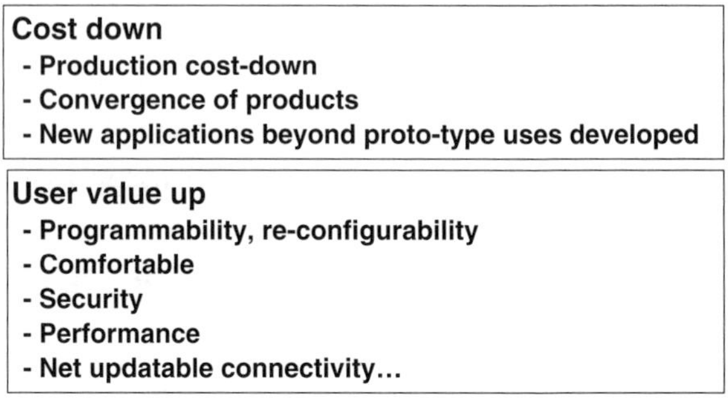

Cost down
- Production cost-down
- Convergence of products
- New applications beyond proto-type uses developed

User value up
- Programmability, re-configurability
- Comfortable
- Security
- Performance
- Net updatable connectivity...

Fig. 7.36 MCU value creation by nonvolatile memory [13]

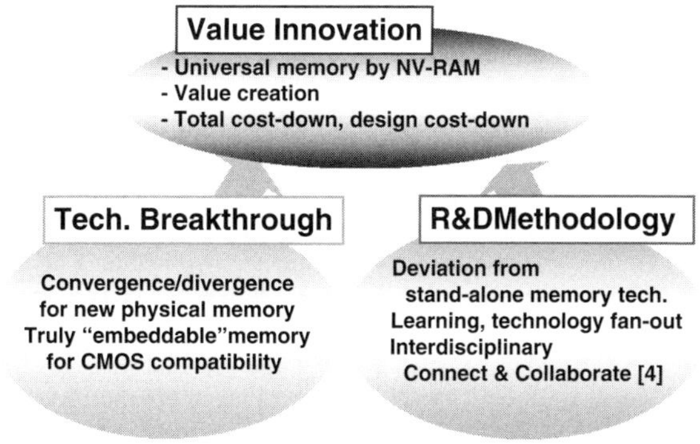

Fig. 7.37 Challenges for embedded nonvolatile memory [6] (© 2008 IEEE)

harmonized for initiating an embedded memory-centric innovation to bring about a drastic economical change.

7.4 Conclusions

Embedded flash memory has inherited much from the stand-alone flash memory technology. Together with the pursuit of truly embeddable flash memory technology, new emerging memory technologies have been explored, some of which have got productized.

Flash-MCU is undergoing a rapid market penetration, which is attributed to the leap in cost/value by the flash-MCU innovation. Emerging nonvolatile memory is required to establish the NV-RAM concept, a key to the next value innovation. Also a truly embeddable nonvolatile memory technology like MRAM may provide a breakthrough in technological limitations for embedded memory. If these two factors are achieved by one technology, this is a focus of convergence necessary for the next innovation.

Embedded nonvolatile memory should move toward the convergence of technology to gain the economy of scale for much diversified application aspects. R&D methodology for embedded memory is required a different learning strategy from that for the stand-alone memory. Moreover, different disciplines like magnetism and phase change material are to be incorporated on the well-established CMOS technology. R&D frameworks are to be established for embedded memory development in value creation, technology breakthrough for embeddable nonvolatile memory, and R&D methodology.

This concludes Chapters 6 and 7 on embedded flash memory and MRAM.

Acknowledgments The author of Chapters 6 and 7 thanks all the members in the Flash-MCU Development Department, Renesas Technology Corp. for their help in providing the data, technical details, and fruitful discussions. These chapters are dedicated to C. H. and late F. A.

References

1. Andre T W, Nahas J J, Subramanian C K, Garni B J, Lin H S, Omair A, Martino W L (2005) A 4-Mb 0.18-μ/m 1T1MTJ toggle MRAM with balanced three input sensing scheme and locally mirrored unidirectional write drivers. IEEE J. Solid-State Circuits 40 1: 301–309
2. DeBrosse J, Arndt C, Barwin C, Bette A, Gogl D, Gow E, Hoenigschmid H, Lammers S, Lamorey M, Lu Y, Maffitt T, Maloney K, Obermeyer W, Sturm A, Viehmann H, Willmott D, Wood M, Gallagher W J, Mueller G, Sitaram A R (2004) A 16 Mb MRAM featuring bootstrapped write drivers. Symp. VLSI Circuits, Dig. Tech. Papers :454–457
3. Durlam M, Naji P, DeHerrera M, Tehrani S, Kerszykowski G, Kyler K (2000) Nonvolatile RAM based on Magnetic Tunnel Junction Elements. Dig Tech Papers ISSCC: 130–131
4. Freedman T L (2005) The World is Flat: A brief history of the twenty-first century. FSG
5. Gogl D, Arndt C, Barwin J C, Bette A, DeBrosse J, Gow E, Hoenigschmid H, Lammers S, Lamorey M, Lu Y, Maffitt T, Maloney K, Obermaier W, Sturm A, Viehmann H, Willmott D, Wood M, Gallagher W J, Mueller G, Sitaram A R (2005) A 16-Mb MRAM Featuring Bootstrapped Write Drivers. IEEE J Solid-State Circuits 40 4: 902–908
6. Hidaka H (2006) Embedded Memory Challenges for Innovations. Proc. Syst. LSI Workshop, IEICE Japan, Nov.27
7. Hidaka H (2007) Embedded NV memory design. ISSCC2007 NV-Memory Forum, Feb. 11
8. Hosomi M, Yamagishi H, Yamamoto T, Bessho K, Higo Y, Yamane K, Yamada H, Shoji M, Hachino H, Fukumoto C, Nagao H, and Kano H (2005) A Novel Nonvolatile Memory with Spin Torque Transfer Magnetization Switching: Spin-RAM. IEDM Tech. Dig. 459–462
9. Iwata Y, Tsuchida K, Inaba T, Shimizu Y, Takizawa R, Ueda Y, Sugibayashi T, Asao Y, Kajiyama T, Hosotani K, Ikegawa S, Kai T, Nakayama M, Tahara S and Yoda H (2006) A 16 Mb MRAM with FORK Wiring Scheme and Burst Modes. Dig. Tech. Papers ISSCC: 477–486
10. Julliere M (1975) Tunneling between ferromagnetic films. Phys. Lett. 54A : 225–226
11. Katti R R (2003) Giant Magnetoresistive Random-Access Memories Based on Current-in-Plane Devices. Proc. IEEE 91 5: 687–702.
12. Kawahara T, Takemura R, Miura K, Hayakawa J, Ikeda S, Lee Y M, Sasaki R, Goto Y, Ito K, Meguro T, Matsukura F, Takahashi H, Matsuoka H, Ohno H (2007) 2 Mb SPRAM (SPin-Transfer Torque RAM) With Bit-by-Bit Bi-Directional Current Write and Parallelizing-Direction Current Read. Dig. Tech. Papers ISSCC :480–481
13. Kim W C and Mauborgne R (2005) Blue Ocean Strategy. HBS Press :16–17
14. Nahas J, Andre T, Subramanian C, Garni B, Lin H, Omair A, and Martino W (2004) A 4 Mb 0.18 μm 1T1MTJ toggle MRAM memory. Dig. Tech. Papers ISSCC: 44–45
15. Nahas J J, Andre T, Subramanian C, Lin H, Alam S M, Papworth K, and Martino W (2007) A 180 Kbit Embeddable MRAM Memory Module. Custom Integrated Circuits Conference:791–794
16. Ohnigian S, Weilerstein I M, Murray D E, and Solomon J (1966) Design of Integrated Selection and Recirculation Circuitry for a High-speed, Low-Power, Magnetic Thin-Film Memory. Dig. Tech. Papers ISSCC:102–103
17. Sakimura N, Sugibayashi T, Honda T, Miura S, Numata H, Hada H, and Tahara S (2003) A 512 Kb Cross- Point Cell MRAM. Dig. Tech. Papers ISSCC:278–279
18. Sakimura N, Sugibayashi T, Honda T, Honjo H, Saito S, Suzuki T, Ishiwata N, and Tahara S (2007) MRAM Cell Technology for Over 500-MHz SoC. IEEE J. Solid-State Circuits 42 4: 830–838
19. Sakimura N, Sugibayashi T, Nebashi R, Honjo H, Saito S, Kato Y, and Kasai N (2007) A 250-MHz 1-Mbit Embedded MRAM Macro Using 2T1MTJ Cell with Bitline Separation and Half-Pitch Shift Architecture. IEEE Asian Solid-State Circuits Conference: 216–219
20. Scheuerlein R, Gallagher W, Parkin S, Lee A, Ray S, Robertazzi R, and Reohr W (2000) A 10 ns read and write non-volatile memory array using a magnetic tunnel junction and FET switch in each cell. Dig. Tech. Papers ISSCC: 128–129

21. Slonczewski J C (1989) Conductance and exchange coupling of two ferromagnets separated by a tunneling barrier. Phys Rev B Volume 39, Number 10: 6995–7002
22. Tanizaki H, Tsuji T, Otani J, Yamaguchi Y, Murai Y, Furuta H, Ueno S, Oishi T, Hayashikoshi M, and Hidaka H (2006) A high-density and high-speed 1T-4MTJ MRAM with Voltage Offset Self-Reference Sensing Scheme. Proc. Tech. Papers Asian Solid-State Circuits Conference: 303–306
23. Tsuji T, Tanizaki H, Ishikawa M, Otani J, Yamaguchi Y, Ueno S, Oishi T, and Hidaka H (2004) A 1.2 V 1 Mbit embedded MRAM core with folded bit-line array architecture. Dig. Tech. Papers, Symp. VLSI Circuits :450–453
24. Ueno S, Eimori T, Kuroiwa T, Furuta H, Tsuchimoto J, Maejima S, Iida S, Ohshita H, Hasegawa S, Hirano S, Yamaguchi T, Kurisu H, Yutani A, Hashikawa N, Maeda H, Ogawa Y, Kawabata K, Okumura Y, Tsuji T, Ohtani J, Tanizaki T, Yamaguchi Y, Ohishi T, Hidaka H, Takenaga T, Beysen S, Kobayashi H, Oomori T, Koga T, Ohji Y (2004) A 0.13 μm MRAM with $0.26 \times 0.44\ \mu m^2$ MTJ optimized on universal MR-RA relation for 1.2 V high-speed operation beyond 143 MHz. Tech. Dig. Int. Electron Device Meeting:579–582

Chapter 8
FeRAM

Shoichiro Kawashima and Jeffrey S. Cross

8.1 Introduction

Ferroelectric materials show spontaneous polarization; FeRAM utilizes the positive and negative polarization direction corresponding to "1" and "0" states for stored data. The basic idea behind FeRAM appeared in 1963 [1] and 1988 [2], however, there have been many scientific and technical improvements needed to convert FeRAM technology into manufactured devices and still further improvements in materials, process fabrication, and circuit architecture are required for further device scaling. This chapter intends to serve as a review primarily focusing on the timeframe from 2000 to 2007. Previously, circuit and architecture of FeRAM devices regarding circuit innovations up to 2000 was summarized by Prof. Sheikholeslami [3]. In this chapter, first we have included an overview of FeRAM-related activities starting with material fundamentals and characteristics, then we discuss design circuit modifications necessary when incorporating characteristics of ferroelectric thin films, and lastly we touch upon future trends and challenges.

8.2 Ferroelectric Materials

This section introduces the origin of spontaneous polarization and basic characteristics of ferroelectric capacitors in Section 8.2.1, then electrical properties, including degradation or problems in FeRAM in Section 8.2.2. The basic terms in FeRAM such as hysteresis, Vc, Pr, P-term, U-term, imprint, fatigue, and their temperature dependencies, which are often used below in the circuit description section will be explained. Understanding of the ferroelectric capacitor response to an electric field, temperature-dependent retention characteristics, aging response, and

S. Kawashima (✉)
Fujitsu Microelectronics Limited, System Micro Division, 1-1 Kamikodanaka 4-chome, Nakahara-ku, Kawasaki, 211-8588, Japan
e-mail: kawashima@jp.fujitsu.com

K. Zhang (ed.), *Embedded Memories for Nano-Scale VLSIs*, Series on Integrated Circuits and Systems, DOI 10.1007/978-0-387-88497-4_8,

switching charge reduction due to cycling are essential to design FeRAM architecture, because like other memories, we are designing memories very close to cell functional limitations.

8.2.1 Fundamentals of Ferroelectric Materials

Because FeRAM process development and electrical characterization of its properties are input to Spice circuit simulations, the actual workings are very close to ferroelectric physics. Within this section, an overview of material science and physics is given in order to understand the ferroelectric materials. Please refer to other prominent textbooks for further information on ferroelectric materials [4–6]. The aim of material science is tailoring a material's properties to fit a specific application. For example, polycarbonate plastics are hard and also very light for airplanes or cars, shape-memory alloys for frames of eyeglasses, or rod antenna used in cellular phones. Overall, there are three major areas of research and development; innovation based upon the periodical atom structure, processing methods for realization of the manufactured goods, and measuring or evaluating properties. Today's science has come to identify molecular structure by utilizing strong intensity X-ray with synchrotron orbital ring, such that we can observe what is going on at the atomic level [7]. And metal organic chemical vapor deposition (MOCVD) facility can grow materials layer-by-layer to form a single crystal with artificial or designed structure. As semiconductor scaling reaches the nanometer-scale level, we are using advanced scientific equipment and are working very closely with molecular sciences. Material science involves the fields of ceramics, metallurgy, semiconductor, polymer, glass, etc. Electro-ceramics includes topics such as dielectrics, optics, magnetics, superconductors, and ferroelectric materials as well as processing and their properties. Porcelains are a traditional style of ceramic; oxygen and metal atoms are baked at a high temperature to create covalent bonding, periodical structure, thus hardened materials can be derived with multiple functions. PZT ($Pb(Zr,Ti)O_3$), also a ceramic, is the most popular ferroelectric material of interest, due to a wide range of applications utilizing its properties such as piezoelectricity or pyroelectricity.

Figure 8.1 shows applications of piezoelectricity, pyroelectricity, and ferroelectricity. Piezoelectricity is defined as charge generation by mechanical fields (pressure), for example, push button electricity generation for an electric spark lighting flame in a cigarette lighter. Reversibly, applying electric field creates shape transformation of the substance, for example, a crystal-earphone inputs electric voltage signal that creates an electric field for a piezoelectric crystal to shape transform thus sounds waves are generated in response to the input electrical signal. By the way, crystal earphones were fabricated from Rochelle salt before 1970s because the single crystal can be easily obtained from super-saturated hot water solution, like growing a large salt crystal or a boric acid crystal in elementary schools. Presently many applications are using PZT as a piezoelectricity material. A quartz oscillator is one of the most popular electronic components that utilize piezoelectricity.

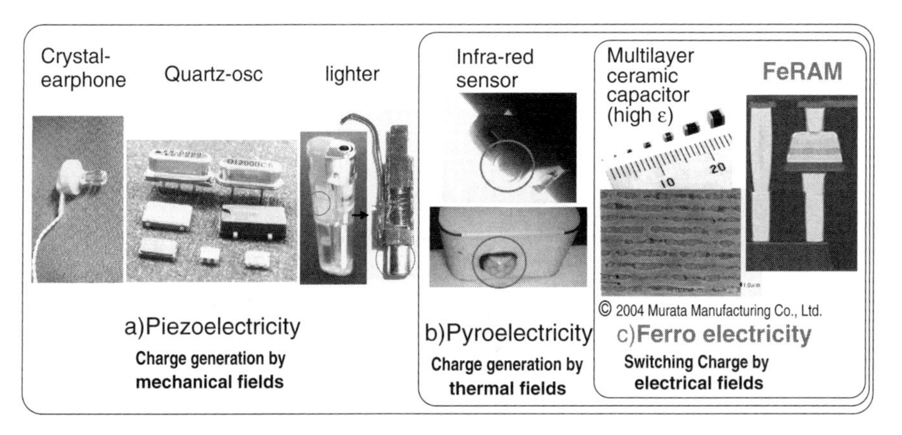

Fig. 8.1 Piezo-, pyro-, and ferroelectricity and applications

However, quartz shows only piezoelectricity, whereas PZT shows also pyroelectricity and ferroelectricity. Pyroelectricity is defined as charge generation by thermal gradients or temperature differences, for example, infrared (IR) detecting sensors utilize it. A pair of IR-LED and PIN photo diode can project IR light and detect reflected IR signals in order to use as a sensor for detecting presence of people, but regarding cost and far range detection capability, sometimes a pyro-sensor is superior and both IR sensors can coexist. $BaTiO_3$ is another popular ferroelectric material, which is mostly used for its high permittivity beyond 10,000 in multilayer ceramic capacitors (MLCC). Recently chip capacitors use $BaTiO_3$ and electrode lamination to achieve even 100 μF in $0.4 \times 0.3 \times 0.2$ mm^3 [8]. In this case, one $BaTiO_3$ layer thickness is about 1 μm thick, which is about five times the thickness used in FeRAM capacitor of 200 nm. There are problems when using $BaTiO_3$ as a capacitor, such as it shows nonlinear capacitance along the applied voltage and applying a reverse polarity is prohibited because it will switch to another polarization direction on the hysteresis loop. Although these are essential ferroelectric material properties, proper usage within a specific range of applied voltage results in high capacitance, for example, in power supply decoupling. A further utilization of ferroelectricity: switching polarized domains or dipoles directions by electrical fields is the core of FeRAM. FeRAM capacitor structure and circuit resembles DRAM's, however, the difference is that the capacitor switching charge emerges when polarization switches for FeRAM. FeRAM stores data by atom location in ferroelectric domains that are nonvolatile and retain their polarized direction even without a power supply, which differentiates it from DRAM. FeRAM capacitors are typically less than 3 μm^2. FeRAM thin films in some cases have columnar grains with crystallization anneals at lower temperatures than processing single crystals preparing for piezoelectricity applications.

The ferroelectric phenomenon is very interesting and complicated if we go into much detail. However, here the mode of operation is limited to illustrating the general idea of what is important for FeRAM. In short, spontaneous polarization

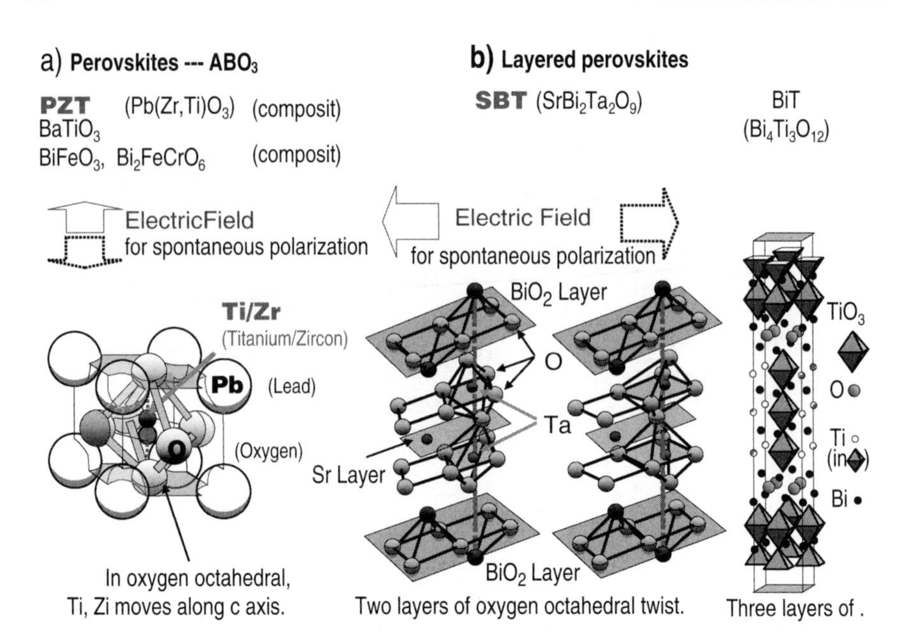

Fig. 8.2 Ferroelectric material molecules used for FeRAM; PZT, SBT, and BiT

(ferroelectricity) comes from ionic crystals exhibiting oriented domains, which respond to an electric field. Oxygen octahedral, a metal atom surrounded by six oxygen atoms (in Fig. 8.2(a)), is the origin of an asymmetrical electric displacement. Oxygen atoms are tightly coupled and the center metal atom may slightly dislocate up or down from the neighboring oxygen atoms which form a tetragonal. Since oxygen atoms are shared with neighboring tetragons, adjacent octahedral oxygen atoms have less energy when the center metal atom moves in a particular direction. Domains are three-dimensional where the center metal atoms in adjourning lattices are oriented in the same direction. The image resembles that of magnetic domains and movement of domain walls. Total summation of ferroelectric domains displacement gives a net spontaneous polarization value.

Currently, popular ferroelectric materials for FeRAM are PZT ($Pb(Zr,Ti)O_3$) and SBT ($SrBi_2Ta_2O_9$). These belong to simple ABO_3 perovskite group(a) and layered perovskite group(b), respectively. Where PZT is a composite material consisting of a solution of $PbZrO_3$ (octahedral containing Zr) and $PbTiO_3$ (octahedral containing Ti) and their relative composition ratio can be varied from 0% Zr to 100% Zr. Since $PbZrO_3$ is orthorhombic and $PbTiO_3$ is tetragonal, their ratio can change 3D unit structure $a1$, $a2$, and c-axis (x, y, height) dimensions and change free energy. Also many dopants substituting into the A(Pb) site and B(Zr/Ti) site are well studied and result in changes in the ferroelectric properties. SBT is a layered perovskite, introduced in the mid-1980s. This kind of layered structure consists of up to five oxygen octahedral layers [9]. The polarization mechanism is different from ABO_3 group and two layers of oxygen octahedral between BiO_2 tightly bond planes, twist

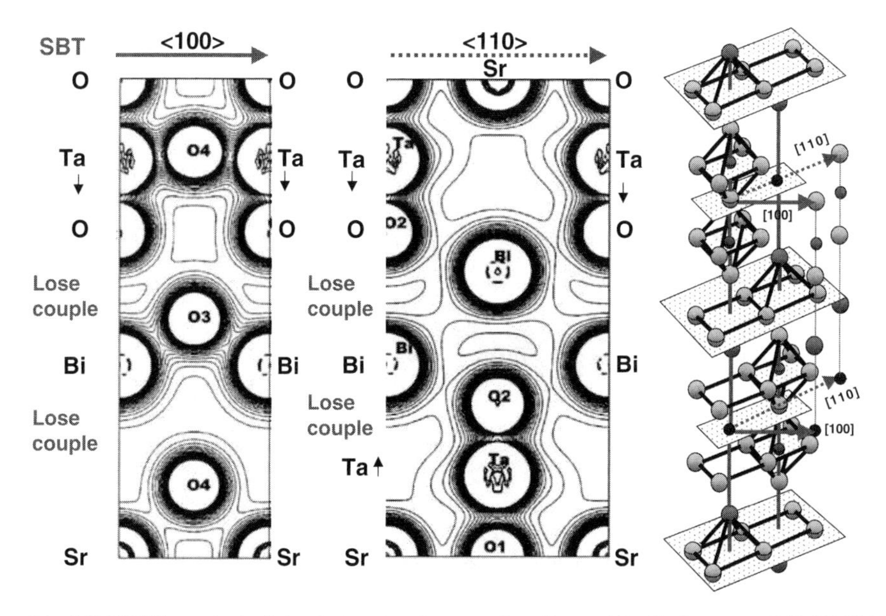

Fig. 8.3 SBT first principal simulation showing atom bonding and Ta displacement [11] (© 2000 APS)

without *c*-axis movement of Ta in SBT [10]. Figure 8.3 shows Ta displacement relative to the BiO_2 layers, so charge displacement (dipole) occurs at each oxygen octahedral [11]. Returning to Fig. 8.2(b), thus the octahedral twist by rotating horizontally in the electric field. BiT ($Bi_4Ti_3O_{12}$) contains three layers of oxygen octahedral [12]. Most descriptions concerning major polarization mechanisms, parallel shift of the octahedral relative to the A site or BiO_2 layers are also said to contribute to the net polarization to some degree. Octahedral are slightly distorted and exhibit asymmetry as the center atom is displaced slightly. BFO ($BiFeO_3$) belongs to ABO_3 group but the octahedral are twisted as shown in Fig. 8.4 [13]. It is not as simple as SiO_2, because it contains many different elements, but by focusing on the oxygen octahedral, it is easy to identify the structure.

Examples of thin film crystals are shown in Fig. 8.5. Theoretical microstructure and formation technology of thin film ferroelectric materials in reality are different from that of a single crystal. A single crystal ingot for piezo applications is available [14]. The Bridgman furnace melts the raw material and then slowly re-crystallizes on a seed crystal while being withdrawn [15]. In contrast, thin films grown on a bottom electrode have slight variations in properties due to material composition variations, added dopants, bottom electrode, crystallization temperature, or processing method such as sol-gel also called chemical solution deposition (CSD), sputter, MOCVD, laser pulse deposition (LPD), etc. Figure 8.5(a) is a relatively thick film with grains diameter of several microns [16]. A cross-sectional schematic shows its grains consist of randomly oriented domains. Applying an electric field, domains closest to the field increase in proportion with domain wall expansion and total

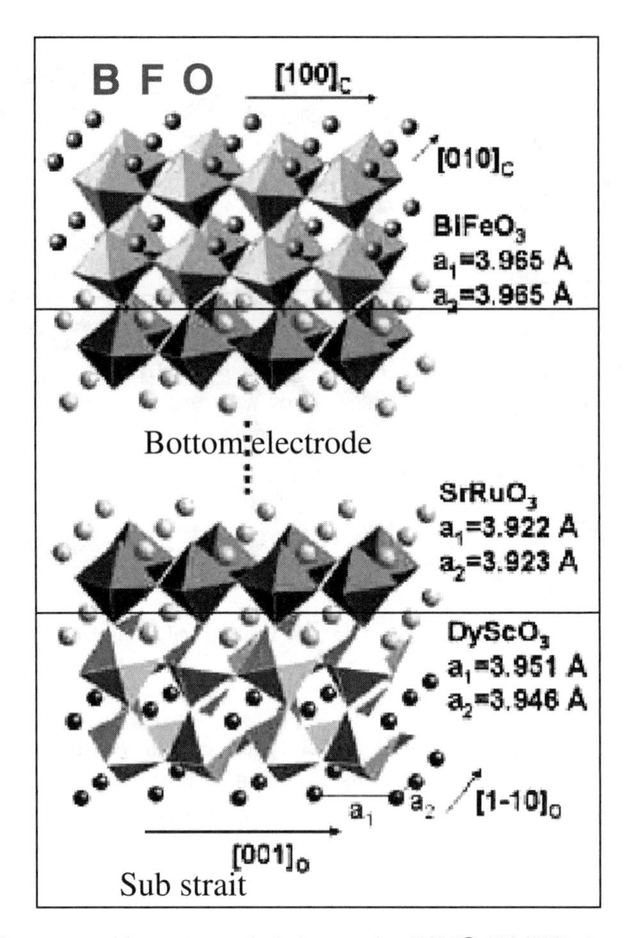

Fig. 8.4 BFO structure with oxygen octahedral expression [13] (© 2006 Wiley)

polarization charge will be less than that of 100% oriented single crystals. Figure 8.5(b) is a thin PZT film with columnar grains where the orientation is influenced by the bottom electrode orientation. It is much denser than thick PZT (a) and has voids between grains. There is an on-going debate on the relationship between the size of columnar grain and Vc (coercive voltage; switching voltage will be described in 8.2.2 detail), larger grains are not always better. In addition, horizontal compressive stress by lattice mismatch of PZT and bottom electrode enlarges vertical length (*c*-axis) of a unit tetragonal. In Fig. 8.5(c) a BFO film grown by LPD is shown, layer-by-layer epitaxial growth. Electron micro-diffraction pattern shows good uniformity of a single crystal [17]. However, LPD is not yet suitable for industrial production, since it shows slow growth rate and the area is limited to small samples. Deposition-processing technologies are reviewed in Chapter 11 of Scott [4].

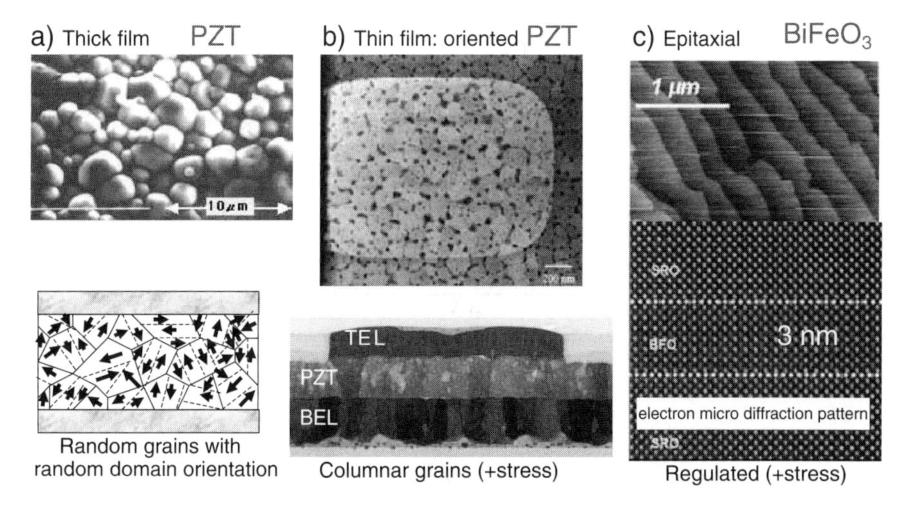

Fig. 8.5 Thin film ferroelectric materials (a) [16] (© 2000 Titech.HST), (c) [17] (© 2007 AIP)

8.2.2 Electrical Characteristics

The FeRAM capacitor is also a two-terminal circuit element. We apply voltage between two terminals but usually we measure charge flow at one terminal. Differences compared to a normal linear capacitor are a ferroelectric capacitor shows a hysteresis loop for electrical charge or Q versus V curve and Q shows a response time, voltage, and retention interval time dependency. Also a ferroelectric capacitor shows a relatively large temperature dependency than do ordinary dielectric capacitors. The hysteresis loop shape is associated with a problem concerning minor-loop operation where insufficient voltage is applied resulting in a partial or incomplete switching of domains. A two-terminal capacitor operation cannot identify what percentage of domains switch under an applied voltage. This phenomenon is something like an AC coupling amplifier that amplifies only differentiation of input signal and one does not know the exact DC input level.

Spontaneous polarization exhibits a hysteresis loop when applying bipolar voltage. The measured charge is composed of domain switching and linear charge. Let us explain the loop shown in Fig 8.6, starting from the –Pr point at the lower y-axis intersection point at 0 V. Pr is referred to as remanent[1] polarization when no voltage is applied across the top and bottom electrode and is the remanent state. The percentage of domain switching to one or the other direction depends on the history of the previously applied maximum voltage (and temperature). In the Fig. 8.6, assuming full polarization at –Pr and in the schematic the three domains have the same direction of spontaneous polarization. Second, with increasing voltage

[1]There are two spellings for remnant or remanent in the literature. I think material scientists prefer remnant.

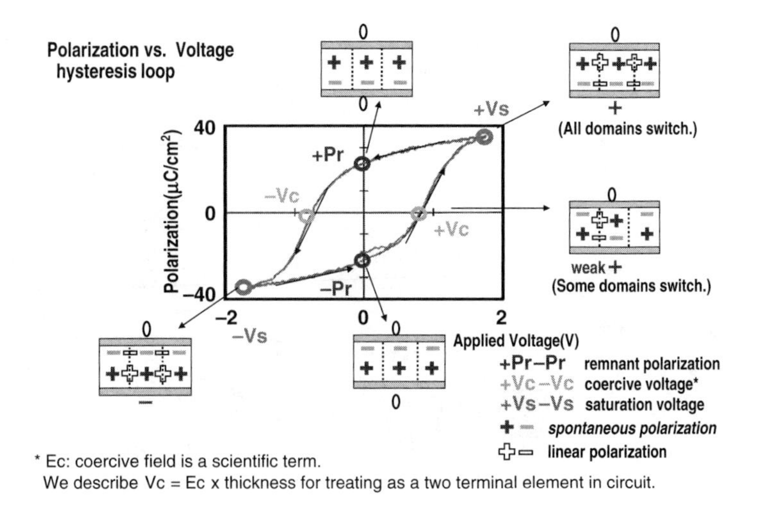

Fig. 8.6 *Q– V* hysteresis curve and Pr, Vc

potential applied at the bottom electrode, linear charge appears and a portion of domains switch to the opposite direction to reach +Vc point at the *x*-axis intersection point. Vc is also called coercive voltage, its name is derived from the switching movement and the capacitance (=d*Q*/d*V*) is at its maximum value. For single crystals, on applying external voltage below Vc, non-switched domains occupy a larger proportion than switched domains so removing the applied electric field is expected to cause a return back to the –Pr position. For thin films with columnar grains, we are not sure this logic is applicable. From a material science viewpoint, Ec, coercive electric field, is a primitive value that does not depend on film thickness. Anyway, we will describe Vc as merely the *x*-axis intersection point when the hysteresis loop is centered on the *Y*-axis. Third, at +Vs, saturation voltage point, all domains switch to the opposite direction and linear charge increases in proportion to the voltage. Applying voltage above +Vs, only linear charge increases until leakage current increases or breakdown occurs. Fourth, at +Pr position, after removing the applied voltage, linear charge decreases and only spontaneous polarization that is switched to reversal at –Pr remains. The response to negative voltage is almost the same despite switching in the opposite direction and goes through –Vc and the lowest voltage –Vs then returns to –Pr. This is the basic description of the operation of a ferroelectric capacitor. However, the hysteresis shape is process and materials dependent and not as sharp as an electrical flip-flop circuit element, thus the above-mentioned phenomena need to be considered with further detailed other effects when designing a robust FeRAM circuit.

For a FeRAM capacitor read operation, raising one terminal potential and keeping the other terminal potential at a fixed potential while detecting charge emerging from polarization switching is necessary. In Fig. 8.7(a), a hysteresis curve with a continuous bipolar triangular voltage appliance for polarization measurement is

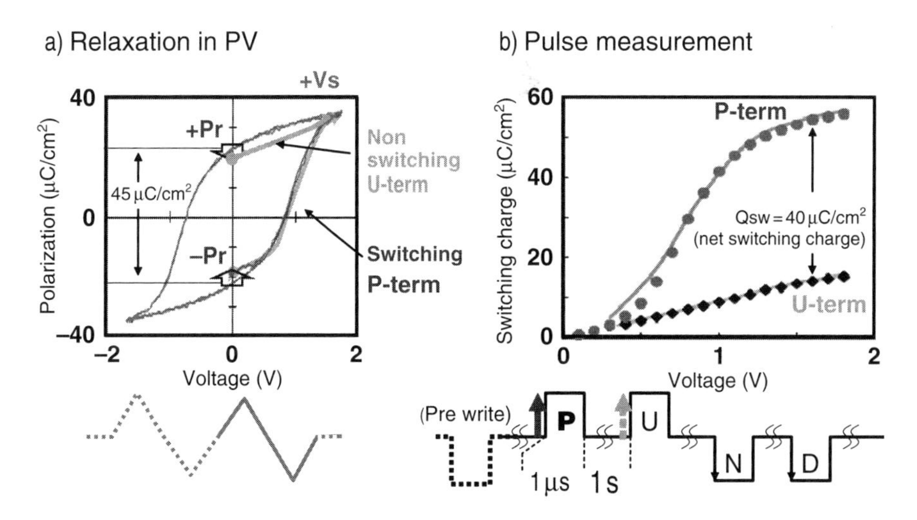

Fig. 8.7 Back switching (relaxation) and pulse measurement

shown. FeRAM operation of reading path from –Pr and +Pr are shown as bold lines. When starting from –Pr, the polarization switches to the opposite direction and simultaneously emits switching and linear charge. We name this path or pulse the P-term. When starting from +Pr, the polarization remains in the same direction thus only linear charge emerges. We name this path or pulse the U-term. Figure 8.7(b) shows pulse train measurement with 1 s intervals. This terminology called PUND is an abbreviation for positive, up, negative, and down pulses. The first positive pulse corresponds to the P-term, switching (and linear) charge, and the second positive pulse corresponds to the U-term, non-switching or linear charge. The value of Qsw=(P-term) – (U-term)=40 $\mu C/cm^2$ is slightly less than 2Pr = (+Pr) – (-Pr)=45 $\mu C/cm^2$. This indicates that some of the domains are back switching or domain relaxation occurs during the wait time of 1 s. Short-term reduction of 2Pr may be explained in terms of internal electric field charge compensation, piezoelectric effect, or domain wall movement, but we will not discuss it here. Thus, a 1 s interval pulse measurement is closer to FeRAM operation than by continuous wave and less charge is expected than from a PV hysteresis curve. Also the pulse measurement shows READ voltage dependence only for the previously fully polarized ferroelectric capacitor.

For FeRAM capacitor writing, the write voltage directly affects the relative proportion of polarization domain directions thus impacting the amount of switching charge that can read by subsequent applied voltage pulse. Figure 8.8 shows minor loop and major or saturated loop of hysteresis. A minor loop results from partial switching when insufficient voltage less than Vs (saturated voltage) is applied. In this case, Pr depends on applied voltage during the previous opposite switching event, that is, writing. A major or saturated loop takes place when sufficient or excess voltage beyond Vs is applied. However, the maximum Pr value is limited

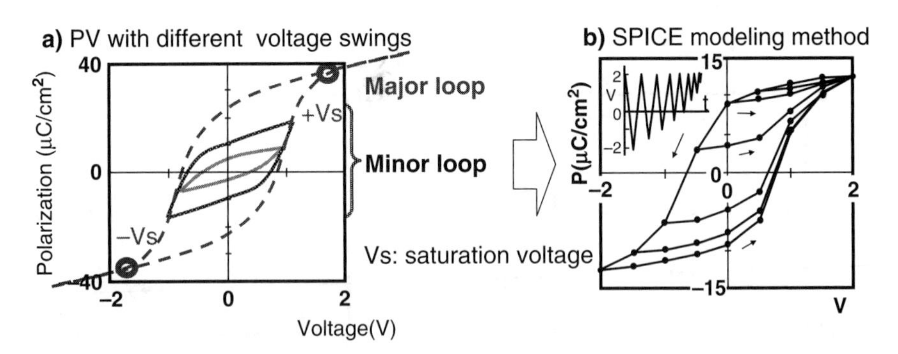

Fig. 8.8 Minor loop and major loop

by the material characteristics. In reality, saturation voltage Vs is difficult to identify, since interface charge injection or capacitor leakage increases when applying higher voltage. Usually we use a term called V90, that is, 90% of the polarization is obtained compared to the polarization at the maximum applied voltage. If we assume a normalized distribution of Vc for domains and parallel capacitance model, roughly Vs = 2Vc. Since reading charge is a function of writing voltage and reading voltage history, it is no longer possible to use a simple graphical approach for PV hysteresis curve and BL capacitance load line. So Spice modeling is important for designing FeRAM in detail.

A parallel capacitance model, which is used for modeling a hysteresis loop, employs a linear capacitor and many (3–32) capacitor elements corresponding to a divided voltage region in Fig. 8.9(a) [18, 19]. C(n) represents the amount of switching charge of a division and Vc(n) represents at what voltage the element shall switch. This is implemented with voltage-controlled current source or voltage-controlled resistor (VCR). Using VCR is preferable because simulation convergence is faster with lower peak current and furthermore, the resistance value can be used to determine the time response. The first step for the fitting is to extract the ferroelectric PV terms by applying asymmetric relatively long pulse trains. Then the second step is to measure the charge with a fast pulse measurements setup for time in nanosecond order versus voltage, then fitting the CR delay with varying resistance values. This fast pulse measurement system has a limitation of 5 ns because of test fixture impedance and stray parasitic capacitance. So the U-term value corresponding to linear charge in the figure is inaccurate for pulses of less than 5 ns. If in the future less than 5 ns access time is required, then a more sophisticated test fixture like using electron beam tester will be needed. The time response results (Fig. 8.9(c)) show that Qsw(=P–U) at 5 ns is 20% lower than that with a 1 μs pulse measurement. In addition, less charge can be obtained when lower voltage is applied, for example, at 1 V, Fig. 8.7(b) shows Qsw=30 μC/cm^2 at the end of 1 μs pulse but in Fig. 8.9(c) Qsw=5 μC/cm^2 at 10 ns. Polarization switching speed is slower around Vc [20]. Several models on switching speed have been developed to model the time

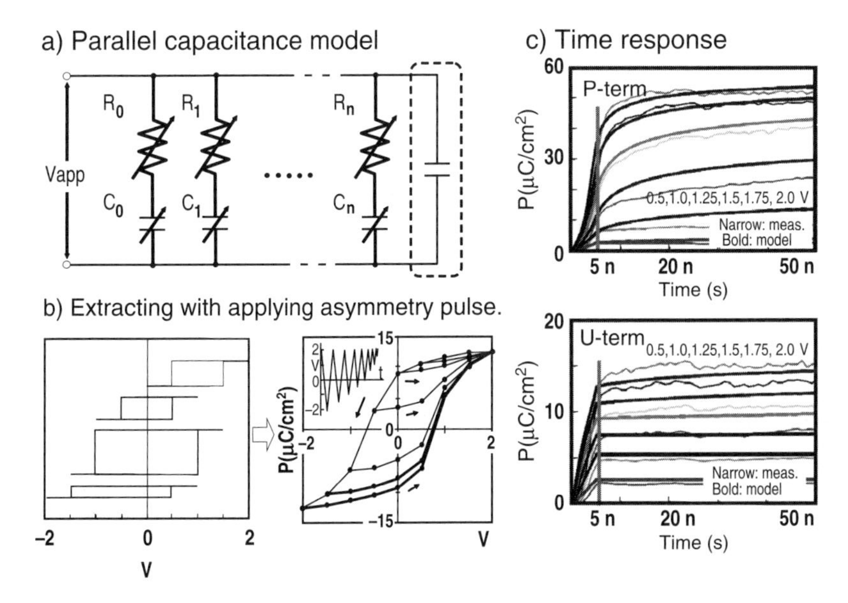

Fig. 8.9 SPICE modeling with parallel capacitance and implemented time response (a), (b) [18] (© 2000 IEEE)

response. This indicates that the read voltage also must be sufficiently larger than Vc when designing a first FeRAM circuit.

Furthermore, Qsw loss is observed as a function of temperature. Figure 8.10 shows P-term and U-term charge first noted at room temperature (RT), and then

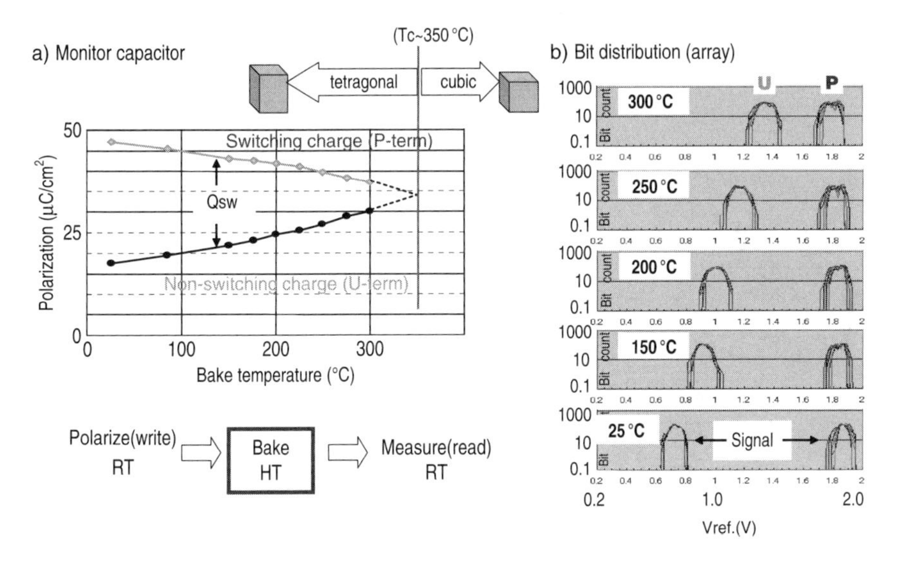

Fig. 8.10 Temperature dependence of P- and U-terms [21] (© 2007 IEICE)

baking at specific temperature and then measuring the retained charge at RT [21]. Since PZT crystal changes from tetragonal to cubic structure above Curie temperature (Tc), the polarization disappears. Upon cooling to RT, random domains form without a preferred polarization direction. The domains only become oriented or aligned upon applying an electric field. Thus, the thermally depolarized state and reduced Pr remains even after cooling. The process is reversible meaning that by rewriting at RT, the polarization can return to its initial value. Tc for PZT capacitors is approximately 350°C and for SBT is 320°C, but Tc can vary with strain from the substrate, dopants and columnar grain size, re-crystallization temperature, etc., for thin ferroelectric films. These values of Tc are lower than that of ceramic materials. Data retention after soldering to a circuit board at 270°C is a technical challenge for FeRAM, but it is possible. After getting through high-temperature wire bonding, plastic molding, and soldering on a circuit board, writing and reading operations at less than 100°C have no impact on retention of data. In Fig. 8.10(a), a pulse measurement of a monitor capacitor written with a 1 μs pulse and in Fig. 8.10(b) a bit-distribution histogram of 2 Kbit cell capacitors on an FeRAM device with applied external reference level to S/A is shown. Both experiments show good agreement in the net reduction of P-U or Qsw versus bake temperature from RT to 300°C.

Another ferroelectric characteristic, which shows a temperature dependency, is the Vc. Fig. 8.11(a) shows a PV hysteresis loop measured at –45°C has larger Vc values than that at 85°C. Since Vs (saturation voltage) is almost 2Vc, at lower temperature the hysteresis loop is in a minor loop (partially switched), which means it is not fully polarized during the write operation. Also while reading, higher Vc results in a smaller P-term charge because of a slower time response. Both Qsw and Vc dependency on temperature are explained qualitatively by energy potential

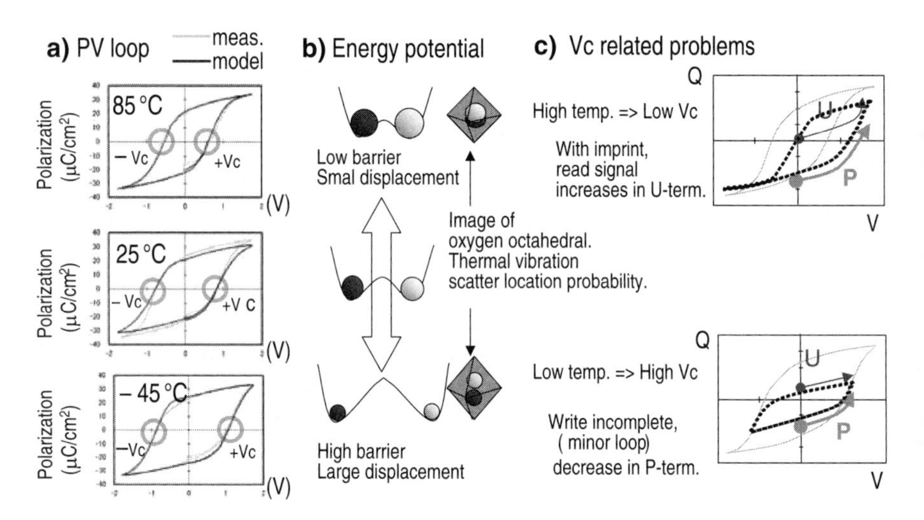

Fig. 8.11 Temperature dependence of Vc

schematic (b). In the oxygen octahedral, there are two possible low-energy positions for Ti or Zr. At higher temperature, atoms are vibrating and their probable location is enlarged as shown schematically by the size of the larger circle. Also the tetragonality decreases slightly lowering the potential barrier height for displacement that is why Vc and Qsw are smaller.

On the contrary at lower temperature, atom vibration is lower and tetragonality slightly increases thus the potential barrier is higher and displacement is larger. This explains why there is a larger Vc and Qsw at lower temperature. Above the Tc, the energy potential is parabolic whereas the lattice is cubic and there is only one possible position for Ti or Zr thus they show no spontaneous polarization. When Vc is low as at a high temperature (c), defects and trapped charge can impact hysteresis horizontal asymmetry as well as back switching, which increases the U-term while decreasing the P-term resulting in lower Qsw. In the case when Vc is high as at a low temperature, a low-voltage writing results in a minor loop, which results in reading a lower Qsw. For a given operating voltage and temperature range, an appropriate Vc value exists in an FeRAM capacitor.

There are further Qsw reduction mechanisms particular to ferroelectric materials that include short-term degradation also called Relaxation (Back switching) and long-term degradation, namely Imprint or aging, Fatigue degradation and hydrogen degradation.

Relaxation was previously mentioned with regards to pulse measurements with delays of 1 s between pulses as shown in Fig. 8.7. Here Fig. 8.12(a2) shows a bit-distribution measurement result of readings immediately after writing and also after waiting 1 s after writing. The P-and U-term peaks and also distributions move closer together after a 1 s time interval. This indicates capacitor or bit-by-bit uniformity is desirable although some process steps may have introduced defects into the capacitors. An extremely bad flyer (outlier) bit may be replaced by redundancy circuitry but uniformity of the entire capacitor distribution is strongly pronounced. Minimization of charge relaxation is strongly desirable such as reducing electrode interface charge traps or by reducing piezoelectricity by introducing dopants as well as through optimization of process steps.

Imprint or aging is typically observed after a write operation then a long-term storage or baking, which results in the hysteresis loop Vc shifting to the left Fig. 8.12(b1) or right direction on the voltage axis (b2), depending upon the data storage state, +Pr or –Pr. The root causes for imprint proposed are charged defects motion during storage and internal electric field dissipation, since internal electric field remains in the presence of spontaneous polarization. Figure 8.12(b) shows imprinted hysteresis curves for the long-term storage state at +Pr (b1) shifts to the left and –Pr (b2) shifts to the right which correspond U-term (0) and P-term (1) storages, respectively.

Explanations for same state and opposite state are referring [22]. Qss: Same-state charge remaining after a memory state is baked at 150°C and read at room temperature. Qos: Opposite-state charge remaining after a memory state is baked at 150°C, reset to the opposite-state and read after a time delay.

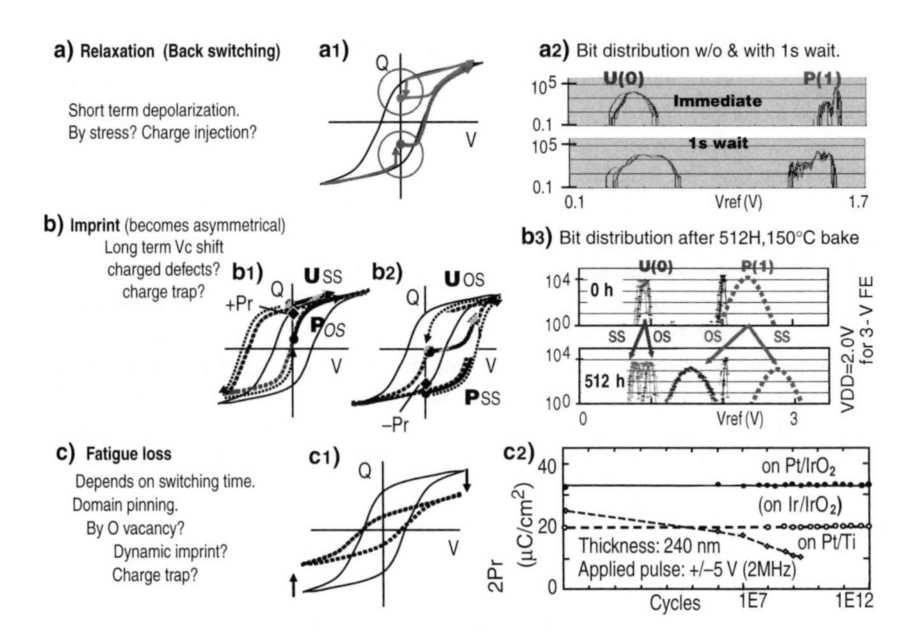

Fig. 8.12 Reliability issues: relaxation/imprint/fatigue (C2) [12]

Thus the left-shifted (b1) case, read out charge for U-term (+Pr to +Vread) defined as Uss and that of for P-term (–Pr to +Vread) defined as Pos decrease. It is noticeable that this leftward shift results in an incomplete switching for a fixed negative writing voltage thus the loop falls into a minor loop that decreases | –Pr | charge. On the contrary, the rightward shift (b2) case, read out charge for U-term defined as Uos and P-term defined as Pss increase. Sometimes the Pss decreases when Vread is low to switch only a portion of the polarization. Thus, an application such as those in ROM uses Pss and Uss and has no data loss. On the contrary, writing opposite (complimentary) data as in the case of RAM where "1" or "0" may involve Pos and Uos, this Vc shift impacts not only semi-static Qsw reduction but also the switching speed which becomes slower with lower voltage FeRAM as imprint progresses. From Fig.(b3) the scattering of Pss and Pos in the bit-distribution measured with VDD of 2.0 V is not seen at 0h because the S/A is designed to saturate P-term, as a result the distribution spread appears narrow. However, after a 512 h bake at 150C, two different distributions for the broad spread of Pos and for saturated narrow Pss are observed. Of course, on applying VDD of 3.0 V, the separation between the Pos distribution other than at high saturation and Uos distribution remains large. So designing for higher voltage reading architecture may reduce imprint's impact. It is noticeable that an imprinted hysteresis curve falls into minor loop at one polarity and the shape transforms to "water flea" shape that has a tail and a round fat belly, as shown in Fig. (b2). The imprint rate (Pos–Uos) shows a linearly decreasing rate with log(time). However, the rate depends on the saturation characteristics of the material as represented by the hysteresis shape such as the slope of Q–V loop and the measurement method such as pulse voltage, pulse measurement speed (switching speed), and measuring temperature (Vc). Also increasing the baking temperature should result in a Qsw reduction because it may decrease the internal electric field and impacts secondary mechanisms although a true imprint charge accumulation model [23] supports higher temperature acceleration.

Fatigue loss (c) that is Qsw reduction with increasing number of switching cycles was major problem for FeRAM with endurance limited to 10^7 cycles until the mid-1990s, but introducing conductive metal oxide such as IrOx electrodes increased PZT fatigue endurance to 10^{12} cycles (c2) [24]. Another way to circumvent fatigue was by introducing a new ferroelectric material for thin film ferroelectric capacitors such as SBT [25], which also exhibits 10^{12} fatigue endurance by the end of 1990s. With these improvements in endurance, FeRAM production reached a commercial level. The root causes of fatigue are still being debated but attributed to domain pinning fixed with charged defects [26], oxygen vacancies, injected interfacial charge, or dynamic imprint that is charge traps [27]. Changing electrode materials improves fatigue indicating that the interface defects in PZT seems to be the primary cause. Although applying higher voltage is beneficial for overcoming imprint, it is detrimental for fatigue.

Relating to above three degradation phenomena of polarization charge, the descriptions of only the root causes were given, now the hydrogen-induced degradation will be mentioned. A degradation of polarization charge is associated with hydrogen intercalation into the ferroelectric material during CMOS integration processing and subsequent loss of polarization. Exposure of the ferroelectric material to hydrogen from the interlayer dielectric and metal layer deposition at elevated temperature is frequent during integration. The ferroelectric capacitor electrodes typically consist of catalytic materials such as Pt or Ir which can dissociate hydrogen or water into atomic hydrogen which diffuses into the ferroelectric material disrupting domain motion which results in device failures [28, 29]. To circumvent this problem, the ferroelectric capacitor is typically encapsulated with hydrogen-resistant diffusion barriers such as AlOx or TiN [30]. Hydrogen degradation of the capacitor is a problem seen only in integrated devices, whereas the other types of degradation are general phenomena seen in all ferroelectrics capacitors.

In summary of this ferroelectric material section, we have seen that PZT and other ferroelectric materials are used widely for its piezo-, pyro-, and ferroelectricity. Ionic crystals contain oxygen octahedral which exhibits oriented domains that respond to an electric field and is the origin of ferroelectricity. We see from an example that a single crystal and thin-film ferroelectric materials and grains are different. From the electrical characteristics, *P–V* hysteresis curve, Vc, Pr terms as well as linear charge and spontaneous charge were explained. Then detailed characteristics of relaxation with 1 s wait pulse measurement, minor loop and major loop are discussed for introduction to the parallel capacitance model including time response for Spice simulation. Further, Qsw reduction and Vc dependency as a function of temperature are explained. Reliability issues including relaxation, imprint, and fatigue are briefly described. About 10 years ago, FeRAM was designed primarily based upon the hysteresis curve at RT, but now we understand that more widely measured characteristics, cf. temperature dependence, time response with read/write voltage, and minor loop are essential especially for designing 1T1C fast FeRAM. Through the use of failure analyses and continued improvements in processing and materials development, technology has resulted in improved device ferroelectric characteristics. A major problem with the Spice modeling is that the electrical properties of thin ferroelectric films depend heavily on the fabrication process and samples vary greatly depending upon the company or university lab where they are fabricated. Generalizations can be made to some degree but some characteristics are solely dependent upon the sample. Knowing these fundamentals is of general interest, but circuit designers want to know how to overcome the above-mentioned issues which is the topic of next section.

8.3 FeRAM Cells and Circuit Technology

In this section, the FeRAM cell structure will be explained. For a smaller footprint unit cell, planar to stack, and then future 3D style cells with higher Qsw and lower Vc ferroelectric material are discussed. Then the circuit schematic of FeRAM cell and BL are reviewed. Basic HiZ read scheme and related design issues, UP-DOWN or UP-only, Plate Line sensing, NRTZ and RTZ writing, and 1/2VDD PL are explained. An important topic of 1T1C reference generation will be discussed based upon four major reference schemes. Advanced topic of redundancy or ECC and power on and off sequences to prevent data loss will be summarized. Then, the essence of Fujitsu's unique BGS architecture will be explained. Finally, other interesting topics on field programmable gate array (FPGA), non-switching latch, and nondestructive read out (NDRO) (and multilevel storage, if possible) will be summarized.

8.3.1 FeRAM Cell Physical and Schematic Structures

The historical FeRAM developments are plotted in Fig. 8.13. Around 1990, there were wide ranging activities on integrating FeRAM, which are not plotted here. The primary problem during this time period was ferroelectric degradation during CMOS integration [6]. The trend toward higher-density devices show process improvements owing to bulk feature size scaling and also owing to cell structure namely a planar capacitor change to a stacked capacitor and circuit improvements of 6T2C cell, 2T2C cell, to 1T1C cell. The details will be discussed later in Figs. 8.15 and 8.18.

Figure 8.14 shows the general trends of process improvements up to 2008. At first, ferroelectric materials were also expected to be integrated into a DRAM capacitor because of its high permittivity. However, DRAM has evolved into 3D structure capacitors with very fine feature sizes, which FeRAM cannot yet achieve. In 1990s, the first priority was to reduce fatigue loss in FeRAM materials. Two different ways

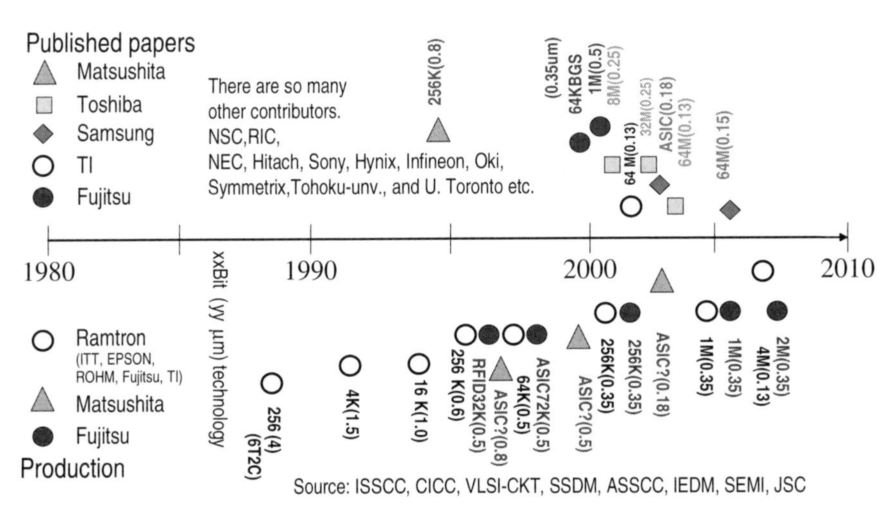

Fig. 8.13 FeRAM development history

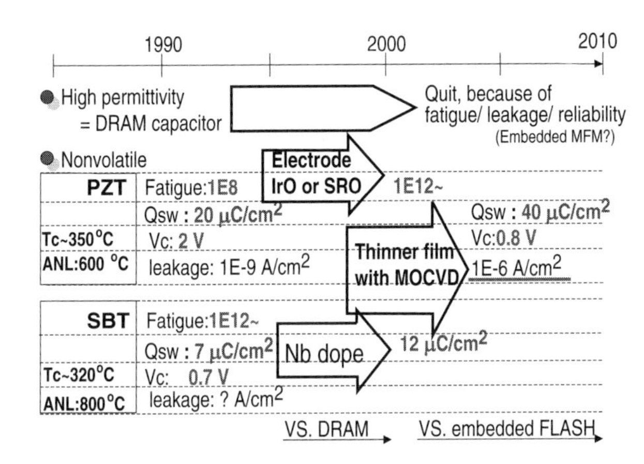

Fig. 8.14 Process improvements in FeRAM

were demonstrated such as change the ferroelectric material to SBT, and the other was by introducing metal oxide IrOx or $SrRuO_3$ thin-film top electrodes for PZT. This fatigue-endurance improvement of more than 10^{12} cycles has helped to realize commercial devices. A second issue is lowering the switching threshold or Vc while also retaining low-leakage current, since VDD of popular ICs was reduced to 3 V (0.35 μm) in the mid-1990s. The third issue is how to achieve a smaller cell footprint, this involves high-temperature etching technology of metal electrodes, Pt or IrOx and ferroelectric material, and increasing Qsw/unit-area for smaller capacitor sizes. Niobium (Nb) is frequently added to SBT to increase Qsw by raising the Curie temperature (Tc) and Vc. MOCVD undoped PZT-based capacitors show nearly twice as much Qsw/unit-area and low Vc of 0.8 V, which makes it an appealing candidate for 0.18 or 0.13 μm devices which operate at less than 1.8 V VDD.

Also investigations for the new ferroelectric materials having lower crystallization temperatures suitable for integrating with a low temperature fine advanced CMOS bulk technology are performed all over the world.

Figure 8.15 shows cross-sectional TEM of three basic structures of ferroelectric capacitors for FeRAM. Planar schematic (a1) is shown in 1996 [31], the capacitor was overlaid on top of MOSFETs. Ferroelectric capacitors are formed over an entire wafer, then patterning of capacitors is done and contact via holes are opened by etching from above for top electrode (TEL) contact and bottom electrode (BEL) contact. Recently, 1T1C planar cell (a2) uses chemical mechanical polishing (CMP) for preparing a flat surface before depositing BEL, which impacts the crystal orientation of the ferroelectric material.

A stack via structure (b) reduces the footprint of a cell and also this type of one-mask and one-step etching for TEL, ferroelectric, and BEL reduces process steps and lowers the manufacturing cost. Investigation-level 3D trench stacked cells are demonstrated in Fig. 8.15(c1) [32] and (c2) [33], which will be expected to increase the net Qsw for a cell capacitor with a given footprint.

Those ferroelectric material growths for cell capacitor are mainly by sol-gel, sputter, or MOCVD method [4]. Sol-gel, CSD, and spin coating methods for fabricating ferroelectric materials are widely used in university laboratories and even for mass production at companies, since the fabrication equipment is rather simple. Films of thicknesses 200 nm can be obtained by multiple spin coating and pre-baking steps. However, in mass production, obtaining the high-quality solutions

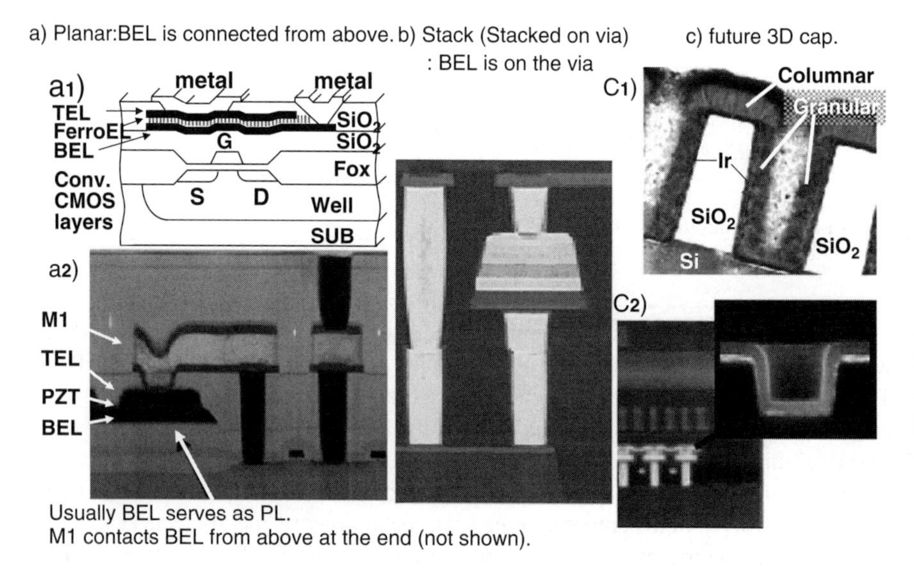

Fig. 8.15 SEM photographs showing planar/stack/3D stack capacitors. (a1) [31] (©1996 IEEE) & [3] (©2000 IEEE), (c1) [32] (©2005 IEEE), (c2) [33]

in mass volume is a key issue. For sputter deposition, once a target composition is fixed, the unit wafer process turn-around time is superior to sol-gel method. In the case of sol-gel and sputtered films, they are initially deposited as amorphous and then heat-treated for crystallization. In the case of MOCVD films due to the high deposition temperature, polycrystalline films grow directly on the BEL. Thus, foreseeing application in 3D structure capacitor, this type of isotropic growth is essential.

Typically ferroelectric crystallization temperature is between 550 and 750°C, so a 450°C metallization process prior to ferroelectric film deposition is impossible. Thus ferroelectric capacitors are integrated under the metal wiring layers of aluminum or copper metallization process. On the contrary, phase change, magnetic (TMR or spin injection), and resistive memories have their memory elements on top of the metal layers because of their low-temperature process.

Basic structure of capacitor over bitline (COB) (Fig. 8.16(a)) and capacitor under bitline (CUB) (b) are shown. DRAM history [34] shows COB obtains smaller cell footprint, but high-temperature treatment of ferroelectric crystallization necessitates using tungsten (W) or other high melting point metal wiring under the ferroelectric. For a high-capacity discrete FeRAM, it may be possible to develop a dedicated process. But for embedded FeRAMs, introducing COB with such W wiring processes is not compatible to a logic process.

Unified BEL structure for a planar cell (Fig. 8.16(c)) and unified TEL structure for a stack cell (d) [33] are referred. Those capacitor electrodes can be used for wiring if additional mask steps are employed.

Basic array structures are summarized in Ref. [3]. PL parallel to WL structure with divided PL is the most popular structure used in commercial devices. Figure 8.17 shows advanced array architectures of staircase WL (a) [21] and chain

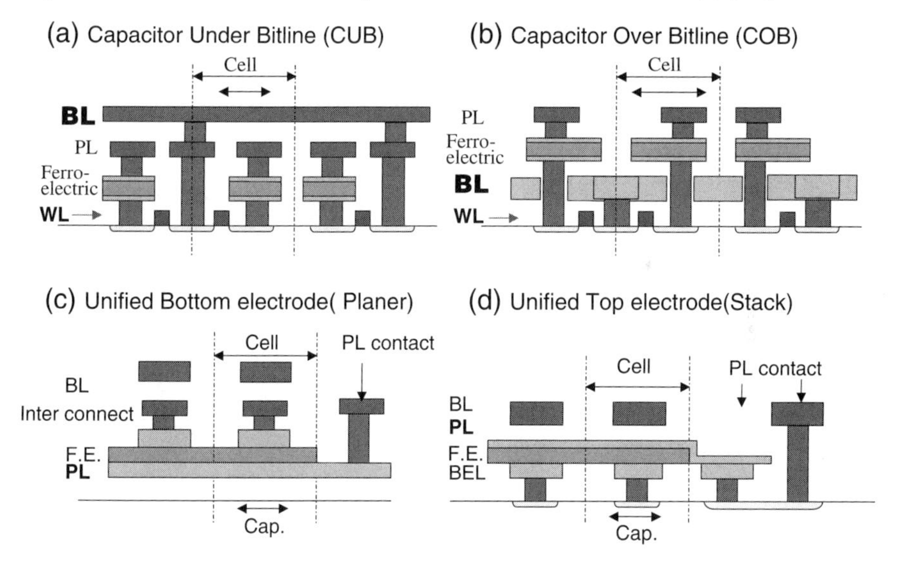

Fig. 8.16 Capacitor array structures

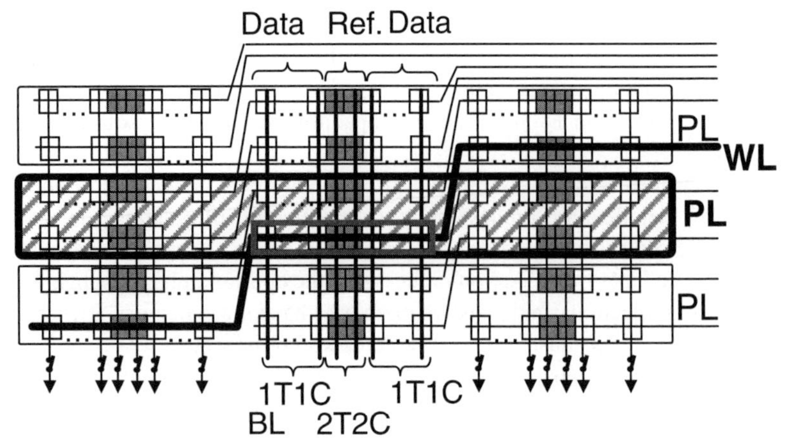

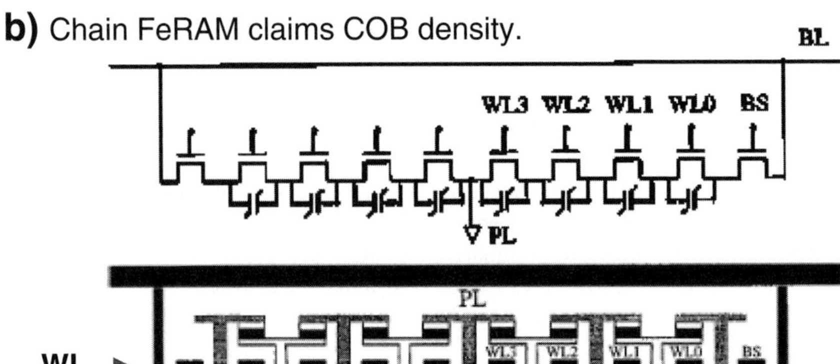

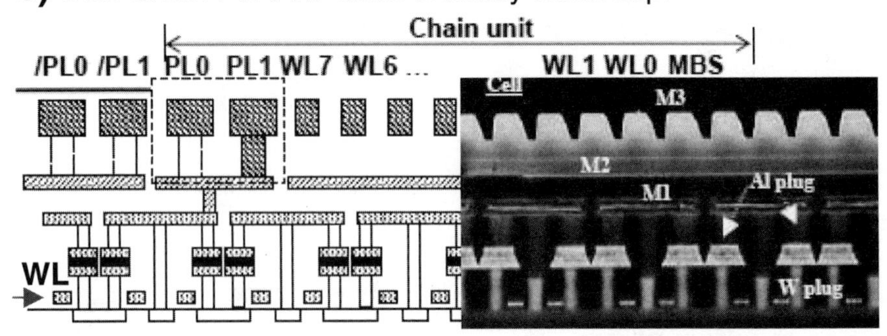

Fig. 8.17 Array structures of staircase WL and chain FeRAM. (**b**) [35] (©1998 IEEE), (**c**) [36] (©2006 IEEE) & [37] (©2004 IEEE)

FeRAM (b) [35] (c) [36]. Those structures reduce accessing cell number for employing a relatively large cell matrix. This is in contrast to small-divided arrays, where the peripheral circuit of row decoders and sense amplifiers occupy a larger area, those large matrix architectures realize around 70% cell area occupation. Accessing a small number of cells results in lower power consumption for PL drives and less fatigue compared with entire row (32–256 cells) activation. The reading method for chain FeRAM resembles that of NAND-FLASH, unselected WLs are at a high level to short circuit TEL and BEL of a cell capacitor. A cell at selected low-level WL (off) is applied TEL to BEL voltage by BL and PL drive source.

8.3.2 Basic Sensing Architectures

In this section, schematic cell structures and basic operating conditions are explained regarding applying capacitor voltage and its relationship on the position within the ferroelectric hysteresis loop. In last part of this section, reference scheme for 1T1C and the essence of Fujitsu bitline ground sensing (BGS) are described.

Figure 8.18 shows the historical improvements in reducing the cell elements and cell footprint. FeRAM started with an SRAM cell with adding two ferroelectric capacitors in 1988 [38]. At the time, since the fatigue endurance was poor (10^6 cycles), a shadow SRAM was designed. Shadow SRAM recalls the nonvolatile ferroelectric data upon power-up and operates as an SRAM during power on with less fatigue by keeping PL at half VDD, and stores data to ferroelectric capacitors by pulsing PL to VDD and GND just before the power supply turns off. Parallel recalling and storing of cells enables a short lead-time overhead for the specific actions. This structure is also often used for the investigation of MRAM, PCRAM, and ReRAM. However, this method has a large SRAM cell footprint. So with the improvements in ferroelectric capacitor, DRAM architectures were adopted for further reduction in the cell footprint. Although this necessitates a restoring time followed by a destructive read after each operating cycles. Thus, the internal operation

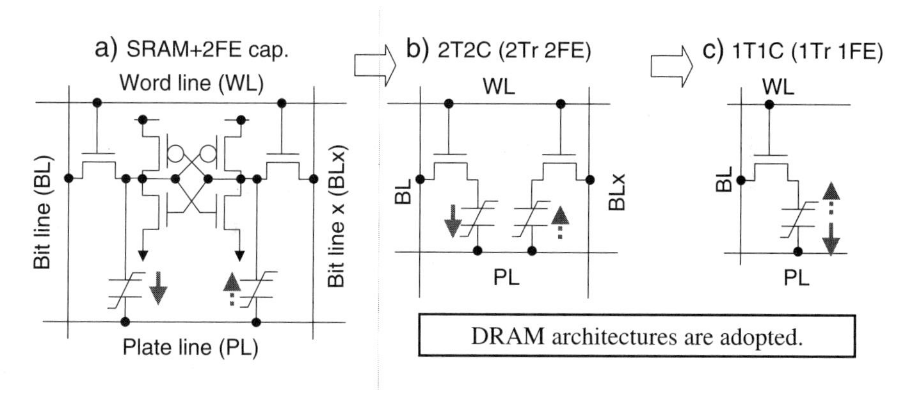

Fig. 8.18 Basic FeRAM cell circuit scheme

is totally clock dependent, with a peripheral interface to mimic SRAM operation. The 2Tr2Cap cell is rather robust because complementary data is stored simultaneously and thermal history (Qsw reduction by elevated temperature) is the same for both capacitors. Thus, the complimentary P and U stored capacitors serve ideal reference voltage generation schemes. So, some embedded FeRAM macros still use this 2T2C type cell even now. The 1Tr1Cap cell achieves the smallest footprint, but in reading, a sense amplifier compares the BL level and a reference level that is preferably mid-point level of P-BL(1) and U-BL(0). Thus, good reference generation method tracing the mid-point though P and U values are impacted by temperature, VDD voltage, or long-term degradations, is very important in this 1T1C design. There is a good cell structure comparison table shown in [39].

Figure 8.19 explains storage capacitance (Cs)/bitline capacitance (Cb) ratio which is namely the stored charge ratio to the bitline capacitance for FeRAM. The term Cs/Cb is identical to DRAM that is used to determine BL signal amplitude. However, in FeRAM operation, Cs is not a linear capacitance but is a summation of charge emitted by non-switching capacitor U-term or by switching capacitor P-term with PL rise. Thus, nonlinear ferroelectric charge can be displayed as *QV* hysteresis curve and linear BL capacitance is transformed to –1/Cbl slope line in graph solution Fig 8.19(b). Since PL pulse voltage Vpl is distributed to the capacitor and the BL respective voltages, Qsw=Cs × Vsc = Cb × Vbl, and Vpl=Vsc + Vbl, where Cs is a nonlinear value. Thus, Vbl=Qsw/Cb, and Vsc=Vpl – Vbl, so Vsc=Vpl – Qsw/Cb, and the relationship of Vsc to Qsw is similar to the hysteresis curve. Hysteresis curve is typically plotted as μC/cm^2, so Cb is also multiplied by the μC/cm^2 unit corresponding to a capacitor layout area of 1 cm^2. In general, as seen in the

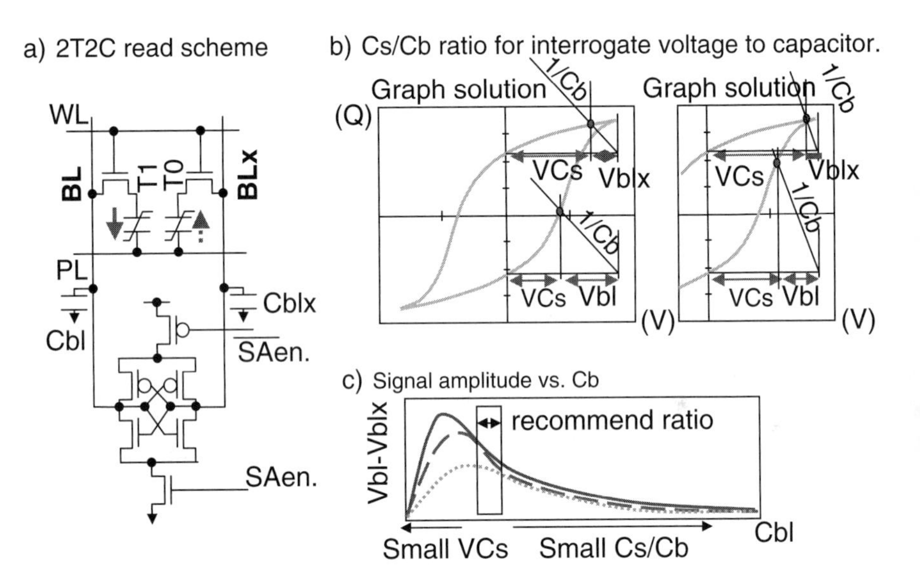

Fig. 8.19 Explanation of Cs/Cb ratio in FeRAM nonlinear Cs ferroelectric capacitor

graphical solution displayed in Fig. 8.19(b), larger Cs/Cb ratio leads to moderate 1/Cb slope and Vbl is larger. Smaller Cs/Cb ratio leads to a steep 1/Cbl slope and Vbl is smaller. But assuming Cb is close to zero, then Vsc is close to zero and small Qsw emerges. Thus as Cb approaches zero signal, the amplitude may also approach zero as shown in the signal amplitude versus amplitude graph in Fig. 8.19(c). A more complicated situation is P- and U-terms have relatively larger temperature dependency and also show end of life Qsw reduction. Thus Cs/Cb ratio is preferably not designed to maximize the signal amplitude but typically designed below the maximum peak on the right side of the slope where the amplitude does not vary greatly when Qsw varies. It is noticeable that the P-term capacitor has smaller Vsc than U-term's. The advantage of this HiZ BL architecture are improving fatigue endurance because some of the polarization remains the same, and self-convergence scheme for P-BL level because if a larger Qsw is emitted then less Vsc is applied. The disadvantage is applied Vsc is always smaller than VDD or Vpl, thus relatively slow switching occurs even when applying full VDD to PL. The graphical solution is based upon a hysteresis loop, which does not account for switching-time response, a Spice simulation using a ferroelectric model including ferroelectric time response would predict more precisely operating signals.

Continuing the discussion from the previous figure, the 2T2C basic read out scheme with WL, PL, and BL signals are shown in Fig. 8.20. The first method (b) is pulsing the PL and after that, a sense amplifier will latch the BL voltage difference to full VDD swing, it is called UP-DOWN, pulsed, or after (pulse) -sensing. The second method (c) is raising the PL voltage and before the PL falls, a sense amplifier will latch the BL voltage difference and then PL voltage decreases to GND level, it is called UP-only, step (pulse), or on-pulse-sensing. The idea of UP-DOWN sensing is subtracting linear capacitance charge while PL falls, which leaves bitlines with net switching charge. It is a robust method if the fabricated capacitance

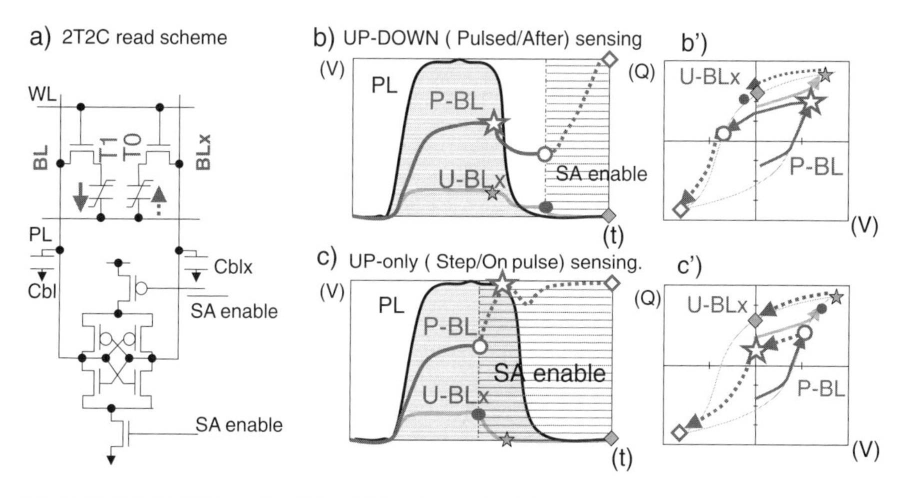

Fig. 8.20 UP–DOWN sensing (b) and UP only sensing (c)

area varies among cell distribution and linear charge varies. In the case when the ferroelectric experiences charge injection at the electrodes interfaces, this can be cancelled and net switching charge can be sensed. The demerit for this approach is a relatively slow sensing time, because PL may have a large load (about few nano coulombs) and cause a delay. Also after the sense amp is enabled, rewriting a P-term by switching to –Vs position will cause a large load in amplifying the BL signal thus a sensing delay will result. UP-only sensing is faster because PL depends upon the rise time delay and sense amp latching will not cause P-term switching thus resulting in smaller capacitance load. However, it requires good process uniformity and less injection charge for ferroelectric capacitors. Usually UP-DOWN sensing uses a sense amp with pMOS latching first then nMOS latching second, because BL levels are lower. For UP-only sensing, pMOS and nMOS latching may be simultaneously applied, because BL levels are near 1/2 VDD and preferable level for amplification. That comparison between two sensing methods appears in a 64 Mb device [40] that uses resistors for switching two testing settings.

Figure 8.21 shows two writing methods and two PL structures with their possible combination. Non-return to zero (NRTZ) (c) is similar to DRAM writing, WL decreases while keeping positive BL level. This leaves positive charge on the TEL with a P-term write (1) cell thus a positive electric field is applied to ferroelectric capacitor for a time period of milliseconds until the charge dissipates via leakage paths. This provides sufficient time for slow switching portion of the ferroelectric capacitor to switch, and net increase of reading switching charge by reducing the imprint effect. Thus NRTZ is rather a robust writing method. With subsequent readings, the positive charge remaining in the cell can be added to real switching charge to be shared with BL or a short time discharge period can be inserted just before PL rising to avoid the DRAM charge influences on the BL levels.

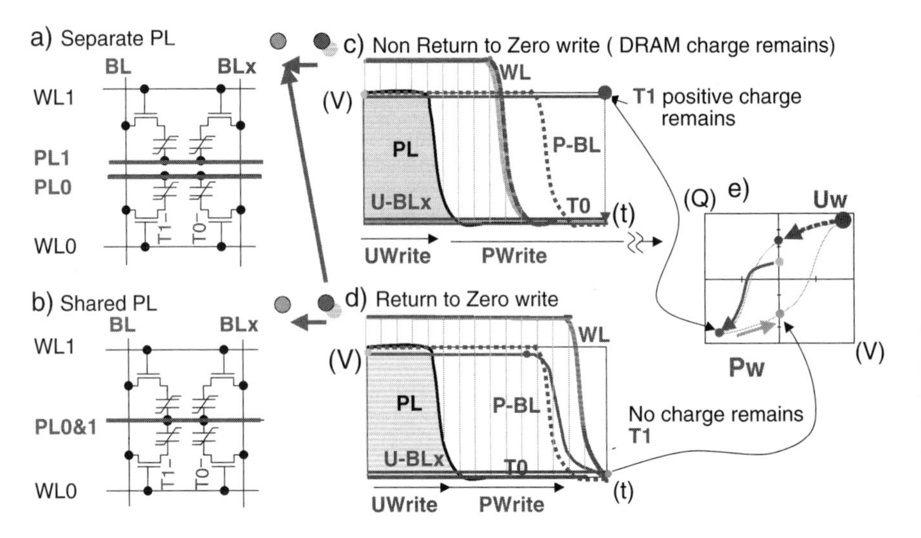

Fig. 8.21 Non-return to zero (c) and return to zero (d) writing method with PL structures

Return to zero writing (RTZ) (d) consists of the WL lowering after BL level decreases to GND. Using this method, no charge remains at TEL, so time for applying electric field to the capacitor is limited and good ferroelectric characteristics are required. However, a shared PL structure (b) can be applied which reduces the cell footprint, since there is no PL to PL separation region as in the case when separate PLs (a) are applied. The problem is that when a shared PL structure with NRTZ is utilized, the remaining VDD voltage for TEL when the WL is off will be boosted to 2 VDD when PL rises to VDD. This results in applying excessive voltage to the nMOS, which are 1.8 V or 3.0 V tolerant when used as WL nMOS (if not a high voltage 5 V tolerant MOS is especially adopted to the WL).

In addition, WL is preferably one Vt boosted higher than VDD for supplying full VDD to a TEL during writing when nMOS WL is adopted.

Figure 8.22(a) explains a problem with half VDD PL scheme and preferable time sequential 0 and 1 write scheme (b). The half VDD PL scheme is identical to a DRAM scheme. If we use mid-level PL, writing can be performed simultaneously for P(1) and U(0) cells by applying VDD and GND swing BLs, respectively. Also approximately Vc of 1/4 VDD or less would be needed for this scheme. Another challenging aspect of this method is the junction or MOS gate leakage at T1 (P-term TEL) node, T1 would eventually reach GND level after a long time that will cause reverse writing (0) to the T1. On the other hand, the time sequential method applies full VDD for writing, thus Vc value of 1/2 VDD is sufficient to write, this is preferable for higher Vc value of PZT. Also there is no problem in the case of higher leakage that causes reverse writing for the P-term. Architectural decisions need depending upon the further considerations for operating VDD and Vc value of targeted ferroelectric capacitor.

To realize 1T1C cell scheme, a good Vref generation circuit that traces process variation, temperature dependency, imprint, and fatigue degradation is ideal. Basically a column reference scheme (one reference cell responds for cells connected to a BL) shown in Fig. 8.23 and row reference scheme (one reference cell responds for cells connected to a WL) shown in Fig. 8.24 are widely used [41]. For a column reference scheme, in Fig. 8.23(a), one reference cell is connected to BLx for generating reference level and compared to P or U levels of cells on BL by a sensing amplifier. Because the reference cell can be accessed 256 times greater than real cells if they have 256 WLs, fatigue degradation for the reference cell would be two orders

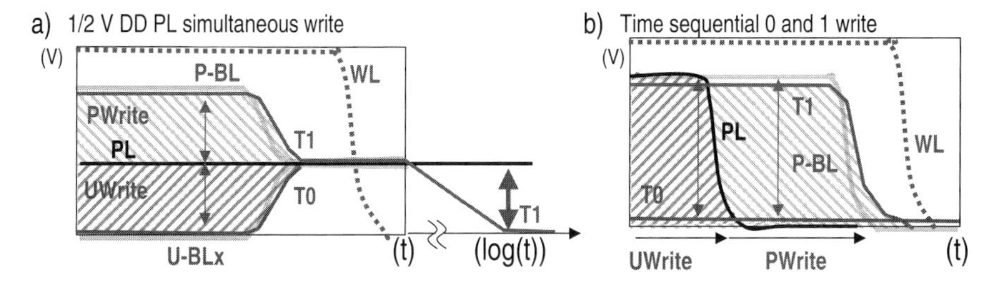

Fig. 8.22 Half VDD PL scheme (**a**) and PL pulsing scheme (**b**) with time sequential 0 and 1 write

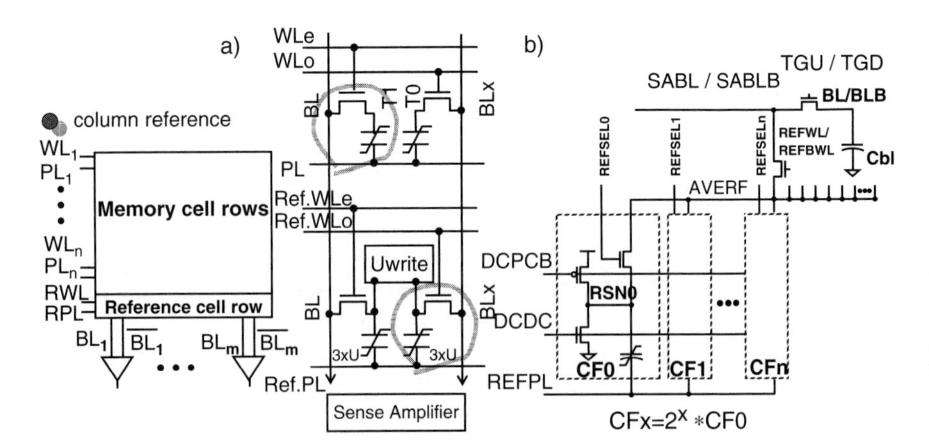

Fig. 8.23 Column reference scheme and reference level generation circuit for 1T1C. (b) [40] (© 2004 IEEE)

higher than data cells. Thus the reference cell is designed as a non-switching operation either a U-term (+Pr to +Vs) or Da term (–Vs to –Pr) which results in unlimited cycles. Capacitor sizing of the reference cell provides level trimming for the reference BLx. The reference WL should be off before the S/A latches and amplifies

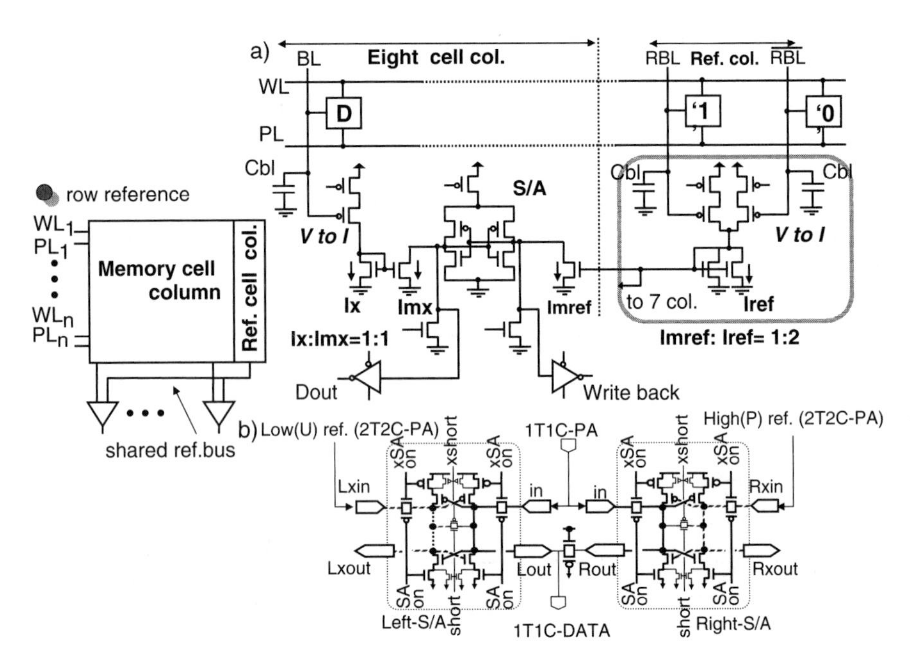

Fig. 8.24 Row reference scheme and reference voltage generation circuits. (a) [41] (©2002 IEEE), (b) [21] (©2007 IEICE)

VDD to GND. The cut off of reference WL prevents the reference cell polarization from switching during full swing amplitude at complementary BLs. A more sophisticated method (b) [40] is a digitally trimmed reference level that is distributed to all reference BLs.

In the row reference scheme shown in Fig. 8.24, accessing time and thermal history of data cells and one or two reference cells are identical. Thus we can use P-term reference cell which experiences the same order of fatigue degradation for all real data cells. Ideally, an averaged BL voltage levels for P- and U-terms reference cells tracks P- or U-term value changes in the data cells. However, the reference voltage to distribute to all the S/As or BLs needs more advanced circuit techniques. For example, P,U references averaging with current mirror and employing current force latch amplifier for every BL in Fig. 8.24(a) [41] and P,U reference averaging with a dual reference voltage latches (b) [42] have been proposed.

Figure 8.25(b) shows bitline ground-level sensing (BGS) [21, 42], in contrast to conventional HiZ BL(a). The aim is mainly to apply full VDD to read both 0 and 1 stored cell capacitors (Vcell), thus effectively reducing the imprint shift and imply higher-speed switching. The disadvantage is a relatively large circuit area compared with a DRAM like sense amp and several clocks needed for pre-charging negative voltages. Figure 8.25(c) compares sensing signal amplitudes for BGS and HiZ impact of bitline capacitance. BGS shows almost no dependency on Cbl thus it enables having 1 K cells on a BL structure.

Figure 8.26 shows other aspects of the BGS. For the BGS, to compensate for having a large area S/A, we reduced the number of S/As to 16 bits + 6 ECC bits, rather than the conventional per BL S/A. As the FeRAM cell has three terminals,

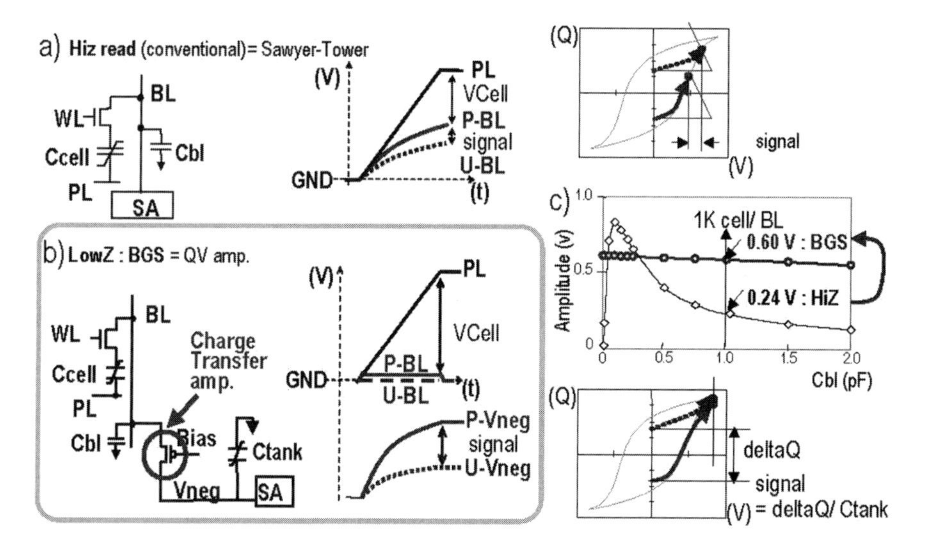

Fig. 8.25 BL reading scheme of conventional HiZ reading (a) and BGS low Z BL reading (b). [42] (© 2002 IEEE)

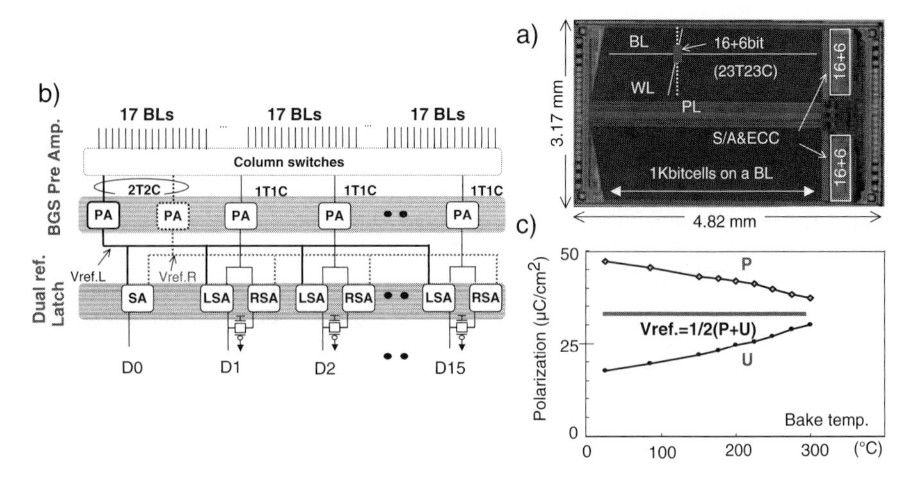

Fig. 8.26 Staircase WL for reducing numbers of active cells and sense amplifier [21] (© 2007 IEICE)

WL, BL, and PL, combination of three can reduce accessing (or destructive reading) cells by the staircase WL, also shown in Fig. 8.17(a). Since 2T2C reference is employed, 23T23C unit (2T2C + 15-1T1C + 6 ECC 1T1C) compose one address word as in Fig. 8.26(a). The 2T2C serves to generate the P- and U-term reference levels and also serves as one data bit, so as to reduce capacitor number. Those reference levels are fed into every dual reference latch amplifier (b) and effective reference level is the average of the two, which dual reference latch S/A is shown in Fig. 8.24(b) and PA is shown in Fig. 8.25(b). Effective reference voltage with the Fig.8.26(b) architecture is averaging of P and U term, which is stable for elevated temperature (c).

8.3.3 Redundancy and Data Protection for FeRAM

Memories beyond 1 Mbits capacity usually employ redundancy architecture for improving yields. How to use ferroelectric capacitor's nonvolatility for redundancy circuits are discussed. Also, data loss likely to occur during power-on/off will be explained.

Figure 8.27 shows redundancy architectures using nonvolatility of ferroelectric capacitor. In Fig. 8.27(b), a direct method of employing ferroelectric nonvolatile latches, which will be used as programmable address pointers for defect cells, is shown [43]. It resembles that FLASHs use programmable elements with FLASH cells. Since this method uses 6T2C cells for the element, a relatively large footprint and well-designed store/recall sequences are needed during power up and down. In Fig. 8.27(a), a schematic containing replacement I/O pointer bits and a spear cell in a large cell matrix is shown [44, 45]. This reduces footprint for elements and

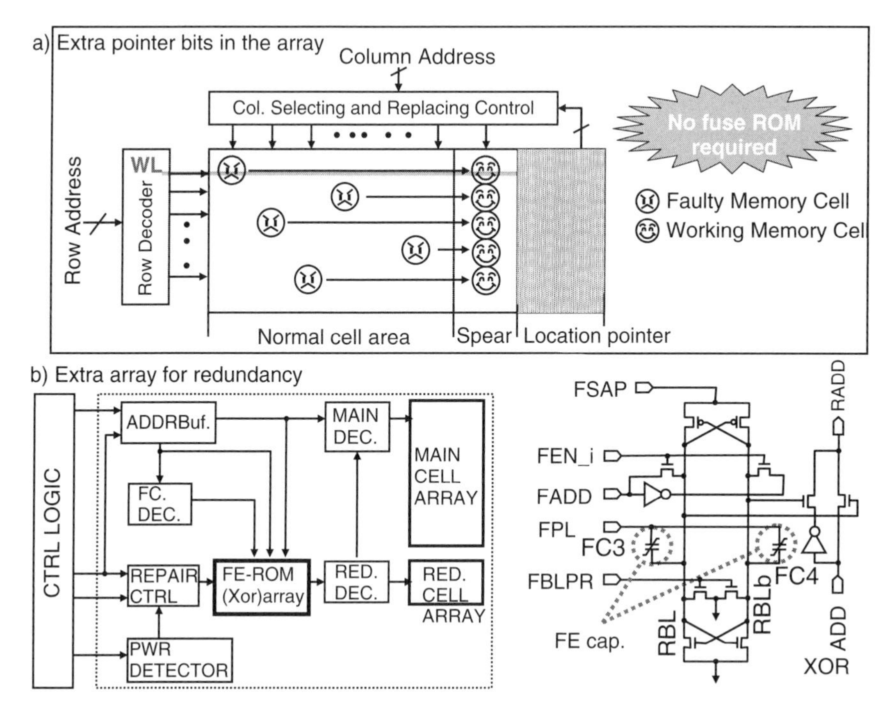

Fig. 8.27 Redundancy, extra pointer bits in array (a) [45] (© 2000 Fujitsu Micro Electronics Limited) and nonvolatile FEcap. element (b) [43] (©2005 IEEE)

eliminates special sequences, but redundancy ability is limited to correct real data cell bit area. However, added extra bits result in the same penalty as employing error correction code (ECC) that also allows for correction to the extra bits.

Figure 8.28(a) shows examples of Hamming ECC architecture applied to FeRAM [21, 36]. By increasing the number of internal word bits accessed, for example, from 16 to 64, one bit ECC can be achieved with a relatively small number of extra bits. However, ECC capability per bit will decrease. So, statistical analysis of defect density and also for ferroelectric cell probability of future anticipated failed bits after 10 years degradation is essential for determining how many ECC bits or redundant spare bits are needed. Usually 1 bit error correction with Hum-

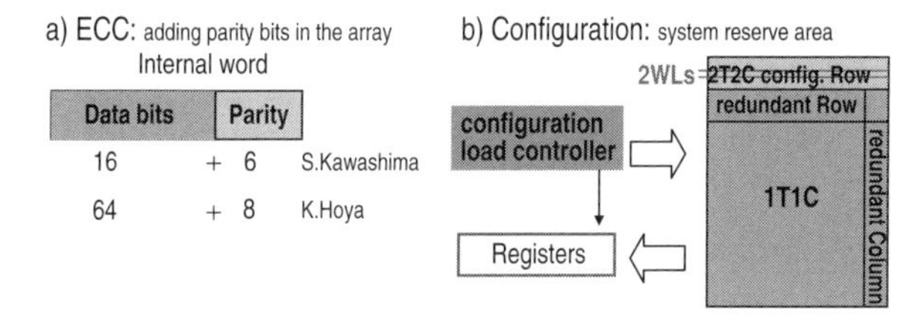

Fig. 8.28 ECC (a) and configuration for FeRAM redundancy (b)

ming code ECC is employed because 2 bits ECC will need much more complicated circuits and cause propagation delays. If we go into block access of serial file application, more sophisticated cyclic or convolution ECC methods, which are applied to digital wireless or optical communications or HDD Read channel PRML or PR4 and are capable of 2 or 3 bit error correction. Such proposals of BCH (Bose, Rey-Chauduri, and Hocquenghem) code or Trellis code for FLASH were pronounced [46, 47]. However, considering the cost considerations of circuit area, access time, and power consumption, there is a debate that 1 bit correction is sufficient combined with enhanced cell array architecture [48, 49].

Another method shown in Fig. 8.28(b) is to access the 2T2C area just after applying power and transferring the settings to conventional volatile logic latches with a sequencer [50]. This method is suitable for a large-capacity FeRAM where the sequencer area penalty is negligible. However, it requires register-setting time each time the device is powered up.

Since the FeRAM write cycle is fast consisting of a destructive read and restore, maintaining nonvolatility during power on and off is relatively more complicated than with FLASH, which has a slow write cycle and nondestructive read. Care needs to be taken with FeRAM as shown in Fig. 8.29(a1), almost the same as is paid to an SRAM transition to a battery backup mode (VDD=1 V) and recovery from the mode. Although many discrete FeRAMs employ low-voltage protection to monitor the VDD level at the supply pin, on a system board, a power on system reset signal can stay active (resetting continues) even after supplying sufficient VDD. Resetting may cause MCU output terminals to stay HiZ or emit low-level pulses that may drive FeRAM's/CS pin and FeRAM will unintentionally operate one cycle thus destroying stored data. Figure 8.29(a2) shows software protection for a parallel discrete part [51] and (b) shows a serial discrete part [52], basically they have the same control as for EEPROM discrete devices. On power-up, the FeRAM starts with write protec-

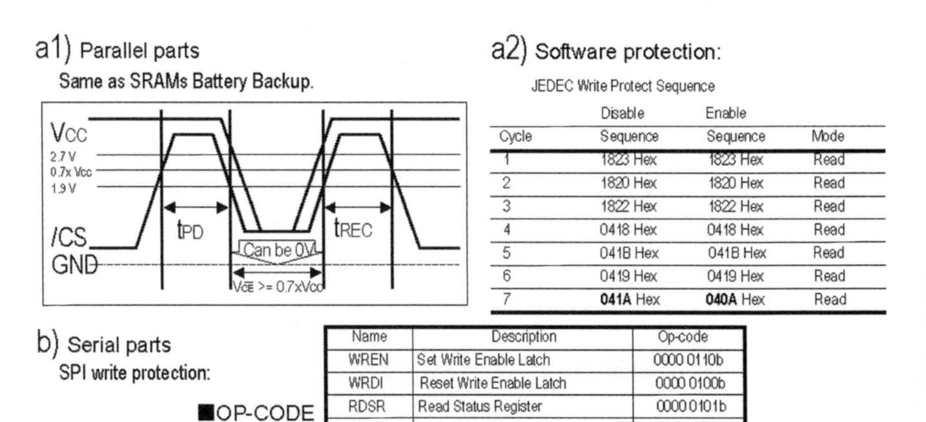

Cycle	Disable Sequence	Enable Sequence	Mode
1	1823 Hex	1823 Hex	Read
2	1820 Hex	1820 Hex	Read
3	1822 Hex	1822 Hex	Read
4	0418 Hex	0418 Hex	Read
5	041B Hex	041B Hex	Read
6	0419 Hex	0419 Hex	Read
7	**041A** Hex	**040A** Hex	Read

Name	Description	Op-code
WREN	Set Write Enable Latch	0000 0110b
WRDI	Reset Write Enable Latch	0000 0100b
RDSR	Read Status Register	0000 0101b
WRSR	Write Status Register	0000 0001b
READ	Read Memory Code	0000 0011b
WRITE	Write Memory Code	0000 0010b

Fig. 8.29 Power-on, power-off protects for parallel parts (a) [51] and serial parts (b) [52]

tion flag up and never writes before feeding with a particular sequences. However, an illegal read cycle shorter than that required for the restore operation also could destroy data. And those illegal short cycles can occur at the onset of asynchronous reset signal. Therefore, it is necessary to add a pull-up resistor to /CS or control pins of FeRAM to VDD and check MCU specifications during or at the onset of reset signal when using discrete FeRAMs. For the embedded FeRAM macro, things are much the same and identical care should be taken.

8.3.4 Circuit Challenges to Fatigue-Free Operation or Nondestructive Read Out

Circuit challenges for fatigue-free operation have been investigated to some degree. The goal here is for a non-destructive read out (NDRO) but an acceptable target is a non-switching read, which can be fatigue-free too but requires a restore operation.

The same result of reducing read voltage for a latch is seen in the 6T4C ferroelectric latch [53] in Fig. 8.30(a). This series connection of the ferroelectric capacitor adds appropriate capacitance at S1 and S2 nodes and each capacitor is applied appropriate higher voltage than if FC3 and FC4 were absent as 6T+2FE.cap. in Fig. 8.18. Or in other words, lower VDD operation for the recall is possible with this scheme. Except the recall and store sequences, PL1 and PL2 are fixed at 1/2 VDD to minimize the applied voltage to interrogate the FE caps, thus a non-switching operation results fatigue-free endurance can be achieved. However, this latch occupies at least as much area as an SRAM cell and additional control circuit for PLs is required. So it is suitable for small bit counts required in high-reliability applications.

Figure 8.30(b) shows operation of non-switching read out [54]. After precharging BLs to VDD and PL at GND, also TEL nodes are GND, then WL goes up and charge distribution from BL to TEL node generates BL voltage amplitude for corresponding U-term(0) or P-term(1). BL stray capacitance or precharge voltage are designed so as to apply less voltage than Vc for P-term capacitor, basically no polarization switching occurs and fatigue-free read time is achieved.

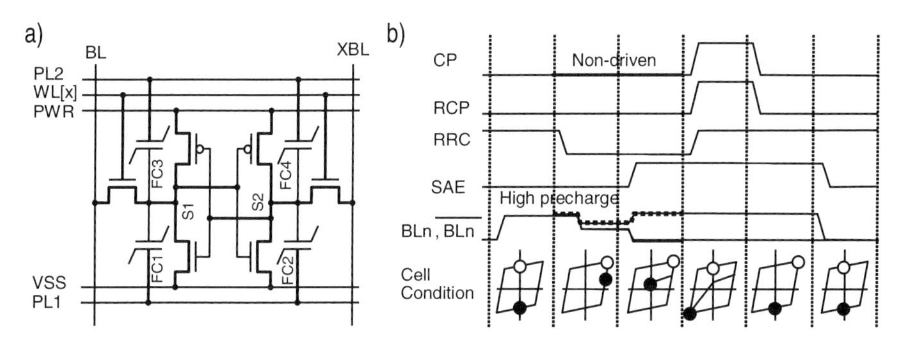

Fig. 8.30 (a) 6T4C ferroelectric latch [53] (© 2003 IEEE) and (b) non-switching READ scheme [54] (© 1997 IEEE)

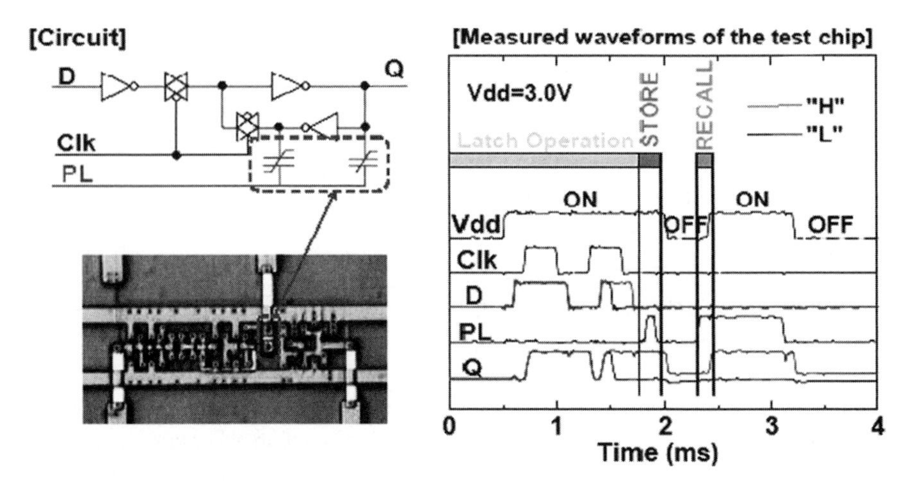

Fig. 8.31 Non-switching FE.cap. data latch schematic and test waveform ([55] © 2003 FED)

Figure 8.31 shows a ferroelectric capacitor base data latch with reduced polarization switching times [55]. It toggles normal PL level VDD or GND with every other power on–off cycle. Storing can be done with positive PL pulse or negative PL pulse (PL voltage decreases then stays at GND afterward power is turned off) depending on a particular normal PL level. During normal operation, both FE caps do not switch so there is no limitation for input switching times thus they are combined with normal logic library with a high-frequency clock. Only store and recall times enforce two cycles of polarization switching cycles but power on–off is not likely to occur up to 10^{12} cycles of the ferroelectric limitation.

The benefits of NDRO are fatigue-free operation and short read cycle time without restoring period and obtaining compatible read performance to FLASH. Figure 8.32 shows a realization of NDRO [56]. Since low voltage is applied to the capacitor it does not cause polarization switching, very small charge can be derived especially for P-term (b). Thus, Vt compensation technique and gain cell structure were employed (a) with a charge-detecting sense amplifier (c), relatively large cell, and slower operation speed. In the future, when BFO or other such high-switching charge materials are introduced into FeRAM, this type of circuit will be of more beneficial.

Although metal-ferroelectric-insulator-semiconductor type FET cells, which use spontaneous polarization electric field to control Vt of the channel seems to be the ideal NDRO ferroelectric cell, overcoming fabrication, write and read disturbs, and interface issues are currently insurmountable. Many attempts have been made to implement non-Si FET with a sandwiched ferroelectric material with some success [57]. However, how to integrate such non-Si FETs on Si VLSIs is still a challenge.

Figure 8.33(a) shows cell distributions for a ternary FeRAM [58] and (b) a 3b/cell NAND [59]. A multivalue storage cell is another topic of interest for smaller die or macro sizes. But previous attempts concluded that the cell Qsw distribution, con-

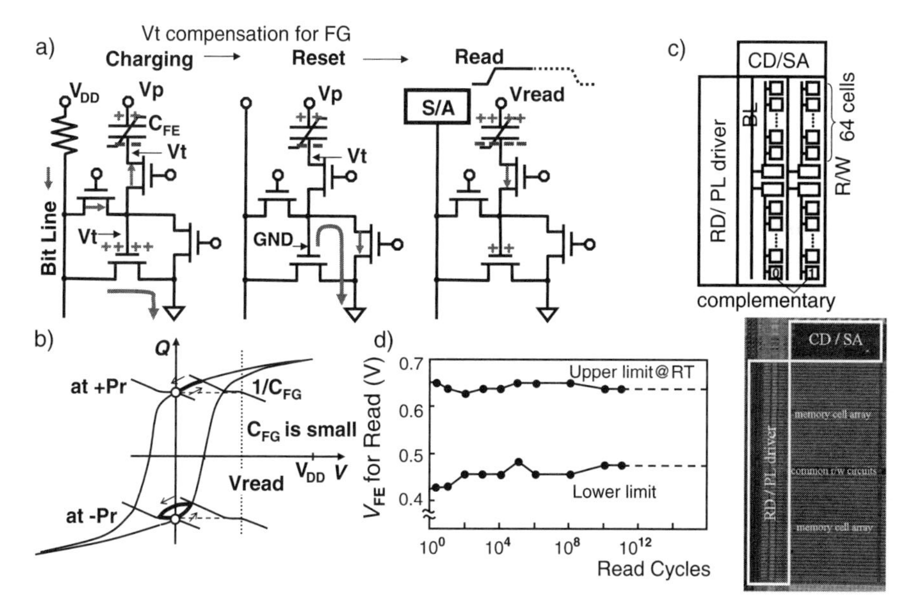

Fig. 8.32 Non-destructive read out (NDRO). (a) circuit, (b) hysteresis loop, (c) layout, (d) endurance results to 10^{11} cycles [56] (© 2005 IEEE)

sidering deviation of TEL area at the moment, is not sufficiently separated and it is not likely to operate even at ternary levels. FLASHs realized 3b/cell with series of pulse injection to setting the precise Vt. On the contrary, if FeRAM can do such trimming by pulse train control, not only the write cycle but also restoring time will be longer, and FeRAM does not differentiate like FLASH by differences in writing and reading speeds.

In addition to this section, sufficient circuit knowledge to realize many reading and writing architectures is evitable. However, the strategy and reason why a particular architecture shall be selected depends upon fully understanding the characteristics of your targeted ferroelectric capacitor where the top-level design architectural

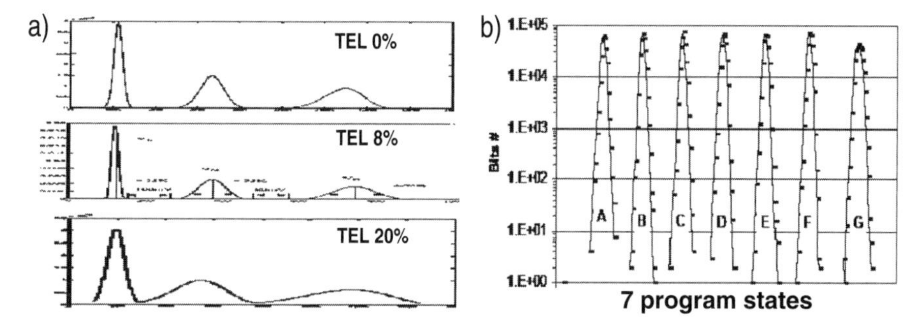

Fig. 8.33 **(a)** BL voltage distributions of a ternary FeRAM cells [58] (©2005 IEEE) and **(b)** Vt for 3b/cell NAND [59] (©2008 IEEE)

decision is the most important. In some cases, it is necessary to test two or three architectures because process development is on-going parallel to circuit design and the ferroelectric characteristics are not fixed. Electrical trimming of the timings, reference level, and even alternate NRTZ/RTZ or UP-only/UP-DOWN sensing eliminates designing many masks for the combination of options at the expense of adding rather complicated support circuits. Since FeRAM has inherent degradation factors, such as imprint and fatigue, reliability testing with 1000 h baking or 10^{10} cycles burn in will take 3 months or more and it will take additional time for testing many different architectures during process development phase. Stand-alone discrete FeRAM aiming for high density with a dedicated process technology and embedded FeRAM with logic-compatible processes forces selecting the appropriate cell structure and even architecture, as is the case for FLASH or DRAM.

8.4 Current FeRAM Markets

In this section, comparisons particularly with FLASH are discussed. As noted above, for low-density devices FeRAMs occupy less area than EEPROMs or FLASH thus it is cost competitive. However, due to larger FeRAM cell area, the Macro size is larger than that of EEPROM or FLASH for some Mbit class NV Macro. Thus, FeRAM is differentiated from other NV memories due to its unique characteristics such as tamper resistance, radiation hardness, and fast write cycle and other advantages depending upon a particular application.

Figure 8.34(a) shows a 4 Kbit embedded FeRAM [60] that occupies less area than previously allocated for an EEPROM macro. Since EEPROM and FLASH macros contain an internal voltage booster or a charge pump for generating high voltage to program and erase, this circuit area occupies a relatively large percentage if the cell area is small. On the contrary, FeRAM operates within VDD and no pumping circuit is involved. Thus, even if one cell footprint is large for FeRAM, macro size

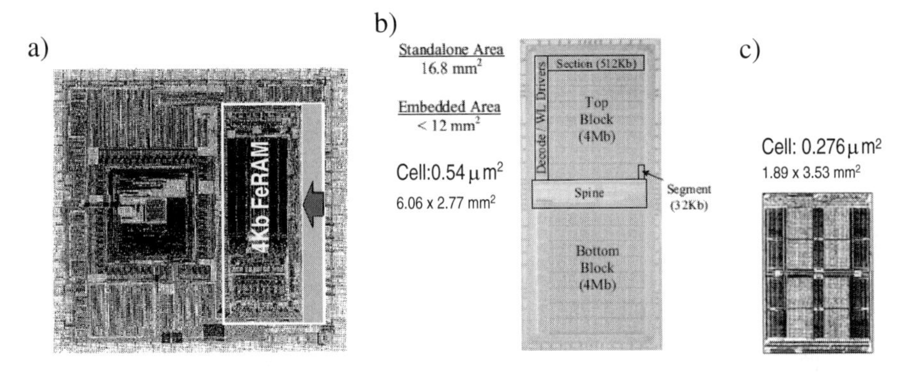

Fig. 8.34 FeRAM in small-density application (a) [60] (© 1996 IEEE) and comparison between FeRAM (b) [50] (© 2005 IEEE) and FLASH (c) [61] (©2004 IEEE) for large-capacity devices

	FUJITSU-FeRAM Product Sputter	FUJITSU-FeRAM Prototype MOCVD	EEPROM	FLASH	DRAM	SRAM
Read cycle	**120 ns**	**50 ns**	<20 ns	<10 ns	<2 ns	<1 ns
Read power/ MHz	**4 mW**	**1.3 mW**	0.5 mW	0.5 mW	1.5 mW	5 mW
Write cycle	**120 ns**	**50 ns**	10 ms	150 μs (erase 2 ms)	<2 ns	<1 ns
Write power/ MHz	**4 mW**	**1.3 mW**	1 mW	1 mW	1.5 mW	5 mW
Write Voltage	**3 V**	**1.8 V**	14 V	5.5 V	1.0 V	0.6 V
Write endurance	**10^{12}**	**10^{14}**	10^5	10^5	∞	∞
volatility	**NVolatile**		NVolatile	NVolatile	Volatile	Volatile
Cell size	**5.4 μm²**	**2.3 μm²**	0.5 μm²	0.016 μm²	0.13 μm²	0.25 μm²
Cell efficiency	**55 %**	**50%**	25%	50%	45%	38%
Technology	**(0.35 μm)**	**(0.18 μm)**	(0.35 μm)	(56 nm)	(65 nm)	(45 nm)

source: ISSCC, IEDM (Rounded off numbers)

Fig. 8.35 A table comparison of FeRAM to other memories features

is smaller. However, an 8 Mbit discrete chip (b) [50] is twice as large as an 8 Mbit (single bit per cell) FLASH (c) [61]. If those multi bit per cell architectures were adopted to FLASH, effective cell size would be approximately half. Thus, large-capacity memory is not yet a cost efficient trait for using FeRAM. However, there are unique advantages of FeRAM that can overcome this inherent area penalty due to the larger cell size.

Figure 8.35 compares FeRAM and other memories. Since other memories utilize much more advanced technology nodes, cell size and power consumption are smaller than FeRAM. However, if based on the same 0.35 μm technology, FeRAM is superior to both. Among nonvolatile memories, FeRAM is superior in write endurance and has faster overwrite speed. The following chart demonstrates these features.

Figure 8.36 shows FeRAM position in (a) capacity to write cycle time and (b) power consumption to write voltage charts. In Fig. 8.36(a), long write cycle time FLASHs are categorized as ROM, while other memories, read and write cycles of same order of time, are categorized as RAM. 2T2C or 1T1C FeRAM exhibit moderate speeds of 50–250 ns and up to 64 Mbit. Much faster approaches are 6T4C or 6T2C FeRAMs that are basically SRAM cells and achieve several nanoseconds cycle times at the expense of a larger cell area and thus limited capacity. Fig. 8.36(b) compares power consumption and write voltage needed for physical cell structure. Since FeRAM write and read are phenomena of polarization switching, only normal VDD voltage and no current for heating or generating magnetic fields are needed, thus FeRAM shows lower power consumption. While DRAM needs refreshing to compensate for charge loss, SRAM dissipates leakage current of MOSs, thus main-

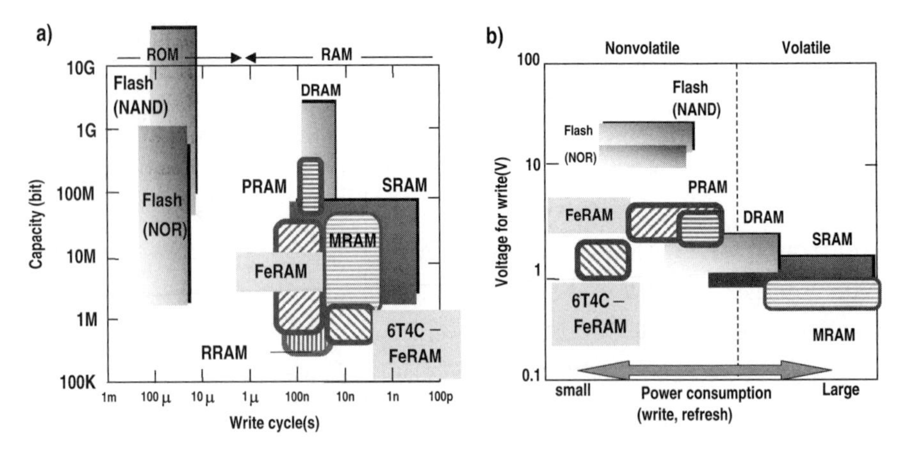

Fig. 8.36 FeRAM comparison to other memories in write cycle time (a) and power consumptions (b)

taining data voltage at the expense of nanometer MOS power consumption. Other NV memories including FeRAM do not dissipate leakage current in the MOSs. So FeRAM is superior due to shorter write cycle time and lower operating power. Other FeRAM unique advantages include tamper resistance and radiation hardness that conventional comparison of memories does not include are described in the following figures.

Security chips also usually require tamper resistance that make it very hard to do reverse engineering. Figure 8.37(a) shows ferroelectric surface that stores "0" and "1" by atomic force microscope, which shows no difference in topography; while charge stored in EEPROM or FLASH is relatively easy to distinguish. For example, scanning nonlinear dielectric microscopy (Fig 8.37(b) [62[MONOS][FG]]) distinguishes the charged and discharged state of FLASH cells. Although scanning probing microscopy can be used to contact a fine TEL node, it is difficult to measure small polarization switch charge by applying a rising pulse because of noise issues and difficulty in reverse engineering sample. Even after applying a pulse, the initially stored data will be overwritten. Thus reading an FeRAM cell makes it very difficult to determine stored data "0" or "1," so using FeRAM itself is also similar to adding another tamper resistance to a security chip.

Airport luggage is routinely exposed to X-rays at security checkpoints. Thanks to recent progress in lower radiation dosage, ISO 400 photographic film is not over exposed these days. But some sterilization methods use radiation exposure such as neutrons to kill microorganisms. In aerospace applications, where exposure to X-rays or neutrons is common, FeRAM is more suitable than FLASH. Because FLASH stores charge, which is likely to be disturbed by high-energy particles. Figure 8.38 shows X-ray dosage and loss of Qsw for SBT (a) [64] and PZT (c) [65]. Those measurements were made after doses that resemble TAG use and special care should be taken for a CMOS circuit if the chip is dosed during VDD on state.

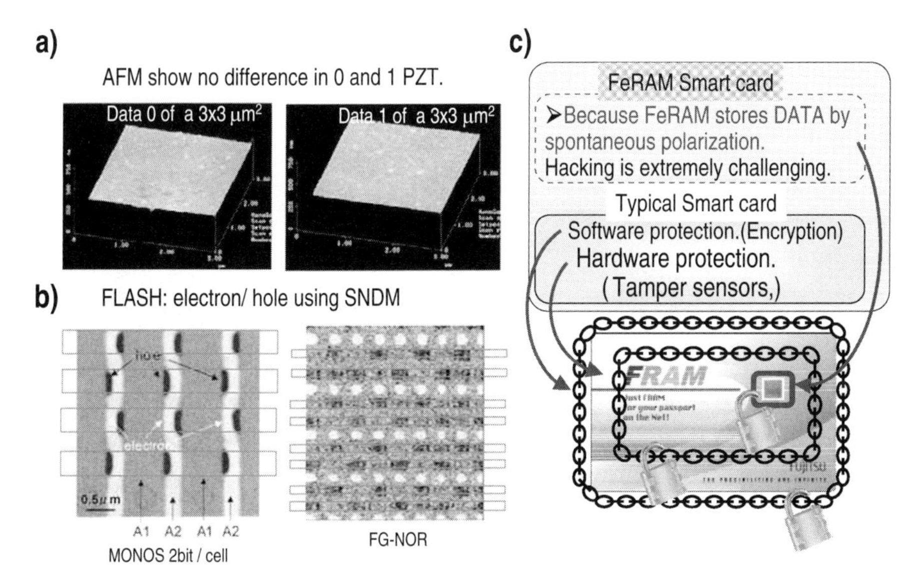

Fig. 8.37 FeRAM tamper resistance. (b) MONOS [62] ©2005 AIP and FG[63] ©2006 Nanotechnology

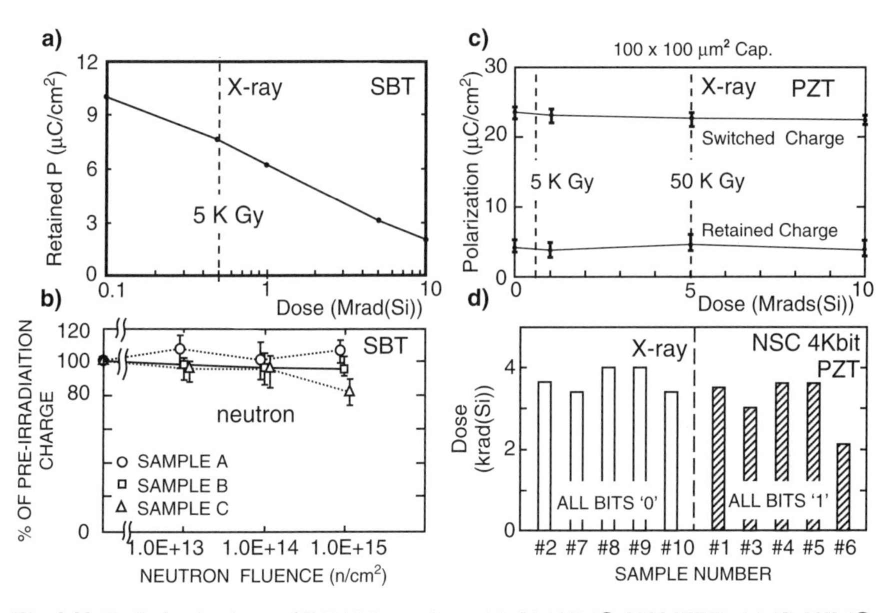

Fig. 8.38 Radiation hardness of FeRAM capacitors. (**a**) (**b**) [64] (© 2000 IEEE), (**c**) (**d**) [65] (© 1991 IEEE)

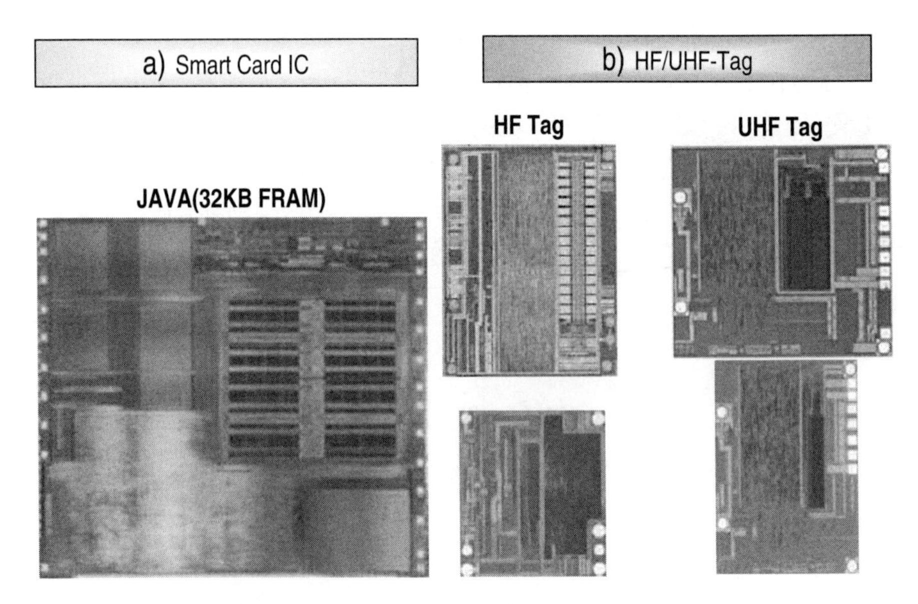

Fig. 8.39 Embedded FeRAM smartcard (a) and TAGs (b)

Figure 8.39 (**a**) shows a smartcard (MCU embedded) and (**b**) TAGs (logic embedded) containing FeRAM macros. The fast writing cycle of FeRAM is of most benefit for applications that handle a large DATA capacity and at high-speed communication that results in a shorter download time. FLASH can write by 2 K bytes block buffer to enhance data rate speed for a file application, but it needs a longer erasing time when overwriting occurs. In contrast, FeRAM can be overwritten address by address in the same cycle time as during reading. For example, smartcard performs JAVA applications that are downloaded to FeRAM. For the radio frequency (RF) passive TAGs with relatively small DATA storage capacity FeRAM, has lower power consumption operation both for reading and writing compared to EEPROM. The low power consumption enables a longer distance communication. Also small-capacity (<256 K bit) FeRAM macro is smaller in area than EEPROM that employs a charge pump circuit.

Small bit FeRAM authentication chips are more cost effective than embedded EEPROM or FLASH. Figure 8.40 shows an example of a rechargeable battery pack, which communicates with the equipment and a battery charger to avoid explosion due to over heating.

Figure 8.41 shows examples of utilizing fast write cycle time of FeRAM. An FeRAM TAG is five times faster in communication with a reader/writer when the protocol is tuned for FeRAM. Using a discrete serial FeRAM for setting trimming data in a production line will shorten presetting data writing time, thus more products can be produced.

In conclusion of this section, table comparison of current FeRAM and other memories are shown. It is different from what was predicted in the 1990s when

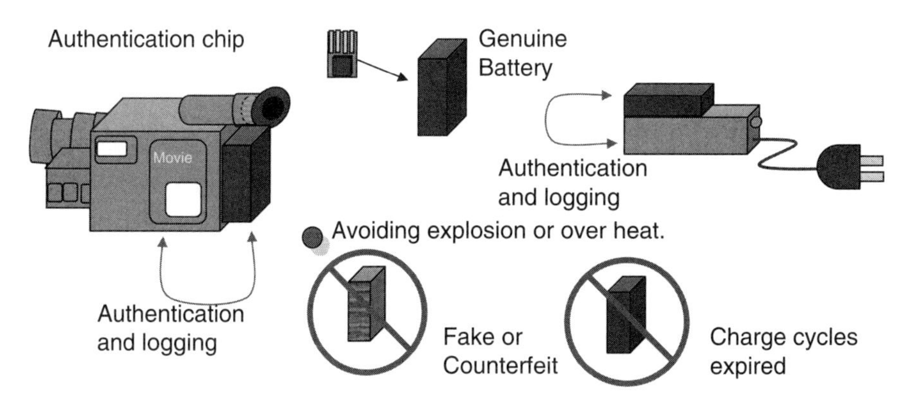

Fig. 8.40 Authentication chip and example of use for use in a battery pack.

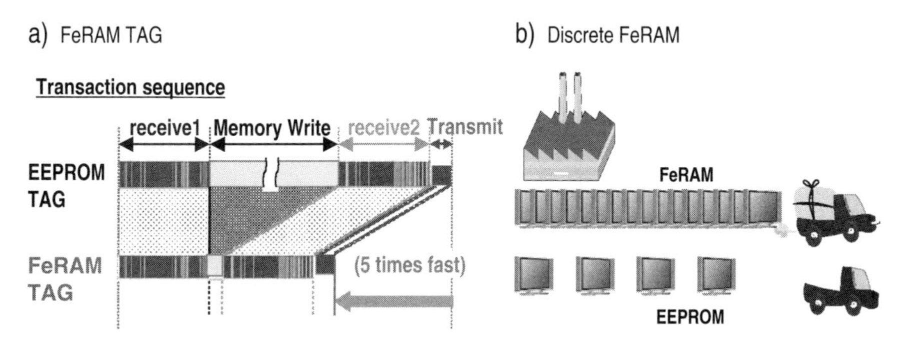

Fig. 8.41 Advantages of FeRAM due to faster write cycle than EEPROMs

the first FeRAM boom occurred and FeRAM was proposed as an ideal or universal memory that would compete with all the memories. Since technology scaling has evolved more quickly than expected, FLASH technology has broken through 0.18 μm technology node owing partially to circuit architectural improvements. However, FeRAM is still alive in niche markets for the reason of its unique characteristics. Though it has taken much more time to develop than has been expected, further improvements are expected resulting in larger memory capacity and higher-endurance applications. Game cartridges actually pulled FeRAM into mass production, authentication chips are a new market, and RF-TAGs or Smartcards are excellent applications for FeRAM taking advantage of fast overwrite, small macro size, and small power consumption, although the market for those RF applications has developed much slower than was expected. It should be also noted that costs concerns are also an issue with large-capacity discrete FeRAM compared to more traditional memories.

8.5 Low-Voltage Challenges and Future Trends

Since embedded memories are usually directly connected to the logic circuit, FeRAM that use same VDD as logic library is preferable. This requires ferroelectric characteristics of Vc to be around 1/2 VDD. So with each generation of technology, 5 V (0.5 μm) to 3 V (0.35 μm) and to 1.8 V (0.18 μm) or future 1.0 V (90 nm), a new ferroelectric capacitor has been developed. In addition, circuit-side improvements for applying higher voltage to the ferroelectric capacitor are required. CMOS thermal budgets are lowered for each technology generation and lower temperature re-crystallization ferroelectric materials are desired.

The ITRS roadmap for FeRAM in 2007 was based upon corporate surveys of FeRAM manufactures. Published papers in the literature seem to use two or three times larger Qsw cell than required based upon the minimum characteristics in the ITRS FeRAM roadmap which are derived from DRAM capacitance characteristics. However, their access time and cycle time trends are almost identical with the roadmap.

Finally, new promising FeRAM applications consist of FeRAM-MCU and SSD (solid state drive using FLASH) controller or SD-card controller are mentioned.

8.5.1 Low-Voltage Challenges

As process technology nodes continue further scaling, LSI operating supply voltage is simultaneously reducing. FeRAM functionality at low-voltage operation mainly depends on a ferroelectric capacitor having low Vc, cell uniformity, and a slow rate of time-dependent degradation toward end of life. Although these are mainly process issues, support from the circuit side by design or architecture is also needed.

Further process improvements are expected to lead to Vc reduction by optimization of the electrode interfaces, thinner ferroelectric films, or Ec reduction by modifying crystal structure. Among them, reducing film thickness is technically the easiest route but with an unexpected side effect of increasing the leakage current. Thus, doping control or two-step crystallization [66] to reduce the leakage current are possible solutions. Interface of the electrode and ferroelectric show Schottky junction characteristics. Thus, Nb doping to PZT or SBT will change the flat band of the ferroelectric and also electrode material choice of Pt, IrOx, or SRO will affect the interface junction voltage. Ec or Vc reduction in the film may also be achieved by controlling the columnar grain size or lattice size by doping, Zr/Ti ratio, or fabricated stress and crystal orientation by seed electrode orientation and lattice size. There are many possibilities for improvements, however, what we can practically control is limited when preparing devices and always side effects will occur. Parameters that can be easily controlled are crystallization temperature, crystallization ambient, deposition rate, multistep deposition, doping elements and quantity, Zr/Ti ratio, electrode material, electrode material lattice dimensions, interface morphology, and physical thickness of ferroelectric material. In general, there is trade-off

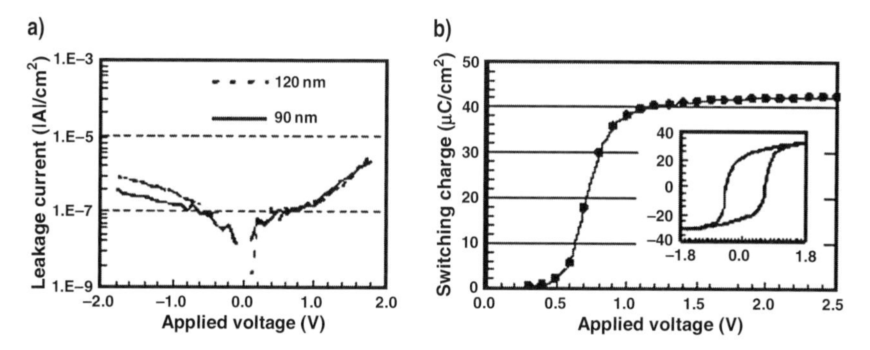

Fig. 8.42 (a) Leakage characteristics of a PZT capacitor and (b) polarization versus voltage plot showing saturation at 1 V [67] (© 2005 IEEE)

between obtaining low Vc with low-leakage current. If leakage current over all cells is uniform, leakage on the order of mA/cm^2 is acceptable for some 10 ns pulse operation, but in reality leakage pathways are dominated by particular cells and averaged leakage values of μA/cm^2 are more acceptable.

An example of a 90 nm thick PZT capacitor [67] is shown in Fig. 8.42. Leakage current is on the order of μA/cm^2, 2Pr of more than 40 μC/cm^2 and Vc of 0.75 V were demonstrated. Also lateral size reduction in the ferroelectric area was shown and the figures are for a 0.45×0.45 μm^2 capacitor array. Thus, even with PZT there is further potential to double the nominal value of Qsw and lower Vc of about one half compared to current production level FeRAM capacitors. These are examples of possible process solutions for low-voltage operation challenges.

In addition, circuit solutions for low-voltage operation challenges are also proposed as in Fig. 8.43. (a): The BGS [21] keeps the BL voltage level at GND and can apply as much voltage as PL level directly to the reading capacitor. The LoZ BL sensing itself is a method for applying higher voltage, moreover when PL boost is adopted, all the boosted voltage is applied to the ferroelectric capacitors. Thus, effectiveness of PL boosting is larger with BGS than conventional HiZ BL scheme. Fig. 8.43(b) [69] for HiZ BL is a method that a coupling capacitor pushes BL level lower for both P-term and U-term BLs then the reading capacitor will experience a higher voltage by the portion that was pushed down. In this method, the voltage sensing will start while "over drive" or BL are being pushed down. In addition, this was applied to a chain FeRAM and the selected WL goes GND during accessing, similar to NAND-FLASH WLs. In Fig. 8.43(c) "zero cancellation" [50] is another BL push-down scheme for HiZ BL. In the method, sensing begins after returning drive pulse that pulls up BL levels. This method seems robust because the mismatch of BL coupling capacitors will be cancelled. In addition, this figure shows "after (PL) pulse" or UP/DOWN sensing timing. For those HiZ BL and push-down schemes, since U-term BLs level must be higher than GND, it implies limitation in increase of P-term capacitor readout voltage.

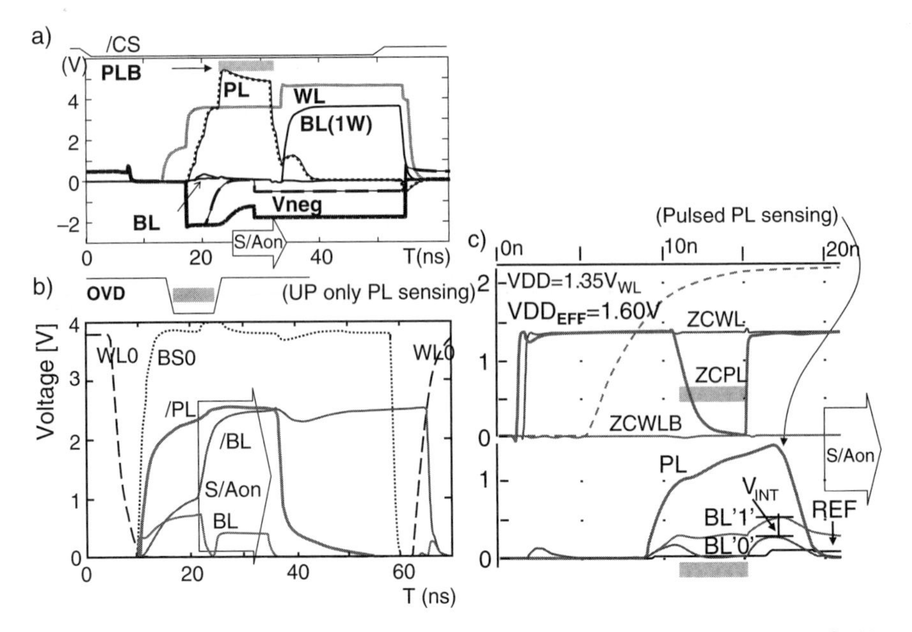

Fig. 8.43 Low-voltage operation circuit proposals for applying higher voltage. (a) [21] (©2007 IEICE), (b) [68] (© 2001 IEEE), and (c) [50] (© 2005 IEEE)

8.5.2 Future Trends

The ITRS 2007 [69] FeRAM roadmap shows more realistic trends rather than the catch up to DRAM story of ITRS 2005. The following two figures compare ITRS 2007 and data from published papers in storage charge trends and in access speed, respectively. Film process methods are sputter or sol-gel spin coating (CSD) and ferroelectric materials are PZT and SBTN for 0.25 μm and 0.35 μm nodes; 0.18 μm and 0.13 μm nodes are utilizing MOCVD process because with its in-situ crystallization thinner ferroelectric films with enhanced properties can be obtained. MOCVD films are mainly nondoped PZT and BiT. Beyond 90 nm nodes, also MOCVD is necessary and 3D capacitors with concave or cylinder shape for reducing cell footprint will be a realistic choice while using PZT or BiT. On the other hand, introducing a new material such as BFO or BFCO that have more than 100 $\mu C/cm^2$ Qsw and still using a simple stack planar capacitor is another option.

Figure 8.44 compares cell switching charge trends of ITRS and published papers for each (half or gate length) technology node. Not only for Fujitsu products but also demonstrated chips are designed with two or three times larger cell Qsw than ITRS charge model predicted. The ITRS maybe assuming a charge model that is derived from DRAM, for example, charge is derived from a simple Cs/Cb expression. In reality, ferroelectric Qsw has a distribution among cells due to process variations, thermal history, and time-dependent reduction by imprint or fatigue. Thus we need larger Qsw margins to account for time-dependent charge loss. A relatively new thin

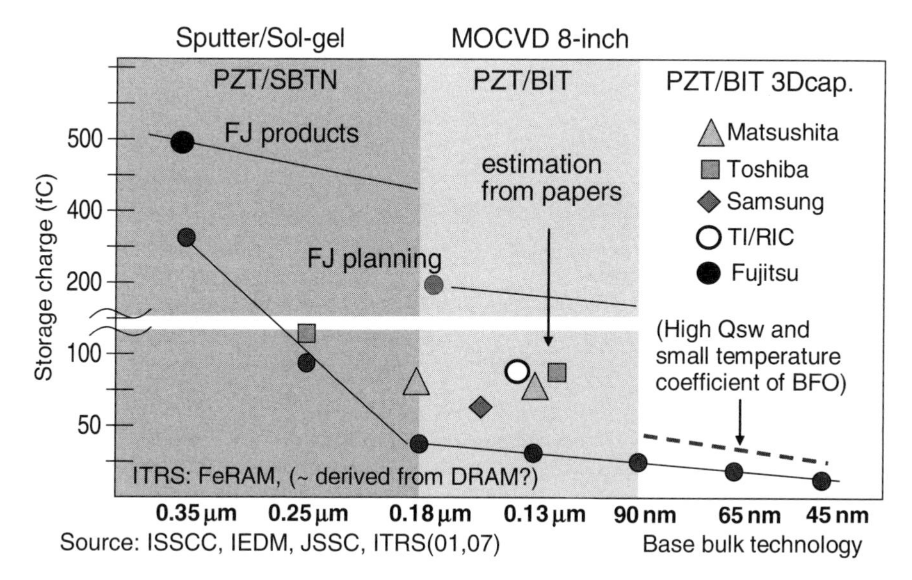

Fig. 8.44 ITRS capacitor storage charge trends for FeRAM cell

film material BFO shows less temperature dependence in Qsw because its Curie temperature is above 820°C, twice that of PZT. So introducing such a material will likely reduce the device gap compared to the roadmap depending upon its reliability characteristics.

Figure 8.45 plots access times and cycle times. Published experimental data are very close to the roadmap under typical voltage and temperature (V–T) conditions. But products show about two times slower values since they are under slowest case V–T conditions and is often the case for other memories. As a whole, access time greater than 10 ns are easy circuit design issues depending on the process node speed.

Figure 8.46 shows new material candidates (a) [70] and electrical characteristics of a BFO(b) (c) [71]. Fig. 8.46(a) shows the plotted relationship between Tc and Pr. Roughly, a higher Tc material results in a higher Pr, although they are for single crystal data and thin films vary to some degree. BFO single crystal films have Qsw > 100 $\mu C/cm^2$ at room temperature, Tc of 820°C, and anneal temperature of 550°C. Matsushita suggests BiT [72] as a successor of SBTN. BiT is said to have Qsw of 25–100 $\mu C/cm^2$, Tc of 700°C, and anneal temperature of 500°C. These large Qsw and low-temperature growth are attractive for more advanced CMOS. BFO shows Vc of 3 V and showed Qsw more than 100 $\mu C/cm^2$ (b), but leakage current is still three orders of magnitude higher than PZT (c). Lowering Vc while keeping leakage current on the orders of $\mu A/cm^2$ is a material-side challenge.

Figure 8.47 shows examples of potential FeRAM applications with regards to bit density and endurance cycle. Embedded FeRAMs typically started with smartcards or TAGs, because endurance was around 10^{10} cycles and integration

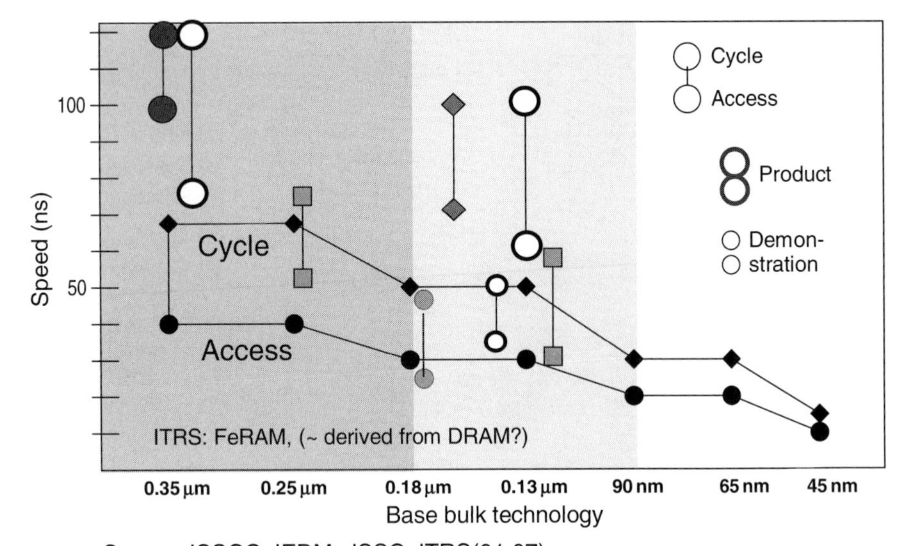

Fig. 8.45 ITRS operation speed trends for FeRAM

density was less than 256 Kbit. Originally, discrete FeRAMs targeted replacement of EEPROMs due to the fast write cycle of FeRAM. As the endurance improved to 10^{12} with the process improvements, application area has increased further because FeRAM has achieved what may be described as "unlimited write cycles." Thus replacement of battery backup SRAM is now possible. Next-generation FeRAM technology will use MOCVD processing for deposition of the ferroelectric film and the endurance is expected to reach 10^{15}. In that case, we can truly say that FeRAM has "infinite write cycles" comparable to SRAMs or DRAMs. In the case of no limitation for endurance, a unified FeRAM memory can be embedded with a MCU thus replacing SRAM, ROM, and FLASH macros. This also allows for greater

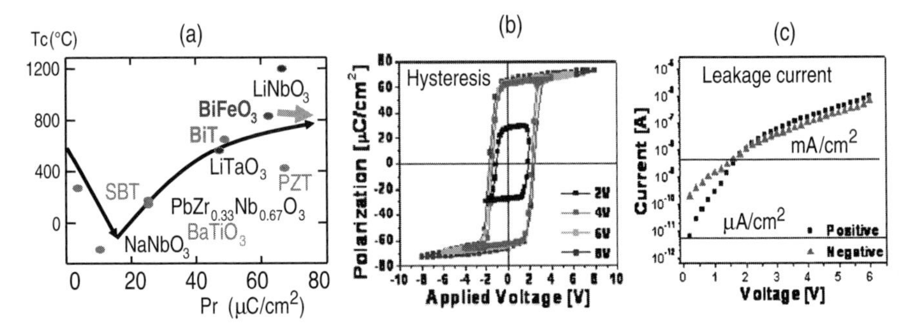

Fig. 8.46 New ferroelectric materials (a) material Curie temperature relationship, (b) BFO thin film capacitor data polarization loop and (c) leakage data (b) (c) [71] ©2007 AIP

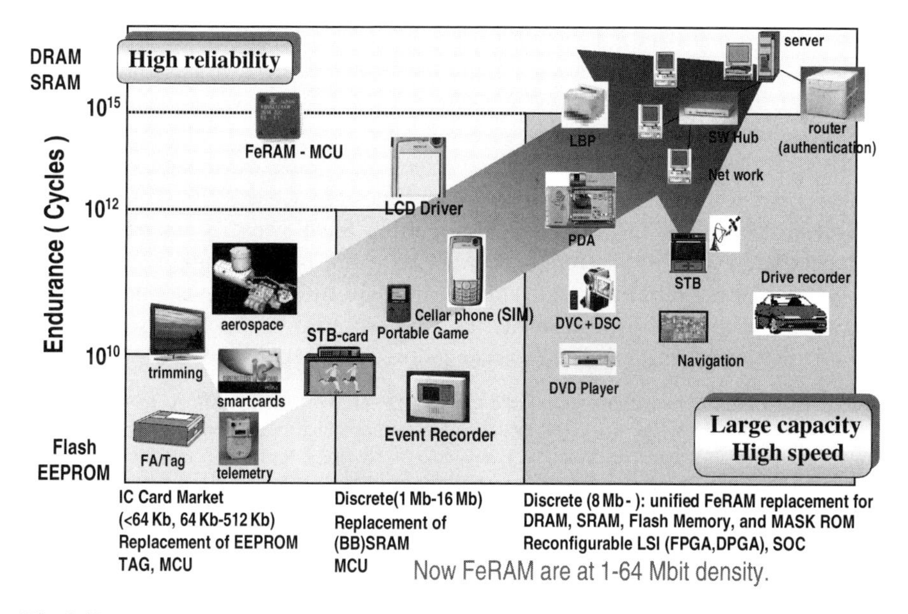

Fig. 8.47 FeRAM growth in applications with increasing density and fatigue endurance

flexibility in memory allocation in the MCU. Reconfigurable LSIs such as field programmable gate array (FPGA) or dynamic PGA (DFPGA) using FeRAM are also of interest. Due to high cost in advanced technology masks, not all the ASICs can be developed with cutting-edge technology. Therefore designs will favor high volume mass production with advanced technology. In our opinion, fewer types of hardware chips will be produced then customized. For example, flexible FeRAM memory partitioning with software and some reconfigure hardware to allow multiple applications with low costs. Recently developed applications complement FLASH usage, such as SD-card or solid state drive (SSD) controller that were not even predicted 10 years ago. Full FeRAM implementation in a SSD is not yet cost effective but composing it mainly by FLASH and utilizing 10^{15} endurance of a few FeRAM cash seems attractive. Greater FeRAM penetration into the PC, cellar phone, and network communication products is anticipated with 10^{15} endurance and more Mbit capacity. However, there is a 1000 times memory density gap comparing FLASH to FeRAM. So FeRAM will not directly compete with cutting-edge high-density FLASH but again differentiate by fast overwrite cycle, 10^{15} endurance, and random access.

In summary of the future trends, process and circuit issues will continue to advance. Process issues such as 90 nm node and advanced technologies, which utilize 300 mm wafers, and ferroelectric deposition with MOCVD will require further development. MOCVD equipment capable to handle 300 mm diameter wafer is needed as well as lowering the ferroelectric crystallization temperature to cope with lower temperature of nm CMOS processes. Further improvements in reliability issues especially fatigue endurance to achieve 10^{15} cycles that is comparable to SRAM, and data retention 100 years at 125°C competitive to FLASH are needed. One potential solution is to change the ferroelectric material to BFO, which shows Qsw of >100 μC/cm^2, less temperature sensitivity Qsw, and below 550°C low crystallization temperature. Regarding circuit issues, developments in multi-level cells for higher-density FeRAM and achieving Non Destructive Read Out both for faster cycle time and for infinite reading times are needed. But these features completely depend on the ferroelectric properties. Small-capacity embedded FeRAMs and large-capacity discrete FeRAMs may differ in application requirements therefore requiring different architectures. There are still many development issues for increasing market share in memories, thus we need greater understanding of ferroelectric materials and related science tailored for FeRAM use.

Acknowledgments The authors would like to thank Dr. H. Arimoto and Mr. A. Inoue for their encouragement regarding the writing of this chapter, also appreciation is extended to Dr. T. Eshita, Dr. H. Tanaka, Dr. J. Watanabe, Dr. K. Sugiyama, and Mr. K. Takai for many useful discussions on material properties as well as the circuit and process engineers of Fujitsu Micro Electronics Ltd. (FML) for realizing BGS devices.

References

1. L. Moll and Y. Tarui, "A new solid state memory resistor," IEEE Trans. Electron Devices, vol. 10, no. 6, p. 338, Sept. 1963.
2. J.T. Evans and R. Womack, "An experimental 512-bit nonvolatile memory with ferroelectric storage cell," IEEE J. Solid-State Circuits, vol. 23, no. 5, pp.1171–1175, Oct. 1998.
3. A. Sheikholeslami and P. Gulak, "A survey of circuit innovations in ferroelectric random-access memories," Proc. IEEE, vol. 88, no. 5, pp.667–789, May 2000.
4. J. F. Scott, "Ferroelectric Memories," Advanced Microelectronics, Publisher: Springer-Verlag Berlin and Heidelberg GmbH & Co.KG, 2000.
5. H. Ishiwara, M. Okuyama, and Y. Arimoto (Eds,) "Ferroelectric random access memories," Fundamentals and applications, Topics in applied physics, vol. 93, Publisher: Springer-Verlag Berlin and Heidelberg GmbH & Co. KG, 2004.
6. J. F. Scott, "Limitations on ULSI-FeRAMs," IEICE Trans. Electron., vol. E81-C, no. 4, pp.477–487, Apr. 1998.
7. S. Kimura http://sunbeam.spring8.or.jp/top/seika/ohp/2001/nec.pdf : Japanese language
8. http://www.murata.co.jp/izumomurata/trend/index4.html Japanese language
9. A. Onodera, S. Mouri, M. Fukunaga, S. Hiramatsu, M. Takesada, and H. Yamashita, "Phase transition in Bi-layered oxides with five perovskite layers," Jpn. J. Appl. Phys., vol. 45, no. 12, pp. 9125–9128, Dec. 2006.
10. H. Yamashita, K. Yoshio, W. Murata, and A. Onodera, "Structural changes and ferroelectricity in Bi-layered SrBi2Ta2O9," Jpn. J. Appl. Phys., vol. 41, pt. 1, no. 11B, pp. 7076–7079, 2002.
11. M.G. Stachiotti, C.O. Rodriguez, C. Ambrosch-Draxl, N. E. Christensen, "Electronic structure and ferroelectricity in SrBi2Ta2O9," Phys. Rev., B61, 14434, 2000. (http://prola.aps.org/abstract/PRB/v61/i21/p14434_1).
12. E. C. Subbarao, "Ferroelectricity in Bi4Ti3O12 and its solid solutions," Phys. Rev., vol. 122, no. 3, pp. 804–807, 1961. (http://prola.aps.org/abstract/PR/v122/i3/p804_1)
13. Ying-Hao Chu, Qian Zhan, L. W. Martin, M. P. Cruz, Pei-Ling Yang, G. W. Pabst, F. Zavaliche, Seung-Yeul Yang, Jing-Xian Zhang, Long-Qing Chen, D. G. Schlom, I.-Nan Lin,Tai-Bor Wu, and R. Ramesh, "Nanoscale domain control in multiferroic $BiFeO_3$ thin films," Adv. Mater., vol. 18, pp. 2307–2311, 2006. Fig. 1(a):Copyright Wiley-VCH Verlag GmbH & Co. KGaA. Reproduced with permission.

14. http://www.jfe-mineral.co.jp/e_mineral/seihin/eseihin034.html
15. Y. Hosono and Y. Yamashita, "High-efficiency piezoelectric single crystals", Toshiba review, vol. 59, no. 10, pp.39–42, 2004 : Japanese language.
16. http://www.hst.titech.ac.jp/~meb/Ceramics/speaker/speaker.html. : Japanese language
17. Y. H. Chu, T. Zhao, M. P. Cruz, Q. Zhan, P. L. Yang, L. W. Martin, M. Huijben, C. H. Yang, F. Zavaliche, H. Zheng, and R. Ramesh, "Ferroelectric size effects in multiferroic $BiFeO_3$ thin films," Appl. Phys. Lett., vol. 90, 252906, 2007.
18. A. Sheikholeslami, P.G. Gulak, H. Takauchi, H. Tamura, H. Yoshioka, and T.Tamura, "A pulse-based, parallel-element macromodel for ferroelectric capacitors," IEEE Trans. Ultrasonics, Ferroelectrics and Frequency Control, vol. 47, no. 4, pp. 784–791, Jul. 2000.
19. J. Chow, A. Sheikholeslami, J. S. Cross, and S. Masui, "A voltage-dependent switching-time (VDST) model of ferroelectric capacitors for low-voltage FeRAM circuits," Digest of Technical Papers, Symp. VLSI Circuits, pp. 448–449, Jun. 2004.
20. A. Gruverman, B. J. Rodriguez, C. Dehoff, J. D. Waldrep, A. I. Kingon, and R. J. Nemanich, "Direct studies of domain switching dynamics in thin film ferroelectric capacitors," Appl. Phys. Lett., vol. 87, 082902, 2005.
21. S. Kawashima, I. Fukushi, K. Morita, K. Nakabayashi, M. Nakazawa, K.Yamane, T. Hirayama, and T. Endo, "A reliable 1T1C FeRAM using a thermal history tracking 2T2C dual reference level technique for a smart card application chip," IEICE Trans. Electron., vol. E90-C, no. 10, pp. 1941–1948, Oct. 2007.
22. F. Chu, G. Fox, and T. Davenport, "Scaled PLZT thin film capacitors with excellent endurance and retention performance," Proc. MRS, Ferroelectric Thin Films IX, Symposium CC , vol. 655, CC1.2, 2001.
23. Y. Shimada, A. Inoue, T. Nasu, Y. Nagano, A. Matsuda, K. Arita, Y. Uemoto, E. Fujii, and T. Otsuki, "Time-dependent leakage current behavior of integrated $Ba_{0.7}Sr_{0.3}TiO_3$ thin film capacitors during stressing," Jpn. J. Appl. Phys., vol. 35, pt. 1, no. 9B, pp. 4919–4924, Sep. 1996.
24. T. Nakamura, Y. Nakao, A. Kamisawa, and H. Takasu, "Preparation of $Pb(Zr,Ti)O_3$ thin films on Ir and IrO_2 electrodes," Jpn. J. Appl. Phys., vol. 33, pt. 1, no. 9B, pp. 5207–5210, Sep. 1994.
25. K. Arita, Y. Shimada, Y. Uemoto, S. Hayashi, M. Azuma, Y. Judai, T.Sumi, E. Fujii, T. Otsuki, L.D. McMillan, and Carlos A. Paz de Araujoz, "Ferroelectric nonvolatile memory technology with bismuth layer-structured ferroelectric materials," IEEE Proc., Tenth Inter. Symp. on Appl. Ferroelectrics, ISAF '96., vol. 1, pp.13–16, Aug. 1996.
26. X. J. Lou, M. Zhang, S. A. T. Redfern, and J. F. Scott, "Fatigue as a local phase decomposition: A switching-induced charge-injection model," Phys. Rev. B vol. 75, 224104, 2007.
27. A. K. Tagantsev, I. Stolichnov, E. L. Colla, and N. Setter, "Polarization fatigue in ferroelectric films: Basic experimental findings, phenomenological scenarios, and microscopic features," J. Appl. Phys., vol. 90, no. 3, pp. 1387–1402, Aug. 2001.
28. S. Aggarwal, S. R. Perusse, C. W. Tipton, R. Ramesh, H. D. Drew, T. Venkatesan, D. B. Romero, V. B. Podobedov, and A. Weber, "Effect of hydrogen on $Pb(Zr,Ti)O_3$-based ferroelectric capacitors," Appl. Phys. Lett.,vol. 73, no. 14, pp. 1973–1975, Oct. 1998.
29. J.S. Cross and M. Tsukada, "Degradation of PLZT capacitors at elevated temperature in deuterium gas accession," Trans. Mater. Res. Soc. Jpn., vol. 28, no. 1, pp.117–120, 2003.
30. Y. Nagano, T. Mikawa, T. Kutsunai, S. Natsume, T. Tatsunari, T. Ito, A. Noma, T. Nasu, S. Hayashi, H. Hirano, Y. Gohou, Y. Judai, and E. Fujii, "Embedded ferroelectric memory technology with completely encapsulated hydrogen barrier structure," IEEE Trans. Semiconductor Manufacturing, vol. 18, no. 1, pp. 49–54, Feb. 2005.
31. E. M. Philofsky, "FRAM-the ultimate memory," Sixth Biennial IEEE International Nonvolatile Memory Tech. Conf. 1996. , pp.99–104, Jun. 1996.

32. June-Mo Koo, Bum-Seok Seo, Sukpil Kim, Sangmin Shin, Jung-Hyun Lee, Hionsuck Baik, Jang-Ho Lee, Jun Ho Lee, Byoung-Jae Bae, Ji-Eun Lim, Dong-Chul Yoo, Soon-Oh Park, Hee-Suk Kim, Hee Han, Sunggi Baik, Jae-Young Choi, Yong Jun Park, and Youngsoo Park, "Fabrication of 3D trench PZT capacitors for 256Mbit FRAM device application," IEEE Internatinal Electron Devices Meeting 2005. Tech. Digest., pp. 351–354, Dec. 2005.
33. Y. Kato, Y. Kaneko, H. Tanaka, K. Kaibara, S. Koyama, K. Isogai, T. Yamada, and Y. Shimada, "Overview and future challenge of ferroelectric random access memory technologies," Jpn. J. Appl. Phys., vol. 46, pt. 1, no. 4B, pp. 2157–2163, 2007.
34. T. Ema, S. Kawanago, T. Nishi, S. Yoshida, H. Nishibe, T. Yabu, Y. Kodama, T. Nakano, M. Taguchi, "3-dimensional stacked capacitor cell for 16 and 64 M DRAMs," IEEE International Electron Devices Meeting 1988, IEDM ′88. Tech. Digest., pp.592–595, 1988.
35. D. Takashima and I. Kunishima, "High-density chain ferroelectric random access memory (Chain FRAM)," IEEE J. Solid-State Circuits, vol. 33, pp. 787–792, May 1998.
36. K. Hoya, D. Takashima, S. Shiratake, R. Ogiwara, T. Miyakawa, H. Shiga, S. Doumae, S. Ohtsuki, Y. Kumura, S. Shuto, T. Ozaki, K. Yamakawa, I. Kunishima, A. Nitayama, and S. Fujii, "A 64 Mb Chain FeRAM with quad-BL architecture and 200 MB/s burst," Dig. Tech. Papers 7.2, IEEE International Solid-State Circuits Conference, pp. 459–466, Feb. 2006
37. H. Kanaya, K. Tomioka, T. Matsushita, M. Omura, T. Ozaki, Y. Kumura, Y. Shimojo, T. Morimoto, O. Hidaka, S. Shuto, H. Koyama, Y. Yamada, K. Osari, N. Tokoh, F. Fujisaki, N. Iwabuchi, N. Yamaguchi, T. Watanabe, M. Yabuki, H. Shinomiya, N. Watanabe, E. Itoh, R. Tsuchiya, K. Yamakawa, K. Natori, S. Yamazaki, K. Nakazawa, D. Takashima, S. Shiratake, S. Ohtsuki, Y. Oowaki, I. Kunishima, and A. Nitayama, " A 0.602 μm^2 nestled ′Chain′ cell structure formed by one-mask etching process for 64 Mbit FeRAM", IEEE Dig. Tech. Papers, Symp. VLSI Technology, pp. 150–151, Jun. 2004.
38. http://www.edn.com/archives/1997/041097/08df_03.htm (1997.)
39. S. Deleonibus (EDT), "Electronic device architectures for the nano-CMOS era," : From ultimate CMOS scaling to beyond CMOS device : Publisher: World Scientific Pub. Co. Inc. to be published 2008/08, -US-ISBN:9789814241281, in "FRAM and MRAM" chaptor by Y. Arimoto.
40. H.P. McAdams, R. Acklin, T. Blake, Xiao-Hong Du, J. Eliason, J. Fong, W. F. Kraus, D. Liu, S. Madan, T. Moise, S. Natarajan, N.Qian, Y. Qiu, K. A. Remack, J. Rodriguez, J. Roscher, A. Seshadri, and S. R. Summerfelt, "A 64-Mb embedded FRAM utilizing a 130-nm 5LM Cu/FSG logic process," IEEE J. Solid-State Circuits, vol. 39, no. 4, pp. 667–677, Apr. 2004.
41. J. Siu, Y. Eslami, A. Sheikholeslami, P. Gulak, T. Endo and S. Kawashima, "A 16 kb 1T1C FeRAM testchip using current-based reference scheme," IEEE Proc. 2002 Custom Integrated Circuits Conference, 7.3.1, pp. 107–110, May 2002.
42. S. Kawashima, T. Endo, A. Yamamoto, K. Nakabayashi, M. Nakazawa, K. Morita, and M. Aoki, "Bitline GND sensing technique for low-voltage operation FeRAM," IEEE J. Solid-State Circuits, vol. 37, no. 5, pp. 592–598, May 2002.
43. B.-J.Min, K.-W. Lee, H.-J. Lee, S.-R. Kim, S.-G. Oh, B.-G. Jeon, H.-H. Yang, M.-K. Kim, S.-H. Cho, H. Cheong, C. Chung, and K. Kim, "An embedded non-volatile FRAM with electrical fuse repair scheme and one time programming scheme for high performance smart cards," Proc. IEEE 2005 Custom Integrated Circuits Conference, P-16-1, pp.255–258, Sept. 2005.
44. C. Ohno, H. Yamazaki, H. Suzuki. E. Nagai, H. Miyazawa, K. Saigoh, T. Yamazaki, Y. Chung, W. Kraus, D. Verhaeghe, G. Argos, J. Walberl, and S. Mitra, "A highly reliable 1T1C 1 Mb FRAM with novel ferro-programmable redundancy scheme," Dig. Tech. Papers IEEE International Solid-State Circuits Conference, 2.5, pp. 36–37, Feb. 2001.
45. http://edevice.fujitsu.com/fj/MARCOM/find/19-4e/pdf/007.pdf
46. Wei Liu, J. Rho, and W. Sung, "Low-power high-throughput BCH error correction VLSI design for multi-level cell NAND Flash memories," IEEE Workshop on Sig. Proc. Sys. Design and Implementation, SIPS ′06, pp. 303–308, Oct. 2006.

47. Fei Sun, S. Devarajan, K. Rose, T. Zhan, "Multilevel flash memory on-chip error correction based on trellis coded modulation," IEEE Proc. 2006 International Symposium on Circuits and Systems, pp.1443–1446, May 2006.
48. S. Gregori, A. Cabrini, O. Khouri, and G. Torelli, "On-chip error correcting techniques for new-generation flash memories," Proc. IEEE, vol. 91, no. 4, pp. 602–616, Apr. 2003.
49. M. Spica and T.M. Mak, "Do we need anything more than single bit error correction (ECC)?" Rec. 2004 International Workshop on Memory Technology, Design and Testing, pp. 111–116, Aug. 2004:
50. J. Eliason, S. Madan, H. McAdams, G. Fox, T. Moise, C. Lin, K. Schwartz, J. Gallia, E. Jabillo, B. Kraus, and S. Summerfelt, "An 8 Mb 1T1C ferroelectric memory with zero cancellation and micro-granularity redundancy," Proc. IEEE 2005 Custom Integrated Circuits Conference, pp. 427–430, Sept. 2005.
51. HM71V832(32 K x8) data sheet, Hitach(1995):http://www.datasheetarchive.com/H-195.htm
52. MB85RS256(SPI256K) data sheet, Fujitsu(2005): http://edevice.fujitsu.com/fj/DATASHEET/e-ds/e513105.pdf
53. S. Masui, T. Ninomiya, M. Oura, W. Yokozeki, K. Mukaida, and S. Kawashima, "A ferroelectric memory-based secure dynamically programmable gate array," IEEE J. Solid-State Circuits, vol. 38, no. 5, pp. 715–725, May 2003.
54. H. Hirano, T. Honda, N. Moriwaki, T. Nakakuma, A. Inoue, G. Nakane, S. Chaya, and T. Sumi, "2-V/100-ns 1T/1C nonvolatile ferroelectric memory architecture with bitline-driven read scheme and nonrelaxation reference cell," IEEE J. Solid-State Circuits, vol. 32, no. 5, pp.649–654, May 1997.
55. H. Takasu, FED Review, vol. 2, no. 7, pp. 1–24, Feb. 2003 ; Japanese language. Also in SS.10, H. Ishiwara, T.Fuchigami et al 'Recent progress in ferroelectric memories,' CMC publishing ISBN4-88231-819-9 (Feb.2004) http://www.cmcbooks.co.jp/books/b0712.php; Japanese language.
56. Y. Kato, T. Yamada, and Y. Shimada, "0.18-μm nondestructive readout FeRAM using charge compensation technique," IEEE Trans. Electron Devices, vol. 52, no. 12, pp.2616–2621, Dec. 2005. Also S. Koyama, Y. Kato, T. Yamada, and Y. Shimada, "Improvement in readout reliability of a nondestructive readout FeRAM by asymmetrical programming," International Meeting for Future of Electron Devices 2004, pp.125–126, Jul. 2004: describes asymmetry rewrite.
57. Y. Kaneko, H. Tanaka, Y. Kato, and Y. Shimada, "Two-dimensional electron gas switching in an ultra thin epitaxial ZnO layer on a ferroelectric gate structure," Ext. Abst. International Conference on Solid State Devices and Materials 2007, J-8-2, pp. 1156–1157, Sept. 2007.
58. K.R. Raiter, and B.F. Cockburn, "An investigation into three-level ferroelectric memory," IEEE International Workshop on Memory Technology, Design, and Testing 2005, pp. 38–43, Aug. 2005
59. Yan Li, S. Lee, Y. Fong, F. Pan, Tien-Chien Kuo, J. Park, T. Samaddar, H. Nguyen, M. Mui, K. Htoo, T. Kamei, M. Higashitani, E. Yero, G. Kwon, P. Kliza, J. Wan, T. Kaneko, H. Maejima, H. Shiga, M. Hamada, N. Fujita, K. Kanebako, E. Tam, A. Koh, I. Lu, C. Kuo, T. Pham, J. Huynh, Q. Nguyen, H. Chibvongodze, M. Watanabe, K. Oowada, G. Shah, B. Woo, R. Gao, J. Chan, J. Lan, P. Hong, L. Peng, D. Das, D. Ghosh, V. Kalluru, S. Kulkarni, R. Cernea, S. Huynh, D. Pantelakis, Chi-Ming Wang, and K. Quader, "A 16 Gb 3b/ cell NAND Flash memory in 56-nm with 8 MB/s write rate," Dig. Tech. Papers IEEE International Solid-State Circuits Conference 2008, pp. 506–632, Feb. 2008.
60. T. Fukushima, A. Kawahara, T. Nanba, M. Matsumoto, T. Nishimoto, N. Ikeda, Y. Judai, T. Sumi, K. Arita, and T. Otsuki, "A microcontroller embedded with 4 Kbit ferroelectric nonvolatile memory," IEEE Dig. Tech. Papers, 1996 Symp. VLSI Circuits, pp. 46–47, Jun. 1996.
61. M.K. Seo, S.H. Sim, Y.H. Sim, M.H. On, S.W. Kim, I.W. Cho, H.S. Lee, G.H. Kim, and M.G. Kim, "A 0.9 V 66 MHz access, 0.13um 8 M(256 K 32) local SONOS embedded flash EEPROM," IEEE Dig. Tech. Papers, 2004 Symp. VLSI Circuits, pp. 68–71, Jun. 2004.

62. K. Honda, S. Hashimoto, and Y. Cho, "Visualization of electrons and holes localized in gate thin film of metal SiO_2–Si_3N_4–SiO_2 semiconductor-type flash memory using scanning nonlinear dielectric microscopy after writing-erasing cycling," Appl. Phys. Lett. 86, 063515, Feb. 2005.
63. K. Honda, S. Hashimoto and Y. Cho "Visualization of charges stored in the floating gate of flash memory by scanning nonlinear dielectric microscopy," Nanotechnology **17** (2006) S185–S188, © 2006 IOP Publishing Ltd Printed in the UK
64. S.C. Philpy, D.A.Kamp, A.D. DeVilbiss, A.F. Isacson, and G.F. Derbenwick, "Ferroelectric memory technology for aerospace applications," IEEE Proc. 2000 Aerospace Conference. vol. 5, pp. 377–383, Mar. 2000.
65. J.M.Benedetto, W.M. De Lancey, T.R. Oldham, J.M. McGarrity, C.W. Tipton, M. Brassington, and D.E. Fisch, "Radiation evaluation of commercial ferroelectric nonvolatile memories," IEEE Trans. Nuclear Science, vol. 38, no. 6, pt. 1, pp. 1410–1414, Dec. 1991.
66. J.H. Park, H.J. Joo, S.K. Kang, Y.M. Kang, H.S. Rhie, B.J. Koo, S.Y. Lee, B.J. Bae, J.E. Lim, H.S. Jeong, and K. Kim, "Fully logic compatible (1.6 V Vcc, 2 additional FRAM masks) highly reliable sub $10F^2$ embedded FRAM with advanced direct via technology and robust 100 nm thick MOCVD PZT technology," IEEE International Electron Devices Meeting Tech. Dig. , pp. 591–594, Dec. 2004.
67. Y. Kumura, T. Ozaki, H. Kanaya, O. Hidaka, Y. Shimojo, S. Shuto, Y. Yamada, K. Tomioka, K. Yamakawa, S. Yamazaki, D. Takashima, T. Miyakawa, S. Shiratake, S. Ohtsuki, I. Kunishima and A. Nitayama, "A $SrRuO_3/IrO_2$ top electrode FeRAM with Cu BEOL process for embedded memory of 130 nm generation and beyond," IEEE Proc. 35th European Solid-State Device Research Conference 2005, Paper 8.B.3, pp.557–560, Sept. 2005.
68. D. Takashima, Y. Takeuchi, T. Miyakawa, Y. Itoh, R. Ogiwara, M. Kamoshida, K. Hoya, S.M. Doumae, T. Ozaki, H. Kanaya, K. Yamakawa, I. Kunishima, and Y.Oowaki, "A 76-mm^2 8-Mb chain ferroelectric memory," IEEE J. Solid State Circuits, vol. 36, no. 11, pp. 1713–1720, Nov. 2001.
69. http://www.itrs.net/reports.html.
70. K. Singh, D. K. Bopardikar, and D.V. Atkare, "A compendium of Tc-Us and Ps-Δz data for displacive ferroelectric," Ferroelectrics, vol. 82, pp. 55–67, Jun. 1988. Publisher: Taylor & Francis.
71. G. W. Pabst, L. W. Martin, Ying-Hao Chu, and R. Ramesh, "Leakage mechanisms in $BiFeO_3$ thin films", Appl. Phys. Lett. 90, 072902, Feb. 2007.
72. Y. Kato, H. Tanaka, K. Isogai, K. Kaibara, Y. Kaneko, Y. Shimada, M. Brubaker, J. Celinska, L.D. McMillan, C.A.P. de Araujo, "Embedded FeRAM challenges in the 65-nm technology node and beyond," IEEE ISAF '06. , pp. 81–84, Aug. 2006.

Chapter 9
Statistical Blockade: Estimating Rare Event Statistics for Memories

Amith Singhee and Rob A. Rutenbar

Abstract As we move deeper into sub-65 nm technology nodes, uncontrollable random parametric variations have become a critical hurdle for achieving high yield. This problem is particularly crippling for *high-replication circuits* (HRCs) – circuits like SRAM cells, nonvolatile memory cells, and other memory cells that are replicated millions of times on the same chip – because of aggressive cell design, the requirement of meeting very high $>5\sigma$ levels of yield and the usual higher sensitivity of such circuits to process variations. However, it has proved difficult to even estimate such high yield values efficiently, making it very difficult for designers to adopt an accurate, variation-aware design methodology. This chapter develops a general statistical methodology to estimate parametric memory yields. The keystone of the methodology is a technique is called *statistical blockade*, which combines Monte Carlo simulation, machine learning, and extreme value theory to simulate very rare failure events and to compute analytical models for the tail distributions of the circuit performance metrics. Several circuit examples are analyzed in detail to enable a deep understanding of the theory and its practical use in a real-world setting. The treatment is directed toward both the memory designer and the EDA engineer.

9.1 Introduction

VLSI technology is moving deep into the nanometer regime, with transistor feature sizes of 45 nm already in widespread production. Computer-aided design (CAD) tools have traditionally kept up with the difficult requirements for handling complex physical effects and multi-million-transistor designs, under the assumption of fixed or deterministic circuit parameters. However, at such small feature sizes, even small variations due to inaccuracies in the manufacturing process can cause large

---

R.A. Rutenbar (✉)
Electrical and Computer Engineering, Carnegie Mellon University, Pittsburgh, PA, USA
e-mail: rutenbar@ece.cmu.edu

K. Zhang (ed.), *Embedded Memories for Nano-Scale VLSIs*, Series on Integrated Circuits and Systems, DOI 10.1007/978-0-387-88497-4_9,

relative variations in the behavior of the circuit. Such variations may be classified into two broad categories, based on the source of variation: (1) systematic variation, and (2) random variation. Systematic variation constitutes the deterministic part of these variations; e.g., proximity-based lithography effects, nonlinear etching effects, etc. [1]. Random variations constitute the unexplained part of the manufacturing variations, and show stochastic behavior; e.g., gate oxide thickness (t_{ox}) variations, poly-Si random crystal orientation (RCO), and random dopant fluctuation (RDF) [2].

Systematic variations are typically pattern dependent and can often be well accounted for in memory cells using more accurate models of the process. The reason for this is that the cells have relatively small geometrical complexity: for example, the ubiquitous 6T SRAM cell contains only six transistors. Random variations, however, cannot simply be accounted for by more accurate models of the physics of the process because of their inherent random nature (until we understand and model the physics well enough to accurately predict the behavior of each ion implanted into the wafer). Further, the impact of variations is widely believed to be roughly inversely proportional to the square root of the transistor area [3], and the transistors in memory cells tend be of close to minimum size. Hence, memory cells are particularly susceptible to these random variations, increasing the need for an efficient yield estimation technique for these cells. The 2007 ITRS[1] roadmap predicts a large 42% variability in the threshold voltage (V_t) of memory devices in the year 2009. It also predicts that almost all of this V_t variability (40%) will result from random doping variability.

The bottom-line impact of these unaccounted for variations then is unpredicted variation in the memory performance and dreaded yield loss. This problem for SRAM is more alarming since SRAM cells are often used for technology development when scaling to the next node. If the SRAM yield is not predicted well in advance, successful technology development may take too long, and may even be difficult to achieve.

Memory designers have often side-stepped the problem of yield estimation by using multiple process and environmental corners with large safety margins. This approach, of course, is unreliable since it does not account for the actual statistics of the memory cell performance metrics. Worse, it usually results in significant over-design, which translates to a squander of chip area and power, both being expensive commodities. Of course, such an approach has been adopted only because it is an engineering solution that has worked in the past. However, it will fail going forward [6, 5]. Pre-Si yield estimation is now critical for memory design, and robust statistical techniques are needed to enable it. In this chapter we will present the mathematical and algorithmic advances that enable rigorous statistical analysis of memory performance.

---

[1] International Technology Roadmap for Semiconductors.

9.2 The Parametric Yield of Memory Arrays

Memory yield requirements are usually specified for the entire array. For example, the designer may be comfortable with an *array failure probability* of one in a thousand; i.e., $F_{\mathrm{f,chip}} \leq 10^{-3}$. However, how does this translate to a yield requirement for the *memory cell*? What is the maximum *cell failure probability*, $F_{f,\mathrm{cell}}$, allowed to satisfy this array failure probability requirement? In this section, we answer this question, starting from an approximate, intuitive analysis, leading up to a more rigorous analysis that accounts for redundancy in the array. Our purpose behind this stage-wise development is to achieve a clear understanding of these statistics.

9.2.1 Simple, Approximate Analysis

Suppose our cell array has N cells, and the chip failure probability specification is $F_{\mathrm{f,chip}}$. For example, the designer may be comfortable with a chip failure rate of one in a thousand; i.e., $F_{\mathrm{f,chip}} \leq 10^{-3}$. A common, "back-of-the-envelope" argument is that a thousand chips will contain $1{,}000N$ cells, and if one out of every $1{,}000N$ cells fails we will have one in every 1,000 arrays failing. Hence, we will need a *cell failure probability*, $F_{f,\mathrm{cell}}$ of $\leq 10^{-3}/N$. In the general case, we can write this simple, approximate relation as

$$F_{f,\mathrm{cell}} \approx \frac{F_{f,\mathrm{chip}}}{N} = F_{\mathrm{simple}} \tag{9.1}$$

Let us now perform an exact analysis, and compare with this approximate estimate of $F_{f,\mathrm{cell}}$.

9.2.2 Accurate Analysis

$F_{\mathrm{f,chip}}$, the chip or array failure probability, is the probability of *one or more* cells failing in any one array. Hence, if we manufacture 10 arrays, and 6 of them have a total of 8 faulty cells, our array failure probability is (approximately) 6/10, and not 8/10. Another, equivalent way to define $F_{\mathrm{f,chip}}$ is the probability of the *worst* cell in an array being faulty. Hence, if we are measuring some performance metric y (e.g., static noise margin), we are really interested in the statistics of the worst value of y from among N values, where N is the number of cells in our array. Let us define this worst value as M_N, and let us assume that by worst, we mean the maximum; i.e., large values of y are bad. If small values of y are bad, then we can use the maximum of $-y$ as the worst value.

$$M_N = \max(Y_1, \ldots, Y_N)$$

where $Y_1, \ldots, Y_N$ are the measured performance metric values for the N cells in an array. Also suppose that the failure threshold for y is y_{f}; i.e., any cell with $y > y_f$ is defined as failing. Then, $F_{f,\mathrm{chip}}$ can be defined as

$$F_{f,\text{chip}} = P(M_N > y_f); \tag{9.2}$$

that is, the probability of the worst case y in the array being greater than the failure threshold y_f. Now,

$$P(M_N > y_f) = 1 - P(M_N \leq y_f),$$

where $P(M_N \leq y_f)$ is the probability of the worst case cell passing: this is the same as the probability of all the cells in the array passing; that is,

$$\begin{aligned} P\left(M_N \leq y\right) &= P\left(Y_1 \leq y_f, \ldots, Y_N \leq y_f\right) \\ &= \Pi_{i=1}^{N} P\left(Y_i \leq y_f\right) = \left[P(Y \leq y_f)\right]^N. \end{aligned} \tag{9.3}$$

Note that here we have made the reasonable assumption that most of the local variability is spatially independant (e.g., RDF). Hence, $P(Y_i \leq y_f)$ is independant of $P(Y_j \leq y_f)$ for $i \neq j$.

Now,

$$P(Y \leq y_f) = 1 - P(Y > y_f), \tag{9.4}$$

where $P(Y_i > y_f)$ is the failure probability of a cell; i.e., $F_{f,\text{cell}}$. Hence, combining all equations from (9.2 to 9.4), we have

$$F_{f,\text{chip}} = 1 - \left[1 - F_{f,\text{cell}}\right]^N, \text{ or} \tag{9.5}$$

$$F_{f,\text{cell}} = 1 - \left[1 - F_{f,\text{chip}}\right]^{\frac{1}{N}} = F_{\text{exact}} \tag{9.6}$$

Comparing this with (9.1), we see that (9.1) is clearly approximate; but how approximate really? Figure 9.1 illustrates this difference graphically. Here, we have plotted the percentage error in F_{simple}

$$\text{error} = 100 \times \frac{F_{\text{simple}} - F_{\text{exact}}}{F_{\text{exact}}},$$

versus the array failure probability $F_{f,\text{chip}}$. We see that F_{simple} underestimates the cell failure probability; that is, it is conservative. For reasonable values of $F_{f,\text{chip}}$, however, the error is small enough to allow casual use of F_{simple}.

9.2.3 Incorporating Redundancy

As array sizes increase, the chances of having a failing cell in one array increase. In reality, this failure can be due to radiation-induced soft errors, manufacturing

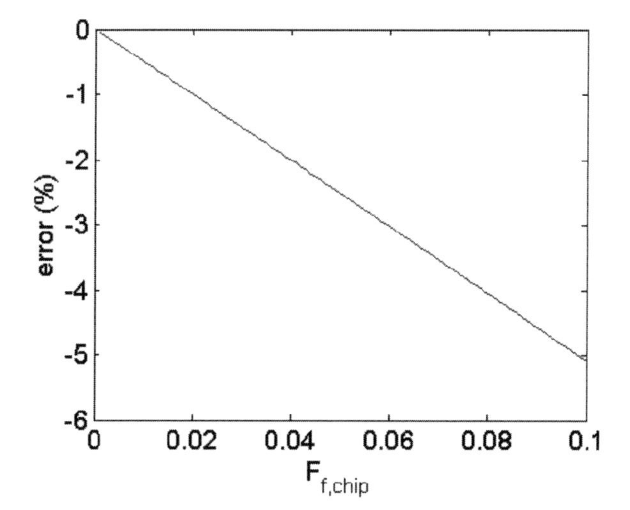

Fig. 9.1 Percentage error in the approximate cell failure probability estimate of (9.1), as the array failure probability $F_{\mathrm{f,chip}}$ increases

defects, or process variation. To counter this increased fault probability, a recent trend is to increase the fault tolerance of the memory by including redundancy in the array. One common approach is to have redundant columns in the array that are used to replace defective columns in the chip (see, for example, [7]). In the simplest case, if there is one redundant column, the chip can tolerate one failing cell. The use of redundancy, of course, changes the statistics of the array failure probability in relation to the cell failure probability. We will now develop a simple model to allow the computation of process-induced cell failure probability $F_{f,\mathrm{cell}}$, given the maximum tolerable array failure probability $F_{\mathrm{f,chip}}$, in the presence of redundancy in the array.

Let us say that the amount of redundancy in the array is r. For the case of column redundancy, this would mean that we have r redundant columns, and can tolerate up to r faulty columns in the array. If each column has n_{c} cells, then we can tolerate up to rn_{c} faulty cells, depending on their spatial configuration. However, under the assumption of spatially independant local variability from the previous section, the probability of having two or more faulty cells in the same column can be shown to be negligibly small. Hence, we consider only the case where all the faulty cells in the array are in different columns. This will help us keep unnecessary and tedious mathematics away, without losing the core insights of the derivation. In this case, the array would operate without faults with up to r faulty cells in it, and the array failure probability is the probability of having more than r faulty cells in it. If the probability of having k faulty cells in an array is denoted by $F_{\mathrm{f},k}$, then we can write the array failure probability $F_{\mathrm{f,chip}}$ as

$$F_{f,\mathrm{chip}} = 1 - \left(F_{f,0} + F_{f,1} + \cdots + F_{f,r}\right). \tag{9.7}$$

Now, all we need is to compute every $F_{f,k}$ in terms of the cell failure probability $F_{\mathrm{f,cell}}$. We can write this $F_{f,k}$, the probability of k cells failing in an array, as

$$F_{f,k} = {}^{N}C_k F_{f,\text{cell}}^k \left(1 - F_{f,\text{cell}}\right)^{N-k}, \tag{9.8}$$

where ${}^{N}C_k$ is the number of combinations of k cells in an array of N cells. For $k = 0$, we get

$$F_{f,0} = \left(1 - F_{f,\text{cell}}\right)^{N}.$$

Hence, if we have zero redundancy, we get from (9.7)

$$F_{f,\text{chip}} = 1 - \left(1 - F_{f,\text{cell}}\right)^{N},$$

which is the same result as (9.5). As we would expect, we get the same result whether we use the worst case argument or zero-redundancy argument, since we are modeling the same statistic. Using (9.8) in (9.7), we can obtain the array failure probability in the presence of redundancy. However, (9.8) is inconvenient because for N large (1 million or more), the ${}^{N}C_k$ term can be inconveniently large, and the $F_{f,\text{cell}}^k$ term can be inconveniently small for floating point arithmetic.

9.2.4 The Poisson Yield Model

Let us now go one step further in simplifying the expression (9.7) to obtain a more popular model of the yield: the *Poisson model*. This model will also not suffer from the unwieldy numerics of (9.8). Let us define λ as the expected number of faulty cells in every N cells we manufacture. Hence,

$$\lambda = F_{f,\text{cell}} N.$$

For example, if $F_{f,\text{cell}}$ is 1 ppm, and N is 10 million, λ will be 1 ppm $\times$ 10 million = 10. Then, we can write (9.8) as

$$\begin{aligned} F_{f,k} &= \frac{N!}{k!(N-k)!} \frac{\lambda^k}{N^k} \left(1 - \frac{\lambda}{N}\right)^{N-k} \\ &= \frac{N!}{(N-k)!N^k} \times \left(1 - \frac{\lambda}{N}\right)^{N} \times \left(1 - \frac{\lambda}{N}\right)^{-k} \times \frac{\lambda^k}{k!}. \end{aligned} \tag{9.9}$$

There are now four terms that are multiplied together. Note that the array size N is very large compared to all relevant values of k and λ. For large values of N, we can approximate the first three terms in this equation as follows:

$$\text{Term1}: \left| \begin{aligned} \lim_{N\to\infty} \tfrac{N!}{(N-k)!N^k} &= \lim_{N\to\infty} \tfrac{N}{N}\tfrac{N-1}{N}\cdots\tfrac{N-k+1}{N} \\ &= \lim_{N\to\infty} 1\left(1-\tfrac{1}{N}\right)\left(1-\tfrac{2}{N}\right)\cdots\left(1-\tfrac{k-1}{N}\right) = 1. \end{aligned} \right.$$

$$\text{Term2}: \left| \begin{aligned} \lim_{N\to\infty}\left(1-\tfrac{\lambda}{N}\right)^N &= \lim_{N\to\infty} 1 + \sum_{i=1}^{\infty} \tfrac{N!}{i!(N-i)!}\tfrac{(-\lambda)^i}{N^i} \\ &= 1 + \sum_{i=1}^{\infty} \tfrac{(-\lambda)^i}{i!}\left(\lim_{N\to\infty}\tfrac{N!}{(N-i)!N^i}\right) \\ &= 1 + \sum_{i=1}^{\infty} \tfrac{(-\lambda)^i}{i!}(1) = e^{-\lambda}. \end{aligned} \right.$$

Note here, that the limit expression of Term 1 emerges in the summation and we have replaced it with the limit value of 1.

$$\text{Term3}: \left| \lim_{N\to\infty}\left(1-\frac{\lambda}{N}\right)^{-k} = (1-0)^{-k} = 1. \right.$$

Using these limiting values for the first three terms in (9.9), we obtain

$$F_{f,k} \approx 1 \times e^{-\lambda} \times 1 \times \frac{\lambda^k}{k!} = \frac{\lambda^k e^{-\lambda}}{k!}.$$

Hence, we see that $F_{f,k}$ follows the well-known *Poisson process*. We can finally write the array failure probability $F_{f,\text{chip}}$ as

$$F_{f,\text{chip}} \approx 1 - \sum_{k=0}^{r} \frac{\lambda^k e^{-\lambda}}{k!} = 1 - \sum_{k=0}^{r} \frac{\left(F_{f,\text{cell}}N\right)^k e^{-F_{f,\text{cell}}N}}{k!}. \tag{9.10}$$

If we have a specification for $F_{f,\text{chip}}$, we can easily compute the specification for $F_{f,\text{cell}}$ numerically.

9.2.4.1 An Example: Quantifying Fault Tolerance with Statistical Analysis

As an example, let us take an array with $N = 10$ million cells. Now, suppose that we require an array yield of no less than 0.9; that is, $F_{f,\text{chip}} = 0.1$ (or less). Using (9.10), we can now easily study the impact of redundancy on the required cell yield: we can compute the cell failure probability $F_{f,\text{cell}}$ for different values of r. The results are shown in Table 9.1. We can see that increasing the redundancy increases the fault tolerance: we can use cells with higher failure probability. However, the real measure of increase in fault tolerance is the percentage increase in cell failure probability allowed by an increase in redundancy. This is shown in column 3, and it is clear that adding redundancy helps, but with diminishing returns. At some point we reach an optimal trade-off point where further increase in fault tolerance is too small to justify the area, delay, and power costs of adding redundant circuitry. Such

Table 9.1 Cell failure probability $F_{f,cell}$, computed from (9.10) for a 10 Mb array and a cell failure probability $F_{f,chip}$ of 0.1. As the fault tolerance (redundancy) increases, we can tolerate much larger cell failure probability, but with diminishing returns

Number of faults tolerable (r)	Maximum allowed $F_{f,cell}$ (ppm)	Percent increase in $F_{f,cell}$
0	0.0105	–
1	0.0532	404.8
2	0.1102	107.2
3	0.1745	58.32
4	0.2433	39.42
5	0.3152	29.57
6	0.3895	23.57
7	0.4656	19.55
8	0.5432	16.67

critical, yield-aware analysis of design trade-offs is only possible with a rigorous statistical analysis methodology.

From the preceding discussion, we now know how to compute the specification on the cell failure probability, given a specification on the array failure probability. Suppose now that we design a cell. How do we estimate its failure probability to check if it meets this criterion? This will be the focus of the rest of the chapter.

9.3 Estimating Failure Statistics with Monte Carlo Simulation

Monte Carlo simulation is the most general and accurate method for computing the failure probability (or yield) of any circuit. This section provides a brief conceptual introduction to Monte Carlo that is immediately relevant to our problem of estimating circuit failure probability. For a more general and detailed discussion on Monte Carlo, see [8].

9.3.1 Process Variation Statistics: Prerequisites for Statistical Analysis

To obtain any statistics of the circuit performance (e.g., yield), some process characterization is required to obtain the statistics of various device or circuit parameters. For example, to estimate the statistics of the static noise margin of an SRAM cell using SPICE simulations, we need to know the statistics of, say, the threshold voltage variation for PMOS and NMOS transistors in the manufacturing process used, due to random dopant fluctuation (RDF). It is usually observed that this variation follows a Gaussian distribution [9]. We name all such parameters as *statistical parameters*. As another example, via or contact resistance may follow some non-Gaussian distribution [10]. These probability distributions have to be estimated. The method of estimation may depend on how mature the technology is. In early stages of technology development, these distributions may be extrapolated from the previous technology, or from TCAD, as in [2] or [11]. In later stages, they may be characterized using hardware measurements, as in [9].

Let us say that we have s statistical parameters for our memory cell. For example, if we are considering only RDF in an 6T SRAM cell, we would have $s = 6$ threshold voltage deviations as the statistical parameters, each having a Gaussian distribution, as per [9]. We denote each parameter by x_i, where $i = 1, \ldots, s$. Together the entire set is denoted by the vector $\mathbf{x} = \{x_1, \ldots, x_s\}$. Each parameter has a probability distribution function (PDF) denoted by $\pi_i(x_i)$, and the joint PDF of $\mathbf{x}$ is denoted by $\pi(\mathbf{x})$.

9.3.2 Monte Carlo Simulation

Given certain a value for each of the s statistical parameter, we can enter these values into a SPICE netlist and run a SPICE simulation to compute the corresponding value of any circuit performance metric. For example, we can compute the static noise margin of a 6T SRAM cell given the six threshold voltage values. Let us say the performance metric is y and we have some specification on the performance metric; for example, the write time for a memory cell must be less than 50 ps. Let us denote this specification value by t, and the requirement by $y \leq t$. Note that here we use $\leq$ without any loss of generality. Now, for certain combinations of values for the statistical parameters, our circuit will fail this specification; that is, y will be greater than t. For example, if the threshold voltage is too large, the access time may exceed the maximum allowed value due to slower devices.

To estimate the yield of the circuit using Monte Carlo simulation, we *randomly* draw values for all the s statistical parameters from their joint PDF. One such set of s random values is called a *sample point*, or simply a *sample*. We generate several such samples until we have at least one sample point for which the circuit fails its specification. In practice, we would want to have obtained several failing points. Suppose that we generate n total points and obtain n_{f} failing points. The *estimate* circuit failure probability is then given by

$$\hat{F}_f = \frac{n_f}{n} \tag{9.11}$$

and the estimate of the circuit yield is $1 - \hat{F}_f$.

Note that if we re-ran a Monte Carlo run with the same number of points, but with a different random number sequence, we would obtain a different estimate of F_{f}. Hence, there is some statistical error in the Monte Carlo estimate, and this error decreases with increasing number of points n [8]. Since, the Monte Carlo estimate is essentially a random variable, we can also associate a probability distribution with it, for a given n. The spread of this PDF decreases with increasing n. Since we run a full SPICE simulation for each sample point, the cost of an n-point Monte Carlo run is roughly the same as n SPICE simulations, assuming that the cost of generating random numbers and using (9.11) is negligible in comparison.

9.3.3 The Problem with Memories

Consider the case of a 1 megabit (Mb) memory array, which has 1 million "identical" instances of a memory cell. To simplify our discussion, we will assume that there is no redundancy in the memory. These cell instances are designed to be identical, but due to manufacturing variations, they usually differ. Suppose we desire a *chip* yield of 99%; that is, no more than 1 chip per 100 should fail. Using (9.6), this translates to a maximum *cell* failure rate of 10.05 per billion or approximately 0.01 ppm.

A failure probability of 0.01 ppm is the same as for a 5.6σ point on a normal distribution, where σ is its standard deviation. If we want to estimate the yield of such a memory cell, a standard Monte Carlo approach would require at least 100 million SPICE simulations on average to obtain just one failing sample point! Even then, the estimate of the yield or failure probability will be suspect because of the lack of statistical confidence, the estimate being computed using only one failing example. Such a large simulation count is usually intractable. This example clearly illustrates the widespread problem with designing robust memories in the presence of process variations: we need to simulate *rare* or *extreme* events and estimate the statistics of these rare events. The problem of simulating and modeling rare events stands for any high-replication circuit (HRC), like SRAM cells, eDRAM cells, and nonvolatile memory cells. In this chapter, we will use mainly SRAMs for illustration, except where some diversion provides insight into the simulation and modeling methods being discussed. All methods developed, however, apply similarly to other memory circuits, like eDRAM and nonvolatile memory.

9.3.4 Methods for Estimating Rare Events Statistics

Monte Carlo simulation would be the ideal technique for reliably estimating the yield, but as we saw in the preceding discussion, it can be prohibitively expensive for HRCs. One avenue of attack is to abandon Monte Carlo. Several analytical and semi-analytical approaches have been suggested to model the behavior of SRAM cells [4, 5, 6] and digital circuits [12] in the presence of process variations. All suffer from approximations necessary to make the problem tractable, or apply to a specific performance metric. References [5] and [12] assume a linear relationship between the statistical variables and the performance metrics (e.g., SNM), and assume that the statistical process parameters are normally distributed. These assumptions result in a normal distribution assumption for the performance metric too, which can suffer from gross errors, especially while modeling rare events: we shall see examples in the results section. When the distribution varies significantly from Gaussian, [5] chooses an F-distribution in an ad hoc manner. Reference [4] presents a complex analytical model limited to a specific long-channel transistor model (the transregional model) and further limited to only static noise margin analysis for the 6T SRAM cell. Reference [6] again models only the SNM for subthreshold SRAM cells under assumptions of independence and identical distribution of the upper and lower SNM, which may not always be valid. All these methods are specific to either one circuit, or one device model, or one performance metric. This is a general problem with analytical methods: they are not generalizable. In this chapter, however, we are interested in general techniques for modeling the statistics of performance metrics of HRCs; techniques that are independent of the circuit, performance metric, and the process statistics.

A different avenue of attack is to modify the Monte Carlo strategy. Reference [13] shows how importance sampling can be used to predict failure probabilities. Recently, [14] applied an efficient formulation of these ideas for modeling rare failure events of single 6T SRAM cells, based on the concept of mixture importance sampling from [15]. The approach uses real SPICE simulations with no approximating equations. However, the method only estimates the failure (exceedance) probability of a *single* threshold value of the performance metric. A re-run is needed to obtain probability estimates for another failure threshold: no complete model of the tail of the distribution is computed. The method also combines all performance metrics to compute a failure probability, given fixed thresholds. Hence, there is no way to obtain separate probability estimates for each metric, other than a separate run per metric.

In this chapter, we go further and discuss a very general and efficient Monte Carlo method that addresses both of the problems previously mentioned: very fast generation of (1) rare event samples, and (2) sound models of the rare event (distribution tail) statistics for any performance metric. We refer to this method as *statistical blockade* (SB). It imposes almost no a priori limitations on the form of the statistics for the statistical parameters, device models, or performance metrics. The method is conceptually simple and employs ideas from two rather non-traditional sources: *extreme value theory* and *machine learning*.

9.3.5 Methods in this Chapter: A Brief Overview

Extreme value theory (EVT) [16] is a branch of probability that studies and optimally quantifies the statistics of, as the name suggests, extreme or rare events. The theory has found wide statistical application in fields such as hydrology [17], insurance [17], and finance [17] among several others: wherever there is a need to estimate the probability of rare events. One of the most consequential applications, and, indeed, one of the driving forces for the development of the theory of extremal statistics, was the Dutch-dike project following the disastrous North Sea flood of 1953 that took over 1,800 human lives. One aspect of the post-flood response was to determine appropriate heights for the sea dikes in the Netherlands, such that the probability of a flood in a year is reduced to some very small amount (e.g., 10^{-4}). Technically, this involved estimating the height of sea-water level corresponding to this probability level – definitely a rare event – using statistical inference based on historical data of sea-water level measurements. Furthermore, the quantile to be estimated was much beyond the available data range. Our problem of estimating extreme quantiles of the SRAM static noise margin, using a limited number of Monte Carlo samples, is similar in flavor to the dike height problem (if not in direct impact on the human condition). Hence, we can employ the same technical tools from EVT for our problem.

However, the yield estimation problem is more "extreme" in the sense of the failure probabilities to be estimated: often 10^{-8}–10^{-9} or smaller. To achieve reliable estimates of these quantiles we will need, again, impractically large Monte Carlo sample sizes. We will tackle this problem by using a *filter* to intelligently simulate only those points that are important; i.e., rare. We note that *generating* each Monte Carlo point is neither challenging, nor expensive relative to *evaluating* it using a SPICE simulation. Hence, we use an appropriate filter to *block* those points that are unlikely to fall in the low-probability tails of the performance metrics. Many points are generated, but only the "rare" events are simulated. Such a partial sampling of the performance distributions fits well with the results from EVT that we exploit. The filter we use is a standard *classifier* from machine learning, and its "blocking" activity gives the method its name of statistical blockade.

In the rest of the chapter, we will review relevant results from EVT, highlighting the limit theorems for the distributions of rare events. Then we will see how we can use these results for statistical inference from data, which in our case is generated from a modified Monte Carlo simulation. Some background on classifiers is discussed, allowing us to develop the statistical blockade framework. We will then extend this framework to handle metrics with conditionals (e.g., max(), min()) that result in disjoint rare event regions in the statistical parameter space, along with a recursion-based extension to produce reliable estimates for extremely rare events (6–8σ). Finally, we will see experimental results demonstrating the effectiveness of statistical blockade on realistic, circuit test cases. These ideas are based on the proposals of [19] and [20].

9.4 Modeling Rare Event Statistics

In Section 9.2, we saw how we can obtain the required yield of a memory cell, given a specification on the array yield. However, once we have a cell design, how do we evaluate its yield, or equivalently, its failure probability? We discussed the difficulties associated with problem in the previous section. The rest of the chapter will focus on solving this problem.

The failure of a reasonable memory cell is a very *rare event*, with probability of much less than 1 ppm. The specific problem here is to be able to estimate the statistics (e.g., failure probability) of such rare events, given the designed circuit.

In this section, we will clearly define what we mean by rare events and what the modeling problem we are interested in is. Then we will review some remarkable, foundational results from extreme value theory that will help us build a theoretically sound model for the statistics of these rare events.

9.4.1 The Tail Modeling Problem

Suppose we want to model the rare event statistics of the write time of an SRAM cell. Figure 9.2 shows an example of the distribution of the write time.

We see that it is skewed to the right with a *heavy* right tail. A typical approach is to run a Monte Carlo with a small sample size (e.g., 1,000) and fit a standard analytical distribution to the data, for example, a normal or a lognormal distribution. Such an approach can be accurate for fitting the "body" of the distribution, but will be grossly inaccurate in the tail of the distribution: the skewness of the actual distribution or the heaviness of its tail will be difficult to match. As a result, any prediction of the statistics of rare events, lying far in the tail, will be very inaccurate.

Let F denote the cumulative distribution function (CDF) of the write time y, and let us define a tail threshold t to mark the beginning of the tail (e.g., the 99th percentile). Let z be the excess over the threshold t. We can write the *conditional* CDF of the tail as

$$F_t(z) = P(Y - t \leq z | Y > t) = \frac{F(z+t) - F(t)}{1 - F(t)}, \tag{9.12}$$

and the overall CDF as

$$F(z+t) = (1 - F(t))\, F_t(z) + F(t). \tag{9.13}$$

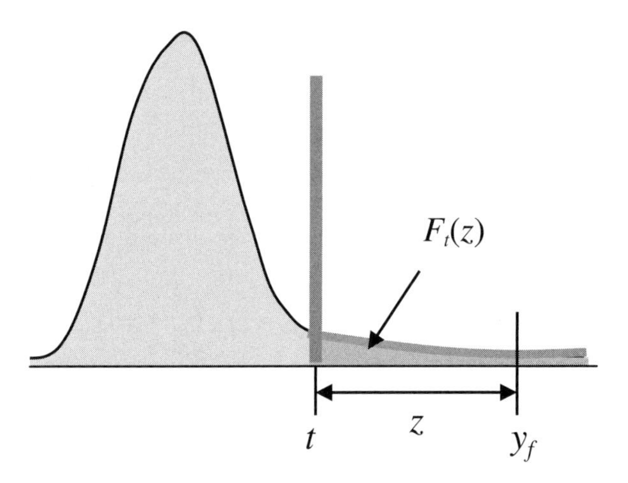

Fig. 9.2 A possible skewed distribution for some SRAM metric (e.g., write time)

If we know $F(t)$ and can estimate the conditional CDF of the tail $F_t(z)$ accurately, we can accurately estimate rare event statistics. For example, the yield for some extreme threshold $y_f > t$ is given as

$$F(y_f) = (1 - F(t))\, F_t(y_f - t) + F(t), \tag{9.14}$$

and the corresponding failure probability $F_f(y_f) = 1 - F(y_f)$ is given as

$$F_f(y_f) = (1 - F(t))\left(1 - F_t(y_f - t)\right). \tag{9.15}$$

$F(t)$ can be accurately estimated using a few thousand simulations, since t is not too far out in the tail. Then, the problem here is to efficiently estimate the conditional tail CDF F_t as a simple analytical form, which can then be used to compute statistical metrics such as (9.14) and (9.15) for rare events. Of course, here we assume that any threshold y_f of interest will be far into the tail, such that $y_f >> t$. This is easily satisfied for any real HRC scenario, for example, our 1 Mb cache example from the beginning of this chapter. We also assume that the extreme values of interest lie only in the *upper* tail of the distribution. This is without any loss of generality, because any lower tail can be converted to the upper tail by replacing $y = -y$. This same approach of fitting a CDF to the exceedances over some threshold has been developed and widely applied by hydrologists under the name of the *peaks over threshold* (POT) method [18]. In their case though, the data is from historical record and not synthetically generated. We now look at some results from extreme value theory, that are directly applicable to the problem of estimating the tail CDF.

9.4.2 Extreme Value Theory: Tail Distributions

Suppose that $Y_1, Y_2, \ldots$ is a sequence of independent, identically distributed random variables from the CDF F. For any sample $Y_1, Y_2, \ldots, Y_N$ of size N, define the *sample maximum* as

$$M_N = \max\left(Y_1, Y_2, \ldots, Y_N\right), \quad N \geq 2. \tag{9.16}$$

The probability of $M_N \leq y$ is the probability of all of $Y_1, Y_2, \ldots, Y_N$ being $\leq y$, as in (9.3), reproduced here:

$$P\left(M_N \leq y\right) = P\left(Y_1 \leq y, \ldots, Y_N \leq y\right) = \Pi_{i=1}^{N} P\left(Y_i \leq y\right) = F^N(y). \tag{9.17}$$

An important result from EVT addresses the question: *What are the possible limiting distributions of M_N as $N \to \infty$?* This result is stated in the following theorem by Fisher and Tippett [21].

Theorem 9.1 [Fisher-Tippett [21]] *If there exist normalizing constants a_N, b_N, and some non-degenerate CDF H, such that*

$$P\left(\frac{M_N - b_N}{a_N} \le y\right) = F^N\left(a_N y + b_N\right) \to H(y) \text{ as} \quad N \to \infty,\ y \in \Re \tag{9.18}$$

then H belongs to the type of one of the following three CDFs:

$$\begin{aligned} &\text{Frechet}: \quad \Phi_\alpha(y) = \begin{cases} 0, & y \le 0 \\ e^{-y^{-\alpha}}, & y > 0 \end{cases}, \quad \alpha > 0 \\ &\text{Weibull}: \quad \Psi_\alpha(y) = \begin{cases} e^{-(-y)^\alpha}, & y \le 0 \\ 1, & y > 0 \end{cases}, \quad \alpha > 0 \\ &\text{Gumbel}: \quad \Lambda(y) = e^{-e^{-y}}, \quad y \in \Re \end{aligned} \tag{9.19}$$

This amazing result formed the foundation of estimation of rare event statistics. Roughly, it says that for a very large class of CDFs, we can model the distribution of the normalized sample maximum M_N as one of three standard distributions: Fréchet, Weibull, and Gumbel. These three CDFs can be combined together into a *generalized extreme value* (GEV) distribution:

$$H_\xi(y) = \begin{cases} e^{-(1-\xi y)^{1/\xi}}, & \xi \ne 0 \\ e^{-e^{-y}}, & \xi = 0 \end{cases} \quad \text{where} \quad 1 - \xi y > 0. \tag{9.20}$$

The three CDFs are obtained as follows:

- $\xi = -\alpha^{-1} < 0$ gives the Fréchet CDF Φ_α,
- $\xi = \alpha^{-1} > 0$ gives the Weibull CDF Ψ_α, and
- $\xi = 0$ gives the Gumbell CDF Λ.

The condition (9.18) is commonly stated as *F lies in the maximum domain of attraction of H*, or $F \in \text{MDA}(H)$. Hence, for non-degenerate H, Theorem 9.1 can be stated succinctly as

$$F \in \text{MDA}(H) \Rightarrow H \text{ is of type } H_\xi.$$

It is interesting to note the similarity between this theorem regarding maxima and the popular central limit theorem (CLT), which provides the limiting distribution for the *sum* of i.i.d. random variables. The most popular form of the CLT is as follows:

Theorem 9.2 [Central limit theorem] *Define $S_N = Y_1 + Y_2 + \ldots + Y_N$ as the sample sum of N i.i.d. random variables from some CDF F. Let $\mu = E(Y)$ be the mean, and $\sigma_2 = E[(y-\mu)^2]$ be the variance for F. If $\sigma < \infty$, then*

Table 9.2 The Fisher–Tippett theorem for maxima is congruent to the central limit theorem for sums

CLT	S_N	$\sigma\sqrt{N}$	μ_N	Φ standard normal
Fisher–Tippett	M_N	a_N	b_N	H_ξ GEV

$$P\left(\frac{S_N - \mu N}{\sigma\sqrt{N}} < y\right) \to \Phi(y) \text{ as } N \to \infty, \;\; y \in \Re, \tag{9.21}$$

where Φ *is the standard normal CDF with mean 0 and variance 1.*

Φ, the standard normal CDF, is not to be confused with Φ_α, the Fréchet CDF. We use this potentially confusing notation for consistency with standard literature. Comparing (9.21) and (9.18), we easily see the parallels between the two theorems, made explicit in Table 9.2.

Of course, the limiting distribution for maxima is more complex than that for sums (of RVs with *finite* variance) because it has an extra parameter ξ, and the normalizing constants have a more complex dependence on the CDF F. Reference [18] provides these constants for some common distribution types of F. Reference [18] also gives a general form of the CLT that handles infinite variance.

The conditions for which $F \in \text{MDA}(H)$ for some non-degenerate H, although tighter than for the general CLT, are quite general for most practical purposes, and known well. Gnedenko [22] provided the first rigorous proof for the Fisher–Tippett theorem, showing conditions on F required for the convergence to each of the three limiting CDFs. For details regarding these conditions, see [23, 18, 16]. Here, we only list some common distributions belonging to $\text{MDA}(H_\xi)$, in Table 9.3, and immediately proceed to the result due to Balkema and de Haan [24] and Pickands [25], that forms the basis for our tail modeling method.

We recall the definition of F_t as the conditional tail CDF for a tail threshold t, as in (9.12). Then, the following is true.

Table 9.3 Some common distributions lying in MDA (H_ξ). For a longer list, see [18]

H_ξ	Distributions in $\text{MDA}(H_\xi)$
$\Phi_{-1/\xi}$	Cauchy Pareto Loggamma
$\Psi_{1/\xi}$	Uniform Beta
Λ	Normal Lognormal Gamma Exponential

Theorem 9.3 [Balkema and de Haan [24] and Pickands [25]] *For every* $\xi \in \Re$, $F \in MDA(H_\xi)$ *if and only if*

$$\lim_{t \to \infty} \sup_{z \geq 0} \left| F_t(z) - G_{\xi,\beta(t)}(z) \right| = 0 \tag{9.22}$$

for some positive function $\beta(t)$*, where* $G_{\xi,\beta}(z)$ *is the generalized Pareto distribution (GPD)*

$$G_{\xi,\beta}(z) = \begin{cases} 1 - \left(1 - \xi \frac{z}{\beta}\right)^{1/\xi}, & \xi \neq 0, \; z \in D(\xi, \beta) \\ 1 - e^{-z/\beta}, & \xi = 0, \; z \geq 0 \end{cases}, \textit{where} \tag{9.23}$$

$$D(\xi, \beta) = \begin{cases} [0, \infty), & \xi \leq 0 \\ [0, \beta/\xi & \xi > 0 \end{cases}$$

In other words, for any distribution F in the maximum domain of attraction of the GEV distribution, the conditional tail distribution F_t converges to a GPD as we move further out in the tail. This is an extremely useful result: it implies that, if we can generate enough points in the tail of a distribution ($y \geq t$), in most practical cases, we can fit the simple, analytical GPD to the data and make predictions further out in the tail. This approach would be independent of the circuit or the performance metric being considered.

Of course, two important questions remain

- How do we efficiently generate a large number of points in the tail ($y \geq t$)?
- How do we fit the GPD to the generated tail points?

We answer the second question next.

9.4.3 Estimating the Tail: Fitting the GPD to Data

For now, let us suppose that we can generate a reasonably large number of points in the tail of our performance distribution. For this we might, theoretically, use standard Monte Carlo simulation with an extremely large sample size, or, more practically, the statistical blockade sampling method we will discuss in Section 9.5.2. Let this data be $\mathbf{Z} = (Z_1, \ldots, Z_n)$, where each Z_i is the *exceedance* over the tail threshold t ($Z_i > 0$, for all i). All Z_i are i.i.d. random variables with common CDF F_t. Then we have the problem of estimating the optimal GPD parameters ξ, β from this tail data, so as to best fit the conditional tail CDF F_t. There are several options; we review three of the most popular ones here. In particular, we focus on methods that require no manual effort and can be completely optimized. For manual methods based on graphical exploration of the data, see [18].

9.4.3.1 Maximum Likelihood Estimation

Maximum likelihood estimation (MLE) is a standard statistical estimation technique that tries to estimate those model parameters (here ξ, β of the GPD) that maximize the "chances" of obtaining the data that we have observed. The probability density function of a GPD $G_{\xi,\beta}$ is given as

$$g_{\xi,\beta}(z) = \begin{cases} \frac{1}{\beta}\left(1 - \xi \frac{z}{\beta}\right)^{1/\xi - 1}, & \xi \neq 0,\ z \in D(\xi,\beta) \\ \frac{1}{\beta} e^{-z/\beta}, & \xi = 0,\ z \geq 0 \end{cases}, \tag{9.24}$$

where $D(\xi, \beta)$ is defined in Theorem 9.3. Recall that all Z_i are i.i.d. random variables with common CDF F_t. We assume that F_t is of the form of a GPD. The *likelihood* (chances) of having seen this data from an underlying GPD is the multivariate probability density associated with it, and is given as

$$\ell(\xi, \beta|\mathbf{Z}) = g_{\xi,\beta}(Z_1, \ldots, Z_n) = \prod_{i=1}^{n} g_{\xi,\beta}(Z_i).$$

Since $\ell(\xi, \beta|\mathbf{Z})$ can be too small for accurate computation with finite accuracy, it is typical to use the *log-likelihood function*

$$\ln(\ell(\xi, \beta|\mathbf{Z})) = \sum_{i=1}^{n} \ln\left(g_{\xi,\beta}(Z_i)\right),$$

which increases monotonically with ℓ. MLE then computes (ξ, β) to maximize this log-likelihood, as

$$(\hat{\xi}, \hat{\beta})_{\text{mle}} = \arg\max_{\xi,\beta} \sum_{i=1}^{n} \ln\left(g_{\xi,\beta}(Z_i)\right). \tag{9.25}$$

Substitution of (9.24) in (9.25) and subsequent algebra allows for a simplification to a one-dimensional search that can be exploited by a careful implementation of a Newton–Raphson algorithm. Details regarding such an implementation are shown in [26].

Smith [27] studies convergence when F_t is not exactly of GPD form, and provides limit results for the distributions of $(\hat{\xi}, \hat{\beta})_{\text{mle}}$ for each of the three cases, $F \in \text{MDA}(\Phi_{-1/\xi})$, $F \in \text{MDA}(\Lambda)$, and $F \in \text{MDA}(\Psi_{1/\xi})$. For $\xi < 1/2$, the MLE estimates are asymptotically normal and efficient (bias = 0) under certain regularity assumptions on F. If (ξ, β) are the exact values to be estimated, then as the sample size $n \to \infty$, the variance of the MLE estimates is given as

$$\text{var}\begin{bmatrix}\hat{\xi} \\ \hat{\beta}\end{bmatrix} \to \frac{1-\xi}{n}\begin{bmatrix}1-\xi & \beta \\ \beta & 2\beta^2\end{bmatrix}, \quad \xi < \frac{1}{2}.$$

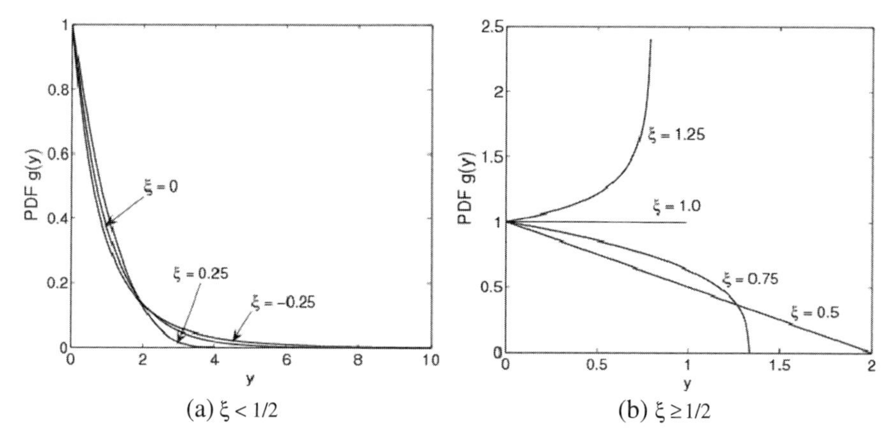

Fig. 9.3 The probability density function for a GPD with $\beta = 1$. We get long unbounded tails for $\xi \leq 0$[2]

When $\xi \geq 1/2$, MLE convergence can be difficult and special techniques are needed, as discussed in [28]. However, $\xi \geq 1/2$ is usually rare, since it corresponds to a finite tail with $g_{\xi,\beta}(z) > 0$ at the endpoint (Fig. 9.3).

9.4.3.2 Moment Matching

An ad hoc way of estimating the GPD parameters is to match the moments of the GPD with the moments of the data, as we now describe. According to [29], the pth moment for the GPD exists if $\xi > -1/p$. In many practical cases we expect finite mean and variance, and hence, existence of the first and second moments. The mean and variance for $G_{\xi,\beta}$ are given as

$$\mu = \frac{\beta}{1+\xi}, \qquad \sigma^2 = \frac{\beta^2}{(1+\xi)^2(1+2\xi)},$$

respectively. Equating these with the *sample* mean and variance, $\hat{\mu}$ and $\hat{\sigma}^2$, respectively, we can compute estimates of ξ and β:

$$\hat{\xi} = \frac{1}{2}\left(\frac{\hat{\mu}^2}{\hat{\sigma}^2} - 1\right), \qquad \hat{\beta} = \frac{\hat{\mu}}{2}\left(\frac{\hat{\mu}^2}{\hat{\sigma}^2} + 1\right).$$

For $\xi > -1/4$, the estimates are asymptotically normal. See [29] for variance estimates for this limit distribution, which are skipped here since this method is not as popular or reliable as the other two methods (MLE and PWM).

[2]Reprinted with permission from *Technometrics*.

9.4.3.3 Probability-Weighted Moment Matching

Probability-weighted moments (PWMs) [30] of a continuous random variable Y with CDF F are generalizations of the standard moments, and are defined as

$$M_{p,r,s} = E[Y^p F^r(Y)(1 - F(Y))^s].$$

The standard pth moment is given by $M_{p,0,0}$. For the GPD, we have a convenient relationship between $M_{1,0,s}$ and (ξ, β), given by

$$m_s = M_{1,0,s} = \frac{\beta}{(1+s)(1+s+\xi)}, \qquad \xi > 0.$$

Then, we can write

$$\beta = \frac{2m_0 m_1}{m_0 - 2m_1}, \qquad \xi = \frac{m_0}{m_0 - 2m_1} - 2.$$

We estimate these PWMs from the data sample, as

$$\hat{m}_s = \frac{1}{n} \sum_{i=1}^{n} (1 - q_i)^s Y_{i,n},$$

where

$$Y_{1,n} \leq Y_{2,n} \leq \cdots \leq Y_{n,n}$$

is the *ordered sample*, and

$$q_i = \frac{i + \gamma}{n + \delta}.$$

Here, γ and δ are fitting parameters. We can use $\gamma = -0.35$ and $\delta = 0$, as suggested for GPD fitting in [29]. The estimates $\hat{\xi}, \hat{\beta}$ converge to the exact values as $n \to \infty$, and are asymptotically normally distributed with covariance given by

$$\operatorname{var}\begin{bmatrix} \hat{\xi} \\ \hat{\beta} \end{bmatrix} \rightarrow \frac{n^{-1}}{(1+2\xi)(3+2\xi)} \times \begin{bmatrix} (1+\xi)(2+\xi)^2(1+\xi+2\xi^2) & \beta(2+\xi)(2+6\xi+7\xi^2+2\xi^3) \\ \beta(2+\xi)(2+6\xi+7\xi^2+2\xi^3) & \beta^2(7+18\xi+11\xi^2+2\xi^3) \end{bmatrix} \tag{9.26}$$

Based on an extensive simulation study, [29] suggests that the PWM method often has lower bias than moment matching and MLE for sample sizes up to 500. Also, the MLE search (9.25) is shown to suffer from some convergence problems when ξ is estimated close to $1/2$. Finally, the study also suggests that PWM matching gives more reliable estimates of the variability of the estimated parameters, as per (9.26). Based on these reasons, we choose PWM matching here.

Once we have estimated a GPD model of the conditional CDF above a threshold t, we can estimate the failure probability for any value y_{f} by substituting the GPD in (9.15) as

$$P(Y > y_f) \approx (1 - F(t)) \left(1 - G_{\xi,\beta}(y_f - t)\right). \tag{9.27}$$

The next section addresses the important remaining question: How do we efficiently generate a large number of points in the tail ($y \geq t$)?

9.5 Statistical Blockade

Before we can introduce the efficient rare event sampling technique, a review of the concept of *classification* from machine learning seems appropriate, given that it will form a cornerstone for the sampling technique. Readers familiar with classifiers may skip to Section 9.5.2.

9.5.1 Classification

Consider the problem of detecting spam in incoming email. A spam detector is some computer program that takes in as inputs, certain features of any new incoming email message, and predicts whether the message is "email" or "spam." The input *features* may be the sender email id, the occurrence of words commonly seen in spam email, etc. We can think of this program as a *function* with these features as inputs and a *categorical* or *discrete* output. The output can assume one of two possible values, "email" or "spam," which we call *classes*. Any such function, that predicts the class of any given input vector, is called a *classifier*, and this act of such prediction is called *classification*. In the general case, we may have any number of classes, and any number of input features. Consider, for example, Fig. 9.4, where there are two input features $\mathbf{x} = (x_1, x_2)$ and three possible classes denoted by ○, Δ, and +.

A classifier would be a function $C(\mathbf{x})$ that, given some input vector $\mathbf{x}$, returns 0, 1, or 2 (for ○, Δ, or +, respectively); i.e., it predicts the class that $\mathbf{x}$ belongs to. Hence, the classifier defines some inter-class boundaries in the input space: Fig. 9.4 also shows a possible set of such boundaries. For a two-class problem with one *linear* boundary, a simple classifier can be based on linear regression as

$$C(\mathbf{x}) = \text{sign}[\mathbf{x}^T \mathbf{w} + b], \tag{9.28}$$

where $\mathbf{w}$ and b are chosen such that the linear function $\mathbf{x}^T\mathbf{w} + b$ is > 0 for any point in one class and <0 for any point in the other class. The sign[$\cdot$] function returns the sign of the linear function, converting its argument from a real variable to a discrete valued variable $\in \{-1,1\}$. The boundary defined by such a linear regression based classifier is

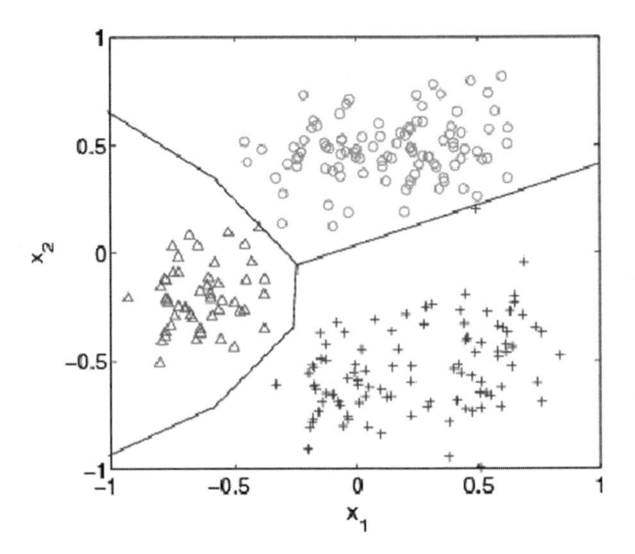

Fig. 9.4 Example with two input features (dimensions) and three classes. The *solid lines* show a possible set of boundaries dividing the classes

$$\mathbf{x}^T\mathbf{w} + b = 0, \tag{9.29}$$

Before any classifier can be used, it has to be "trained;" i.e., optimal values for its parameters have to be determined. In our linear regression example, the parameters are the elements of $\mathbf{w}$, and b. Such training starts from a training sample of points for which the class values are known, and then computes the parameters so as to minimize the error between the classifier predictions and the actual class values for all the training points. Denote the vector of all classifier parameters by $\mathbf{p}$. Let $(\mathbf{x}_1, y_1), (\mathbf{x}_2, y_2), \ldots, (\mathbf{x}_n, y_n)$ be the n training points, where y_i is the class of the ith point $\mathbf{x}_i$. We compute $\mathbf{p}$ such that

$$\min_{\mathbf{p}} \text{Error}\left(\{C(\mathbf{x}_i), y_i\}_{i=1}^n\right),$$

where the precise definition of Error depends on the particular classifier and optimization method used. We now review one particularly successful type of classifier, called *support vector machines*, which we use for an experimental implementation of statistical blockade in this chapter.

9.5.1.1 Support Vector Classifier

A support vector machine (SVM) classifier uses the *optimal separating hyperplane* as the decision boundary between classes in the input space. We provide an introductory discussion of the basic ideas behind SVMs in this section. SVMs enjoy extensive application for statistical inference in a wide variety of problem domains; see [31], for example. There are good reasons for this widespread popularity of SMVs. The basic idea is intuitive and simple, and it allows for classifiers with very good generalizability (low overfitting), relative to many other competing approaches [31, 32]. Reference [31] provides a good tutorial of SVMs in the classification context.

Basic SVM separates two classes with a linear boundary (the optimal separating hyperplane), although, it has been generalized to easily handle nonlinear boundaries and multiple (≥ 3) classes, as discussed in [31] and [32], for instance. Here we restrict our discussion to the two-class, linear classifier that we use for our implementation of the statistical blockade method. First, let us assume that the training points are separable with a linear boundary: there exists a hyperplane that can completely divide the two classes without any errors. This case is shown in Fig. 9.5. We call this hyperplane, the *separating hyperplane*.

We start with a linear classification rule, as in (9.28)

$$C(\mathbf{x}) = \text{sign}[\mathbf{x}^T \mathbf{w} + b]$$

with the separating hyperplane given by

$$S \,:\, \mathbf{x}^T \mathbf{w} + b = 0.$$

Recall that the training data consists of n pairs $(\mathbf{x}_1, y_1), (\mathbf{x}_2, y_2), \ldots, (\mathbf{x}_n, y_n)$, where $y_1 \in \{-1,1\}$. If we can correctly orient this separating hyperplane S; i.e., there are no training errors, then we expect that

$$\begin{aligned} \mathbf{x}_i^T \mathbf{w} + b &> 0 \text{ whenever } y_i = 1, \text{ and} \\ \mathbf{x}_i^T \mathbf{w} + b &< 0 \text{ whenever } y_i = -1. \end{aligned}$$

Then,

$$y_i(\mathbf{x}_i^T \mathbf{w} + b) > 0, \ \forall i.$$

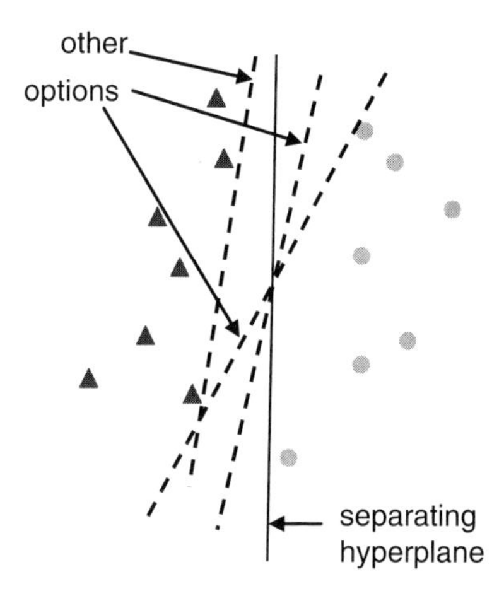

Fig. 9.5 Example of data separable with a hyperplane. In this case, the hyperplane is a *straight line*. Multiple options for the separating hyperplane are shown as *dashed straight lines*

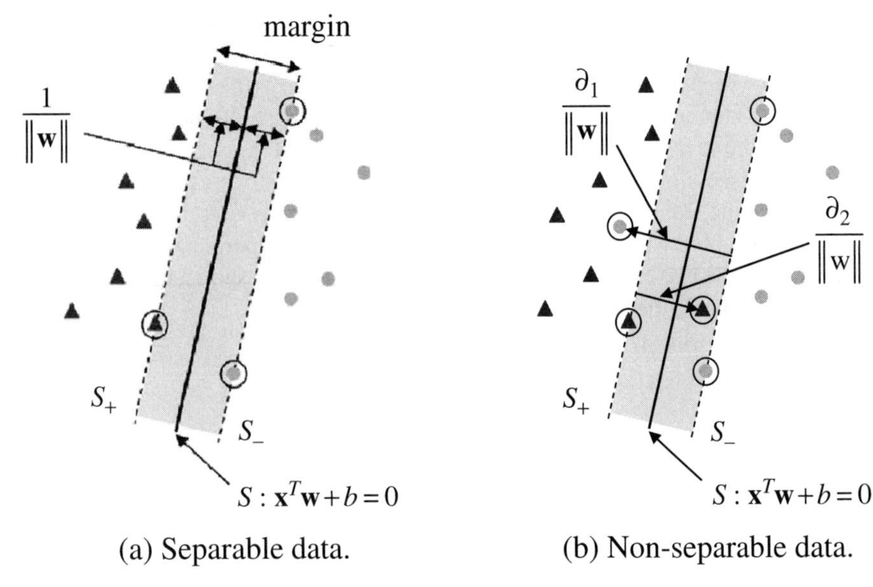

Fig. 9.6 Optimal separating hyperplane maximizes the margin between the nearest points from the two classes, shown here for both separable and non-separable data. The support vectors lie on the margin hyperplanes, and are *circled* in both cases

Even with this condition, there are several choices for the separating hyperplane, as shown in Fig. 9.5: which is the best choice? The best choice is called, understandably, the *optimal* separating hyperplane, and it maximizes the distance to the point nearest to it from either class, as shown in Fig. 9.6(a). For any such hyperplane, let d_+ and d_- denote the distance to the nearest point from classes 1 and –1, respectively.

We define the *margin* of the separating hyperplane as $d_+ + d_-$. Suppose, we always center the hyperplane in the margin, so that $y_i(\mathbf{x}_i{}^T\mathbf{w} + b) = 1$ for the nearest points in either class. The signed perpendicular distance between any point $\mathbf{x}$ and the hyperplane is given by

$$\frac{\mathbf{x}^T\mathbf{w} + b}{\|\mathbf{w}\|}.$$

Hence,

$$d_+ = d_- = \frac{1}{\|\mathbf{w}\|},$$

giving us a margin of $2\|\mathbf{w}\|^{-1}$. The optimal separating hyperplane is obtained by maximizing this margin as

$$\min_{\mathbf{w},b} \tfrac{1}{2} \|\mathbf{w}\|^2 \text{ subject to } y_i(\mathbf{x}_i^T \mathbf{w} + b) \geq 1, \; i = 1, \ldots, n. \tag{9.30}$$

The Lagrangian formulation of this optimization problem is given as

$$\min_{\mathbf{w},b} \tfrac{1}{2} \|\mathbf{w}\|^2 - \sum_{i=1}^{n} \mu_i \left[y_i(\mathbf{x}_i^T \mathbf{w} + b) - 1 \right],$$

where μ_i are the Lagrangian multipliers. At the solution, the derivatives of the objective are zero, giving us

$$\mathbf{w} = \sum_{i=1}^{n} \mu_i y_i \mathbf{x}_i .$$

The multipliers μ_i are nonzero only for those points that satisfy the *equality* in the constraints in (9.30); i.e., for points exactly on the margin hyperplanes (S_+, S_-). Hence, the orientation of the separating hyperplane is determined only by those points that lie on the margin hyperplanes, giving them the name of *support points*. Figure 9.6(a) shows the support points as circled.

All of this has development assumed that the training points are separable using a hyperplane. Of course, in many practical situations this is not true: even the optimal separating hyperplane can suffer from misclassifications due to "overlap" in the classes. Figure 9.6(b) shows an example. The complete linear SVM formulation accounts for these cases by including positive "slack" variables δ_i, $i = 1, \ldots, n$ in the constraints of (9.30) as

$$\begin{aligned} & y_i(\mathbf{x}_i^T \mathbf{w} + b) \geq 1 - \delta_i, \; i = 1, \ldots, n \\ & \delta_i \geq 0, \; i = 1, \ldots, n. \end{aligned} \tag{9.31}$$

For an error to occur on point i, $\delta_i > 1$, hence $\Sigma\delta_i$ is an upper bound on the number of misclassifications. Since we wish to minimize the training error *and* maximize the margin size, a natural choice for the optimization problem is

$$\min_{\mathbf{w},b} \tfrac{1}{2} \|\mathbf{w}\|^2 + \gamma \sum_{i=1}^{n} \delta_i \text{ subject to (9.31)}, \tag{9.32}$$

where γ is a user-supplied tuning parameter. Note that the distance of a misclassified point from its margin hyperplane is $\delta_i||\mathbf{w}||^{-1}$, as shown in Fig. 9.6(b). All points on their margin hyperplane or on the wrong side of it are called the *support vectors*, since they alone determine the estimates of $\mathbf{w}$ and b, in a manner similar to the separable case. This optimization problem is easiest to solve in its dual form. We refer the reader to [31, 32] for further details on these aspects, and generalizations of SVMs to nonlinear boundaries and multiple classes. SVMs are a popular, well-researched classification strategy, and optimized software implementations are readily available; for example, SVM^{light} [33] and WEKA [34].

9.5.2 The Statistical Blockade Algorithm

We are now almost ready to synthesize all the pieces of the statistical blockade algorithm: only a mapping of the theory in the foregoing sections to various aspects of the high-replication circuit problem is needed. For this, let any circuit performance metric, or simply, output y be computed as

$$y = f_{\text{sim}}(\mathbf{x}), \tag{9.33}$$

Here, $\mathbf{x}$ is a point in the statistical parameter (e.g., V_{t}, t_{ox}) space, or simply, the input space, and f_{sim} includes expensive SPICE simulation. We assume that y has some probability distribution F, with an extended tail. Suppose, we define a large tail threshold t for y, then from the developments in Section 9.4.2 we know that we can approximate the conditional tail CDF F_t by a generalized Pareto distribution $G_{\xi,\beta}$. Section 9.4.3 shows how we can estimate the GPD parameters (ξ, β) from data drawn from the tail distribution. We now discuss an efficient tail sampling strategy that will generate the tail points for fitting this GPD.

Corresponding to the tail of output distribution, we expect a "tail region" in the input space: any statistical parameter values drawn from this tail region will give an output value $y > t$. Figure 9.7 shows an example of such a tail region for two inputs. The rest of the input space is called the "body" region, corresponding to the body of the output distribution F. In Fig. 9.7 these two regions are separated by a dashed line.

The key idea behind the efficient sampling technique is to identify the tail region and simulate only those Monte Carlo points that are likely to lie in this tail region. Here, we exploit the common fact that *generating* the random values for a Monte

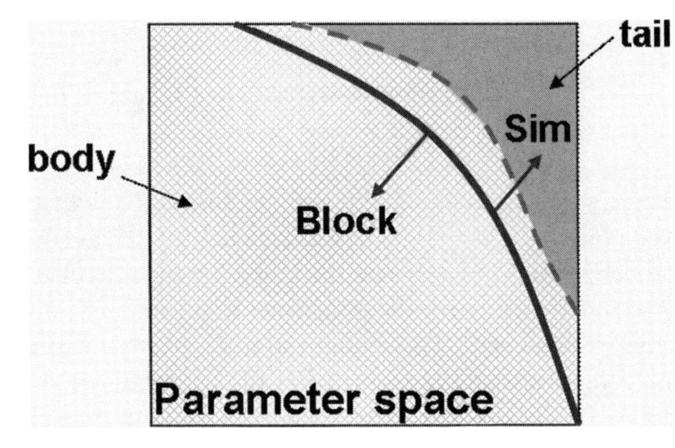

Fig. 9.7 The tail and body regions in the statistical parameter space. The *dashed line* is the exact tail region boundary for tail threshold t. The *solid line* is the relaxed boundary modeled by the classifier for a classification threshold $t_{\text{c}} < t$

Carlo sample point is very cheap compared to actually *simulating* the point as in (9.33). Hence, if we generate points as in standard Monte Carlo, but *block* – not simulate – those points that are unlikely to fall in the tail region, we can drastically cut down the total time spent. This reduction in time spent is drastic because we are trying to simulate only the rare events, which by definition constitute a very small percentage of the total Monte Carlo sample size. As might be obvious to the discerning reader, we can use a classifier to distinguish the tail and body regions, and to block out the body points. For any candidate point in the input space, generated from standard Monte Carlo, the classifier can predict its membership in either the "body" or the "tail" classes. Only the "tail" points are then simulated.

To build this model of the tail region boundary, the classifier can be trained with a small (e.g., 1,000 points) training set of simulated Monte Carlo sample points. However, it is difficult, if not impossible to build an *exact* model of the boundary in general. Misclassifications, at least on points unseen during training, is unavoidable. Hence, we relax the accuracy requirement to allow for classification error. This is done by building the classification boundary at a *classification threshold* t_c that is less than the tail threshold t. Since we have assumed that only the upper (right) tail is relevant, the tail region corresponding to t will be a subset of the tail region corresponding to t_c, if $t_c < t$. This will help to ensure that, even if the classifier is imperfect, it is unlikely that it will misclassify points in the true tail region (for t). The relaxed boundary corresponding to such a t_c is shown as the solid line in Fig. 9.7. The statistical blockade algorithm is then as in Algorithm 9.1. The algorithm derives its name from the blocking activity of the classifier. This classifier is also referred to as the *blockade filter* and its blocking activity as *blockade filtering*.

The thresholds $t = p_t$th percentile and $t_c = p_c$-th percentile are estimated from the small initial Monte Carlo run, which also gives the n_0 training points for the classifier. Typical values for these constants are shown in Algorithm 9.1. The function *MonteCarlo*(n) generates n points in the statistical parameter space, which are stored in the $n \times s$ matrix $\mathbf{X}$, where s is the input dimensionality. Each row of $\mathbf{X}$ is a point in s dimensions. $\mathbf{y}$ is a vector of output values computed from simulations. The

Algorithm 9.1 The statistical blockade algorithm for efficiently sampling rare events and estimating their probability distribution

Require: training sample size n_0 (e.g., 1,000)
total sample size n
percentages p_t (e.g., 99%), p_c (e.g., 97%)

1. $\mathbf{X} = \text{MonteCarlo}(n_0)$
2. $\mathbf{y} = f_{\text{sim}}(\mathbf{X})$
3. $t = \text{Percentile}(\mathbf{y}, p_t)$
4. $t_c = \text{Percentile}(\mathbf{y}, p_c)$
5. $C = \text{BuildClassifier}(\mathbf{X}, \mathbf{y}, t_c)$ // C is a classifier
6. $\mathbf{y} = f_{\text{sim}}(\text{Filter}(C, \text{MonteCarlo}(n)))$
7. $\mathbf{y}_{\text{tail}} = \{y_i \in \mathbf{y} : y_i > t\}$
8. $(\xi, \beta) = \text{FitGPD}(\mathbf{y}_{\text{tail}} - t)$

function *BuildClassifier*(**X**, **y**, t_c) trains and returns a classifier using the training set (**X**, **y**) and classification threshold t_c. The function *Filter*(*C*, **X**) blocks the points in **X** classified as "body" by the classifier *C*, and returns only the points classified as "tail". *FitGPD*($\mathbf{y}_{\text{tail}} - t$) computes the parameters (ξ, β) for the best GPD approximation $G_{\xi,\beta}$ to the conditional CDF of the exceedances of the tail points in $\mathbf{y}_{\text{tail}}$ over t. We can then use this GPD model to compute statistical metrics for rare events, for example, the failure probability for some threshold y_f, as in (9.27). This sampling procedure is also illustrated in Fig. 9.8.

9.5.2.1 Note on Choosing and Unbiasing the Classifier

The algorithm places no restrictions on the choice of classifier. Here, we will use support vector machines for our experiments. We now make some practical observations here that are relevant for the choice of classifier. High-replication circuits naturally tend to be small, relatively simple circuits. It is highly unlikely that a complex, large circuit will be replicated thousands to millions of times on the same chip. This level of replication often naturally coincides with simple functionality. As a result, we often do not expect to see drastically nonlinear boundaries for the tail regions of these circuits. Nor do we expect to see very complex topologies of the tail regions. These considerations, along with the safety margin awarded by a classification threshold t_c less than t, led us to use linear SVMs. Indeed, linear SVMs suffer minimally from overfitting issues and from the complex parameter selection problems of nonlinear, kernel-based SVMs. As we shall see with experiments in Sections 9.5.3 and 9.6.4, this choice does result in an effective implementation of statistical blockade. For cases where a strongly nonlinear boundary exists, a linear

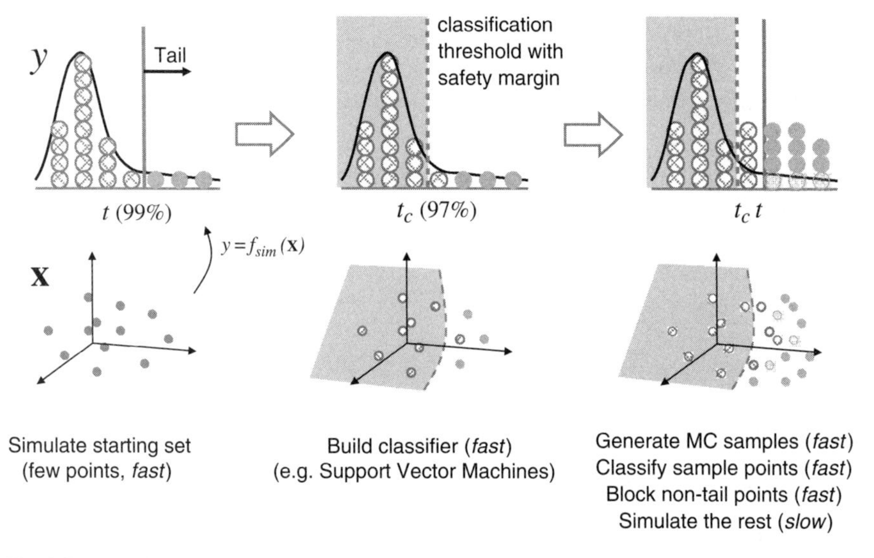

Fig. 9.8 The efficient tail (rare event) sampling method of statistical blockade

classifier may not suffice, and more sophisticated classification techniques of the type discussed in [32] may be required. The statistical blockade framework, however, should not need any fundamental change.

An important technical point to note about the classifier construction is as follows. The training set will typically have many more body points than tail points. Hence, even if all or most of the tail points are misclassified, the training error will be low as long as most of the body points are correctly classified. This will result in a classifier that is biased to allow more misclassifications of points in the tail region. However, we need to minimize misclassification of tail points to avoid distorting the statistics of the simulated tail points. Hence, we need to reverse bias the classification error.

Using the technique proposed in [35], we penalize misclassifications of tail points more than misclassifications of body points. In the context of SVMs, let us rewrite the training objective function (9.32) as

$$\min_{\mathbf{w},b} \frac{1}{2} \|\mathbf{w}\|^2 + \gamma_+ \sum_{i\,:\,y_i=1} \delta_i + \gamma_- \sum_{i\,:\,y_i=-1} \delta_i \text{ subject to (9.31)},$$

where γ_+ and γ_- are possibly different penalty factors for the two classes ("tail" and "body" in our case). If, as in [35], we choose

$$\frac{\gamma_+}{\gamma_-} = \frac{\text{Number of ` } - \text{ ' training points}}{\text{Number of ` } + \text{ ' training points}} = \frac{\text{Number of `body' points}}{\text{Number of `tail' points}},$$

we can obtain an unbiased classifier. Any other choice of classifier (instead of SVMs) will also require such asymmetric penalties during training.

9.5.3 Experimental Results

Let us now apply the statistical blockade method to three test cases to illustrate its use:

1) a 6T SRAM cell,
2) a complete 64-bit SRAM column with write driver, and
3) a master-slave flip-flop with the scan chain component.

We construct each blockade filter using a training sample set of $n_0 = 1{,}000$ points from a standard Monte Carlo run. The filter is an SVM classifier built using the 97th percentile of each relevant performance metric as the classification threshold t_c. The tail threshold is defined as the 99th percentile. In all cases the rare event statistical metric we compute is the failure probability $F_f(y_f)$ for any failure threshold y_f, using the GPD fit to the tail defined by the tail threshold t. We represent this failure probability as the equivalent quantile y_σ on the standard normal distribution:

$$y_\sigma = \Phi^{-1}\left(1 - F_f(y_f)\right) = \Phi^{-1}\left(F(y_f)\right),$$

where Φ is the standard normal CDF. For example, a failure probability of $F_f = 0.00135$ implies a cumulative probability of $F = 1 - F_f = 0.99865$. The equivalent point on a standard normal, having the same cumulative probability, is $y_\sigma = 3$. In other words, any y_f with a failure probability of 0.00135 is a "3σ" point.

We can compute $F_f(y_f)$, and hence y_σ, in three different ways:

I. *Empirically*: Run a large Monte Carlo run where all points are fully simulated; i.e., with no use of blockade filtering or EVT. Say we use a sample size of n_{MC} (e.g., 1 million), giving us n_{MC} values y_i, $i = 1, \ldots, n_{\mathrm{MC}}$. Then we can empirically compute $F_f(y_f)$ as

$$F_f(y_f) \approx \frac{\left|\{y_i \,:\, y_i > y_f\}\right|}{n_{\mathrm{MC}}}.$$

Of course, for any $y_f > \max(\{y_1, \ldots, y_{n_{\mathrm{MC}}}\})$, we will get the same estimate of 0 failure probability, and $y_\sigma = \infty$, since there are no points beyond this y_f to give us any information about such rare events. Hence, the prediction power of the empirical method is limited by the Monte Carlo sample size.

II. *Using GPD model, with no blockade filtering*: We can run a full Monte Carlo run with no filtering, as in the empirical estimation case, but then fit a GPD to the points in the tail, defined by the tail threshold t. These are the points $\{y_i : y_i > t\}$. Using this GPD, $G_{\xi,\beta}$, in (9.27), which we reproduce here for convenience,

$$F_f(y_f) = P(Y > y_f) \approx (1 - F(t))\left(1 - G_{\xi,\beta}(y_f - t)\right), \tag{9.34}$$

we can compute the failure probability. $F(t)$ can be estimated empirically with good accuracy. The GPD model extends the prediction power all the way to ∞. Of course, the confidence in the prediction would decrease as we move to very high values of y_f.

III. *Using statistical blockade*: Here we use the complete statistical blockade flow, where only candidate tail points identified by the blockade filter are simulated, and a GPD tail model is estimated from the actual tail points $y > t$. Here, too, we use (9.34), but the points used to estimate (ξ, β) are obtained from blockade filtering. Further, we use a Monte Carlo sample size that is much smaller than for method II, to test statistical blockade in a practical setting, where we want to use as small a sample size as possible.

For all the test cases we will compare the predictions of y_σ from these three methods. Method II will give us the most accurate estimates, since it uses a large number of points and no filtering. In some cases, we will also see estimates computed using a Gaussian distribution fit to highlight the error in such an approach. Let us now look at the test circuits in more detail, along with the results we obtain.

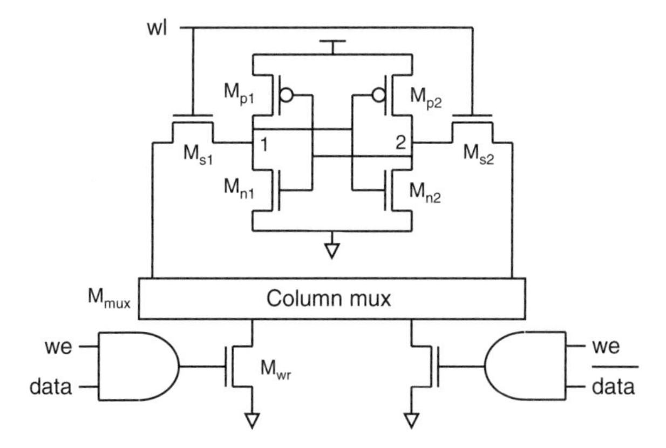

Fig. 9.9 A 6T SRAM cell with write driver and column mux

9.5.3.1 6T SRAM Cell

The first test case is a standard 6T SRAM cell with bit-lines connected to a column multiplexer and a non-restoring write driver, shown in Fig. 9.9.

Here, we use the Cadence 90 nm Generic PDK library for the device models, and model RDF on every transistor as independent, normally distributed threshold voltage (V_t) variation per transistor. The V_t standard deviation is taken to be

$$\sigma(\mathrm{V_t}) = \frac{5000}{\sqrt{WL}}\mathrm{mV},$$

where W and L are the transistor width and length, respectively, in nanometers. We also include a global gate oxide thickness variation, also normally distributed and with a standard deviation of 2% of the nominal thickness. This gives us a total of nine statistical parameters. The metric we are measuring here is the write time τ_{w}: the time between the wordline going high, to the non-driven cell node (node 2) transitioning. Here, "going high" and "transitioning" imply crossing 50% of the full voltage change.

For methods I and II, we use $n_{\mathrm{MC}} = 1$ million Monte Carlo points. For statistical blockade (method III), 100,000 Monte Carlo points are filtered through the classifier, generating 4,379 tail candidates. On simulating these 4,379 points, 978 true tail points ($\tau_{\mathrm{w}} > t$) were obtained, which were then used to compute a GPD model for the tail conditional CDF. Table 9.4 shows a comparison of the y_σ values estimated by the three different methods. We can see a close match between the predictions by the accurate method II and statistical blockade, method III. Figure 9.10 compares the conditional tail CDFs computed from the empirical method and from statistical blockade, showing a good match.

Some observations highlighting the efficiency of statistical blockade can be made immediately.

Table 9.4 Prediction of failure probability as y_σ by methods I, II, and III, for a 6T SRAM cell. The number of simulations for statistical blockade includes the 1,000 training points. The write time values are in "fanout of 4 delay" units

τ_w (y_f) (FO4)	(I) Standard Monte Carlo	(II) GPD *no* blockade filter	(III) Statistical blockade
2.4	3.404	3.408	3.379
2.5	3.886	3.886	3.868
2.6	4.526	4.354	4.352
2.7	∞	4.821	4.845
2.8	∞	5.297	5.356
2.9	∞	5.789	5.899
3.0	∞	6.310	6.493
Number of simulations	1,000,000	1,000,000	5,379

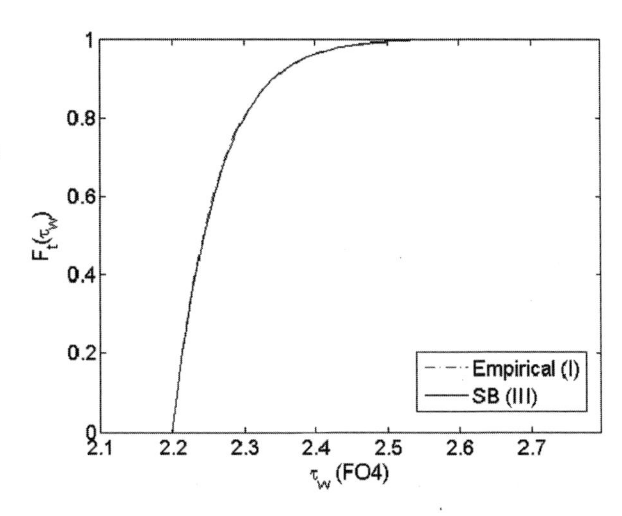

Fig. 9.10 Comparison of GPD tail model from statistical blockade (5,379 simulations) and the empirical tail CDF (1 million simulations) for the write time of the 6T SRAM cell

- The empirical method fails beyond 2.6 FO4, corresponding to about 1 ppm circuit failure probability, because there are no points generated by the Monte Carlo run so far out in the tail.
- Fitting a GPD model to the tail points (method II) allows us to make predictions far out in the tail, even though we have no points that far out.
- Using blockade filtering, coupled with the GPD tail model, we can drastically reduce the number of simulations (from 1 million to 5,379) with very small change to the tail model.

Of course, the tail model cannot be relied on too far out from the available data, as suggested by the increased discrepancy between methods II and III for the largest τ_w values. We further discuss and attack this problem in Section 9.6.

9.5.3.2 64-bit SRAM Column

The next test case is a 64-bit SRAM column, with a non-restoring write driver and column multiplexer (Fig. 9.11). Only one cell is being accessed, while all the other wordlines are turned off. Random threshold variation on all 402 transistors (including the write driver and column mux) are considered, along with a global gate oxide variation. The device and variation models are the same 90 nm technology as for the SRAM cell we just saw. In scaled technologies, leakage current is no longer negligible [36]. Hence, process variations on transistors that are meant to be inaccessible (or off) can also impact the overall behavior of a circuit. This test case allows us to see the impact of variations in the leakage current passing through the 63 off cells, along with variations in the write driver. Since the BSIM3v3 models [37] are used, the gate leakage is not well modeled, but the drain leakage is.

Once again we measure the write time, in this case from the wordline wl_0 to node 2, for falling node 2. The number of statistical parameters is 403. Building a reliable classifier with only $n_0 = 1{,}000$ points in 403 dimensional space is nearly impossible because of the *curse of dimensionality* [32]. However, we can reduce the dimensionality by choosing only those dimensions (statistical parameters) that have

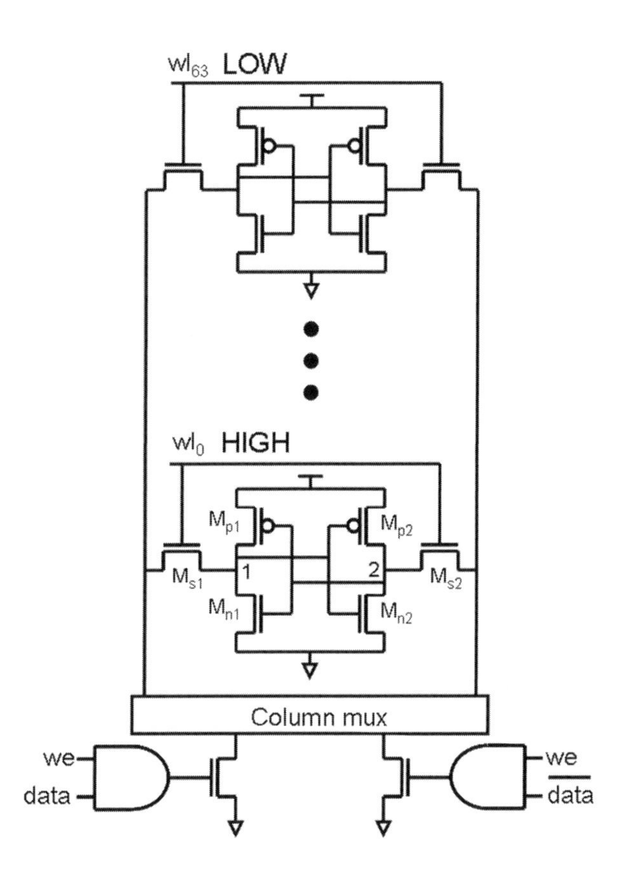

Fig. 9.11 A 64-bit SRAM column with write driver and column multiplexer

a significant impact on the write time. This problem of detecting the "important" variables appears in a wide variety of problems. For a relevant discussion in the context of circuits, see [23].

Here we use a simple measure of impact: the Spearman's rank correlation coefficient, ρ_S (9.35), between each statistical parameter and the circuit performance metric to quantitatively estimate the strength of their relationship. Spearman's rank correlation between two variables x, y, given the sample set $\{x_j, y_j\}$ for $j = 1, \ldots, n$, is given by

$$\rho_S(x, y) = \frac{\sum_{j=1}^{n}(P_j - P)(Q_j - Q)}{\sqrt{\sum_{j=1}^{n}(P_j - P)^2}\sqrt{\sum_{j=1}^{n}(Q_j - Q)^2}}, \tag{9.35}$$

where P_j and Q_j are the *ranks* of x_j and y_j in the sample set, as shown by the example in Table 9.5. To compute the rank of, say x_j, we sort all the x values in increasing order and take the position of x_j in this sorted list as its rank. P and Q denote the means of the ranks. Hence, ρ_S is just Pearson's linear correlation on the ranks. However, this measure of correlation does not assume linearity like the latter, and hence, gives more relevant estimates of the sensitivities. For classification, only parameters with $|\rho_S| > 0.1$ are used, reducing the dimensionality to only 11.

Figure 9.12 shows the sorted magnitudes of the 403 rank correlation values: we can see that only a handful of the statistical parameters have significant correlation with the write time. The transistors (the threshold voltages) chosen by this method are

- the pull-down and output transistors in the active write-driver AND gate,
- the bitline pull-down transistors, and
- all transistors in the active 6T cell, except for M_{p2} (since node 2 is being pulled down in this case).

We can see that this selection coincides with a designer's intuition of the devices that would have the most impact on the write time.

y_σ is computed for increasing failure thresholds, using all three methods. We use $n_{MC} = 100{,}000$ simulated Monte Carlo points for methods I and II. For statistical blockade, method III, we filter these 100,000 points through the classifier in reduced dimensions, giving 5,314 candidate tail point. On simulation, we finally

Table 9.5 Example illustrating the concept of ranks for Spearman's rank correlation. The rank of a value is its position in a sorted list of its class; for example, 0.76 is third in the list of x values sorted in increasing order

x	P	y	Q
0.1	2	101	2
–0.1	1	89	1
0.89	4	130	3
0.76	3	132	4

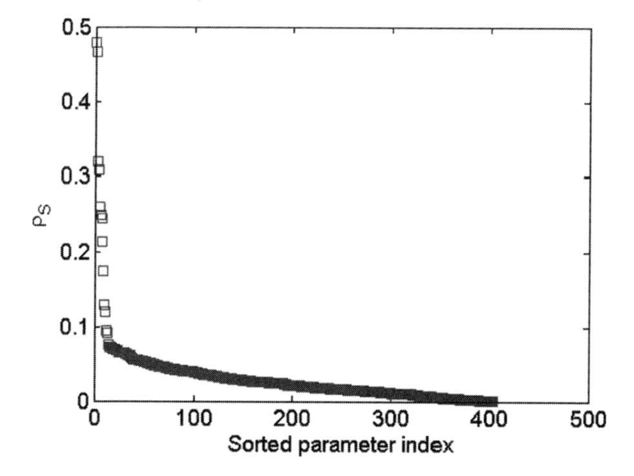

Fig. 9.12 Magnitudes of rank correlation between the statistical parameters and the write time of the SRAM column. Only a few parameters have a strong relationship with the write time

Table 9.6 Prediction of failure probability as y_σ by methods I, II, III and by Gaussian approximation, for the SRAM column. The number of simulations for statistical blockade includes the 1,000 training samples. The write time values are in "fanout of 4" delay units

τ_w (y_f)(FO4)	(I) Standard Monte Carlo	(II) GPD *no* blockade filter	(III) Statistical blockade (100 K)	(III) Statistical blockade (20 K)	Gaussian approximation
2.7	2.966	2.986	3.010	2.990	3.364
2.8	3.367	3.73	3.390	3.425	3.898
2.9	3.808	3.743	3.747	3.900	4.432
3.0	∞	4.101	4.088	4.448	4.966
3.1	∞	4.452	4.416	5.138	5.499
3.2	∞	4.799	4.736	6.180	6.033
3.3	∞	5.147	5.049	–	6.567
3.4	∞	5.496	5.357	–	7.100
Number of simulatios	100,000	100,000	6,314	2,046	20,000

obtain 1,077 true tail points. Table 9.6 compares the predictions by these three methods. We can see the close match between the accurate method II and statistical blockade, even though the total number of simulations is reduced from 100,000 to 6,314. The empirical method, once again, falls short of our needs, running out of data beyond $\tau_w = 2.9$ FO4. Figure 9.13 graphically shows the agreement between the conditional tail models extracted empirically and using statistical blockade.

Let us now further reduce the Monte Carlo sample size for statistical blockade, to see if the simulation cost can be further reduced while maintaining accuracy. We use statistical blockade on only 20,000 Monte Carlo points, giving 1,046 filtered candidate tail points and 218 true tail points. However, the predictions (column 5) show large errors compared to our reference, method II. This suggests that a tail sample of only 218 is not sufficient to obtain a reliable model.

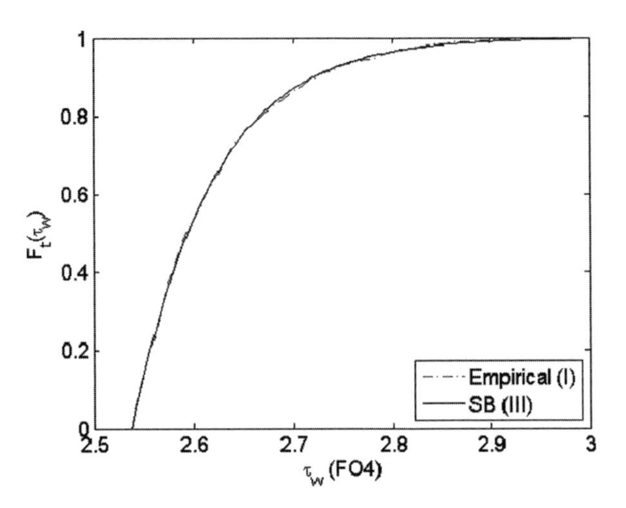

Fig. 9.13 GPD model of SRAM column write time from statistical blockade (6,314 simulations) compared with the empirical conditional CDF (100,000 simulations)

We also use a Gaussian fit to 20,000 simulated Monte Carlo points for estimating y_σ. It is clear from the table that, in this case, a Gaussian fit underestimates the failure probability, with the error increasing as we move to rarer events. Comparing the statistics for the SRAM column in Table 9.6 with the statistics for the SRAM cell in Table 9.4, we can see that the distribution of write time has a larger spread for the SRAM column than for the SRAM cell. For example, the 4.8σ point for the SRAM cell is 2.7 FO4, while for the SRAM column, it is 3.2 FO4. The reason for this increased spread is that the variations in the leakage current of the entire column contribute significantly to the variation of the performance of any single cell. This observation suggests that, in general, simulating variations in a single circuit, without modeling variations in its environment circuitry can lead to large errors in the estimated statistics.

9.5.3.3 Master–Slave Flip-Flop with Scan Chain

This last test case is a master–slave flip-flop with the scan chain component (MSFF) (Fig. 9.14).It may seem surprising to see a flip-flop test case in a book on embedded memory. However, this test case will help us to highlight certain subtleties in the statistical blockade framework which any practical implementation must account for. The discussion will attempt to focus more on these subtleties, without involving details of the flip-flop operation. Flip-flops are ubiquitous in digital circuits, and can be highly replicated in large chips. There may be situations where a rare event study for flip-flops may be relevant.

This circuit is implemented in 45 nm technology using the predictive technology models (PTM) from [38]. Once again, we include normally distributed V_t variation per transistor, with a relatively large standard deviation of

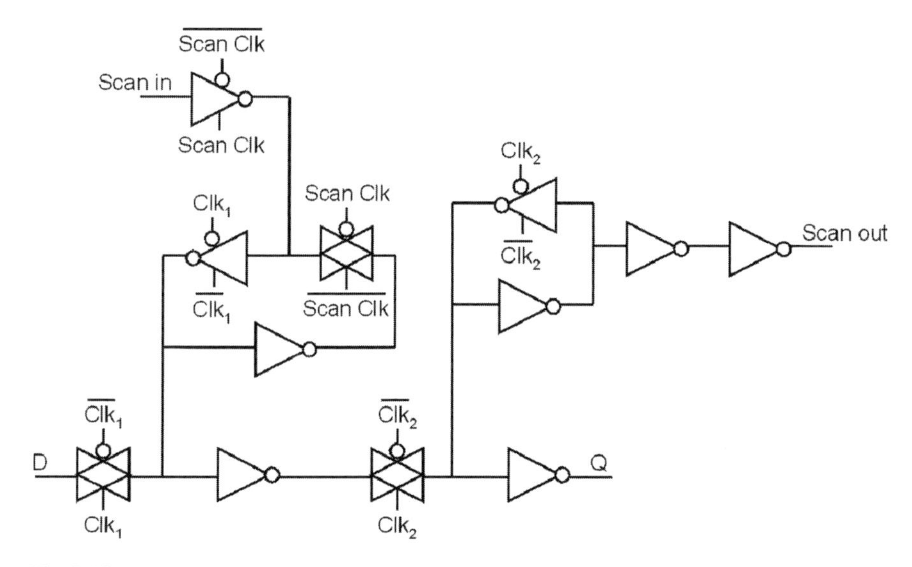

Fig. 9.14 A master–slave flip-flop with the scan chain component

$$\sigma(\mathrm{V_t}) = \frac{13.5\mathrm{V_{t0}}}{\sqrt{WL}},$$

where W and L are the transistor width and length, respectively, in nanometers, and V_{t0} is the nominal V_t in the PTM. Normally distributed global gate oxide thickness variation is also considered, with a standard deviation of 2% of the nominal value.

For this circuit, we are measuring the clock-output delay τ_{cq}. The flip-flop has a peculiarity in its rare event behavior. For large deviations in the statistical parameters, the flip-flop reaches metastable behavior and, using standard circuit simulators, we fail to see the flip-flop output converge to the value at the input. This leads to an undefined clock-output delay for some Monte Carlo points. We reject any such points without replacement in this experiment. Although these rejected points are also rare events, they distort the smoothness of the tail that is required to apply the EVT limit theorem, if not rejected. This still allows us to test the speed and tail modeling efficiency of statistical blockade, since we use the same rejection method across all estimation methods. In practice, the fraction of such undefined delay events can be estimated from the simulated points in statistical blockade and added to the failure probability estimated using the GPD model, to give the overall failure probability.

For methods I and II, we simulate $n_{MC} = 500{,}000$ Monte Carlo points. For statistical blockade (method III), we filter 100,000 Monte Carlo points to obtain 7,785 candidate tail points which, on simulating, yield 692 tail points. Note that here we have ignored any tail points for which the flip-flop output did not capture the input value. Figure 9.15 compares the conditional tail CDFs from the empirical and statistical blockade models.

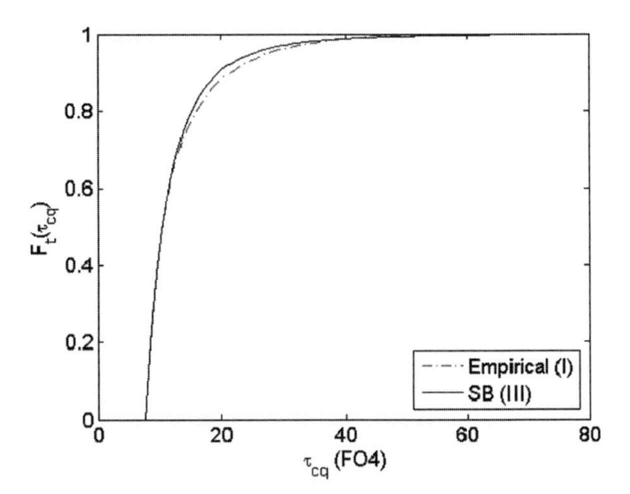

Fig. 9.15 GPD model of MSFF delay from statistical blockade (8,785 simulations) compared with the empirical conditional CDF (500,000 simulations)

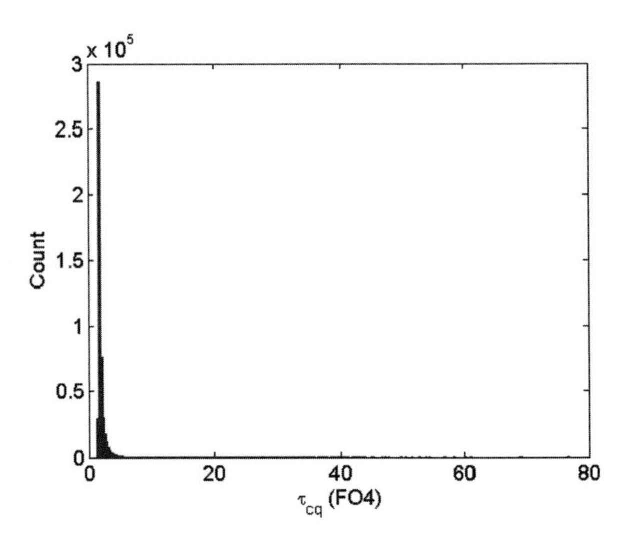

Fig. 9.16 Histogram of clock-output delay of the MSFF, showing a long heavy tail and high skewness

We also compare predictions from these methods with those from a Gaussian fit to 20,000 simulated Monte Carlo points. Figure 9.16 shows a histogram of the delay values obtained from the 500,000-point Monte Carlo run. The extreme skewness and the heavy tail of the histogram suggest that a Gaussian fit would be grossly inaccurate. Table 9.7 shows the estimates of y_σ computed by these four methods: we can clearly see the gross errors in the Gaussian estimates.

In this case, we also see some discrepancy between the empirical and GPD-predicted values that is larger than in the cases of the SRAM cell and column. There can be two reasons for this

Table 9.7 Prediction of failure probability as y_σ by methods I, II, III and by Gaussian approximation, for the MSFF. The number of simulations for statistical blockade includes the 1,000 training samples. The delay values are in "fanout of 4" delay units

τ_{cq} (y_f) (FO4)	(I) Standard Monte Carlo	(II) GPD no blockade filter	(III) Statistical blockade	Gaussian approximation
30	3.424	3.466	3.431	22.127
40	3.724	3.686	3.661	30.050
50	4.008	3.854	3.837	37.974
60	4.219	3.990	3.978	45.898
70	4.607	4.102	4.095	53.821
80	∞	4.199	4.195	61.745
90	∞	4.283	4.282	69.669
Number of simulatios	500,000	500,000	8,785	20,000

1) Due to the quite heavy tail, slight variations in the chosen tail samples can cause significant variations in the model.
2) The tail threshold of $t =$ the 99th percentile might not be large enough to fit a GPD with near exactness; that is, the tail conditional CDF might not have converged to the GPD form.

It turns out that the actual reason is the second one. We further explore, and address, this problem in Section 9.6.2.

The GPD fits do, however, capture the heavy tail of the distribution. To see this, compare Table 9.7 with the results for the SRAM cell in Table 9.4. A 20% increase in the SRAM write time, from 2.5 FO4 to 3 FO4, results in an increase of 2.424 in y_σ, while a similar percentage increase in the MSFF delay, from 50 FO4 to 60 FO4, increases y_σ by only 0.136, even though the increases are from similar probability levels: 3.886σ for the SRAM cell, and 3.854σ for the MSFF.

In summary of these results, we see that statistical blockade provides an efficient way of sampling rare events and modeling their statistics. However, there are some issues that need to be addressed. First, the predictions may not be reliable for events that are very far out in the tail. Second, we saw some notable discrepancy between the empirical and GPD tail models for the case of the flip-flop: the exact reason for this is not yet obvious. The next section explores these issues in more detail and shows enhancements to the statistical blockade method to address them.

9.6 Making Statistical Blockade Practical

Although statistical blockade provides us an effective method for sampling rare events and modeling their statistics, there are still some practical issues left unresolved by the algorithm in Section 9.5.2. We saw a glimpse of some of these issues in the results presented in Section 9.5.3. Let us look at these in more detail now.

9.6.1 Conditionals and Disjoint Tail Regions

9.6.1.1 The Problem

SRAM performance metrics are often computed for two states of the SRAM cell: while storing a 1, and while storing a 0. The final metric value is then a maximum or a minimum of the vales for these two states. The presence of such *conditionals* (max, min) can result in *disjoint* tail regions in the statistical parameter space, making it difficult to use a single classifier to define the boundary of the tail region. Let us look at an example to illustrate this problem.

Consider the 6T SRAM cell. With technology scaling reaching nanometer feature sizes, subthreshold and gate leakage have become very significant. In particular, for the large memory blocks seen today the standby power consumption due to leakage can be intolerably high. Supply voltage (V_{dd}) scaling [39] is a powerful technique to reduce this leakage, whereby the supply voltage is reduced when the memory bank is not being accessed. However, lowering V_{dd} also makes the cell unstable, ultimately resulting in data loss at some threshold value of V_{dd}, known as the *data retention voltage* or DRV. Hence, the DRV of an SRAM cell is the lowest supply voltage that still preserves the data stored in the cell. DRV can be computed as follows:

$$\mathrm{DRV} = \max(\mathrm{DRV}_0,\ \mathrm{DRV}_1), \tag{9.36}$$

where DRV_0 is the DRV when the cell is storing a 0, and DRV_1 is the DRV when it is storing a 1. If the cell is balanced (symmetric), with identical left and right halves, then $\mathrm{DRV}_0 = \mathrm{DRV}_1$. However, if there is any mismatch due to process variations, they become unequal. This creates the situation where the standard statistical blockade classification technique would fail because of the presence of disjoint tail regions.

Suppose we run a 1,000-point Monte Carlo, varying all the mismatch parameters in the SRAM cell according to their statistical distributions. This would give us distributions of values for DRV_0, DRV_1, and DRV. In certain parts of the mismatch parameter space $\mathrm{DRV}_0 > \mathrm{DRV}_1$, and in other parts, $\mathrm{DRV}_0 < \mathrm{DRV}_1$. This is clearly illustrated by Fig. 9.18: let us see how. First, we extract the direction in the parameter space that has maximum influence on DRV_0. This direction is shown as $\mathbf{w}_{1,\mathrm{DRV0}}$ in the two-dimensional parameter space example of Fig. 9.17. As we move along this vector in the statistical parameter space, the corresponding DRV_0 increases from low to high (from bad to good). [3] Other directions would show changes in DRV_0 that are no greater than the change along this direction. The figure plots the simulated DRV_0 and DRV_1 values from a 1,000-point Monte Carlo run, along this direction. It is clear that the two DRV measures are inversely related: one decreases as the other increases.

[3] Here we extract this vector using the SiLVR tool of Ref. [40], so that $\mathbf{w}_{1,\mathrm{DRV0}}$ is essentially the projection vector of the first latent variable of DRV_0.

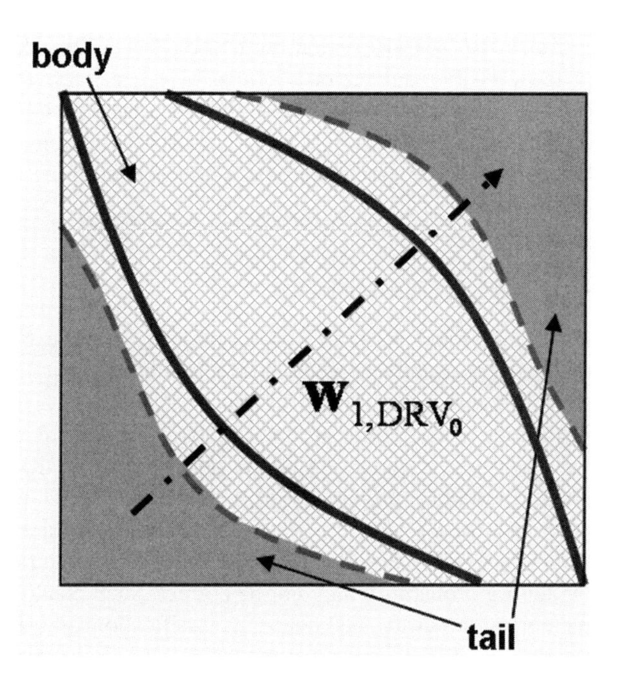

Fig. 9.17 A circuit metric (e.g., DRV) with two disjoint tail regions. The tail regions are separated from the body region by *dashed lines*. $w_{1,DRV0}$ is the direction of maximum variation of the circuit metric

Now, let us take the maximum as in (9.36), and choose the classification threshold t_c equal to the 97th percentile. Then we pick out the worst 3% points from the classifier training data and plot them against the same latent variable in Fig. 9.18, as squares. Note that we have not trained the classifier yet, we are just looking at the points that the classifier would have to classify as being in the tail. We can clearly see that these points (squares) lie in two disjoint parts of the parameter space.

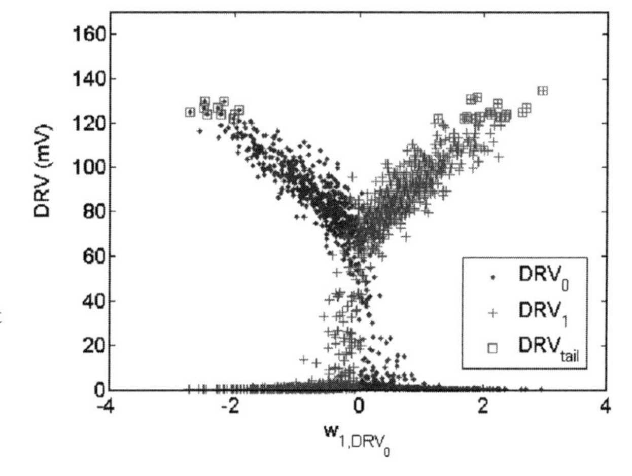

Fig. 9.18 Behavior of DRV_0 and DRV_1 along the direction of maximum variation in DRV_0. The worst 3% DRV values are plotted as squares, showing the disjoint tail regions (along this direction in the parameter space)

Since the true tail region defined by the tail threshold $t > t_c$ will be a subset of the classifier tail region (defined by t_c), it is obvious that the true tail region consists of two disjoint regions of the parameter space. This is illustrated in our two-dimensional example in Fig. 9.17. The dark tail regions on the top-right and bottom-left corners of the parameter space correspond to the large DRV values shown as squares in Fig. 9.18.

Asymmetrical Cell Topologies

For symmetrical memory cell topologies, like the 6T SRAM cell, the disjoint regions will usually be symmetrical around the origin in the parameter space for a two argument conditional like in (9.36). Hence, a simple solution would be to just multiply the failure probability of one argument by two. In the case of DRV, we would just compute the failure probability for DRV_0 and multiply it by two to obtain the failure probability of the overall DRV. However, for asymmetrical circuits, like the 8T SRAM cell of [41], that has a separate read port from one of the storage nodes, the disjoint regions may not be symmetrical and a simple doubling of the failure probability may not be correct. Such conditionals are very common for memory cell metrics, and hence, a classification strategy for such cases is essential for practical use of statistical blockade. We now describe such a strategy.

9.6.1.2 The Solution

Instead of building a single classifier for the tail of in (9.36), let us build two separate classifiers, one for the 97th percentile ($t_{c,DRV0}$) of DRV_0, and another for the 97th percentile ($t_{c,\,DRV1}$) of DRV_1. The generated Monte Carlo samples can then be filtered through both these classifiers: points classified as "body" by *both* the classifiers will be blocked, and the rest will be simulated. In the general case for arbitrary number of arguments in the conditional, let the circuit metric be given as

$$y = \max(y_0, y_1, \ldots). \tag{9.37}$$

Note that a min() operator can easily be converted to max() by negating the arguments. The resulting general algorithm is then as follows:

1) Perform initial sampling to generate training data to build the classifiers, and estimate tail and classification thresholds, t_i and $t_{c,i}$, respectively, for each y_i, $i = 0, 1, \ldots$. Also estimate the tail threshold t for y.
2) For each argument, y_i, $i = 0, 1, \ldots$, of the conditional (9.37), build a classifier C_i at a classification threshold $t_{c,i}$ that is less than the corresponding tail threshold t_i.
3) Generate more points using Monte Carlo, but block the points classified as "body" by *all* the classifiers. Simulate the rest and compute y for the simulated points.

Hence, in the case of Fig. 9.17, we build a separate classifier for each of the two boundaries. The resulting classification boundaries are shown as solid lines. From the resulting simulated points, those with $y > t$ are chosen as tail points for further analysis; e.g., for computing a GPD model for the tail distribution of y.

Note that this same algorithm can also be used for the case of multiple circuit metrics. Each metric would have its own thresholds and its own classifier, just like each argument in (9.37), the only difference being that we would not be computing any conditional.

9.6.2 Extremely Rare Events and Their Statistics

9.6.2.1 Extremely Rare Events

The GPD tail model can be used to make predictions regarding rare events that are farther out in the tail than any of the data we used to compute the GPD model. Indeed, this is the compelling reason for adopting the GPD model. However, as suggested by common intuition and the results presented in Section 9.5.3, we expect the statistical confidence in the estimates to decrease as we predict farther out in the tail. Equivalently, the variance of the predictions will probably increase as we move out in the tail. We can estimate this confidence or variance in two different ways:

1) *Empirically*: Suppose we run 50 runs of Monte Carlo with n_{MC} samples each and compute a GPD tail model from each run, using points that exceed some fixed threshold t. This gives us 50 slightly different pairs of the GPD parameters (ξ, β), one for each of 50 GPD models so computed. Then, we can estimate variance and confidence intervals of any statistical metric using the 50 estimates obtained from these tail models.
2) *Using asymptotic variance*: Section 9.4.3 gives us expressions for the asymptotic covariance matrix $\Sigma_{\xi,\beta}$ of ($\hat{\xi}$, $\hat{\beta}$) estimated using probability-weighted moment matching (see Eq. (9.26)). We can replace the exact values of (ξ, β) with the estimated ($\hat{\xi}$, $\hat{\beta}$) to obtain an approximate covariance matrix $\hat{\Sigma}_{\xi,\beta}$. For reasonable large number of tail samples used in the estimation, we can assume normal distribution of these estimates with mean ($\hat{\xi}$, $\hat{\beta}$) and covariance $\hat{\Sigma}_{\xi,\beta}$. Then, we can sample the GPD parameters from this distribution to compute different estimates of some statistical metric, which can be used to compute the variance. Here we need to build only one GPD model using a single Monte Carlo run of n_{MC} points. However, because of the assumption of the onset of asymptotic normal distribution and the approximation of the covariance matrix, we expect this method to show some error.

We use both these methods to compute 95% confidence intervals for the estimation of the $m\sigma$ point of the SRAM cell write time, where $m \in [3, 6]$. For the empirical method we use 50 Monte Carlo runs of $n_{\mathrm{MC}} = 100{,}000$ points each, and

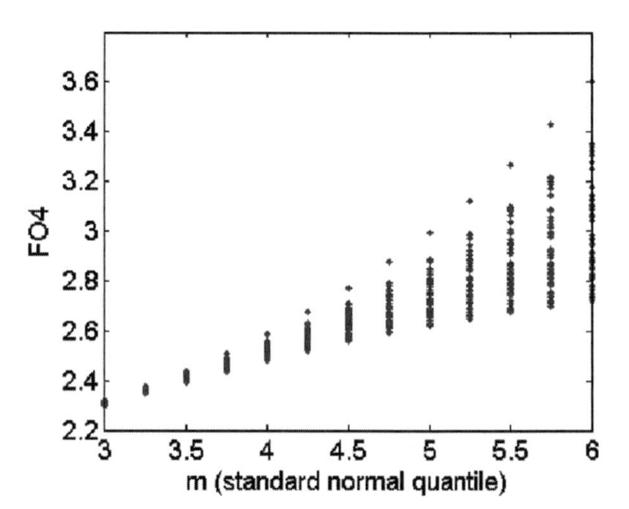

Fig. 9.19 The spread of $m\sigma$ point estimates across 50 runs of statistical blockade

compute GPD models with $t =$ the 99-percentile write time. This gives us 50 different estimates of the $m\sigma$ point. These estimates are shown in Fig. 9.19.

As expected, the spread of the estimates increases as we extrapolate further with the GPD model. We then compute 95% confidence intervals of the $m\sigma$ point estimates using these 50 models. Say we have 50 estimates $y_i(m)$, $i = 1, \ldots, 50$ for the $m\sigma$ point. From these we can empirically estimate the 97.5 percentile and 2.5 percentile points, $y_{97.5\%}(m)$ and $y_{2.5\%}(m)$, respectively. A 95% confidence interval width $\kappa_{95\%}(m)$ can then be computed as

$$\kappa_{95\%}(m) = y_{97.5\%}(m) - y_{2.5\%}(m).$$

We can express this interval as a percentage of the mean of the estimates:

$$\kappa'_{95\%}(m) = \frac{\kappa_{95\%}(m)}{\frac{1}{n}\sum_{i=1}^{50} y_i(m)}.$$

We also compute similar 95% confidence intervals using the second method, using 10,000 pairs of GPD parameter values sampled from the normal distribution with mean $(\hat{\xi}, \hat{\beta})$ and covariance $\hat{\Sigma}_{\xi,\beta}$. In this case we express the confidence interval as a percentage of the estimate $y(m)$ computed using the single GPD model $G_{\hat{\xi},\hat{\beta}}$. Figure 9.20 shows these percentage confidence intervals.

Although there is some mismatch in the magnitudes of the two estimates of the 95% confidence interval, we see a common trend: the statistical confidence decreases as we move out in the tail. To keep the error within 5% with a confidence of 95% we should not be predicting farther than 4.28σ. For 10% error, we can go out to 4.95σ. Of course these numbers will change from circuit to circuit and

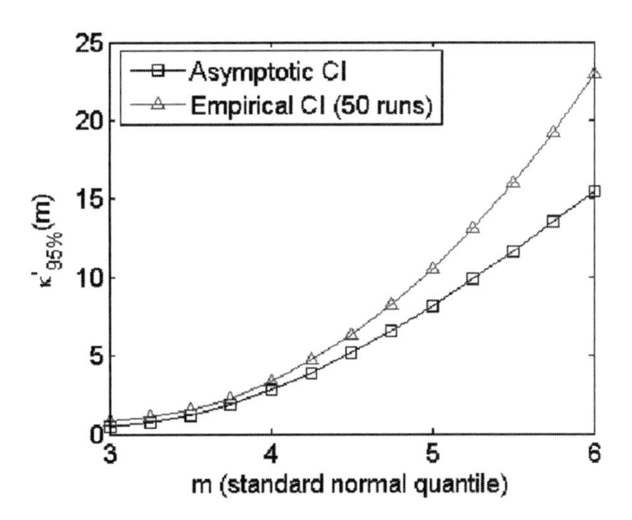

Fig. 9.20 95% confidence intervals as a percentage of the mean value (empirical method) or the single estimate (asymptote method)

performance metric to performance metric. The general inference is that we should not rely on the GPD tail model too far out from our data.

9.6.2.2 The Reason for Error in the MSFF Tail Model

Here we return to our MSFF test circuit from Section 9.5.3, where we saw some discrepancy between the empirical and GPD estimates for the failure probability expressed as y_σ. We will try to develop an explanation for this undesirable, although small, discrepancy. For this purpose we call on a common tool of graphical exploration of statistical data: the *sample mean excess plot*. Reference [18] reviews some properties of the mean excess plot. Here we focus on its properties in relation to the generalized Pareto distribution. The *mean excess function* for a given threshold y_f is defined as

$$e(y_f) = E(y - y_f | y > y_f);$$

that is, the mean of exceedances over y_f. Plotting $e(y_f)$ against y_f gives us the mean excess plot. The *sample* mean excess function is the sample version of $e(y_f)$. For a given sample $\{y_i : i = 1, \ldots, n\}$, it is defined as

$$e_n(y_f) = \frac{\sum_{i=1}^{n} (y_i - y_f)_+}{\left|\{y_i : y_i > y_f\}\right|}, \text{ where } (\bullet)_+ = \max(\bullet,\ 0);$$

that is, the sample mean of only the exceedances over y_f. A plot of $e_n(y_f)$ against y_f gives us the sample mean excess plot. The mean excess function of a GPD $G_{\xi,\beta}$ can be shown (see [18]) to be a straight line given by

$$e(y_f) = \frac{\beta - \xi y_f}{1 - \beta}, \text{ for } y_f \in D(\xi, \beta),$$

where $D(\xi, \beta)$ is as defined in Theorem 9.3. Hence, if the sample mean excess function of any data sample starts to follow roughly a straight line from some threshold, then it is an indication that the exceedances over that threshold follow a GPD. In fact, this feature of the mean excess plot can be employed to manually estimate an appropriate tail threshold.

Let us now look at the sample mean excess plot of the MSFF tail data ($\tau_{cq} \geq$ 99-percentile delay) from the 500,000-point Monte Carlo run. This is shown in Fig. 9.21. The plot suggests a good reason for the observed discrepancy in the estimated failure probabilities. It is clear from the plot that the tail defined by the $t = 99$ point has not converged close to a GPD form. Hence, the discrepancy could be a result of choosing a tail threshold that is not large enough. To test this, let us choose a threshold $t = 3\sigma$ point and fit the GPD model to exceedances over this t. Figure 9.21 suggests that this should show a better fit, since the sample mean excess function seems to be roughly a straight line from the 3σ threshold. The predictions of this new GPD model are shown in Table 9.8. We also reproduce columns 2 and 3 of Table 9.7 for comparison. As expected, we see more accurate predictions.

9.6.2.3 The Problem

For both the issues discussed above, a solution is to sample further out in the tail and use a higher tail threshold for building the GPD model of the tail. This is, of course, "easier said than done." Suppose we wish to support our GPD model with data up to the 6σ point. The failure probability of a 6σ value is roughly 1 part per *billion*, corresponding to a 99% chip yield requirement for a 10 Mb cache.

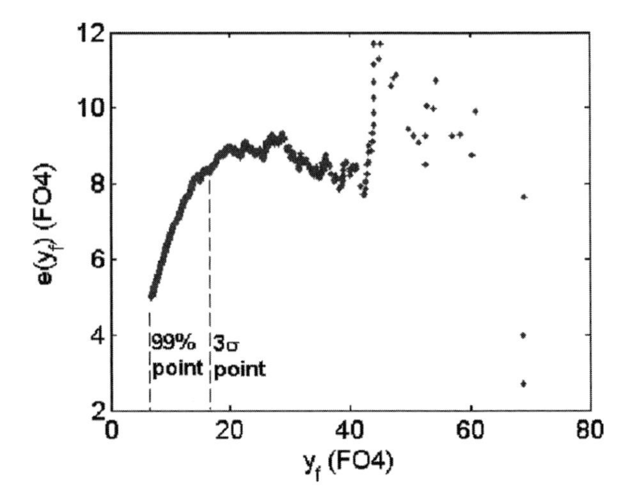

Fig. 9.21 A sample mean excess plot for the MSFF circuit, showing the 99th percentile and 3σ tail thresholds

Table 9.8 Prediction of failure probability as y_σ using a GPD model (method II of Section 9.5.3) with the tail threshold t at the 99th percentile and at the 3σ point

τ_{cq} (y_f) (FO4)	(I) Standard Monte Carlo	(II) GPD at $t = 99$th percentile	(II) GPD at $t = 3\sigma$ point
30	3.424	3.466	3.443
40	3.724	3.686	3.729
50	4.008	3.854	3.978
60	4.219	3.990	4.198
70	4.607	4.102	4.396
80	∞	4.199	4.574
90	∞	4.283	4.737

This is definitely not an impractical requirement. However, for a 99% tail threshold, even a perfect classifier ($t_c = t$) will only reduce the number of simulations to an extremely large 10 million. If we decide to use a 99.9999% threshold, the number of simulations will be reduced to a more practical 1,000 tail points (with a perfect classifier). However, we will need to simulate an extremely large number of points ($\geq$ 1 million) to generate a classifier training set with at least one point in the tail region. In both cases, the circuit simulation counts are too high. We now describe a recursive formulation of statistical blockade that reduces this count drastically.

9.6.3 A Recursive Formulation of Statistical Blockade

Let us first assume that there are no conditionals. For a tail threshold equal to the ath percentile, let us represent it as t^a, and the corresponding classification threshold as $t_c{}^a$. For this threshold, build a classifier C^a and generate sufficient points beyond the tail threshold, $y > t^a$, so that a *higher* percentile (t^b, $t_c{}^b$, $b > a$) can be estimated. For this new, higher threshold $t_c{}^b$, a new classifier C^b is trained and a new set of tail points ($y > t^b$) is generated. This new classifier will block many more points than C^a, significantly reducing the number of simulations. This procedure is repeated to push the threshold out more until the tail region of interest is reached. The complete algorithm is shown in Algorithm 9.2.

The arguments to the algorithm are formulated a little differently from the basic statistical blockade algorithm (Algorithm 9.1). Instead of passing the tail and classification threshold probabilities (p_t, p_c), we pass a tail sample size n_t and a classification threshold probability function $p_c(p)$. The former is the number of tail points to be used finally to compute the GPD tail model. The latter is a function that returns the classification threshold probability for a given tail threshold probability. It is implicitly a function also of the type of classifier being used, since the error in the classifier will determine the appropriate safety margin.

The functions that appear also in Algorithm 9.1 do the same work here, hence, we do not reiterate their description. $\mathbf{f}_{\text{sim}}$ now returns multiple outputs: it computes the values of all the arguments of the conditional in $y = \max(y_0, y_1, \ldots)$. For example, in

Algorithm 9.2 The general recursive statistical blockade algorithm for efficient sampling of *extremely* rare events, in the presence of conditional induced disjoint tail regions

---

Require: initial sample size n_0 (e.g., 1,000);
total sample size n; tail sample size n_t;
function $p_c(p)$, $p \in (0,100)$;
performance metric function $y = \max(y_0, y_1, \ldots)$

---

1. $\mathbf{X} = \text{MonteCarlo}(n_0)$
2. $n' = n_0$
3. $n_c = \max(n_t, 1{,}000)$ // Classifier training set size at least 1,000
4. $\mathbf{Y} = \mathbf{f}_{\text{sim}}(\mathbf{X})$ // Simulate the initial Monte Carlo sample set
5. $\mathbf{y}_{\text{tail},i} = \mathbf{Y}_{\bullet,i}$, $i = 0, 1, \ldots$ // The ith column of $\mathbf{Y}$ contains values for y_i in $y = \max(y_0, y_1, \ldots)$
6. $\mathbf{X}_{\text{tail},i} = \mathbf{X}$, $i = 0, 1, \ldots$
7. **while** $n' < n$
8. $\Delta n = \min(100n', n) - n'$ // Number of points to filter in this recursion step
9. $p_t = \frac{100\Delta n}{n' + \Delta n}$ // The tail threshold is the p_tth percentile
10. $n' = n' + \Delta n$ // Total number of points filtered by the end of this recursion stage
11. $\mathbf{X} = \text{MonteCarloNext}(\Delta n)$ // The next Δn points in the Monte Carlo sequence
12. **forall** $i : y_i$ is an argument in $y = \max(y_0, y_1, \ldots)$
13. $(\mathbf{X}_{\text{tail},i}, \mathbf{y}_{\text{tail},i}) = \text{GetWorst}(n_c, \mathbf{X}_{\text{tail},i}, \mathbf{y}_{\text{tail},i})$ // Get the n_t worst points
14. $t = \text{Percentile}(\mathbf{y}_{\text{tail},i}, p_t)$
15. $t_c = \text{Percentile}(\mathbf{y}_{\text{tail},i}, p_c(p_t))$
16. $C^i = \text{BuildClassifier}(\mathbf{X}_{\text{tail},i}, \mathbf{y}_{\text{tail},i}, t_c)$
17. $(\mathbf{X}_{\text{tail},i}, \mathbf{y}_{\text{tail},i}) = \text{GetGreaterThan}(t, \mathbf{X}_{\text{tail},i}, \mathbf{y}_{\text{tail},i})$ // Get the points with $y_i > t$
18. $\mathbf{X}_{\text{cand},i} = \text{Filter}(C^i, \mathbf{X})$ // Candidate tail points for y_i
19. **endfor**
20. $\mathbf{X} = \begin{bmatrix}\mathbf{X}^T_{\text{cand},0} & \mathbf{X}^T_{\text{cand},1} & \cdots\end{bmatrix}^T$ // Union of *all* candidate tail points
21. $\mathbf{Y} = \mathbf{f}_{\text{sim}}(\mathbf{X})$ // Simulate all candidate tail points
22. $\mathbf{y}_{\text{cand},i} = \{\mathbf{Y}_{j,i} : \mathbf{X}_{j,\bullet} \in \mathbf{X}_{\text{cand},i}\}$, $i = 0, 1, \ldots$ // Extract the tail points for y_i
23. $\mathbf{y}_{\text{tail},i} = \begin{bmatrix}\mathbf{y}^T_{\text{tail},i} & \mathbf{y}^T_{\text{cand},i}\end{bmatrix}^T$, $\mathbf{X}_{\text{tail},i} = \begin{bmatrix}\mathbf{X}^T_{\text{tail},i} & \mathbf{X}^T_{\text{cand},i}\end{bmatrix}^T$, $i = 0, 1, \ldots$ // All tail points till now
24. **endwhile**
25. $\mathbf{y}_{\text{tail}} = \text{MaxOverRows}([\mathbf{y}_{\text{tail},0}\ \mathbf{y}_{\text{tail},1} \ldots])$ // compute the conditional
26. $\mathbf{y}_{\text{tail}} = \text{GetWorst}(n_t, \mathbf{y}_{\text{tail}})$
27. $(\xi, \beta) = \text{FitGPD}(\mathbf{y}_{\text{tail}} - \min(\mathbf{y}_{\text{tail}}))$

---

the case of DRV, it will return the values of DRV_0 and DRV_1. These values, for any one Monte Carlo point, are stored in one row of the result matrix $\mathbf{Y}$. The function *MonteCarloNext*(Δn) returns the *next* Δn points in the sequence of points generated till now. The function *GetWorst*(n, $\mathbf{X}$, $\mathbf{y}$) returns the n worst values in the vector $\mathbf{y}$ and the corresponding rows of the matrix $\mathbf{X}$. This functionality naturally extends to the two argument *GetWorst*(n, $\mathbf{y}$). *GetGreaterThan*(t, $\mathbf{X}$, $\mathbf{y}$) returns the elements of $\mathbf{y}$ that are greater than t, along with the corresponding rows of $\mathbf{X}$.

The function $p_c(p)$ is not easy to determine, hence we also present a less general version as Algorithm 9.3, which can be used immediately by any practitioner. Here, we restrict the total sample size n to be some power of 100, times 1,000:

$$n = 100^j \cdot 1000, \quad j = 0, 1, \ldots \tag{9.38}$$

Algorithm 9.3 The recursive statistical blockade algorithm with fixed sequences for the tail and classification thresholds: $t = 99\%, 99.99\%, 99.9999\%, \ldots$ points, and $t_c = 97\%, 99.97\%, 99.9997\%, \ldots$ points. The total sample size is given by (9.38)

Require: initial sample size n_0 (e.g., 1,000);
total sample size n; tail sample size n_t;
performance metric function $y = \max(y_0, y_1, \ldots)$

---

1. $\mathbf{X} = \text{MonteCarlo}(n_0)$
2. $n' = n_0$
4. $\mathbf{Y} = \mathbf{f}_{\text{sim}}(\mathbf{X})$ // Simulate the initial Monte Carlo sample set
5. $\mathbf{y}_{\text{tail},i} = \mathbf{Y}_{\bullet,i}, i = 0, 1, \ldots$ // The ith column of $\mathbf{Y}$ contains values for y_i in $y = \max(y_0, y_1, \ldots)$
6. $\mathbf{X}_{\text{tail},i} = \mathbf{X}, i = 0, 1, \ldots$
7. **while** $n' < n$
8. $\Delta n = 99n'$ // Number of points to filter in this recursion step
10. $n' = n' + \Delta n$ // Total number of points filtered by the end of this recursion stage
11. $\mathbf{X} = \text{MonteCarloNext}(\Delta n)$ // The next Δn points in the Monte Carlo sequence
12. **forall** $i : y_i$ is an argument in $y = \max(y_0, y_1, \ldots)$
13. $(\mathbf{X}_{\text{tail},i}, \mathbf{y}_{\text{tail},i}) = \text{GetWorst}(1{,}000, \mathbf{X}_{\text{tail},i}, \mathbf{y}_{\text{tail},i})$ // Get the 1,000 worst points
14. $t = \text{Percentile}(\mathbf{y}_{\text{tail},i}, 99)$
15. $t_c = \text{Percentile}(\mathbf{y}_{\text{tail},i}, 97)$
16. $C^i = \text{BuildClassifier}(\mathbf{X}_{\text{tail},i}, \mathbf{y}_{\text{tail},i}, t_c)$
17. $(\mathbf{X}_{\text{tail},i}, \mathbf{y}_{\text{tail},i}) = \text{GetGreaterThan}(t, \mathbf{X}_{\text{tail},i}, \mathbf{y}_{\text{tail},i})$ // Get the points with $y_i > t$
18. $\mathbf{X}_{\text{cand},i} = \text{Filter}(C^i, \mathbf{X})$ // Candidate tail points for y_i
19. **endfor**
20. $\mathbf{X} = \begin{bmatrix}\mathbf{X}^T_{\text{cand},0} & \mathbf{X}^T_{\text{cand},1} & \cdots\end{bmatrix}^T$ // Union of *all* candidate tail points
21. $\mathbf{Y} = \mathbf{f}_{\text{sim}}(\mathbf{X})$ // Simulate all candidate tail points
22. $\mathbf{y}_{\text{cand},i} = \{\mathbf{Y}_{j,i} : \mathbf{X}_{j,\bullet} \in \mathbf{X}_{\text{cand},i}\}, i = 0, 1, \ldots$ // Extract the tail points for y_i
23. $\mathbf{y}_{\text{tail},i} = \begin{bmatrix}\mathbf{y}^T_{\text{tail},i} & \mathbf{y}^T_{\text{cand},i}\end{bmatrix}^T, \mathbf{X}_{\text{tail},i} = \begin{bmatrix}\mathbf{X}^T_{\text{tail},i} & \mathbf{X}^T_{\text{cand},i}\end{bmatrix}^T, i = 0, 1, \ldots$ // All tail points till now
24. **endwhile**
25. $\mathbf{y}_{\text{tail}} = \text{MaxOverRows}([\mathbf{y}_{\text{tail},0}\ \mathbf{y}_{\text{tail},1} \ldots])$ // compute the conditional
26. $\mathbf{y}_{\text{tail}} = \text{GetWorst}(n_t, \mathbf{y}_{\text{tail}})$
27. $(\xi, \beta) = \text{FitGPD}(\mathbf{y}_{\text{tail}} - \min(\mathbf{y}_{\text{tail}}))$

Also, we fix $p_t = 99\%$ and $p_c = 97\%$. This will always give us 1,000 tail points to fit the GPD. The tail threshold t moves with every recursion step as

$$t = 99\text{th percentile}, 99.99\text{th percentile}, 99.9999\text{th percentile}, \ldots,$$

and the classification threshold as

$$t_c = 97\text{th percentile}, 99.97\text{th percentile}, 99.9997\text{th percentile}, \ldots$$

The algorithms presented here are in iterative form, rather than recursive form. To see how the recursion works, suppose we want to estimate the 99.9999% tail. To generate points at and beyond this threshold, we first estimate the 99.99% point and use a classifier at the 99.97% point to generate these points efficiently. To build this classifier in turn, we first estimate the 99% point and use a classifier at the 97%

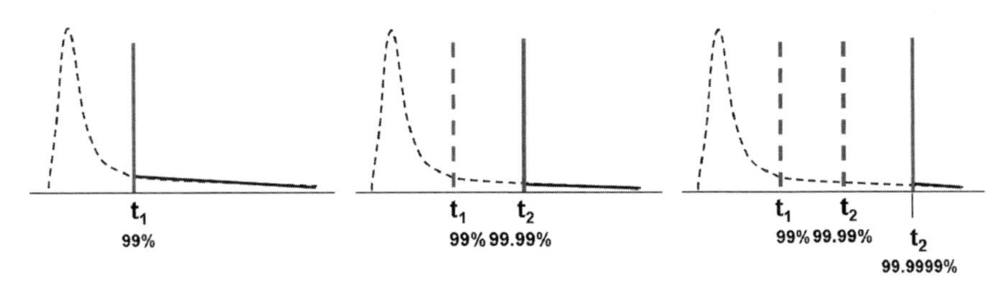

Fig. 9.22 Recursive formulation of statistical blockade as in Algorithm 9.3

point. Figure 9.22 illustrates this recursion on the PDF of any one argument in the conditional (9.37).

9.6.4 Experimental Results

We now test the recursive statistical blockade method on another SRAM cell test case, where we compute the DRV as in (9.36). In this case the SRAM cell is implemented in an industrial 90 nm process. Wang et al. [42] develop an analytical model for predicting the CDF of the DRV that uses not more than 5,000 Monte Carlo points. The CDF is given as

$$F(y) = 1 - \mathrm{erfc}(y_0) + \frac{1}{4}\mathrm{erfc}^2(y_0), \text{ where } y_0 = \frac{\mu_0 + k(y - V_0)}{\sqrt{2}\sigma_0}, \tag{9.39}$$

where y is the DRV value and erfc() is the complementary error function [43]. k is the sensitivity of the SNM of the SRAM cell to the supply voltage, computed using a DC sweep. μ_0 and σ_0 are the mean and standard deviation of the SNM (SNM_0), for a user-defined supply voltage V_0. SNM_0 is the SNM of the cell while storing a 0. These statistics are computed using a short Monte Carlo run of 1,500–5,000 sample points. We direct the reader to [42] for complete details regarding this analytical model of the DRV distribution. The qth quantile can be estimated as

$$\mathrm{DRV}(q) = \frac{1}{k}\left(\sqrt{2}\sigma_0 \mathrm{erfc}^{-1}\left(2 - 2\sqrt{q}\right) - \mu_0\right) + V_0. \tag{9.40}$$

Here $\mathrm{DRV}(q)$ is the supply voltage V_{dd} such that

$$P(\mathrm{DRV}(q)) \leq V_{\mathrm{dd}} = q.$$

Let us now compute the DRV quantiles as $m\sigma$ points, such that q is the cumulative probability for the value m from a standard normal distribution. We will use five different methods to estimate the DRV quantiles for $m \in [3, 8]$:

1) *Analytical*: Use Eq. (9.40).

2) *Recursive statistical blockade without the GPD model*: Algorithm 9.3 is run for $n = 1$ billion. This results in three recursion stages, corresponding to total sample sizes of $n' = 100{,}000$, 10 million, and 1 billion Monte Carlo points. The worst DRV value for these three recursion stages are estimates of the 4.26σ, 5.2σ, and 6σ points, respectively.
3) *GPD model from recursive statistical blockade*: The 1,000 tail points from the last recursion stage of the recursive statistical blockade run are used to fit a GPD model, which is then used to predict the DRV quantiles.
4) *Normal*: A normal distribution is fit to data from a 1,000 point Monte Carlo run, and used to predict the DRV quantiles.
5) *Lognormal*: A lognormal distribution is fit to the same set of 1,000 Monte Carlo points, and used for the predictions.

The results are shown in Fig. 9.23. From the plots in the figure, we can immediately see that the recursive statistical blockade estimates are very close to the estimates from the analytical model. This shows the efficiency of the recursive formulation in reducing the error in predictions for events far out in the tail.

Table 9.9 shows the number of circuit simulations performed at each recursion stage. The total number of circuit simulations is 41,721. This is not small, but in comparison to standard Monte Carlo (1 billion simulations), and basic, non-recursive statistical blockade (approximately, 30 million with $t_c = 97$th percentile) it is extremely fast. About 41,721 simulations for DRV computation of a 6T SRAM cell can be completed in several hours on a single computer today. With the advent of multi-core processors, the total simulation time can be drastically reduced with proper implementation.

Note that we can extend the prediction power to 8σ with the GPD model, without any additional simulations. Standard Monte Carlo would need over 1.5 quadrillion

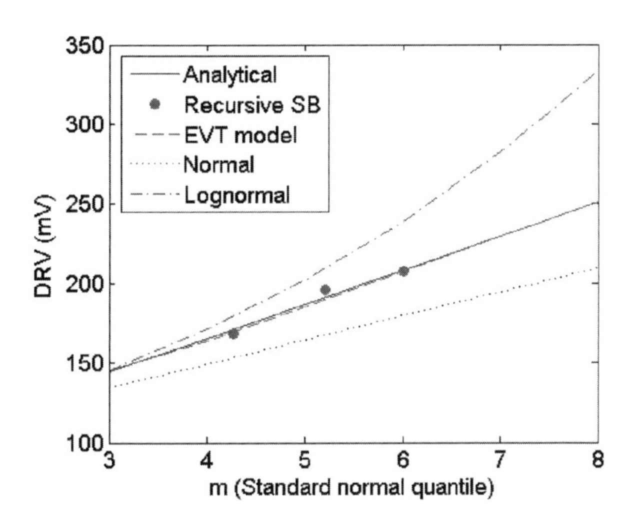

Fig. 9.23 Estimates of DRV quantiles from five estimation methods. The GPD model closely fits the analytical model (9.39). The *solid circles* show the worst DRV values from the three recursion stages of statistical blockade sampling. The normal and lognormal models are quite inaccurate

Table 9.9 Number of circuit simulation needed by recursive statistical blockade to generate a 6σ point

Recursion stage	Number of simulations
Initial	1,000
1	11,032
2	14,184
3	15,505
Total	41,721
Speedup over Monte Carlo	23,969×
Speedup over statistical blockade	719×

circuit simulations to generate a single 8σ point. For this case, the speedup over standard Monte Carlo is extremely large. As expected, the normal and lognormal fits show large errors. The normal fit is unable to capture the skewness of the DRV distribution. On the other hand, the lognormal distribution has a heavier tail than the DRV distribution.

References

1. P. Gupta and F.-L. Heng, *Toward a systematic-variation aware timing methodology*, Proc. IEEE/ACM Design Automation Conf., June 2004.
2. M. Hane, T. Ikezawa, and T. Ezaki, *Atomistic 3d process/device simulation considering gate line-edge roughness and poly-si random crystal orientation effects*, Proc. IEEE Int. Electron Devices Meeting, 2003.
3. M. J. M. Pelgrom, A. C. J. Duinmaijer, and A. P. G. Welbers, *Matching properties of MOS transistors*, IEEE J. Solid-State Circuits, 24(5): 1433–1440, 1989.
4. A. J. Bhavnagarwala, X. Tang, and J. D. Meindl, *The impact of intrinsic device fluctuations on CMOS SRAM cell stability*, IEEE J. Solid-State Circuits, 36(4): 658–665, 2001.
5. S. Mukhopadhyay, H. Mahmoodi, and K. Roy, *Statistical design and optimization of SRAM cell for yield enhancement*, Proc. IEEE/ACM Int. Conf. on CAD, 2004.
6. B. H. Calhoun and A. Chandrakasan, *Analyzing static noise margin for sub-threshold SRAM in 65 nm CMOS*, Proc. Eur. Solid State Cir. Conf., 2005.
7. B. Joshi, R. K. Anand, C. Berg, J. Cruz-Rios, A. Krishnamurthi, N. Nettleton, S. Nguyen, J. Reaves, J. Reed, A. Rogers, S. Rusu, C. Tucker, C. Wang, M. Wong, D. Yee, and J.-H. Chang, *A BiCMOS 50 MHz cache controller for a superscalar microprocessor*, Int. Solid State Cir. Conf., 1992.
8. G. S. Fishman, *A First Course in Monte Carlo*. Duxbury, 2006.
9. K. Agarwal, F. Liu, C. McDowell, S. Nassif, K. Nowka, M. Palmer, D. Acharyya, and J. Plusquellic, *A test structure for characterizing local device mismatches*, Symp. on VLSI Circuits Dig. Tech. Papers, 2006.
10. H. Chang, V. Zolotov, S. Narayan, and C. Visweswariah, *Parameterized block-based statistical timing analysis with non-Gaussian parameters, nonlinear delay functions*, Proc. IEEE/ACM Design Autom. Conf., 2005.
11. T. Ezaki, T. Izekawa, and M. Hane, *Investigation of random dopant fluctuation induced device charestistics variation for sub-100 nm CMOS by using atomistic 3D process/device simulator*, Proc. IEEE Int. Electron Devices Meeting, 2002.
12. H. Mahmoodi, S. Mukhopadhyay, and K. Roy, *Estimation of delay variations due to random-dopant fluctuations in nanoscale CMOS circuits*, IEEE J. Solid State Cir., 40(3): 1787–1796, 2005.

13. D. Hocevar, M. Lightner, and T. Trick, *A study of variance reduction techniques for estimating circuit yields*, IEEE Trans. Computer-Aided Design, 2(3): 279–287, 1983.
14. R. Kanj, R. Joshi, and S. Nassif, *Mixture importance sampling and its application to the analysis of SRAM designs in the presence of rare event failure*, Proc. IEEE/ACM Design Autom. Conf., 2006.
15. T. C. Hesterberg, *Advances in Importance Sampling*, Dept. of Statistics, Stanford University, 1998, 2003.
16. S. I. Resnick, *Extreme Values, Regular Variation and Point Processes*, Springer, New York, 1987.
17. L. de Haan, *Fighting the arch-enemy with mathematics*, Statist. Neerlandica, 44: 45–68, 1990.
19. P. Embrechts, C. Klüppelberg, and T. Mikosch, *Modelling Extremal Events for Insurance and Finance*, Springer-Verlag, Berlin, 4th ed., 2003.
19. A. Singhee and R. A. Rutenbar, *Statistical Blockade: a novel method for very fast Monte Carlo simulation of rare circuit events, and its application*, Proc. Design Autom. Test Europe, 2007.
20. A. Singhee, J. Wang, B. H. Calhoun, and R. A. Rutenbar, *Recursive Statistical Blockade: an enhanced technique for rare event simulation with application to SRAM circuit design*, Proc. Int. Conf. VLSI Design, 2008.
21. R. A. Fisher and L. H. C. Tippett, *Limiting forms of the frequency distribution of the largest or smallest member of a sample*, Proc. Cambridge Philos. Soc., 24: 180–190, 1928.
22. B. Gnedenko, *Sur la distribution limite du terme maximum d'une aleatoire*, Ann. Math., 44(3): 423–453, 1943.
23. A. Singhee, *Novel Algorithms for Fast Statistical Analysis of Scaled Circuits*, PhD Thesis, Electrical and Computer Engg., Carnegie Mellon University, 2007.
24. A. A. Balkema and L. de Haan, *Residual life time at great age*, Ann. Prob., 2(5): 792–804, 1974.
25. J. Pickands III, *Statistical inference using extreme order statistics*, Ann. Stats., 3(1): 119–131, 1975.
26. S. D. Grimshaw, *Computing maximum likelihood estimates for the generalized Pareto distribution*, Technometrics, 35(2): 185–191, 1993.
27. R. L. Smith, *Estimating tails of probability distributions*, Ann. Stats., 15(3): 1174–1207, 1987.
28. R. L. Smith, *Maximum likelihood estimation in a class of non-regular cases*, Biometrika, 72: 67–92, 1985.
29. J. R. M. Hosking and J. R. Wallis, *Parameter and quantile estimation for the generalized Pareto distribution*, Technometrics, 29(3): 339–349, 1987.
30. J. R. M. Hosking, *The theory of probability weighted moments*, IBM Research Report, RC12210, 1986.
31. C. J. C. Burges, *A tutorial on support vector machines for pattern recognition*, Data Mining and Knowledge Discovery, 2(2): 121–167, 1998.
32. T. Hastie, R. Tibshirani and J. Friedman, *The Elements of Statistical Learning: Data Mining, Inference, and Prediction*, Springer, 2001.
33. T. Joachims, *Making large-scale SVM learning practical*, In B. Schölkopf, C. Burges, and A. Smola, editors, Advances in Kernel Methods – Support Vector Learning. MIT Press, 1999.
34. I. H. Witten and E. Frank, *Data Mining: Practical Machine Learning Tools and Techniques*, Morgan Kaufmann, San Francisco, 2nd edition, 2005.
35. K. Morik, P. Brockhausen, and T. Joachims, *Combining statistical learning with a knowledge-based approach – a case study in intensive care monitoring*, Proc. 16th Intn'l. Conf. Machine Learning, 1999.
36. R. Rao, A. Srivastava, D. Blaauw, and D. Sylvester, *Statistical analysis of subthreshold leakage current for VLSI circuits*, IEEE Trans. VLSI Sys., 12(2): 131–139, 2004.
37. W. Liu, X. Jin, J. Chen, M.-C. Jeng, Z. Liu, Y. Cheng, K. Chen, M. Chan, K. Hui, J. Huang, R. Tu, P. Ko, and C. Hu, *BSIM 3v3.2 Mosfet Model Users' Manual*, Univ. California, Berkeley, Tech. Report No. UCB/ERL M98/51, 1988.

38. http://www.eas.asu.edu/~ptm/
39. R. K. Krishnamurthy, A. Alvandpour, V. de, and S. Borkar, *High-performance and low-power challenges for sub-70 nm microprocessor circuits*, Proc. Custom Integ. Circ. Conf., 2002.
40. A. Singhee and R. A. Rutenbar, *Beyond low-order statistical response surfaces: latent variable regression for efficient, highly nonlinear fitting*, Proc. IEEE/ACM Design Autom. Conf., 2007.
41. L. Chang, D. M. Fried, J. Hergenrother, J. W. Sleight, R. H. Dennard, R. K. Montoye, L. Sekaric, S. J. McNab, A. W. Topol, C. D. Adams, K. W. Guarini, and W. Haensch, *Stable SRAM Cell Design for the 32 nm Node and Beyond*, Symp. VLSI Tech. Dig. Tech. Papers, 128–129, 2005.
42. J. Wang, A. Singhee, R. A. Rutenbar, and B. H. Calhoun, *Modeling the minimum standby supply voltage of a full SRAM array*, Proc. Europ. Solid State Cir. Conf., 2007.
43. W. H. Press, B. P. Flannery, A. A. Teukolsky, and W. T. Vetterling, *Numerical Recipes in C: The Art of Scientific Computing*, Cambridge University Press, 2nd edition, 1992.

Index

Continued from page ii
SAT-Based Scalable Formal Verification Solutions
Malay Ganai and Aarti Gupta
ISBN 978-0-387-69166-4, 2007

Ultra-Low Voltage Nano-Scale Memories
Kiyoo Itoh, Masashi Horiguchi and Hitoshi Tanaka
ISBN 978-0-387-33398-4, 2007

Routing Congestion in VLSI Circuits: Estimation and Optimization
Prashant Saxena, Rupesh S. Shelar, Sachin Sapatnekar
ISBN 978-0-387-30037-5, 2007

Ultra-Low Power Wireless Technologies for Sensor Networks
Brian Otis and Jan Rabaey
ISBN 978-0-387-30930-9, 2007

Sub-Threshold Design for Ultra Low-Power Systems
Alice Wang, Benton H. Calhoun and Anantha Chandrakasan
ISBN 978-0-387-33515-5, 2006

High Performance Energy Efficient Microprocessor Design
Vojin Oklibdzija and Ram Krishnamurthy (Eds.)
ISBN 978-0-387-28594-8, 2006

Abstraction Refinement for Large Scale Model Checking
Chao Wang, Gary D. Hachtel, and Fabio Somenzi
ISBN 978-0-387-28594-2, 2006

A Practical Introduction to PSL
Cindy Eisner and Dana Fisman
ISBN 978-0-387-35313-5, 2006

Thermal and Power Management of Integrated Systems
Arman Vassighi and Manoj Sachdev
ISBN 978-0-387-25762-4, 2006

Leakage in Nanometer CMOS Technologies
Siva G. Narendra and Anantha Chandrakasan
ISBN 978-0-387-25737-2, 2005

Statistical Analysis and Optimization for VLSI: Timing and Power
Ashish Srivastava, Dennis Sylvester, and David Blaauw
ISBN 978-0-387-26049-9, 2005

Printed in the United States of America